CORE MATHEMATICS FOR IGCSE

SECOND EDITION

Ric Pimentel and Terry Wall

HODDER EDUCATION
AN HACHETTE UK COMPANY

Hachette's policy is to use papers that are natural, renewable and recyclable products and made from wood grown in sustainable forests. The logging and manufacturing processes are expected to conform to the environmental regulations of the country of origin.

Orders: please contact Bookpoint Ltd, 130 Milton Park, Abingdon, Oxon OX14 4SB. Telephone: (44) 01235 827720. Fax: (44) 01235 400454. Lines are open 9.00–5.00, Monday to Saturday, with a 24-hour message answering service. Visit our website at www.hoddereducation.co.uk

First published in 1997
by John Murray (Publishers) Ltd

This second edition first published 2006
by Hodder Education,
an Hachette UK company,
338 Euston Road,
London NW1 3BH.

Impression number	10 9 8 7
Year	2011

Cover photo: R.Mercer/Robert Harding Picture Library.
Typeset in 10/12pt Times by Pantek Arts Ltd, Maidstone, Kent.
Printed in India.

A catalogue record for this title is available from the British Library.

ISBN: 978 0340 908 129

CONTENTS

INTRODUCTION

This textbook covers the core curriculum of the IGCSE examination – up to grade C. For the convenience of students and teachers, it has been arranged by chapter in almost the same order as the syllabus. It is not intended that the chapters will necessarily be taught in the same order as they occur in the book. For example, many teachers and students may prefer to begin with Chapter 12, Use of an electronic calculator.

We have kept our explanations quite short. Teachers may wish to expand these and to use further examples if appropriate.

Some of the chapters, particularly the algebra and geometry chapters, are quite long. It may prove useful for these to be approached in sections, interspersed with other work.

This textbook is meant to be a core text in both senses. We would expect that it would be supplemented by practical work, especially in the use of measures, in estimation, and in geometry and trigonometry. We would also hope that investigations and practical topics would form a part of the mathematics course.

There are normally two student assessments at the end of each chapter, but where there is a considerable amount of new work introduced within that chapter, there are four student assessments to ensure complete coverage. The student assessments are to help determine how clearly the work in the chapter has been understood; some questions require short answers and some require longer answers, but working should always be shown.

This new edition also includes a section at the end of each topic where Information and Communication Technology (ICT) can be used to solve problems in a different, and more efficient, way.

Students from many countries will be studying from this IGCSE Mathematics textbook. It is interesting that, whatever language you speak, mathematics, and algebra in particular, is a universal language which is understood in every country on Earth.

We hope that all students will approach the work in this book in a positive way, expecting both difficulty and enjoyment. We hope that you all gain a sense of achievement, and that this achievement is reflected in the highest grade in the IGCSE Mathematics examination of which you are capable.

Ric Pimentel
Terry Wall

NUMBER

1 NUMBERS AND SEQUENCES

Identify and use natural numbers, integers, prime numbers, square numbers, common factors and common multiples; use rational and irrational numbers; continue a given number sequence; recognise patterns in sequences and relationships between different sequences; generalise to simple algebraic statements (including expressions for the *n*th term) relating to such sequences.

Whole numbers such as 1, 2, 3, 4, etc are called **natural numbers**.

Integers can be positive, negative or zero. Examples of integers are +7, +12, −3 and −5.

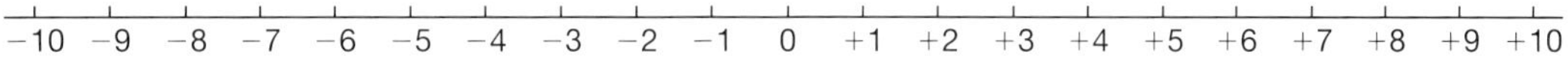

Integers can be shown on the number line above.

Left to right is the positive direction. Right to left is the negative direction.

Adding integers

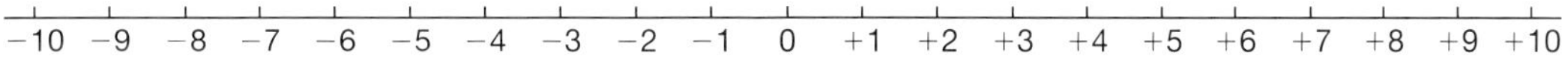

Worked examples

a) Use the number line above to add (+4) and (−2). Start at (+4) and move 2 in the negative direction.

$$(+4) + (-2) = (+2)$$

b) Use the number line above to add (−5) and (+7). Start at (−5) and move 7 in the positive direction.

$$(-5) + (+7) = (+2)$$

c) Use the number line above to add (−1) and (−3). Start at −1 and move 3 in the negative direction.

$$(-1) + (-3) = (-4)$$

Exercise 1.1

Draw a number line in your book from −8 to +8. Use it to answer the questions below:

1. a) (+5) + (−2) b) (+3) + (−3) c) (+8) + (−5)
d) (+6) + (−5) e) (+2) + (−4) f) (+1) + (−6)

2. a) (−5) + (+3) b) (−8) + (+7) c) (−2) + (−6)
d) (−1) + (+7) e) (−2) + (−4) f) (−4) + (+5)

3. a) (−3) + (−3) b) (−3) + (−2) c) (−2) + (−6)
d) (−4) + (−1) e) (−4) + (−2) f) (−3) + (−5)

4. a) (−3) + (+3) + (−1)
b) (−4) + (−4) + (+6)
c) (+1) + (−4) + (−3)
d) (−2) + (−2) + (−3)
e) (+5) + (−8) + (−3)
f) (−1) + (−3) + (+7) + (−2)

Subtracting integers

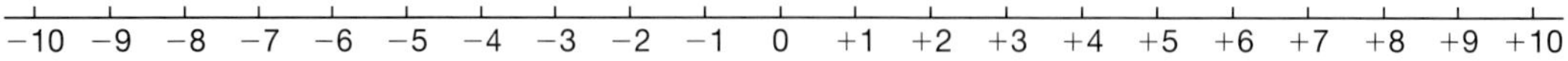

Worked examples

a) Use the number line above to calculate (+6) − (+2).

Always start at the number being subtracted.
Starting at (+2), (+6) is 4 in the positive direction.
So (+6) − (+2) = (+4).

b) Use the number line above to calculate (+2) − (+5).

Start at the number being subtracted, (+5). From +5 to +2 is 3 in the negative direction.

(+2) − (+5) = (−3)

c) Use the number line above to calculate (−2) − (+3).

Starting at (+3), (−2) is 5 in the negative direction.

(−2) − (+3) = (−5)

d) Use the number line above to calculate (−3) − (−2).

Starting at (−2), (−3) is 1 in the negative direction.

(−3) − (−2) = (−1)

e) Use the number line above to calculate (−3) − (−6).

Starting at (−6), (−3) is 3 in the positive direction.

(−3) − (−6) = (+3)

Exercise 1.2

Use a number line like the one on page 1 to help with the following calculations:

1. a) (+4) − (+3) b) (+5) − (+2) c) (+7) − (+5)
d) (+4) − (+1) e) (+8) − (+2) f) (+6) − (+3)
g) (+7) − (+7) h) (+3) − (+1) i) (+4) − (+2)
j) (+5) − (+5)

2. a) $(+2)-(+3)$ b) $(+4)-(+5)$ c) $(+3)-(+6)$
d) $(+2)-(+7)$ e) $(+4)-(+5)$ f) $(+1)-(+2)$
g) $(+3)-(+3)$ h) $(+2)-(+4)$ i) $(+5)-(+6)$
j) $(+2)-(+8)$

3. a) $(-2)-(+4)$ b) $(-3)-(+5)$ c) $(-6)-(+8)$
d) $(-1)-(+6)$ e) $(-3)-(+3)$ f) $(-5)-(+7)$
g) $(-1)-(+2)$ h) $(-2)-(+4)$ i) $(-4)-(+7)$
j) $(-5)-(+5)$

4. a) $(-8)-(-4)$ b) $(-2)-(-1)$ c) $(-5)-(-2)$
d) $(-7)-(-6)$ e) $(-6)-(-3)$ f) $(-7)-(-5)$
g) $(-2)-(-2)$ h) $(-4)-(-3)$ i) $(-5)-(-1)$
j) $(-8)-(-6)$

5. a) $(-2)-(-3)$ b) $(-1)-(-4)$ c) $(-3)-(-7)$
d) $(-5)-(-6)$ e) $(-4)-(-8)$ f) $(-1)-(-3)$
g) $(-2)-(-6)$ h) $(-4)-(-4)$ i) $(-3)-(-3)$
j) $(-6)-(-7)$

Multiplying integers

$(+3)\times(+2)$ means 3 lots of $+2$ or
$(+2)+(+2)+(+2)=(+6)$
$(+3)\times(-2)$ means 3 lots of -2 or
$(-2)+(-2)+(-2)=(-6)$
$(-3)\times(+2)$ means 2 lots of -3
or $(-3)+(-3)=(-6)$

Exercise 1.3

1. Calculate:
a) $(+5)\times(-4)$ b) $(+7)\times(-3)$ c) $(+8)\times(+2)$
d) $(+9)\times(-5)$ e) $(+8)\times(-5)$ f) $(+7)\times(-6)$

2. Calculate:
a) $(-3)\times(+4)$ b) $(-7)\times(+2)$ c) $(-8)\times(+2)$
d) $(-5)\times(+3)$ e) $(-6)\times(+9)$ f) $(-8)\times(+8)$

3. Copy and complete the multiplication table below:

×	–3	–2	–1	0	+1	+2	+3
+3		–6					+9
+2							
+1					+1		
0	0			0			0
–1						–2	
–2							
–3							

4. Use the table in Q.3 to read off:
a) $(+3)\times(+2)$ b) $(-2)\times(+1)$ c) $(+3)\times(-2)$
d) $(-2)\times(-3)$ e) $(-3)\times(-1)$ f) $(-2)\times(-2)$

When multiplying:
A positive number by a positive number, the result is positive.
A negative number by a positive number, the result is negative.
A positive number by a negative number, the result is negative.
A negative number by a negative number, the result is positive.

5. Calculate:
a) $(-3) \times (+7)$ b) $(+6) \times (-5)$ c) $(-8) \times (-5)$
d) $(-3) \times (-5)$ e) $(-4) \times (+9)$ f) $(-7) \times (-6)$

The rules for division are the same as for multiplication.

6. Calculate:
a) $(+15) \div (+3)$ b) $(+15) \div (-3)$ c) $(-15) \div (+3)$
d) $(-15) \div (-3)$ e) $(-28) \div (-7)$ f) $(-45) \div (+9)$
g) $(-18) \div (+2)$ h) $(-36) \div (+12)$ i) $(-81) \div (-9)$
j) $(+48) \div (-6)$

7. Find the missing number in these divisions:
a) $(-20) \div (\ldots) = (-2)$ b) $(-30) \div (\ldots) = (-6)$
c) $(+15) \div (\ldots) = (+3)$ d) $(+18) \div (\ldots) = (+3)$
e) $(+25) \div (\ldots) = (-5)$ f) $(-20) \div (\ldots) = (+4)$
g) $(-12) \div (\ldots) = (+3)$ h) $(-6) \div (\ldots) = (-2)$
i) $(+18) \div (\ldots) = (-3)$ j) $(+36) \div (\ldots) = (-6)$

8. Find the missing numbers to make the statement below true:
a) $(\ldots) \times (+5) = (+15)$ b) $(\ldots) \times (-3) = (-15)$
c) $(\ldots) \times (-4) = (-20)$ d) $(-7) \times (\ldots) = (-21)$
e) $(+5) \times (\ldots) = (-40)$ f) $(-7) \times (\ldots) = (+21)$
g) $(-6) \times (\ldots) = (-42)$ h) $(+5) \times (\ldots) = (-40)$
i) $(+5) \times (\ldots) = (-5)$ j) $(-3) \times (\ldots) = (-3)$
k) $(+10) \times (\ldots) = (+100)$

9. a) The table gives pairs of numbers x and y which add together to make 10. That is, $x + y = 10$.

Copy and complete the table.

x	+5	+4	+3	+2	+1	0	−1	−2	−3	−4	−5
y											

b) If $m + n = 7$, copy and complete the table:

m	+5	+4	+3	+2	+1	0	−1	−2	−3	−4	−5
n											

c) If $p + q = -2$, copy and complete the table:

p	+5	+4	+3	+2	+1	0	−1	−2	−3	−4	−5
q											

d) If $xy = +12$, copy and complete the table (xy means x multiplied by y):

x	+4	+3	+2	+1	−1	−2	−3	−4
y								

e) If $xy = -20$, copy and complete the table:

x	+5	+4	+2	+1	−1	−2	−4	−5
y								

Prime numbers

A prime number is one whose only factors are 1 and itself. (Note that 1 is not a prime number.)

Exercise 1.4

In a 10 by 10 square, write the numbers 1 to 100.

Cross out number 1.

Cross out all the even numbers after 2 (these have 2 as a factor).

Cross out every third number after 3 (these have 3 as a factor).

Continue with 5, 7, 11 and 13, then list all the prime numbers less than 100.

Square numbers

In a 10 by 10 square write the numbers 1 to 100.

Shade in 1 and then 2×2, 3×3, 4×4, 5×5 etc.
These are the square numbers.
3×3 can be written 3^2 (you say three squared)
7×7 can be written 7^2 (the 2 is called an index; plural indices)

Cube numbers

$3 \times 3 \times 3$ can be written 3^3 (you say three cubed)
$5 \times 5 \times 5$ can be written 5^3
$2 \times 2 \times 2 \times 5 \times 5$ can be written $2^3 \times 5^2$

Write the following using indices:

1. a) 9×9
b) 12×12
c) 8×8
d) $7 \times 7 \times 7$
e) $4 \times 4 \times 4$
f) $3 \times 3 \times 2 \times 2 \times 2$
g) $5 \times 5 \times 5 \times 2 \times 2$
h) $4 \times 4 \times 3 \times 3 \times 2 \times 2$

Factors

The factors of 12 are all the numbers which will divide exactly into 12,

i.e. 1, 2, 3, 4, 6 and 12.

Exercise 1.5 List all the factors of the following numbers:

a) 6	b) 9	c) 7	d) 15
e) 24	f) 36	g) 35	h) 25
i) 42	j) 100		

Prime factors

The factors of 12 are 1, 2, 3, 4, 5 and 12.

Of these, 2 and 3 are prime numbers, so 2 and 3 are the prime factors of 12.

Exercise 1.6 List the prime factors of the following numbers:

a) 15	b) 18	c) 24	d) 16
e) 20	f) 13	g) 33	h) 35
i) 70	j) 56		

An easy way to find prime factors is to divide by the prime numbers in order, smallest first.

Worked examples

a) Find the prime factors of 18 and express them as a product of prime numbers:

2	18
3	9
3	3
	1

$18 = 2 \times 3 \times 3$ or 2×3^2

b) Find the prime factors of 24 and express them as a product of prime numbers:

2	24
2	12
2	6
3	3
	1

$24 = 2 \times 2 \times 2 \times 3$ or $2^3 \times 3$

c) Find the prime factors of 75 and express them as a product of prime numbers:

3	75
5	25
5	5
	1

$75 = 3 \times 5 \times 5$ or 3×5^2

Exercise 1.7 Find the prime factors of the following numbers and express them as a product of prime numbers:

a) 12 b) 32 c) 36 d) 40
e) 44 f) 56 g) 45 h) 39
i) 231 j) 63

Highest common factor

The prime factors of 12 are $2 \times 2 \times 3$.
The prime factors of 18 are $2 \times 3 \times 3$.
So the highest common factor (HCF) can be seen by inspection to be 2×3, i.e. 6.

Exercise 1.8

1. Find the HCF of the following numbers:
a) 8, 12 b) 10, 25 c) 12, 18, 24
d) 15, 21, 27 e) 36, 63, 108 f) 22, 110
g) 32, 56, 72 h) 39, 52 i) 34, 51, 68
j) 60, 144

Multiples

Multiples of 5 are 5, 10, 15, 20, etc.
The lowest common multiple (LCM) of 2 and 3 is 6, since 6 is the smallest number divisible by 2 and 3.
The LCM of 3 and 5 is 15.
The LCM of 6 and 10 is 30.

2. Find the LCM of the following:
a) 3, 5 b) 4, 6 c) 2, 7 d) 4, 7
e) 4, 8 f) 2, 3, 5 g) 2, 3, 4 h) 3, 4, 6
i) 3, 4, 5 j) 3, 5, 12

3. Find the LCM of the following:
a) 6, 14 b) 4, 15 c) 2, 7, 10 d) 3, 9, 10
e) 6, 8, 20 f) 3, 5, 7 g) 4, 5, 10 h) 3, 7, 11
i) 6, 10, 16 j) 25, 40, 100

NB: All diagrams are not drawn to scale.

Rational and irrational numbers

A **rational number** is any number which can be expressed as a fraction. Examples of some rational numbers and how they can be expressed as a fraction are shown below:

$$0.2 = \tfrac{1}{5} \quad 0.3 = \tfrac{3}{10} \quad 7 = \tfrac{7}{1} \quad 1.53 = \tfrac{153}{100} \quad 0.\dot{2} = \tfrac{2}{9}$$

An **irrational number** cannot be expressed as a fraction. Examples of irrational numbers include:

$$\sqrt{2}, \quad \sqrt{5}, \quad 6 - \sqrt{3}, \quad \pi$$

In summary:
Rational numbers include:

- whole numbers,
- fractions,
- recurring decimals,
- terminating decimals.

Irrational numbers include:

- the square root of any number other than square numbers,
- a decimal which neither repeats nor terminates (e.g. π).

Exercise 1.9

1. For each of the numbers shown below state whether it is rational or irrational:

a) 1.3	b) $0.\dot{6}$	c) $\sqrt{3}$
d) $-2\frac{3}{5}$	e) $\sqrt{25}$	f) $\sqrt[3]{8}$
g) $\sqrt{7}$	h) 0.625	i) $0.\dot{1}\dot{1}$

2. For each of the numbers shown below state whether it is rational or irrational:

a) $\sqrt{2} \times \sqrt{3}$	b) $\sqrt{2} + \sqrt{3}$	c) $(\sqrt{2} \times \sqrt{3})^2$
d) $\dfrac{\sqrt{8}}{\sqrt{2}}$	e) $\dfrac{2\sqrt{5}}{\sqrt{20}}$	f) $4 + (\sqrt{9} - 4)$

3. In each of the following decide whether the quantity required is rational or irrational. Give reasons for your answer.

a)

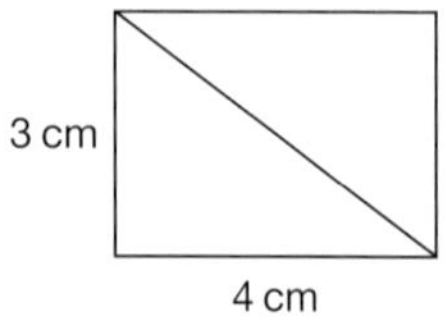

The length of the diagonal

b)

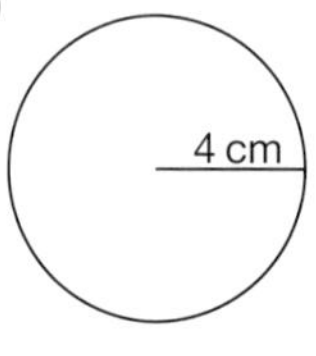

The circumference of the circle

c)

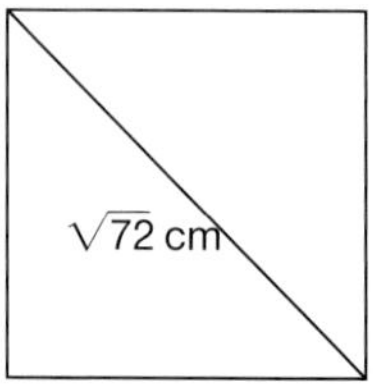

The side length of the square

d)

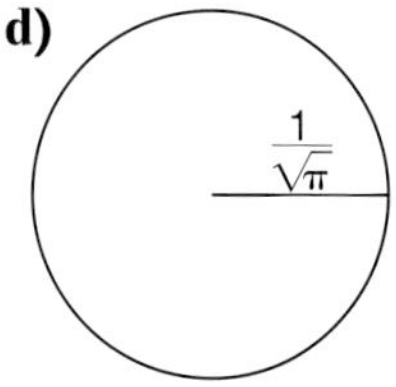

The area of the circle

Sequences

A **sequence** is an ordered set of numbers. Each number in a sequence is known as a **term**. The terms of a sequence form a pattern. For the sequence of numbers

2, 5, 8, 11, 14, 17, ...

the difference between successive terms is +3. The term-to-term rule is therefore + 3.

Worked examples

a) Below is a sequence of numbers.

5, 9, 13, 17, 21, 25, ...

i) What is the term-to-term rule for the sequence?

The term-to-term rule is + 4.

ii) Calculate the 10th term of the sequence.
Continuing the pattern gives:

5, 9, 13, 17, 21, 25, 29, 33, 37, 41, ...

Therefore the 10th term is 41.

b) Below is a sequence of numbers.

1, 2, 4, 8, 16, ...

i) What is the term-to-term rule for the sequence?

The term-to-term rule is $\times$ 2.

ii) Calculate the 10th term of the sequence.
Continuing the pattern gives:

1, 2, 4, 8, 16, 32, 64, 128, 256, 512, ...

Therefore the 10th term is 512.

Exercise 1.10

For each of the sequences given below:

i) State a rule to describe the sequence.
ii) Calculate the 10th term.

a) 3, 6, 9, 12, 15, ...
b) 8, 18, 28, 38, 48, ...
c) 11, 33, 55, 77, 99, ...
d) 0.7, 0.5, 0.3, 0.1, ...
e) $\frac{1}{2}, \frac{1}{3}, \frac{1}{4}, \frac{1}{5}, \ldots$
f) $\frac{1}{2}, \frac{2}{3}, \frac{3}{4}, \frac{4}{5}, \ldots$
g) 1, 4, 9, 16, 25, ...
h) 4, 7, 12, 19, 28, ...
i) 1, 8, 27, 64, ...
j) 5, 25, 125, 625, ...

Sometimes the pattern in a sequence of numbers is not obvious. By looking at the differences between successive terms a pattern is often found.

Worked examples

a) Calculate the 8th term in the sequence

8, 12, 20, 32, 48, ...

The pattern in this sequence is not immediately obvious, so a row for the differences between successive terms can be constructed.

	8		12		20		32		48
1st differences		4		8		12		16	

The pattern in the differences row is + 4 and this can be continued to complete the sequence to the 8th term.

	8		12		20		32		48		68		92		120
1st differences		4		8		12		16		20		24		28	

b) Calculate the 8th term in the sequence

	3		6		13		28		55
1st differences		3		7		15		27	

The row of first differences is not sufficient to spot the pattern, so a row of 2nd differences is constructed.

	3		6		13		28		55
1st differences		3		7		15		27	
2nd differences			4		8		12		

The pattern in the 2nd differences row can be seen to be + 4. This can now be used to complete the sequence.

	3		6		13		28		55		98		161		248
1st differences		3		7		15		27		43		63		87	
2nd differences			4		8		12		16		20		24		

Exercise 1.11

For each of the sequences given below calculate the next two terms:

a) 8, 11, 17, 26, 38, ...
b) 5, 7, 11, 19, 35, ...
c) 9, 3, 3, 9, 21, ...
d) −2, 5, 21, 51, 100, ...
e) 11, 9, 10, 17, 36, 79, ...
f) 4, 7, 11, 19, 36, 69, ...
g) −3, 3, 8, 13, 17, 21, 24, ...

The *n*th term

So far the method used for generating a sequence relies on knowing the previous term in order to work out the next one. This method works but can be a little cumbersome if the 100th term is needed and only the first five terms are given! A more efficient rule is one which is related to a term's position in a sequence.

Worked examples

a) For the sequence shown below give an expression for the nth term.

Position	1	2	3	4	5	n
Term	3	6	9	12	15	?

By looking at the sequence it can be seen that the term is always $3 \times$ position.

Therefore the nth term can be given by the expression $3n$.

b) For the sequence shown below give an expression for the nth term.

Position	1	2	3	4	5	n
Term	2	5	8	11	14	?

You will need to spot similarities between sequences. The terms of the above sequence are the same as the terms in example a) above but with 1 subtracted each time.

The expression for the nth term is therefore $3n - 1$.

Exercise 1.12

For each of the following sequences:

i) Write down the next two terms.
ii) Give an expression for the nth term.

a) 5, 8, 11, 14, 17, ...
b) 5, 9, 13, 17, 21, ...
c) 4, 9, 14, 19, 24, ...
d) 8, 10, 12, 14, 16, ...
e) 1, 8, 15, 22, 29, ...
f) 0, 4, 8, 12, 16, 20, ...
g) 1, 10, 19, 28, 37, ...
h) 15, 25, 35, 45, 55, ...
i) 9, 20, 31, 42, 53, ...
j) 1.5, 3.5, 5.5, 7.5, 9.5, 11.5, ...
k) 0.25, 1.25, 2.25, 3.25, 4.25, ...
l) 0, 1, 2, 3, 4, 5, ...

Exercise 1.13

For each of the following sequences:

i) Write down the next two terms.
ii) Give an expression for the nth term.

a) 2, 5, 10, 17, 26, 37, ...
b) 8, 11, 16, 23, 32, ...
c) 0, 3, 8, 15, 24, 35, ...
d) 1, 8, 27, 64, 125, ...
e) 2, 9, 28, 65, 126, ...
f) 11, 18, 37, 74, 135, ...
g) −2, 5, 24, 61, 122, ...
h) 2, 6, 12, 20, 30, 42, ...

Student Assessment 1

For Q.1 to 7, use a number line if necessary:

1. a) $(+8) + (-3)$ b) $(+6) + (-5)$ c) $(+7) + (-2)$
2. a) $(-5) + (+2)$ b) $(-7) + (+3)$ c) $(-11) + (+1)$
3. a) $(-4) + (-5)$ b) $(-8) + (-9)$ c) $(-8) + (-8)$
4. a) $(+3) - (+2)$ b) $(+12) - (+3)$ c) $(+16) - (+14)$
5. a) $(+4) - (+8)$ b) $(+7) - (+12)$ c) $(+3) - (+9)$
6. a) $(-8) - (-6)$ b) $(-4) - (-2)$ c) $(-9) - (-1)$
7. a) $(-5) - (-12)$ b) $(-8) - (-15)$ c) $(-1) - (-12)$
8. Write each of the following numbers as a product of prime factors:
   a) 60 b) 63 c) 64
9. Find the highest common factor (HCF) of the following pairs of numbers:
   a) 16, 30 b) 12, 22 c) 24, 48
10. Find the lowest common multiple (LCM) of the following pairs of numbers:
    a) 4, 7 b) 6, 10 c) 12, 18
11. a) $(+8) \times (-4)$ b) $(-7) \times (+9)$ c) $(-5) \times (-9)$
12. a) $(-36) \div (+6)$ b) $(+63) \div (-9)$ c) $(-49) \div (-7)$
13. If $x + y = 7$, copy and complete the table:

x	4	3	2	1	0	−1	−2	−3	−4
y									

14. If $xy = 24$, copy and complete the table:

x	+8	+6	+4	+2	−2	−4	−6	−8
y								

15. If $\frac{x}{y} = 2$, copy and complete the table:

x	20	10	6	2	−2	−4	−12
y							

Student Assessment 2

Use a number line if necessary in Q.1 to 7:

1. a) $(+11) + (-3)$ b) $(-10) + (+2)$ c) $(-7) + (-8)$
2. a) $(-8) + (+4)$ b) $(+10) + (+2)$ c) $(-7) + (+5)$

3. a) $(-6) + (-6)$ b) $(-4) + (-3)$ c) $(-5) + (-4)$

4. a) $(-2) + (+1)$ b) $(+6) - (+3)$ c) $(+12) - (+11)$

5. a) $(+2) - (+6)$ b) $(+11) - (+15)$ c) $(+3) - (+12)$

6. a) $(-5) - (-4)$ b) $(-17) - (-9)$ c) $(-18) - (-11)$

7. a) $(-8) - (-10)$ b) $(-6) - (-11)$ c) $(-7) - (-7)$

8. Write each of the following numbers as a product of prime factors:
a) 24 b) 45 c) 50

9. Find the highest common factor (HCF) of the following pairs of numbers:
a) 18, 48 b) 24, 60 c) 51, 68

10. Find the lowest common multiple (LCM) of the following pairs of numbers:
a) 6, 7 b) 9, 15 c) 40, 50

11. a) $(-6) \times (+3)$ b) $(+9) \times (-7)$ c) $(-6) \times (-8)$

12. a) $(+24) \div (-2)$ b) $(-54) \div (+6)$ c) $(-36) \div (-12)$

13. If $x + y = -1$, copy and complete the table:

x	3	2	1	0	−1	−2	−3
y							

14. If $xy = 12$, copy and complete the table:

x	+6	+3	+1	−1	−2	−4	−12
y							

15. If $\frac{x}{y} = -5$, copy and complete the table:

x	+20	+10	+5	−15	−25	−50
y						

Student Assessment 3

1. State whether the following numbers are rational or irrational:
a) 1.5 b) $\sqrt{7}$ c) $0.\dot{7}$
d) $0.\dot{7}\dot{3}$ e) $\sqrt{121}$ f) π

2. For each of the sequences given below:
i) Calculate the next two terms.
ii) Explain the pattern in words.
a) 9, 18, 27, 36, ... b) 54, 48, 42, 36, ...
c) 18, 9, 4.5, ... d) 12, 6, 0, −6, ...
e) 216, 125, 64, ... f) 1, 3, 9, 27, ...

3. For each of the sequences shown below, give an expression for the nth term:
   a) 6, 10, 14, 18, 22, ...
   b) 13, 19, 25, 31, ...
   c) 3, 9, 15, 21, 27, ...
   d) 4, 7, 12, 19, 28, ...
   e) 0, 10, 20, 30, 40, ...
   f) 0, 7, 26, 63, 124, ...

Student Assessment 4

1. Show, by expressing them as fractions or whole numbers, that the following numbers are rational:
   a) 0.625
   b) $\sqrt[3]{27}$
   c) 0.44
2. For each of the sequences given below:
   i) Calculate the next two terms.
   ii) Explain the pattern in words.
   a) 6, 12, 18, 24, ...
   b) 24, 21, 18, 15, ...
   c) 10, 5, 0, ...
   d) 16, 25, 36, 49, ...
   e) 1, 10, 100, ...
   f) 1, $\frac{1}{2}$, $\frac{1}{4}$, $\frac{1}{8}$, ...
3. For each of the sequences shown below, give an expression for the nth term:
   a) 3, 5, 7, 9, 11, ...
   b) 7, 13, 19, 25, 31, ...
   c) 8, 18, 28, 38, ...
   d) 1, 9, 17, 25, ...
   e) −4, 4, 12, 20, ...
   f) 2, 5, 10, 17, 26, ...

2 SQUARES, SQUARE ROOTS AND CUBES

Calculate squares, square roots, cubes and cube roots of numbers.

NB: All diagrams are not drawn to scale.

This is a square of side 1 cm.

This is a square of side 2 cm.
It has four squares of side 1 cm in it.

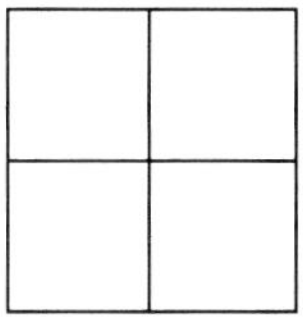

Exercise 2.1

Calculate how many squares of side 1 cm there would be in squares of side:

a) 3 cm	b) 5 cm	c) 8 cm	d) 10 cm
e) 11 cm	f) 12 cm	g) 7 cm	h) 13 cm
i) 15 cm	j) 20 cm		

In index notation, the square numbers are 1^2, 2^2, 3^2, 4^2, etc.
4^2 is read as '4 squared'.

Worked example

This square is of side 1.1 units.
Its area is 1.1×1.1 unit^2.
$A = 1 \times 1 = 1$
$B = 1 \times 0.1 = 0.1$
$B = 1 \times 0.1 = 0.1$
$C = 0.1 \times 0.1 = 0.01$
Total $= 1.21$ units^2

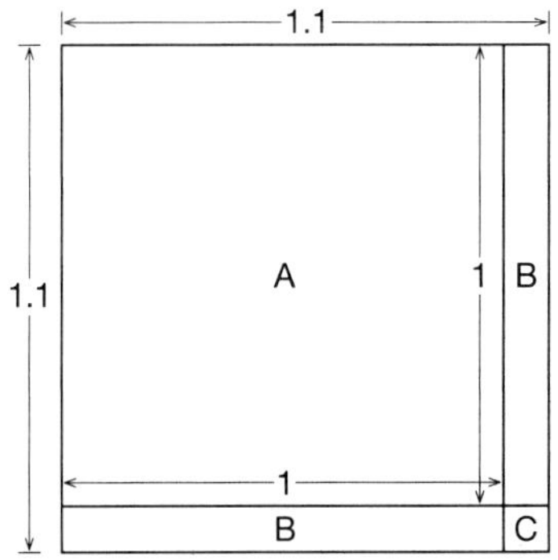

Exercise 2.2

1. Draw diagrams and use them to find the area of squares of side:

a) 2.1 units	b) 3.1 units	c) 1.2 units
d) 2.2 units	e) 2.5 units	f) 1.4 units

2. Use long multiplication to work out the area of squares of side:

a) 2.4	b) 3.3	c) 2.8	d) 6.2
e) 4.6	f) 7.3	g) 0.3	h) 0.8
i) 0.1	j) 0.9		

3. Check your answers to Q.1 and 2 by using the x^2 key on a calculator.

Using a graph

4. Copy and complete the table for the equation $y = x^2$.

x	0	1	2	3	4	5	6	7	8
y				9				49	

Plot the graph of $y = x^2$. Use your graph to find the value of the following:

a) 2.5^2 b) 3.5^2 c) 4.5^2 d) 5.5^2
e) 7.2^2 f) 6.4^2 g) 0.8^2 h) 0.2^2
i) 5.3^2 j) 6.3^2

5. Check your answers to question 4 by using the x^2 key on a calculator.

Square roots

The square on the left contains 16 squares. It has sides of length 4 units.

So the square root of 16 is 4.
This can be written as $\sqrt{16} = 4$.

Note that $4 \times 4 = 16$ so $\sqrt{16}$ is 4
but also $-4 \times -4 = 16$ so $\sqrt{16}$ is also -4.
So $\sqrt{81}$ is ± 9 i.e. $+9$ or -9.

Note that -16 has no square root since any integer squared is positive.

Exercise 2.3

1. Find the following square roots:

a) $\sqrt{25}$ b) $\sqrt{9}$ c) $\sqrt{49}$ d) $\sqrt{100}$
e) $\sqrt{121}$ f) $\sqrt{169}$ g) $\sqrt{0.01}$ h) $\sqrt{0.04}$
i) $\sqrt{0.09}$ j) $\sqrt{0.25}$

2. Use the $\sqrt{\ }$ key on your calculator to check your answers to Q.1.

3. Calculate the following square roots:

a) $\sqrt{\frac{1}{9}}$ b) $\sqrt{\frac{1}{16}}$ c) $\sqrt{\frac{1}{25}}$ d) $\sqrt{\frac{1}{49}}$
e) $\sqrt{\frac{1}{100}}$ f) $\sqrt{\frac{4}{9}}$ g) $\sqrt{\frac{9}{100}}$ h) $\sqrt{\frac{49}{81}}$
i) $\sqrt{2\frac{7}{9}}$ j) $\sqrt{6\frac{1}{4}}$

Exercise 2.4

Using a graph

1. Copy and complete the table below for the equation $y = \sqrt{x}$.

x	0	1	4	9	16	25	36	49	64	81	100
y											

Plot the graph of $y = \sqrt{x}$. Use your graph to find the values of the following square roots:

a) $\sqrt{70}$ b) $\sqrt{40}$ c) $\sqrt{50}$ d) $\sqrt{90}$
e) $\sqrt{35}$ f) $\sqrt{45}$ g) $\sqrt{55}$ h) $\sqrt{60}$
i) $\sqrt{2}$ j) $\sqrt{3}$ k) $\sqrt{20}$ l) $\sqrt{30}$
m) $\sqrt{12}$ n) $\sqrt{75}$ o) $\sqrt{115}$

2. Check your answers to Q.1 above by using the $\sqrt{\ }$ key on a calculator.

Cubes of numbers

The cube below has sides of 1 unit and occupies 1 cubic unit of space.

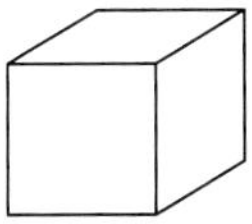

The cube below has sides of 2 units and occupies 8 cubic units of space. (That is, $2 \times 2 \times 2$.)

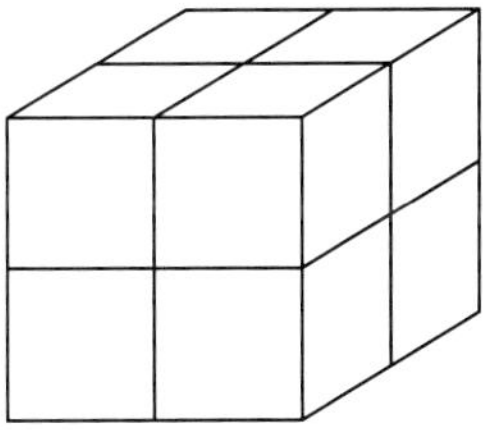

Exercise 2.5

1. How many cubic units would be occupied by cubes of side:
a) 3 units b) 5 units c) 10 units
d) 4 units e) 9 units f) 100 units

In index notation, the cube numbers are 1^3, 2^3, 3^3, 4^3, etc. 4^3 is read as '4 cubed'.

Some calculators have an x^3 key. On others, to find a cube you multiply the number by itself three times.

2. Copy and complete the table below:

Number	1	2	3	4	5	6	7	8	9	10
Cube			27							

3. Use a calculator to find the following:

a) 11^3 b) 0.5^3 c) 1.5^3 d) 2.5^3
e) 20^3 f) 30^3 g) $3^3 + 2^3$ h) $(3 + 2)^3$
i) $7^3 + 3^3$ j) $(7 + 3)^3$

$\sqrt[3]{\ }$ is read as 'the cube root of . . .'.
$\sqrt[3]{64}$ is 4, since $4 \times 4 \times 4 = 64$.

Note that $\sqrt[3]{64}$ is not -4
since $-4 \times -4 \times -4 = -64$
but $\sqrt[3]{-64}$ is -4.

4. Find the following cube roots:
a) $\sqrt[3]{8}$ b) $\sqrt[3]{125}$ c) $\sqrt[3]{27}$ d) $\sqrt[3]{0.001}$
e) $\sqrt[3]{0.027}$ f) $\sqrt[3]{216}$ g) $\sqrt[3]{1000}$ h) $\sqrt[3]{1\,000\,000}$
i) $\sqrt[3]{-8}$ j) $\sqrt[3]{-27}$ k) $\sqrt[3]{-1000}$ l) $\sqrt[3]{-1}$

Student Assessment 1

1. Without using a calculator, find the value of:
a) 7^2 b) 12^2 c) $(0.1)^2$ d) $(0.3)^2$

2. Draw a square of side 4.1 units. Use it to find $(4.1)^2$.

3. Use long multiplication to find the value of:
a) $(2.5)^2$ b) $(3.3)^2$ c) $(0.25)^2$

4. Copy and complete the table below and plot a graph of $y = x^2$.

x	1	2	3	4	5	6	7	8
y								

Plot the graph of $y = x^2$. Use your graph to find the value of:
a) $(3.6)^2$ b) $(6.5)^2$ c) $(7.5)^2$ d) $(5.3)^2$

5. Calculate the following square roots without using a calculator:
a) $\sqrt{49}$ b) $\sqrt{144}$ c) $\sqrt{0.64}$
d) $\sqrt{6\frac{1}{4}}$ e) $\sqrt{\frac{81}{100}}$ f) $\sqrt{2\frac{28}{36}}$

6. Without using a calculator, find:
a) 2^3 b) 5^3 c) 10^3

7. Without using a calculator, find:
a) $\sqrt[3]{64}$ b) $\sqrt[3]{125}$ c) $\sqrt[3]{1000}$

Student Assessment 2

1. Find the value of:
a) 9^2 b) 15^2 c) $(0.2)^2$ d) $(0.7)^2$

2. Draw a square of side 2.5 units. Use it to find $(2.5)^2$.

3. Use long multiplication to find:
a) $(3.5)^2$ b) $(4.1)^2$ c) $(0.15)^2$

4. Copy and complete the table below for $y = \sqrt{x}$.

x	0	1	4	9	16	25	36	49
y								

Plot the graph of $y = \sqrt{x}$. Use your graph to find:

a) $\sqrt{7}$ b) $\sqrt{30}$ c) $\sqrt{45}$

5. Without using a calculator, find:

a) $\sqrt{225}$ b) $\sqrt{0.01}$ c) $\sqrt{0.81}$

d) $\sqrt{\frac{9}{25}}$ e) $\sqrt{5\frac{4}{9}}$ f) $\sqrt{2\frac{23}{49}}$

6. Without using a calculator, find:

a) 4^3 b) $(0.1)^3$ c) $(\frac{2}{3})^3$

7. Without using a calculator, find:

a) $\sqrt[3]{27}$ b) $\sqrt[3]{1\,000\,000}$ c) $\sqrt[3]{\frac{64}{125}}$

3 DIRECTED NUMBERS

Use directed numbers in practical situations.

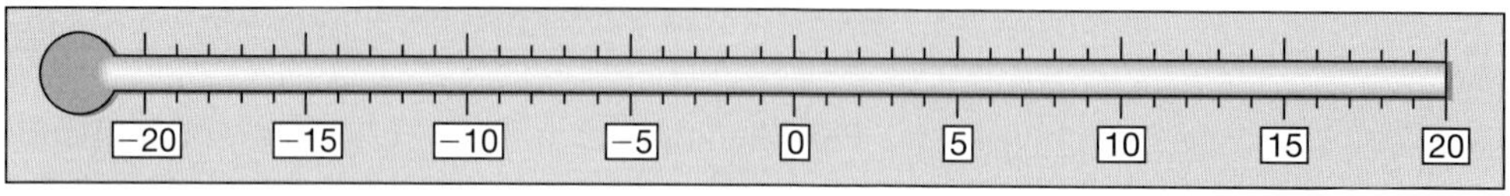

Worked example The diagram above shows the scale of a thermometer. The temperature at 4 a.m. was −3 °C. By 9 a.m. the temperature had risen by 8 °C. What was the temperature at 9 a.m.?

$$(-3)^\circ + (8)^\circ = (5)^\circ$$

Exercise 3.1

Find the new temperature if:

1. a) The temperature was −5 °C, and rises 9 °C.
b) The temperature was −12 °C, and rises 8 °C.
c) The temperature was +14 °C, and falls 8 °C.
d) The temperature was −3 °C, and falls 4 °C.
e) The temperature was −7 °C, and falls 11 °C.
f) The temperature was 2 °C, it falls 8 °C, then rises 6 °C.
g) The temperature was 5 °C, it falls 8 °C, then falls a further 6 °C.
h) The temperature was −2 °C, it falls 6 °C, then rises 10 °C.
i) The temperature was 20 °C, it falls 18 °C, then falls a further 8 °C.
j) The temperature was 5 °C below zero and falls 8 °C.

2. Mark lives in Canada. Every morning before school he reads a thermometer to find the temperature in the garden. The thermometer below shows the results for 5 days in winter.

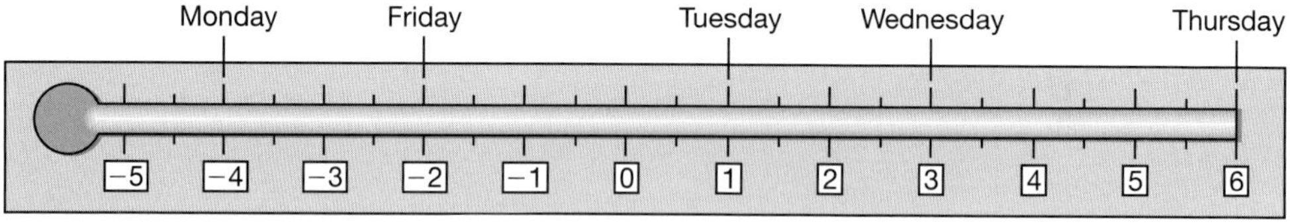

Find the change in temperature between:
a) Monday and Friday
b) Monday and Thursday
c) Tuesday and Friday
d) Thursday and Friday
e) Monday and Tuesday.

3. The highest temperature ever recorded was in Libya. It was 58 °C. The lowest temperature ever recorded was −88 °C in Antarctica. What is the temperature difference?

4. Julius Caesar was born in 100 BC (or BCE) and was 56 years old when he died. In what year did he die?

5. Marcus Flavius was born in 20 BC and died in AD 42 (or CE). How old was he when he died?

6. Rome was founded in 753 BC. Constantinople fell to Mehmet Sultan Ahmet in AD 1453, ending the Roman Empire in the East. For how many years did the Roman Empire last?

7. My bank account shows a credit balance of £105. Describe my balance as a positive or negative number after each of these transactions is made in sequence:
a) rent £140
b) car insurance £283
c) 1 week's salary £230
d) food bill £72
e) credit transfer £250

8. A lift in the Empire State Building in New York has stopped somewhere close to the halfway point. Call this 'floor zero'. Show on a number line the floors it stops at as it makes the following sequence of journeys:
a) up 75 floors
b) down 155 floors
c) up 110 floors
d) down 60 floors
e) down 35 floors
f) up 100 floors

9. A hang-glider is launched from a mountainside in the Swiss Alps. It climbs 650 m and then starts its descent. It falls 1220 m before landing.
a) How far below its launch point was the hang-glider when it landed?
b) If the launch point was at 1650 m above sea level, at what height above sea level did it land?

10. The average noon temperature in Sydney in January is $+32$ °C. The average midnight temperature in Boston in January is -12 °C. What is the temperature difference between the two cities?

11. The temperature in London on New Year's day is -2 °C. The temperature in Moscow on the same day is -14 °C. What is the temperature difference between the two cities?

12. The temperature inside a freezer is -8 °C. To defrost it, the temperature is allowed to rise by 12 °C. What will the temperature be after this rise?

13. A plane flying at 8500 m drops a sonar device onto the ocean floor. If the sonar falls a total of 10 200 m, how deep is the ocean at this point?

14. The roof of an apartment block is 130 m above ground level. The car park beneath the apartment is 35 m below ground level. How high is the roof above the floor of the car park?

15. A submarine is at a depth of 165 m. If the ocean floor is 860 m from the surface, how far is the submarine from the ocean floor?

Student Assessment 1

Date	Event
2900 BC (BCE)	Great Pyramid built
1650 BC	Rhind Papyrus written
540 BC	Pythagoras born
300 BC	Euclid born
290 AD (CE)	Lui Chih calculated π as 3.14
1500 AD	Leonardo da Vinci born
1900 AD	Albert Einstein born
1998 AD	Fermat's last theorem proven

The table above shows dates of some significance to mathematics. Use the table to answer questions 1–6 below.

1. How many years before Einstein was born was the Great Pyramid built?

2. How many years before Leonardo was born was Pythagoras born?

3. How many years after Lui Chih's calculation of π was Fermat's last theorem proved?

4. How many years were there between the births of Euclid and Einstein?

5. How long before Fermat's last theorem was proved was the Rhind Papyrus written?

6. How old was the Great Pyramid when Leonardo was born?

7.

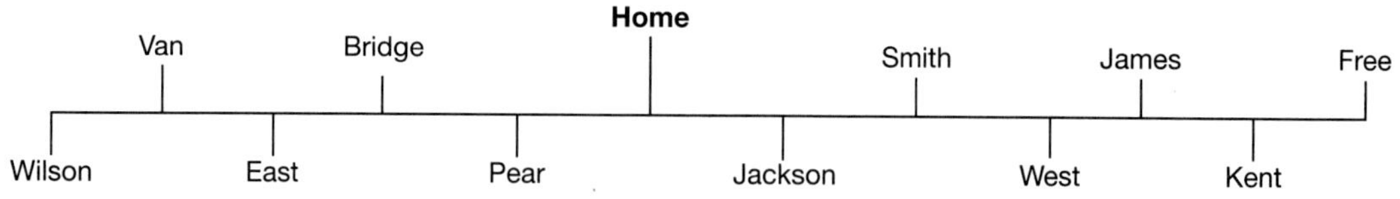

A bus route runs past Danny's house. Each stop is given the name of a street. From Home to Smith Street is the positive direction. Find where Danny is after the stages of these journeys from Home:

a) $+4-3$
b) $+2-5$
c) $+2-7$
d) $-3-2$
e) $-1-1$
f) $+6-8+1$
g) $-1+3-5$
h) $-2-2+8$
i) $+1-3+5$
j) $-5+8-1$

8. Using the diagram from Q.7, and starting from Home each time, find the missing stages in these journeys if they end at the stop given:

a) $+3+?$	Pear	b) $+6+?$	Jackson
c) $-1+?$	Van	d) $-5+?$	James
e) $+5+?$	Home	f) $?-2$	Smith
g) $?+2$	East	h) $?-5$	Van
i) $?-1$	East	j) $?+4$	Pear

Student Assessment 2

Date	Event
1200 BC (BCE)	End of the Hittite Empire in Turkey
300 BC	Ptolemy rules Egypt
969 AD (CE)	Fatimids found Cairo
1258 AD	Mongols destroy Baghdad
1870 AD	Suez Canal opens

Some significant dates in the history of the Middle East are shown in the table above. Use the table to answer Q.1–4 below.

1. How long after the end of the Hittite Empire did Ptolemy rule Egypt?

2. How many years before the destruction of Baghdad by Mongols was Cairo founded?

3. How many years were there between Ptolemy's rule and the opening of the Suez Canal?

4. How many years were there between the end of the Hittite Empire and the founding of Cairo?

5. My bank statement for seven days in October is shown below:

Date	Payments (£)	Receipts (£)	Balance (£)
01/10			200
02/10	284		(a)
03/10		175	(b)
04/10	(c)		46
05/10		(d)	120
06/10	163		(e)
07/10		28	(f)

Copy and complete the statement by entering the amounts (a) to (f).

6. Stops on an underground railway are given the colours of the rainbow:

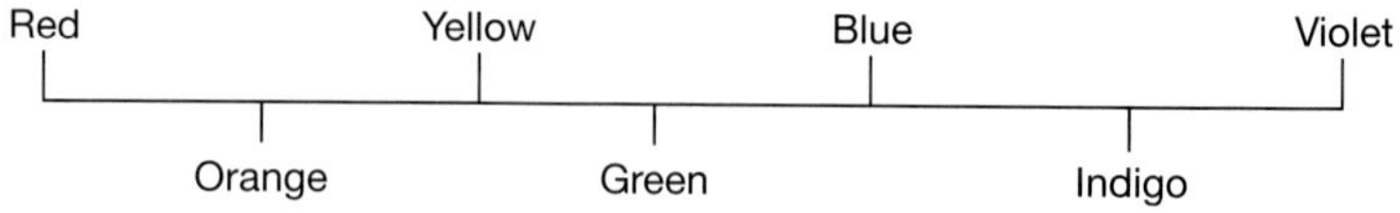

If I start at Green Station, where do I finish up if I make the following moves? (Positive moves are towards Violet and negative moves are towards Red.)

a) $+2-3$ b) $+3-2$ c) $+1-4$
d) $-1-2$ e) $+1-3$ f) $-3+5$
g) $-1-1+3$ h) $-2+5-2$ i) $+3-6+1$
j) $-1+4-2$ k) $-3+6-1$ l) $+3-6+3$

7. The noon and midnight temperatures, in degrees Celsius, in the Sahara during one week in January are shown below:

	Noon	**Midnight**	**Range**
Sunday	+30	0	(a)
Monday	+28	(b)	32
Tuesday	+26	−4	(c)
Wednesday	+32	(d)	34
Thursday	+33	+3	(e)
Friday	+34	(f)	30
Saturday	+28	−1	(g)

Copy and complete the above chart by putting the correct values for (a) to (g).

4 VULGAR AND DECIMAL FRACTIONS, AND PERCENTAGES

Use the language and notation of simple vulgar and decimal fractions and percentages; recognise equivalence and convert between these forms.

A single unit can be broken into equal parts called fractions, e.g. $\frac{1}{2}$, $\frac{1}{3}$, $\frac{1}{6}$.

If, for example, the unit is broken into ten equal parts and three parts are then taken, the fraction is written as $\frac{3}{10}$. That is, three parts out of ten parts.

In the fraction $\frac{3}{10}$:

The ten is called the **denominator**.

The three is called the **numerator**.

A **proper fraction** has its numerator less than its denominator, e.g. $\frac{3}{4}$.

An **improper fraction** has its numerator more than its denominator, e.g. $\frac{9}{2}$.

A **mixed number** is made up of a whole number and a proper fraction, e.g. $4\frac{1}{5}$.

Exercise 4.1

1. Copy the following fractions and indicate which is the numerator and which the denominator.

   a) $\frac{2}{3}$ b) $\frac{15}{22}$ c) $\frac{4}{3}$ d) $\frac{5}{2}$

2. Draw up a table with three columns headed 'proper fraction', 'improper fraction' and 'mixed number'. Put each of the numbers below into the appropriate column.

   a) $\frac{2}{3}$ b) $\frac{15}{22}$ c) $\frac{4}{3}$ d) $\frac{5}{2}$
   e) $1\frac{1}{2}$ f) $2\frac{3}{4}$ g) $\frac{7}{4}$ h) $\frac{7}{11}$
   i) $7\frac{1}{4}$ j) $\frac{5}{6}$ k) $\frac{6}{5}$ l) $1\frac{1}{5}$
   m) $\frac{1}{10}$ n) $2\frac{7}{8}$ o) $\frac{5}{3}$

Worked examples

a) Find $\frac{1}{5}$ of 35.
This means 'divide 35 into 5 equal parts'.
$\frac{1}{5}$ of 35 is $35/5 = 7$.

b) Find $\frac{3}{5}$ of 35.
Since $\frac{1}{5}$ of 35 is 7, $\frac{3}{5}$ of 35 is 7×3.
That is, 21.

Exercise 4.2

1. Evaluate the following:

   a) $\frac{1}{5}$ of 40 b) $\frac{3}{5}$ of 40 c) $\frac{1}{9}$ of 36 d) $\frac{5}{9}$ of 36
   e) $\frac{1}{8}$ of 72 f) $\frac{7}{8}$ of 72 g) $\frac{1}{12}$ of 60 h) $\frac{5}{12}$ of 60
   i) $\frac{1}{4}$ of 8 j) $\frac{3}{4}$ of 8

2. Evaluate the following:
a) $\frac{3}{4}$ of 12 b) $\frac{4}{5}$ of 20 c) $\frac{4}{9}$ of 45 d) $\frac{5}{8}$ of 64
e) $\frac{3}{11}$ of 66 f) $\frac{9}{10}$ of 80 g) $\frac{5}{7}$ of 42 h) $\frac{8}{9}$ of 54
i) $\frac{7}{8}$ of 240 j) $\frac{4}{5}$ of 65

To change a mixed number to an improper fraction:

Worked examples

a) Change $2\frac{3}{4}$ to an improper fraction.

$$1 = \frac{4}{4}$$
$$2 = \frac{8}{4}$$
$$2\frac{3}{4} = \frac{8}{4} + \frac{3}{4}$$
$$= \frac{11}{4}$$

b) Change $3\frac{5}{8}$ into an improper fraction.

$$3\frac{5}{8} = \frac{24}{8} + \frac{5}{8}$$
$$= \frac{24+5}{8}$$
$$= \frac{29}{8}$$

Exercise 4.3

1. Change the following mixed numbers into improper fractions:
a) $4\frac{2}{3}$ b) $3\frac{3}{5}$ c) $5\frac{7}{8}$ d) $2\frac{5}{6}$
e) $8\frac{1}{2}$ f) $9\frac{5}{7}$ g) $6\frac{4}{9}$ h) $4\frac{1}{4}$
i) $5\frac{4}{11}$ j) $7\frac{6}{7}$ k) $4\frac{3}{10}$ l) $11\frac{3}{13}$

To change an improper fraction into a mixed number:

Worked example

Change $\frac{27}{4}$ into a mixed number.

$$\frac{27}{4} = \frac{24+3}{4}$$
$$= \frac{24}{4} + \frac{3}{4}$$
$$= 6\frac{3}{4}$$

2. Change the following improper fractions into mixed numbers.
a) $\frac{29}{4}$ b) $\frac{33}{5}$ c) $\frac{41}{6}$ d) $\frac{54}{8}$
e) $\frac{49}{9}$ f) $\frac{16}{12}$ g) $\frac{66}{7}$ h) $\frac{33}{10}$
i) $\frac{19}{2}$ j) $\frac{73}{12}$

Decimals

H	T	U.	$\frac{1}{10}$	$\frac{1}{100}$	$\frac{1}{1000}$
		3.	2	7	
		0.	0	3	8

3.27 is 3 units, 2 tenths and 7 hundredths

i.e. $3.27 = 3 + \frac{2}{10} + \frac{7}{100}$

0.038 is 3 hundredths and 8 thousandths

i.e. $0.038 = \frac{3}{100} + \frac{8}{1000}$

Note that 2 tenths and 7 hundredths is equivalent to 27 hundredths

i.e. $\frac{2}{10} + \frac{7}{100} = \frac{27}{100}$

and that 3 hundredths and 8 thousandths is equivalent to 38 thousandths

i.e. $\frac{3}{100} + \frac{8}{1000} = \frac{38}{1000}$

Exercise 4.4

1. Make a table similar to the one at the beginning of this section. List the digits in the following numbers in their correct position:

a) 6.023 b) 5.94 c) 18.3
d) 0.071 e) 2.001 f) 3.56

2. Write the following fractions as decimals:

a) $4\frac{5}{10}$ b) $6\frac{3}{10}$ c) $17\frac{8}{10}$ d) $3\frac{7}{100}$
e) $9\frac{27}{100}$ f) $11\frac{36}{100}$ g) $4\frac{6}{1000}$ h) $5\frac{27}{1000}$
i) $4\frac{356}{1000}$ j) $9\frac{204}{1000}$

3. Evaluate the following without using a calculator:

a) $2.7 + 0.35 + 16.09$ b) $1.44 + 0.072 + 82.3$
c) $23.8 - 17.2$ d) $16.9 - 5.74$
e) $121.3 - 85.49$ f) $6.03 + 0.5 - 1.21$
g) $72.5 - 9.08 + 3.72$ h) $100 - 32.74 - 61.2$
i) $16.0 - 9.24 - 5.36$ j) $1.1 - 0.92 - 0.005$

Percentages

A fraction whose denominator is 100 can be expressed as a percentage.

$\frac{29}{100}$ can be written as 29%

$\frac{45}{100}$ can be written as 45%

Exercise 4.5

1. Write the following fractions as percentages:

a) $\frac{30}{100}$ b) $\frac{42}{100}$ c) $\frac{63}{100}$ d) $\frac{5}{100}$

By using equivalent fractions to change the denominator to 100, other fractions can be written as percentages.

Worked example Change $\frac{3}{5}$ to a percentage.

$\frac{3}{5} = \frac{3}{5} \times \frac{20}{20} = \frac{60}{100}$

$\frac{60}{100}$ can be written as 60%

2. Express each of the following as a fraction with denominator 100, then write them as percentages:

a) $\frac{29}{50}$ b) $\frac{17}{25}$ c) $\frac{11}{20}$ d) $\frac{3}{10}$
e) $\frac{23}{25}$ f) $\frac{19}{50}$ g) $\frac{3}{4}$ h) $\frac{2}{5}$

3. Copy and complete the table of equivalents below.

Fraction	Decimal	Percentage
$\frac{1}{10}$		
	0.2	
		30%
$\frac{4}{10}$		
	0.5	
		60%
	0.7	
$\frac{4}{5}$		
	0.9	
$\frac{1}{4}$		
		75%

Student Assessment 1

1. Copy the following numbers. Circle improper fractions and underline mixed numbers:

a) $3\frac{1}{2}$ b) $\frac{2}{5}$ c) $\frac{7}{5}$ d) $\frac{8}{9}$
e) $1\frac{3}{4}$

2. Evaluate the following:

a) $\frac{1}{5}$ of 60 b) $\frac{3}{5}$ of 55 c) $\frac{2}{7}$ of 21 d) $\frac{3}{4}$ of 120

3. Change the following mixed numbers to improper fractions:

a) $2\frac{1}{2}$ b) $3\frac{6}{7}$ c) $5\frac{11}{12}$

4. Change the following improper fractions to mixed numbers:

a) $\frac{22}{7}$ b) $\frac{36}{5}$ c) $\frac{67}{9}$

5. Copy the following set of equivalent fractions and fill in the missing numerators:

$\frac{5}{6} = \frac{}{12} = \frac{}{24} = \frac{}{60} = \frac{}{108}$

6. Write the following fractions as decimals:

a) $\frac{27}{100}$ b) $\frac{105}{1000}$ c) $\frac{7}{100}$ d) $\frac{87}{1000}$

7. Write the following as percentages:

a) $\frac{3}{10}$ b) $\frac{29}{100}$ c) $\frac{1}{2}$ d) $\frac{7}{10}$
e) $\frac{4}{5}$ f) $2\frac{19}{100}$ g) $\frac{6}{100}$ h) $\frac{3}{4}$
i) 0.31 j) 0.07 k) 3.4 l) 2

Student Assessment 2

1. Copy the following numbers. Circle improper fractions and underline mixed numbers:

a) $\frac{3}{11}$ b) $5\frac{3}{4}$ c) $\frac{27}{8}$ d) $\frac{3}{7}$

2. Evaluate the following:

a) $\frac{1}{3}$ of 63 b) $\frac{3}{8}$ of 72 c) $\frac{2}{5}$ of 55 d) $\frac{3}{13}$ of 169

3. Change the following mixed numbers to improper fractions:

a) $2\frac{3}{5}$ b) $3\frac{4}{9}$ c) $5\frac{5}{8}$

4. Change the following improper fractions to mixed numbers:

a) $\frac{33}{5}$ b) $\frac{48}{9}$ c) $\frac{67}{11}$

5. Copy the following set of equivalent fractions and fill in the missing numerators:

$\frac{2}{3} = \frac{}{6} = \frac{}{12} = \frac{}{18} = \frac{}{27} = \frac{}{30}$

6. Write the following fractions as decimals:

a) $\frac{35}{100}$ b) $\frac{275}{1000}$ c) $\frac{675}{100}$ d) $\frac{35}{1000}$

7. Write the following as percentages:

a) $\frac{3}{5}$ b) $\frac{49}{100}$ c) $\frac{1}{4}$ d) $\frac{9}{10}$
e) $1\frac{1}{2}$ f) $3\frac{27}{100}$ g) $\frac{5}{100}$ h) $\frac{7}{20}$
i) 0.77 j) 0.03 k) 2.9 l) 4

5 ORDERING

Order quantities by magnitude; demonstrate familiarity with the symbols $=$, $\neq$, $>$, $<$, $\geqslant$, $\leqslant$.

The following symbols have a specific meaning in mathematics:

$=$	is equal to
$\neq$	is not equal to
$>$	is greater than
$\geqslant$	is greater than or equal to
$<$	is less than
$\leqslant$	is less than or equal to

$x \geqslant 3$ states that x is greater than or equal to 3, i.e. x can be 3, 4, 4.2, 5, 5.6, etc.

$3 \leqslant x$ states that 3 is less than or equal to x, i.e. x is still 3, 4, 4.2, 5, 5.6, etc.

Therefore:

$5 > x$ can be rewritten as $x < 5$, i.e. x can be 4, 3.2, 3, 2.8, 2, 1, etc.

$-7 \leqslant x$ can be rewritten as $x \geqslant -7$, i.e. x can be -7, -6, -5, etc.

These inequalities can also be represented on a number line:

$x < 5$

$x \geqslant -7$

Note that ○⟶ implies that the number is not included in the solution whilst ●⟶ implies that the number is included in the solution.

Worked examples

a) Write $a > 3$ in words:

a is greater than 3.

b) Write 'x is greater than or equal to 8' using appropriate symbols:

$x \geqslant 8$

c) Write 'V is greater than 5, but less than or equal to 12' using the appropriate symbols:

$5 < V \leqslant 12$

Exercise 5.1

1. Write the following in words:

a) $a < 7$ b) $b > 4$ c) $c \neq 8$

d) $d \leqslant 3$ e) $e \geqslant 9$ f) $f \leqslant 11$

2. Rewrite the following, using the appropriate symbols:

a) a is less than 4
b) b is greater than 7
c) c is equal to or less than 9
d) d is equal to or greater than 5
e) e is not equal to 3
f) f is not more than 6
g g is not less than 9
h) h is at least 6
i) i is not 7
j) j is not greater than 20

3. Write the following in words:

a) $5 < n < 10$
b) $6 \leqslant n \leqslant 15$
c) $3 \leqslant n < 9$
d) $8 < n \leqslant 12$

4. Write the following using the appropriate symbols:

a) p is more than 7, but less than 10
b) q is less than 12, but more than 3
c) r is at least 5, but less than 9
d) s is greater than 8, but not more than 15

Worked examples

a) The maximum number of players from one football team allowed on the pitch at any one time is eleven. Represent this information:

i) as an inequality,
ii) on a number line.

i) Let the number of players be represented by the letter n. n must be less than or equal to 11. Therefore $n \leqslant 11$.

ii)

b) The maximum number of players from one football team allowed on the pitch at any one time is eleven. The minimum allowed is seven players. Represent this information:

i) as an inequality,
ii) on a number line.

i) Let the number of players be represented by the letter n. n must be greater than or equal to 7, but less than or equal to 11. Therefore $7 \leqslant n \leqslant 11$.

ii)

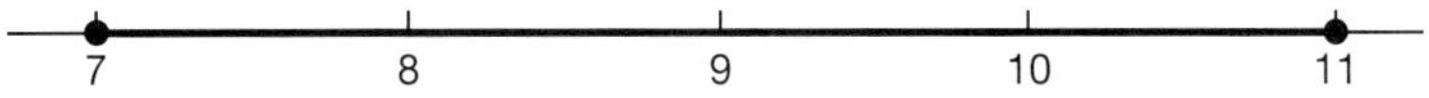

Exercise 5.2

1. Copy each of the following statements, and insert one of the symbols =, > or < into the space to make the statement correct:

a) $7 \times 2 \ldots 8 + 7$
b) $6^2 \ldots 9 \times 4$
c) $5 \times 10 \ldots 7^2$
d) 80 cm . . . 1 m
e) 1000 litres . . . 1 m^3
f) $48 \div 6 \ldots 54 \div 9$

2. Represent each of the following inequalities on a number line, where x is a real number:

a) $x < 2$ b) $x \geqslant 3$
c) $x \leqslant -4$ d) $x \geqslant -2$
e) $2 < x < 5$ f) $-3 < x < 0$
g) $-2 \leqslant x < 2$ h) $2 \geqslant x \geqslant -1$

3. Write down the inequalities which correspond to the following number lines:

a)

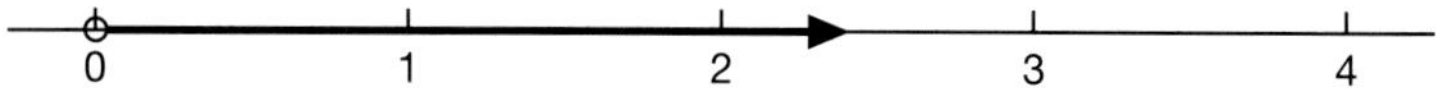

b)

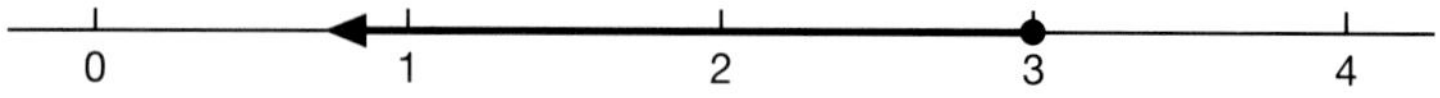

c)

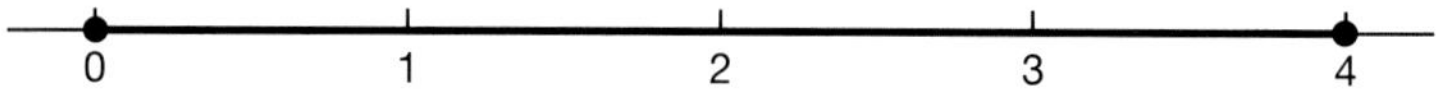

d)

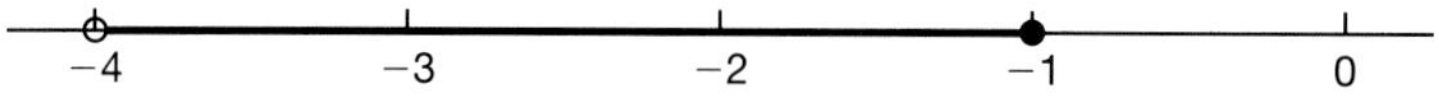

4. Write the following sentences using inequality signs:
   a) The maximum capacity of an athletics stadium is 20 000 people.
   b) In a class, the tallest pupil is 180 cm and the shortest is 135 cm.
   c) Five times a number plus three is less than 20.
   d) The maximum temperature in May was 25 °C.
   e) A farmer has between 350 and 400 apples on each tree in his orchard.
   f) In December, temperatures in Kenya were between 11 °C and 28 °C.

Exercise 5.3

1. Write the following decimals in order of magnitude, starting with the largest:

   0.45 0.405 0.045 4.05 4.5

2. Write the following decimals in order of magnitude, starting with the smallest:

   6.0 0.6 0.66 0.606 0.06 6.6 6.606

3. Write the following decimals in order of magnitude, starting with the largest:

   0.906 0.96 0.096 9.06 0.609 0.690

4. Write the following fractions in order of magnitude, starting with the smallest:

   $\frac{1}{3}$ $\frac{1}{4}$ $\frac{1}{2}$ $\frac{2}{5}$ $\frac{3}{10}$ $\frac{3}{4}$

5. Write the following fractions in order of magnitude, starting with the largest:

   $\frac{1}{2}$ $\frac{1}{3}$ $\frac{6}{13}$ $\frac{4}{5}$ $\frac{7}{18}$ $\frac{2}{19}$

6. Write the following fractions in order of magnitude, starting with the smallest:

   $\frac{3}{4}$ $\frac{3}{5}$ $\frac{2}{3}$ $\frac{4}{7}$ $\frac{5}{9}$ $\frac{1}{2}$

Exercise 5.4

1. Write the following lengths in order of magnitude, starting with the smallest:

   0.5 km 5000 m 15 000 cm $\frac{2}{5}$km 750 m

2. Write the following lengths in order of magnitude, starting with the smallest:

   2 m 60 cm 800 mm 180 cm 0.75 m

3. Write the following masses in order of magnitude, starting with the largest:

   4 kg 3500 g $\frac{3}{4}$kg 700 g 1 kg

4. Write the following volumes in order of magnitude, starting with the smallest:

   1 litre 430 ml 800 cm^3 120 cl 150 cm^3

Student Assessment 1

1. Write the following in words:
   a) $m = 7$ b) $n \neq 10$
   c) $e < 4$ d) $f \leqslant 6$
   e) $x \geqslant 12$ f) $y > -1$

2. Rewrite the following using the appropriate symbols:
   a) *a* is greater than 5 b) *b* is greater than or equal to 6
   c) *c* is at least 4 d) *d* is not equal to 14

3. Illustrate each of the following inequalities on a number line:
   a) $d < 4$ b) $e \geqslant 6$
   c) $2 < f < 5$ d) $-3 \leqslant g < 1$

4. Illustrate the information in each of the following statements on a number line:
   a) The temperature in a greenhouse must be kept between 18 °C and 28 °C.
   b) The pressure in a car tyre should be less than 2.3 bars, and equal to or more than 1.9 bars.

5. Write the following decimals in order of magnitude, starting with the smallest:

   0.99 9.99 9.09 9 0.09 0.9

Student Assessment 2

1. Write the following in words:
   a) $p \neq 2$ b) $q > 0$
   c) $r \leqslant 3$ d) $s = 2$
   e) $t \geqslant 1$ f) $u < -5$

2. Rewrite the following using the appropriate symbols:
   a) *a* is less than 2 b) *b* is less than or equal to 4
   c) *c* is equal to 8 d) *d* is not greater than 0

3. Illustrate each of the following inequalities on a number line:
   a) $j > 2$ b) $k \leqslant 16$
   c) $-2 \leqslant L \leqslant 5$ d) $3 \leqslant m < 7$

4. Illustrate the information in each of the following statements on a number line:
   a) A ferry can carry no more than 280 cars.
   b) The minimum temperature overnight was 4 °C.

5. Write the following masses in order of magnitude, starting with the largest:

   900 g 1 kg 1800 g 1.09 kg 9 g

Student Assessment 3

1. Insert one of the symbols =, > or < into the space to make the following statements correct:

a) $8 \times 5 \dots 5 \times 8$ b) $9^2 \dots 100 - 21$

c) 45 cm . . . 0.5 m d) Days in June . . . 31 days

2. Illustrate the following information on a number line:

a) The most number of days in a month is 31.

b) A month has at least 28 days.

c) A cake takes between $1\frac{1}{2}$ and $1\frac{3}{4}$ hours to bake.

d) An aeroplane will land between 2.40 p.m. and 2.45 p.m.

3. Write the following sentences using inequalities:

a) Not more than 52 people can be carried on a bus.

b) The minimum temperature tonight will be 11 °C.

c) There are between 24 and 38 pupils in a class.

d) Three times a number plus six is less than 50.

e) The minimum reaction time for an alarm system is 0.03 seconds.

4. Illustrate the following inequalities on a number line:

a) $x > 5$ b) $y \leqslant -3$

c) $-3 < x < -1$ d) $-2 \leqslant y < 1$

5. Write the following fractions in order of magnitude, starting with the largest:

$\frac{1}{6}$ $\frac{2}{3}$ $\frac{7}{12}$ $\frac{13}{18}$ $\frac{6}{7}$

Student Assessment 4

1. Insert one of the symbols =, > or < into the space to make the following statements correct:

a) $4 \times 2 \dots 2^3$ b) $6^2 \dots 2^6$

c) 850 ml . . . 0.5 litre d) Days in May . . . 30 days

2. Illustrate the following information on a number line:

a) The temperature during the day reached a maximum of 35 °C.

b) There were between 20 and 25 pupils in a class.

c) The world record for the 100 m sprint is under 10 seconds.

d) Doubling a number and subtracting four gives an answer greater than 16.

3. Write the information on the following number lines as inequalities:

a)

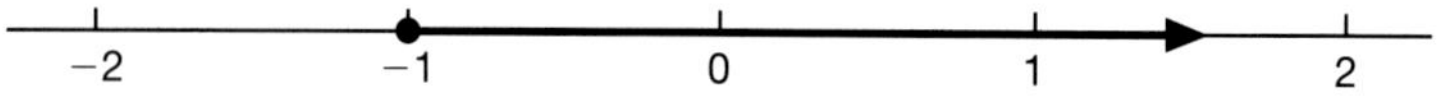

b)

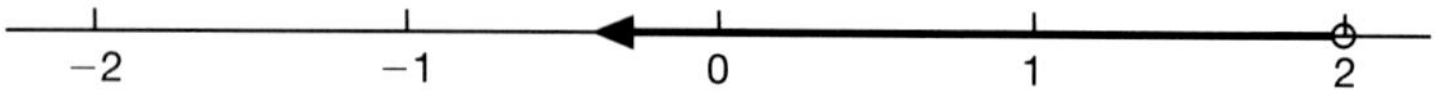

c)

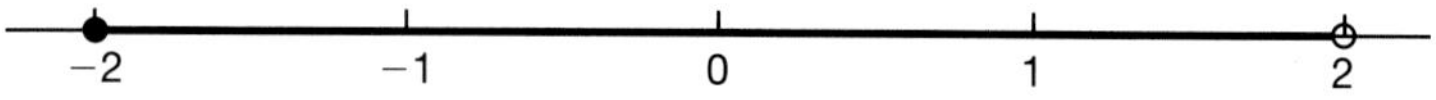

d)

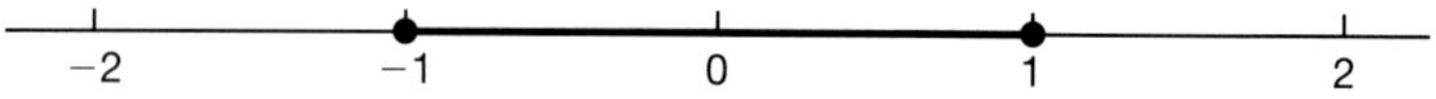

4. Illustrate the following inequalities on a number line:

a) $x \geqslant 3$ b) $m < 4$

c) $0 < x < 4$ d) $-3 \leqslant x < 1$

5. Write the following fractions in order of magnitude, starting with the smallest:

$\frac{4}{7}$ $\frac{3}{14}$ $\frac{9}{10}$ $\frac{1}{2}$ $\frac{2}{5}$

6 STANDARD FORM

Use the standard form $A \times 10^n$ where n is a positive or negative integer, and $1 \leqslant A < 10$.

Standard form is also known as standard index form or sometimes as scientific notation. It involves writing large numbers or very small numbers in terms of powers of 10.

A positive index

$$100 = 1 \times 10^2$$
$$1000 = 1 \times 10^3$$
$$10\,000 = 1 \times 10^4$$
$$3000 = 3 \times 10^3$$

For a number to be in standard form it must take the form $A \times 10^n$ where the index n is a positive or negative integer and A must lie in the range $1 \leqslant A < 10$.

e.g. 3100 can be written in many different ways:

3.1×10^3 $\quad 31 \times 10^2$ $\quad 0.31 \times 10^4$ $\quad$ etc.

However, only 3.1×10^3 satisfies the above conditions and therefore is the only one which is written in standard form.

Worked examples

a) Write 72 000 in standard form.

7.2×10^4

b) Of the numbers below, ring those which are written in standard form:

(4.2×10^3 circled) $\quad 0.35 \times 10^2$ $\quad 18 \times 10^5$ $\quad$ (6×10^3 circled) $\quad 0.01 \times 10^1$

c) Multiply the following and write your answer in standard form:

600×4000
$= 2\,400\,000$
$= 2.4 \times 10^6$

d) Multiply the following and write your answer in standard form:

$(2.4 \times 10^4) \times (5 \times 10^7)$
$= 12 \times 10^{11}$
$= 1.2 \times 10^{12}$ when written in standard form

e) Divide the following and write your answer in standard form:

$(6.4 \times 10^7) \div (1.6 \times 10^3)$
$= 4 \times 10^4$

f) Add the following and write your answer in standard form:

$(3.8 \times 10^6) + (8.7 \times 10^4)$

Changing the indices to the same value gives the sum:

$(380 \times 10^4) + (8.7 \times 10^4)$
$= 388.7 \times 10^4$
$= 3.887 \times 10^6$ when written in standard form

g) Subtract the following and write your answer in standard form:

$(6.5 \times 10^7) - (9.2 \times 10^5)$

Changing the indices to the same value gives:

$(650 \times 10^5) - (9.2 \times 10^5)$
$= 640.8 \times 10^5$
$= 6.408 \times 10^7$ when written in standard form

Exercise 6.1

1. Deduce the value of n in the following:
a) $79\,000 = 7.9 \times 10^n$ b) $53\,000 = 5.3 \times 10^n$
c) $4\,160\,000 = 4.16 \times 10^n$ d) 8 million $= 8 \times 10^n$
e) 247 million $= 2.47 \times 10^n$ f) $24\,000\,000 = 2.4 \times 10^n$

2. Write the following numbers in standard form:
a) 65 000 b) 41 000 c) 723 000
d) 18 million e) 950 000 f) 760 million
g) 720 000 h) $\frac{1}{4}$million

3. Write the numbers below which are written in standard form:

26.3×10^5 2.6×10^7 0.5×10^3 8×10^8
0.85×10^9 8.3×10^{10} 1.8×10^7 18×10^5
3.6×10^6 6.0×10^1

4. Multiply the following and write your answers in standard form:
a) 400×2000 b) 6000×5000 c) $75\,000 \times 200$
d) $33\,000 \times 6000$ e) 8 million $\times$ 250 f) $95\,000 \times 3000$
g) 7.5 million $\times$ 2 h) 8.2 million $\times$ 50
i) $300 \times 200 \times 400$ j) $(7000)^2$

5. Multiply the following and write your answers in standard form:
a) $(4 \times 10^3) \times (2 \times 10^5)$ b) $(2.5 \times 10^4) \times (3 \times 10^4)$
c) $(1.8 \times 10^7) \times (5 \times 10^6)$ d) $(2.1 \times 10^4) \times (4 \times 10^7)$
e) $(3.5 \times 10^4)(4 \times 10^7)$ f) $(4.2 \times 10^5)(3 \times 10^4)$
g) $(2 \times 10^4)^2$ h) $(4 \times 10^8)^2$

6. Find the value of the following and write your answers in standard form:
a) $(8 \times 10^6) \div (2 \times 10^3)$ b) $(8.8 \times 10^9) \div (2.2 \times 10^3)$
c) $(7.6 \times 10^8) \div (4 \times 10^7)$ d) $(6.5 \times 10^{14}) \div (1.3 \times 10^7)$
e) $(5.2 \times 10^6) \div (1.3 \times 10^6)$ f) $(3.8 \times 10^{11}) \div (1.9 \times 10^3)$

7. Find the value of the following and write your answers in standard form:
   a) $(3 \times 10^4) \times (6 \times 10^5) \div (9 \times 10^5)$
   b) $(6.5 \times 10^8) \div (1.3 \times 10^4) \times (5 \times 10^3)$
   c) $(18 \times 10^3) \div 900 \times 250$
   d) $27\,000 \div 3000 \times 8000$
   e) $4000 \times 8000 \div 640$
   f) $2500 \times 2500 \div 1250$

Exercise 6.2

1. Which of the following are not in standard form?
   a) 6.2×10^5 b) 7.834×10^{16}
   c) 8.0×10^5 d) 0.46×10^7
   e) 82.3×10^6 f) 6.75×10^1

2. Write the following numbers in standard form:
   a) 600 000 b) 48 000 000
   c) 784 000 000 000 d) 534 000
   e) 7 million f) 8.5 million

3. Write the following in standard form:
   a) 68×10^5 b) 720×10^6
   c) 8×10^5 d) 0.75×10^8
   e) 0.4×10^{10} f) 50×10^6

4. Multiply the following and write your answers in standard form:
   a) 200×3000 b) 6000×4000
   c) 7 million $\times$ 20 d) 500 $\times$ 6 million
   e) 3 million $\times$ 4 million f) 4500×4000

5. Light from the Sun takes approximately 8 minutes to reach Earth. If light travels at a speed of 3×10^8 m/s, calculate to three significant figures (s.f.) the distance from the Sun to the Earth.

6. Find the value of the following and write your answers in standard form:
   a) $(4.4 \times 10^3) \times (2 \times 10^5)$ b) $(6.8 \times 10^7) \times (3 \times 10^3)$
   c) $(4 \times 10^5) \times (8.3 \times 10^5)$ d) $(5 \times 10^9) \times (8.4 \times 10^{12})$
   e) $(8.5 \times 10^6) \times (6 \times 10^{15})$ f) $(5.0 \times 10^{12})^2$

7. Find the value of the following and write your answers in standard form:
   a) $(3.8 \times 10^8) \div (1.9 \times 10^6)$
   b) $(6.75 \times 10^9) \div (2.25 \times 10^4)$
   c) $(9.6 \times 10^{11}) \div (2.4 \times 10^5)$
   d) $(1.8 \times 10^{12}) \div (9.0 \times 10^7)$
   e) $(2.3 \times 10^{11}) \div (9.2 \times 10^4)$
   f) $(2.4 \times 10^8) \div (6.0 \times 10^3)$

8. Find the value of the following and write your answers in standard form:

a) $(3.8 \times 10^5) + (4.6 \times 10^4)$
b) $(7.9 \times 10^7) + (5.8 \times 10^8)$
c) $(6.3 \times 10^7) + (8.8 \times 10^5)$
d) $(3.15 \times 10^9) + (7.0 \times 10^6)$
e) $(5.3 \times 10^8) - (8.0 \times 10^7)$
f) $(6.5 \times 10^7) - (4.9 \times 10^6)$
g) $(8.93 \times 10^{10}) - (7.8 \times 10^9)$
h) $(4.07 \times 10^7) - (5.1 \times 10^6)$

9. The following list shows the distance of some of the planets of the Solar system from the Sun.

Jupiter	778 million km
Mercury	58 million km
Mars	228 million km
Uranus	2870 million km
Venus	108 million km
Pluto	5920 million km
Earth	150 million km
Saturn	1430 million km

Write each of the distances in standard form and then arrange them in order of magnitude, starting with the distance of the planet closest to the Sun.

A negative index

A negative index is used when writing a number between 0 and 1 in standard form.

e.g.
$$\begin{aligned} 100 &= 1 \times 10^2 \\ 10 &= 1 \times 10^1 \\ 1 &= 1 \times 10^0 \\ 0.1 &= 1 \times 10^{-1} \\ 0.01 &= 1 \times 10^{-2} \\ 0.001 &= 1 \times 10^{-3} \\ 0.0001 &= 1 \times 10^{-4} \end{aligned}$$

Note that A must still lie within the range $1 \leqslant A < 10$.

Worked examples

a) Write 0.0032 in standard form.

3.2×10^{-3}

b) Write the following numbers in order of magnitude, starting with the largest:

3.6×10^{-3} 5.2×10^{-5} 1×10^{-2} 8.35×10^{-2} 6.08×10^{-8}

8.35×10^{-2} 1×10^{-2} 3.6×10^{-3} 5.2×10^{-5} 6.08×10^{-8}

Exercise 6.3

1. Copy and complete the following so that the answers are correct (the first question is done for you):

a) 0.0048 $= 4.8 \times 10^{-3}$
b) 0.0079 $= 7.9 \times \ldots$
c) 0.000 81 $= 8.1 \times \ldots$
d) 0.000 009 $= 9 \times \ldots$
e) 0.000 000 45 $= 4.5 \times \ldots$
f) 0.000 000 003 24 $= 3.24 \times \ldots$
g) 0.000 008 42 $= 8.42 \times \ldots$
h) 0.000 000 000 403 $= 4.03 \times \ldots$

2. Write the following numbers in standard form:
a) 0.0006 b) 0.000 053
c) 0.000 864 d) 0.000 000 088
e) 0.000 000 7 f) 0.000 414 5

3. Write the following numbers in standard form:
a) 68×10^{-5} b) 750×10^{-9}
c) 42×10^{-11} d) 0.08×10^{-7}
e) 0.057×10^{-9} f) 0.4×10^{-10}

4. Deduce the value of n in each of the following cases:
a) $0.000\,25 = 2.5 \times 10^n$ b) $0.003\,57 = 3.57 \times 10^n$
c) $0.000\,000\,06 = 6 \times 10^n$ d) $0.004^2 = 1.6 \times 10^n$
e) $0.000\,65^2 = 4.225 \times 10^n$ f) $0.0002^n = 8 \times 10^{-12}$

5. Write these numbers in order of magnitude, starting with the largest:

3.2×10^{-4} 6.8×10^5 5.57×10^{-9} 6.2×10^3
5.8×10^{-7} 6.741×10^{-4} 8.414×10^2

Student Assessment 1

1. Write the following numbers in standard form:
   a) 8 million b) 0.000 72
   c) 75 000 000 000 d) 0.0004
   e) 4.75 billion f) 0.000 000 64

2. Write the following numbers in order of magnitude, starting with the smallest:

   6.2×10^7 5.5×10^{-3} 4.21×10^7 4.9×10^8
   3.6×10^{-5} 7.41×10^{-9}

3. Write the following numbers:
   a) in standard form,
   b) in order of magnitude, starting with the largest.

   6 million 820 000 0.0044 0.8 52 000

4. Deduce the value of n in each of the following:
   a) $620 = 6.2 \times 10^n$ b) $555\,000\,000 = 5.55 \times 10^n$
   c) $0.000\,45 = 4.5 \times 10^n$ d) $500^2 = 2.5 \times 10^n$
   e) $0.0035^2 = 1.225 \times 10^n$ f) $0.04^3 = 6.4 \times 10^n$

5. Write the answers to the following calculations in standard form:
   a) $4000 \times 30\,000$ b) $(2.8 \times 10^5) \times (2.0 \times 10^3)$
   c) $(3.2 \times 10^9) \div (1.6 \times 10^4)$ d) $(2.4 \times 10^8) \div (9.6 \times 10^2)$

6. The speed of light is 3×10^8 m/s. Venus is 108 million km from the Sun. Calculate the number of minutes it takes for sunlight to reach Venus.

7. A star system is 500 light years away from Earth. If the speed of light is 3×10^5 km/s, calculate the distance the star system is from Earth. Give your answer in kilometres and written in standard form.

Student Assessment 2

1. Write the following numbers in standard form:
   a) 6 million b) 0.0045
   c) 3 800 000 000 d) 0.000 000 361
   e) 460 million f) 3

2. Write the following numbers in order of magnitude, starting with the largest:

   3.6×10^2 2.1×10^{-3} 9×10^1 4.05×10^8
   1.5×10^{-2} 7.2×10^{-3}

3. Write the following numbers:
   a) in standard form,
   b) in order of magnitude, starting with the smallest.

   15 million 430 000 0.000 435 4.8 0.0085

4. Deduce the value of n in each of the following:

a) $4750 = 4.75 \times 10^n$
b) $6\,440\,000\,000 = 6.44 \times 10^n$
c) $0.0040 = 4.0 \times 10^n$
d) $1000^2 = 1 \times 10^n$
e) $0.9^3 = 7.29 \times 10^n$
f) $800^3 = 5.12 \times 10^n$

5. Write the answers to the following calculations in standard form:

a) $50\,000 \times 2400$
b) $(3.7 \times 10^6) \times (4.0 \times 10^4)$
c) $(5.8 \times 10^7) + (9.3 \times 10^6)$
d) $(4.7 \times 10^6) - (8.2 \times 10^5)$

6. The speed of light is 3×10^8 m/s. Jupiter is 778 million km from the Sun. Calculate the number of minutes it takes for sunlight to reach Jupiter.

7. A star is 300 light years away from Earth. If the speed of light is 3×10^5 km/s, calculate the distance from the star to Earth. Give your answer in kilometres and written in standard form.

7 THE FOUR RULES

Use the four rules for calculations with whole numbers, decimal fractions and vulgar (and mixed) fractions, including correct ordering of operations and use of brackets.

Calculations with whole numbers

Addition, subtraction, multiplication and division are mathematical operations.

Long multiplication

When carrying out long multiplication, it is important to remember place value.

Worked example 184×37

```
     1 8 4
  ×    3 7
  --------
   1 2 8 8     (184 × 7)
   5 5 2 0    (184 × 30)
  --------
   6 8 0 8    (184 × 37)
```

Short division

Worked example $453 \div 6$

```
     7  5 r3
  6 |4  5 ³3
```

It is usual, however, to give the final answer in decimal form rather than with a remainder. The division should therefore be continued:

$453 \div 6$

```
     7  5 .  5
  6 |4  5 ³3 . ³0
```

Long division

Worked example Calculate $7184 \div 23$ to one decimal place (d.p.).

```
          3 1 2 . 3 4
     ----------------
  23 | 7  1 8 4 . 0 0
       6  9
       ----
          2 8
          2 3
          ---
            5 4
            4 6
            ---
              8 0
              6 9
              ---
              1 1 0
                9 2
              -----
                1 8
```

Therefore $7184 \div 23 = 312.3$ to 1 d.p.

Mixed operations

When a calculation involves a mixture of operations, the order of the operations is important. Multiplications and divisions are done first, whilst additions and subtractions are done afterwards. To override this, brackets need to be used.

Worked examples

a)
$$\begin{aligned} & 3 + 7 \times 2 - 4 \\ = {} & 3 + 14 - 4 \\ = {} & 13 \end{aligned}$$

b)
$$\begin{aligned} & (3 + 7) \times 2 - 4 \\ = {} & 10 \times 2 - 4 \\ = {} & 20 - 4 \\ = {} & 16 \end{aligned}$$

c)
$$\begin{aligned} & 3 + 7 \times (2 - 4) \\ = {} & 3 + 7 \times (-2) \\ = {} & 3 - 14 \\ = {} & -11 \end{aligned}$$

d)
$$\begin{aligned} & (3 + 7) \times (2 - 4) \\ = {} & 10 \times (-2) \\ = {} & -20 \end{aligned}$$

Exercise 7.1

1. Evaluate the answer to each of the following:
a) $3 + 5 \times 2 - 4$ b) $6 + 4 \times 7 - 12$
c) $3 \times 2 + 4 \times 6$ d) $4 \times 5 - 3 \times 6$
e) $8 \div 2 + 18 \div 6$ f) $12 \div 8 + 6 \div 4$

2. Copy these equations and put brackets in the correct places to make them correct:
a) $6 \times 4 + 6 \div 3 = 20$ b) $6 \times 4 + 6 \div 3 = 36$
c) $8 + 2 \times 4 - 2 = 12$ d) $8 + 2 \times 4 - 2 = 20$
e) $9 - 3 \times 7 + 2 = 44$ f) $9 - 3 \times 7 + 2 = 54$

3. Without using a calculator, work out the solutions to the following multiplications:
a) 63×24 b) 531×64
c) 785×38 d) 164×253
e) 144×144 f) 170×240

4. Work out the remainders in the following divisions:
a) $33 \div 7$ b) $68 \div 5$
c) $72 \div 7$ d) $430 \div 9$
e) $156 \div 5$ f) $687 \div 10$

5. a) The sum of two numbers is 16, their product is 63. What are the two numbers?
b) When a number is divided by 7 the result is 14 remainder 2. What is the number?
c) The difference between two numbers is 5, their product is 176. What are the numbers?
d) How many 9s can be added to 40 before the total exceeds 100?
e) A length of rail track is 9 m long. How many complete lengths will be needed to lay 1 km of track?
f) How many 35 cent stamps can be bought for 10 dollars?

6. Work out the following long divisions to 1 d.p.
a) $7892 \div 7$ b) $45\,623 \div 6$
c) $9452 \div 8$ d) $4564 \div 4$
e) $7892 \div 15$ f) $79\,876 \div 24$

Fractions

Equivalent fractions

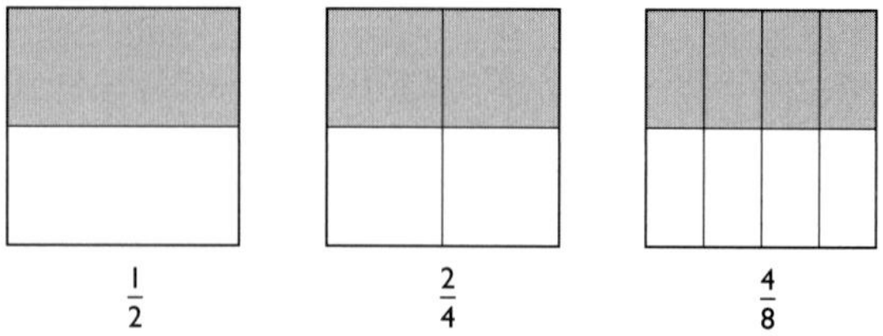

It should be apparent that $\frac{1}{2}$, $\frac{2}{4}$ and $\frac{4}{8}$ are equivalent fractions. Similarly, $\frac{1}{3}$, $\frac{2}{6}$, $\frac{3}{9}$ and $\frac{4}{12}$ are equivalent, as are $\frac{1}{5}$, $\frac{10}{50}$ and $\frac{20}{100}$.

Exercise 7.2

1. Copy the following sets of equivalent fractions and fill in the blanks:

   a) $\frac{1}{4} = \frac{2}{\ } = \frac{\ }{16} = \frac{\ }{64} = \frac{3}{\ }$

   b) $\frac{2}{5} = \frac{4}{\ } = \frac{\ }{20} = \frac{\ }{50} = \frac{16}{\ }$

   c) $\frac{3}{8} = \frac{6}{\ } = \frac{\ }{24} = \frac{15}{\ } = \frac{\ }{72}$

   d) $\frac{\ }{7} = \frac{8}{14} = \frac{12}{\ } = \frac{\ }{56} = \frac{36}{\ }$

   e) $\frac{5}{\ } = \frac{\ }{27} = \frac{20}{36} = \frac{\ }{90} = \frac{55}{\ }$

2. Express the following fractions in their lowest terms.
   e.g. $\frac{12}{16} = \frac{3}{4}$

   a) $\frac{5}{10}$ b) $\frac{7}{21}$ c) $\frac{8}{12}$
   d) $\frac{16}{36}$ e) $\frac{75}{100}$ f) $\frac{81}{90}$

3. Write the following improper fractions as mixed numbers.
   e.g. $\frac{15}{4} = 3\frac{3}{4}$

   a) $\frac{17}{4}$ b) $\frac{23}{5}$ c) $\frac{8}{3}$
   d) $\frac{19}{3}$ e) $\frac{12}{3}$ f) $\frac{43}{12}$

4. Write the following mixed numbers as improper fractions.
   e.g. $3\frac{4}{5} = \frac{19}{5}$

   a) $6\frac{1}{2}$ b) $7\frac{1}{4}$ c) $3\frac{3}{8}$
   d) $11\frac{1}{9}$ e) $6\frac{4}{5}$ f) $8\frac{9}{11}$

Addition and subtraction of fractions

For fractions to be either added or subtracted, the denominators need to be the same.

Worked examples

a) $\frac{3}{11} + \frac{5}{11} = \frac{8}{11}$

b) $\frac{7}{8} + \frac{5}{8} = \frac{12}{8}$

c) $\frac{1}{2} + \frac{1}{3}$
$= \frac{3}{6} + \frac{2}{6} = \frac{5}{6}$

d) $\frac{4}{5} - \frac{1}{3}$
$= \frac{12}{15} - \frac{5}{15} = \frac{7}{15}$

When dealing with calculations involving mixed numbers, it is sometimes easier to change them to improper fractions first.

e) $5\frac{3}{4} - 2\frac{5}{8}$
$= \frac{23}{4} - \frac{21}{8}$
$= \frac{46}{8} - \frac{21}{8}$
$= \frac{25}{8} = 3\frac{1}{8}$

f) $1\frac{4}{7} + 3\frac{3}{4}$
$= \frac{11}{7} + \frac{15}{4}$
$= \frac{44}{28} + \frac{105}{28}$
$= \frac{149}{28} = 5\frac{9}{28}$

Exercise 7.3

Evaluate each of the following and write the answer as a fraction in its simplest form:

1. a) $\frac{3}{5} + \frac{4}{5}$ b) $\frac{3}{11} + \frac{7}{11}$
c) $\frac{2}{3} + \frac{1}{4}$ d) $\frac{3}{5} + \frac{4}{9}$
e) $\frac{8}{13} + \frac{2}{5}$ f) $\frac{1}{2} + \frac{2}{3} + \frac{3}{4}$

2. a) $\frac{1}{8} + \frac{3}{8} + \frac{5}{8}$ b) $\frac{3}{7} + \frac{5}{7} + \frac{4}{7}$
c) $\frac{1}{3} + \frac{1}{2} + \frac{1}{4}$ d) $\frac{1}{5} + \frac{1}{3} + \frac{1}{4}$
e) $\frac{3}{8} + \frac{3}{5} + \frac{3}{4}$ f) $\frac{3}{13} + \frac{1}{4} + \frac{1}{2}$

3. a) $\frac{3}{7} - \frac{2}{7}$ b) $\frac{4}{5} - \frac{7}{10}$
c) $\frac{8}{9} - \frac{1}{3}$ d) $\frac{7}{12} - \frac{1}{2}$
e) $\frac{5}{8} - \frac{2}{5}$ f) $\frac{3}{4} - \frac{2}{5} + \frac{7}{10}$

4. a) $\frac{3}{4} + \frac{1}{5} - \frac{2}{3}$ b) $\frac{3}{8} + \frac{7}{11} - \frac{1}{2}$
c) $\frac{4}{5} - \frac{3}{10} + \frac{7}{20}$ d) $\frac{9}{13} + \frac{1}{3} - \frac{4}{5}$
e) $\frac{9}{10} - \frac{1}{5} - \frac{1}{4}$ f) $\frac{8}{9} - \frac{1}{3} - \frac{1}{2}$

5. a) $2\frac{1}{2} + 3\frac{1}{4}$ b) $3\frac{3}{5} + 1\frac{7}{10}$
c) $6\frac{1}{2} - 3\frac{2}{5}$ d) $8\frac{5}{8} - 2\frac{1}{3}$
e) $5\frac{7}{8} - 4\frac{3}{4}$ f) $3\frac{1}{4} - 2\frac{5}{9}$

6. a) $2\frac{1}{2} + 1\frac{1}{4} + 1\frac{3}{8}$ b) $2\frac{4}{5} + 3\frac{1}{8} + 1\frac{3}{10}$
c) $4\frac{1}{2} - 1\frac{1}{4} - 3\frac{5}{8}$ d) $6\frac{1}{2} - 2\frac{3}{4} - 3\frac{2}{5}$
e) $2\frac{4}{7} - 3\frac{1}{4} - 1\frac{3}{5}$ f) $4\frac{7}{20} - 5\frac{1}{2} + 2\frac{2}{5}$

Multiplication and division of fractions

Worked examples

a) $\frac{3}{4} \times \frac{2}{3}$
$= \frac{6}{12}$
$= \frac{1}{2}$

b) $3\frac{1}{2} \times 4\frac{4}{7}$
$= \frac{7}{2} \times \frac{32}{7}$
$= \frac{224}{14}$
$= 16$

The reciprocal of a number is obtained when 1 is divided by that number. The reciprocal of 5 is $\frac{1}{5}$, the reciprocal of $\frac{2}{5}$ is $\frac{5}{2}$, etc.

Dividing fractions is the same as multiplying by the reciprocal.

c) $\frac{3}{8} \div \frac{3}{4}$
$= \frac{3}{8} \times \frac{4}{3}$
$= \frac{12}{24}$
$= \frac{1}{2}$

d) $5\frac{1}{2} \div 3\frac{2}{3}$
$= \frac{11}{2} \div \frac{11}{3}$
$= \frac{11}{2} \times \frac{3}{11}$
$= \frac{3}{2}$

Exercise 7.4

1. Write the reciprocal of each of the following:
   a) $\frac{3}{4}$ b) $\frac{5}{9}$ c) 7
   d) $\frac{1}{9}$ e) $2\frac{3}{4}$ f) $4\frac{5}{8}$

2. Write the reciprocal of each of the following:
   a) $\frac{1}{8}$ b) $\frac{7}{12}$ c) $\frac{3}{5}$
   d) $1\frac{1}{2}$ e) $3\frac{3}{4}$ f) 6

3. Evaluate the following:
   a) $\frac{3}{8} \times \frac{4}{9}$ b) $\frac{2}{3} \times \frac{9}{10}$ c) $\frac{5}{7} \times \frac{4}{15}$
   d) $\frac{3}{4}$ of $\frac{8}{9}$ e) $\frac{5}{6}$ of $\frac{3}{10}$ f) $\frac{7}{8}$ of $\frac{2}{5}$

4. a) $\frac{5}{8} \div \frac{3}{4}$ b) $\frac{5}{6} \div \frac{1}{3}$ c) $\frac{4}{5} \div \frac{7}{10}$
   d) $1\frac{2}{3} \div \frac{2}{5}$ e) $\frac{3}{7} \div 2\frac{1}{7}$ f) $1\frac{1}{4} \div 1\frac{7}{8}$

5. Evaluate the following:
   a) $\frac{3}{4} \times \frac{4}{5}$ b) $\frac{7}{8} \times \frac{2}{3}$
   c) $\frac{3}{4} \times \frac{4}{7} \times \frac{3}{10}$ d) $(\frac{4}{5} \div \frac{2}{3}) \times \frac{7}{10}$
   e) $\frac{1}{2}$ of $\frac{3}{4}$ f) $4\frac{1}{2} \div 3\frac{1}{9}$

6. Evaluate the following:
   a) $(\frac{3}{8} \times \frac{4}{5}) + (\frac{1}{2}$ of $\frac{3}{5})$ b) $(1\frac{1}{2} \times 3\frac{3}{4}) - (2\frac{3}{5} \div 1\frac{1}{2})$
   c) $(\frac{3}{5}$ of $\frac{4}{9}) + (\frac{4}{9}$ of $\frac{3}{5})$ d) $(1\frac{1}{3} \times 2\frac{5}{8})^2$

Changing a fraction to a decimal

To change a fraction to a decimal, divide the numerator by the denominator.

Worked examples

a) Change $\frac{5}{8}$ to a decimal.

$$\begin{array}{r} 0.6\;2\;5 \\ 8\,\overline{)\,5.0\;{}^{2}0\;{}^{4}0} \end{array}$$

b) Change $2\frac{3}{5}$ to a decimal.
This can be represented as $2 + \frac{3}{5}$.

$$\begin{array}{r} 0.6 \\ 5\,\overline{)\,3.0} \end{array}$$

Therefore $2\frac{3}{5} = 2.6$

Exercise 7.5

1. Change the following fractions to decimals:
   a) $\frac{3}{4}$ b) $\frac{4}{5}$ c) $\frac{9}{20}$
   d) $\frac{17}{50}$ e) $\frac{1}{3}$ f) $\frac{3}{8}$
   g) $\frac{7}{16}$ h) $\frac{2}{9}$ i) $\frac{7}{11}$

2. Change the following mixed numbers to decimals:
   a) $2\frac{3}{4}$ b) $3\frac{3}{5}$ c) $4\frac{7}{20}$
   d) $6\frac{11}{50}$ e) $5\frac{2}{3}$ f) $6\frac{7}{8}$
   g) $5\frac{9}{16}$ h) $4\frac{2}{9}$ i) $5\frac{3}{7}$

Changing a decimal to a fraction

Changing a decimal to a fraction is done by knowing the 'value' of each of the numbers in any decimal.

Worked examples

a) Change 0.45 from a decimal to a fraction.

units	.	tenths	hundredths
0	.	4	5

0.45 is therefore equivalent to 4 tenths and 5 hundredths, which in turn is the same as 45 hundredths.
Therefore $0.45 = \frac{45}{100} = \frac{9}{20}$

b) Change 2.325 from a decimal to a fraction.

units	.	tenths	hundredths	thousandths
2	.	3	2	5

Therefore $2.325 = 2\frac{325}{1000} = 2\frac{13}{40}$

Exercise 7.6

1. Change the following decimals to fractions:

a) 0.5 b) 0.7 c) 0.6
d) 0.75 e) 0.825 f) 0.05
g) 0.050 h) 0.402 i) 0.0002

2. Change the following decimals to mixed numbers:

a) 2.4 b) 6.5 c) 8.2
d) 3.75 e) 10.55 f) 9.204
g) 15.455 h) 30.001 i) 1.0205

Student Assessment 1

1. Evaluate the following:
   a) $5 + 8 \times 3 - 6$ b) $15 + 45 \div 3 - 12$

2. The sum of two numbers is 21 and their product is 90. What are the numbers?

3. How many seconds are there in $2\frac{1}{2}$ hours?

4. Work out 851×27.

5. Work out $6843 \div 19$ giving your answer to 1 d.p.

6. Copy these equivalent fractions and fill in the blanks:

$$\frac{8}{18} = \frac{\quad}{9} = \frac{16}{\quad} = \frac{56}{\quad} = \frac{\quad}{90}$$

7. Evaluate the following:
   a) $3\frac{3}{4} - 1\frac{11}{16}$ b) $4\frac{4}{5} \div \frac{8}{15}$

8. Change the following fractions to decimals:
   a) $\frac{2}{5}$ b) $1\frac{3}{4}$
   c) $\frac{9}{11}$ d) $1\frac{2}{3}$

9. Change the following decimals to fractions. Give each fraction in its simplest form.
   a) 4.2 b) 0.06
   c) 1.85 d) 2.005

Student Assessment 2

1. Evaluate the following:
   a) $6 \times 4 - 3 \times 8$ b) $15 \div 3 + 2 \times 7$

2. The product of two numbers is 72, and their sum is 18. What are the two numbers?

3. How many days are there in 42 weeks?

4. Work out 368×49.

5. Work out $7835 \div 23$ giving your answer to 1 d.p.

6. Copy these equivalent fractions and fill in the blanks:

$$\frac{24}{36} = \frac{\quad}{12} = \frac{4}{\quad} = \frac{\quad}{30} = \frac{60}{\quad}$$

7. Evaluate the following:
   a) $2\frac{1}{2} - \frac{4}{5}$ b) $3\frac{1}{2} \times \frac{4}{7}$

8. Change the following fractions to decimals:
   a) $\frac{7}{8}$ b) $1\frac{2}{5}$
   c) $\frac{8}{9}$ d) $3\frac{2}{7}$

9. Change the following decimals to fractions. Give each fraction in its simplest form.
   a) 6.5 b) 0.04
   c) 3.65 d) 3.008

8 ESTIMATION

Give approximations to specified numbers of significant figures (s.f.) and decimal places (d.p.) and round off answers to reasonable accuracy in the context of a given problem.

NB: All diagrams are not drawn to scale.

In many instances exact numbers are not necessary or even desirable. In those circumstances approximations are given. The approximations can take several forms. The common types of approximation are dealt with below.

Rounding

If 28 617 people attend a gymnastics competition, this figure can be reported to various levels of accuracy.

To the nearest 10 000, this figure would be rounded up to 30 000.
To the nearest 1000, the figure would be rounded up to 29 000.
To the nearest 100, the figure would be rounded down to 28 600.

In this type of situation it is unlikely that the exact number would be reported.

Exercise 8.1

1. Round the following numbers to the nearest 1000:

a) 68 786	b) 74 245	c) 89 000
d) 4020	e) 99 500	f) 999 999

2. Round the following numbers to the nearest 100:

a) 78 540	b) 6858	c) 14 099
d) 8084	e) 950	f) 2984

3. Round the following numbers to the nearest 10:

a) 485	b) 692	c) 8847
d) 83	e) 4	f) 997

Decimal places

A number can also be approximated to a given number of decimal places (d.p.). This refers to the quantity of numbers written after a decimal point.

Worked examples

a) Write 7.864 to 1 d.p.
The answer needs to be written with one number after the decimal point. However, to do this, the second number after the decimal point also needs to be considered. If it is 5 or more then the first number is rounded up.

i.e. 7.864 is written as 7.9 to 1 d.p.

b) Write 5.574 to 2 d.p.
The answer here is to be given with two numbers after the decimal point. In this case the third number after the decimal point needs to be considered. As the third number after the decimal point is less than 5, the second number is not rounded up.

i.e. 5.574 is written as 5.57 to 2 d.p.

Exercise 8.2

1. Give the following to 1 d.p.

a) 5.58	b) 0.73	c) 11.86
d) 157.39	e) 4.04	f) 15.045
g) 2.95	h) 0.98	i) 12.049

2. Give the following to 2 d.p.

a) 6.473	b) 9.587	c) 16.476
d) 0.088	e) 0.014	f) 9.3048
g) 99.996	h) 0.0048	i) 3.0037

Significant figures

Numbers can also be approximated to a given number of significant figures (s.f.). In the number 43.25 the 4 is the most significant figure as it has a value of 40. In contrast, the 5 is the least significant as it only has a value of 5 hundredths.

Worked examples

a) Write 43.25 to 3 s.f.
Only the three most significant numbers are written, however the fourth number needs to be considered to see whether the third number is to be rounded up or not.

i.e. 43.25 is written as 43.3 to 3 s.f.

b) Write 0.0043 to 1 s.f.
In this example only two numbers have any significance, the 4 and the 3. The 4 is the most significant and therefore is the only one of the two to be written in the answer.

i.e. 0.0043 is written as 0.004 to 1 s.f.

Exercise 8.3

1. Write the following to the number of significant figures written in brackets:

a) 48 599 (1 s.f.)	b) 48 599 (3 s.f.)	c) 6841 (1 s.f.)
d) 7538 (2 s.f.)	e) 483.7 (1 s.f.)	f) 2.5728 (3 s.f.)
g) 990 (1 s.f.)	h) 2045 (2 s.f.)	i) 14.952 (3 s.f.)

2. Write the following to the number of significant figures written in brackets:

a) 0.085 62 (1 s.f.)	b) 0.5932 (1 s.f.)	c) 0.942 (2 s.f.)
d) 0.954 (1 s.f.)	e) 0.954 (2 s.f.)	f) 0.003 05 (1 s.f.)
g) 0.003 05 (2 s.f.)	h) 0.009 73 (2 s.f.)	i) 0.009 73 (1 s.f.)

Appropriate accuracy

In many instances calculations carried out using a calculator produce answers which are not whole numbers. A calculator will give the answer to as many decimal places as will fit on its screen. In most cases this degree of accuracy is neither necessary nor desirable. Unless specifically asked for, answers should not be given to more than two decimal places. Indeed, one decimal place is usually sufficient.

Worked example

Calculate $4.64 \div 2.3$ giving your answer to an appropriate degree of accuracy.

The calculator will give the answer to $4.64 \div 2.3$ as 2.017 391 3. However the answer given to 1 d.p. is sufficient. Therefore $4.64 \div 2.3 = 2.0$ (1 d.p.).

Estimating answers to calculations

Even though many calculations can be done quickly and effectively on a calculator, often an estimate for an answer can be a useful check. This is found by rounding each of the numbers in such a way that the calculation becomes relatively straightforward.

Worked examples

a) Estimate the answer to 57×246.

Here are two possibilities:
i) $60 \times 200 = 12\,000$,
ii) $50 \times 250 = 12\,500$.

b) Estimate the answer to $6386 \div 27$.

$6000 \div 30 = 200$

Exercise 8.4

1. Calculate the following, giving your answer to an appropriate degree of accuracy:

a) 23.456×17.89 b) 0.4×12.62 c) 18×9.24
d) $76.24 \div 3.2$ e) 7.6^2 f) 16.42^3
g) $\dfrac{2.3 \times 3.37}{4}$ h) $\dfrac{8.31}{2.02}$ i) $9.2 \div 4^2$

2. Without using a calculator, estimate the answers to the following:

a) 62×19 b) 270×12 c) 55×60
d) 4950×28 e) 0.8×0.95 f) 0.184×475

3. Without using a calculator, estimate the answers to the following:

a) $3946 \div 18$ b) $8287 \div 42$ c) $906 \div 27$
d) $5520 \div 13$ e) $48 \div 0.12$ f) $610 \div 0.22$

4. Without using a calculator, estimate the answers to the following:

a) $78.45 + 51.02$ b) $168.3 - 87.09$ c) 2.93×3.14

d) $84.2 \div 19.5$ e) $\dfrac{4.3 \times 752}{15.6}$ f) $\dfrac{(9.8)^3}{(2.2)^2}$

5. Using estimation, identify which of the following are definitely incorrect. Explain your reasoning clearly.

a) $95 \times 212 = 20\,140$
b) $44 \times 17 = 748$
c) $689 \times 413 = 28\,457$
d) $142\,656 \div 8 = 17\,832$
e) $77.9 \times 22.6 = 2512.54$
f) $\dfrac{84.2 \times 46}{0.2} = 19\,366$

6. Estimate the shaded areas of the following shapes. Do *not* work out an exact answer.

a)

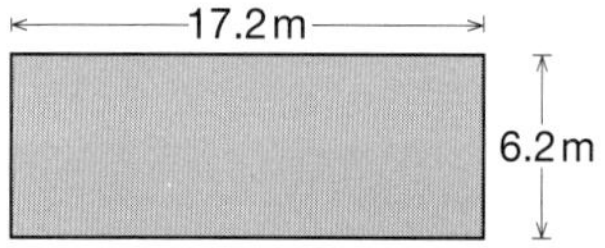

b)

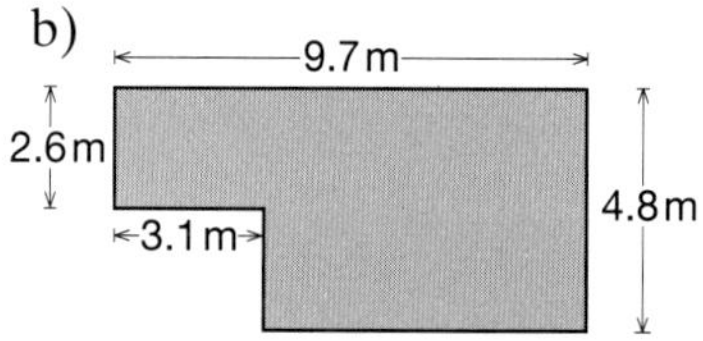

c)

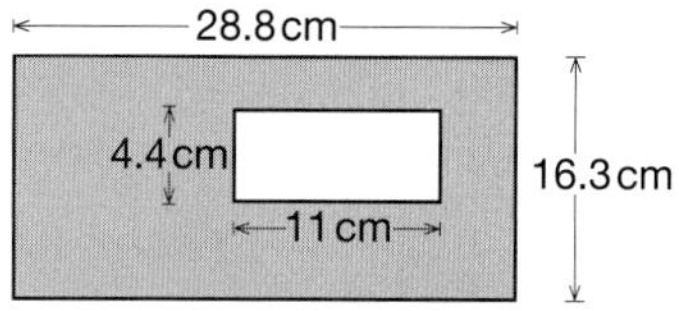

7. Estimate the volume of each of the solids below. Do *not* work out an exact answer.

a)

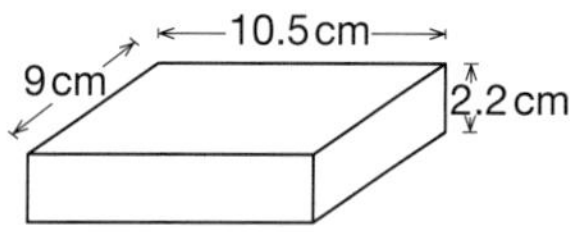

c)

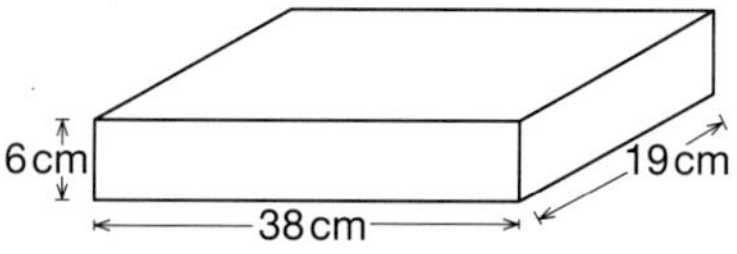

b)

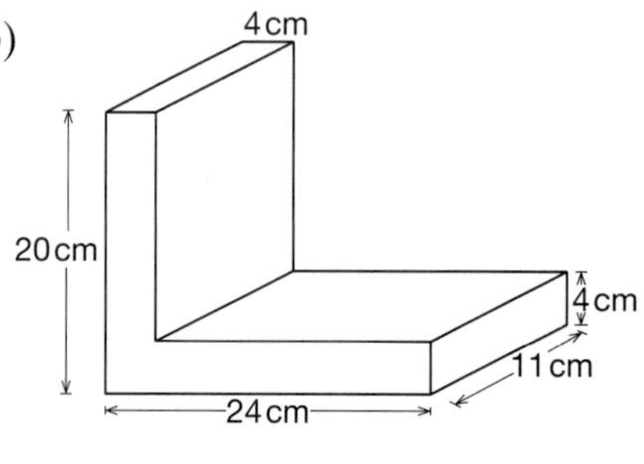

Student Assessment 1

1. Round off the following numbers to the degree of accuracy shown in brackets:
   a) 2841 (nearest 100) b) 7286 (nearest 10)
   c) 48 756 (nearest 1000) d) 951 (nearest 100)

2. Round off the following numbers to the number of decimal places shown in brackets:
   a) 3.84 (1 d.p.) b) 6.792 (1 d.p.)
   c) 0.8526 (2 d.p.) d) 1.5849 (2 d.p.)
   e) 9.954 (1 d.p.) f) 0.0077 (3 d.p.)

3. Round off the following numbers to the number of significant figures shown in brackets:
   a) 3.84 (1 s.f.) b) 6.792 (2 s.f.)
   c) 0.7765 (1 s.f.) d) 9.624 (1 s.f.)
   e) 834.97 (2 s.f.) f) 0.004 51 (1 s.f.)

4. 1 mile is 1760 yards. Estimate the number of yards in 11.5 miles.

5. Estimate the shaded area of the figure below:

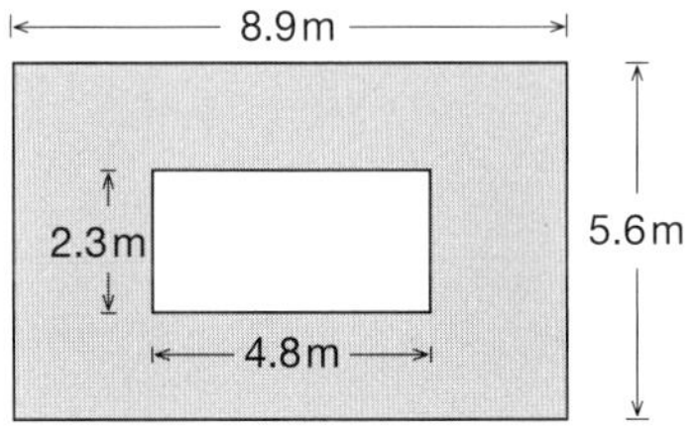

6. Estimate the answers to the following. Do *not* work out exact answers.

   a) $\dfrac{5.3 \times 11.2}{2.1}$ b) $\dfrac{(9.8)^2}{(4.7)^2}$ c) $\dfrac{18.8 \times (7.1)^2}{(3.1)^2 \times (4.9)^2}$

7. A cuboid's dimensions are given as 12.32 cm by 1.8 cm by 4.16 cm. Calculate its volume, giving your answer to an appropriate degree of accuracy.

Student Assessment 2

1. Round off the following numbers to the degree of accuracy shown in brackets:
   a) 6472 (nearest 10) b) 88 465 (nearest 100)
   c) 64 785 (nearest 1000) d) 6.7 (nearest 10)

2. Round off the following numbers to the number of decimal places shown in brackets:
   a) 6.78 (1 d.p.) b) 4.438 (2 d.p.)
   c) 7.975 (1 d.p.) d) 63.084 (2 d.p.)
   e) 0.0567 (3 d.p.) f) 3.95 (2 d.p.)

3. Round off the following numbers to the number of significant figures shown in brackets:

a) 42.6 (1 s.f.) b) 5.432 (2 s.f.)

c) 0.0574 (1 s.f.) d) 48 572 (2 s.f.)

e) 687 453 (1 s.f.) f) 687 453 (3 s.f.)

4. 1 mile is 1760 yards. Estimate the number of yards in 19 miles.

5. Estimate the area of the figure below:

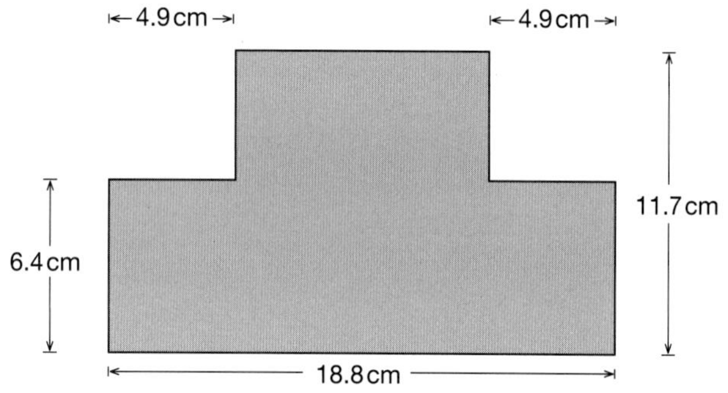

6. Estimate the answers to the following. Do *not* work out exact answers.

a) $\dfrac{3.9 \times 26.4}{4.85}$ b) $\dfrac{(3.2)^3}{(5.4)^2}$ c) $\dfrac{2.8 \times (7.3)^2}{(3.2)^2 \times 6.2}$

7. A cuboid's dimensions are given as 3.973 m by 2.4 m by 3.16 m. Calculate its volume, giving your answer to an appropriate degree of accuracy.

9 LIMITS OF ACCURACY

Give appropriate upper and lower bounds for data to a specified accuracy (e.g. measured lengths).

Numbers can be written to different degrees of accuracy. For example, 4.5, 4.50 and 4.500, although appearing to represent the same number, do not. This is because they are written to different degrees of accuracy.

4.5 is rounded to one decimal place and therefore could represent any number from 4.45 up to but not including 4.55. On a number line this would be represented as:

As an inequality where x represents the number, 4.5 would be expressed as:

$$4.45 \leqslant x < 4.55$$

4.45 is known as the **lower bound** of 4.5, whilst 4.55 is known as the **upper bound**.

4.50 on the other hand is written to two decimal places and only numbers from 4.495 up to but not including 4.505 would be rounded to 4.50. This therefore represents a much smaller range of numbers than those which would be rounded to 4.5. Similarly the range of numbers being rounded to 4.500 would be even smaller.

Worked example A girl's height is given as 162 cm to the nearest centimetre.

i) Work out the lower and upper bounds within which her height can lie.

Lower bound = 161.5 cm
Upper bound = 162.5 cm

ii) Represent this range of numbers on a number line.

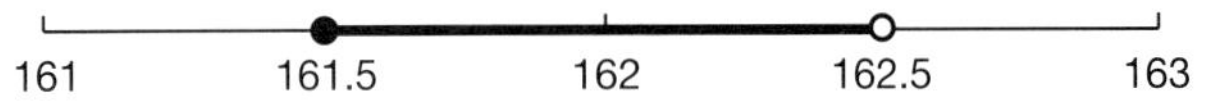

iii) If the girl's height is h cm, express this range as an inequality.

$$161.5 \leqslant h < 162.5$$

Exercise 9.1

1. Each of the following numbers is expressed to the nearest whole number.

i) Give the upper and lower bounds of each.
ii) Using x as the number, express the range in which the number lies as an inequality.

a) 6 b) 83 c) 152
d) 1000 e) 0 f) −4

2. Each of the following numbers is correct to one decimal place.
   i) Give the upper and lower bounds of each.
   ii) Using x as the number, express the range in which the number lies as an inequality.
   a) 3.8 b) 15.6 c) 1.0
   d) 10.0 e) 0.3 f) −0.2

3. Each of the following numbers is correct to two significant figures.
   i) Give the upper and lower bounds of each.
   ii) Using x as the number, express the range in which the number lies as an inequality.
   a) 4.2 b) 0.84 c) 420
   d) 5000 e) 0.045 f) 25 000

4. The mass of a sack of vegetables is given as 5.4 kg.
   a) Illustrate the lower and upper bounds of the mass on a number line.
   b) Using M kg for the mass, express the range of values in which M must lie as an inequality.

5. At a school sports day, the winning time for the 100 m race was given as 11.8 seconds.
   a) Illustrate the lower and upper bounds of the time on a number line.
   b) Using T seconds for the time, express the range of values in which T must lie as an inequality.

6. The capacity of a swimming pool is given as 620 m^3 correct to two significant figures.
   a) Calculate the lower and upper bounds of the pool's capacity.
   b) Using x cubic metres for the capacity, express the range of values in which x must lie as an inequality.

7. A farmer measures the dimensions of his rectangular field to the nearest 10 m. The length is recorded as 630 m and the width is recorded as 400 m.
   a) Calculate the lower and upper bounds of the length.
   b) Using W metres for the width, express the range of values in which W must lie as an inequality.

Exercise 9.2

1. Each of the following numbers is expressed to the nearest whole number.
   i) Give the upper and lower bounds of each.
   ii) Using x as the number, express the range in which the number lies as an inequality.
   a) 8 b) 71 c) 146
   d) 200 e) 1 f) −1

2. Each of the following numbers is correct to one decimal place.
   i) Give the upper and lower bounds of each.
   ii) Using x as the number, express the range in which the number lies as an inequality.

a) 2.5	b) 14.1	c) 2.0
d) 20.0	e) 0.5	f) −0.5

3. Each of the following numbers is correct to two significant figures.
   i) Give the upper and lower bounds of each.
   ii) Using x as the number, express the range in which the number lies as an inequality.

a) 5.4	b) 0.75	c) 550
d) 6000	e) 0.012	f) 10 000

4. The mass of a sack of vegetables is given as 7.8 kg.
   a) Illustrate the lower and upper bounds of the mass on a number line.
   b) Using M kg for the mass, express the range of values in which M must lie as an inequality.

5. At a school sports day, the winning time for the 100 m race was given as 12.1 s.
   a) Illustrate the lower and upper bounds of the time on a number line.
   b) Using T seconds for the time, express the range of values in which T must lie as an inequality.

6. The capacity of a swimming pool is given as 740 m^3 correct to two significant figures.
   a) Calculate the lower and upper bounds of the pool's capacity.
   b) Using x cubic metres for the capacity, express the range of values in which x must lie as an inequality.

7. A farmer measures the dimensions of his rectangular field to the nearest 10 m. The length is recorded as 570 m and the width is recorded as 340 m.
   a) Calculate the lower and upper bounds of the length.
   b) Using W metres for the width, express the range of values in which W must lie as an inequality.

Student Assessment 1

1. The following numbers are expressed to the nearest whole number. Illustrate on a number line the range in which each must lie.
   a) 7
   b) 40
   c) 0
   d) −200

2. The following numbers are expressed correct to two significant figures. Representing each number by the letter x, express the range in which each must lie using an inequality.
   a) 210
   b) 64
   c) 3.0
   d) 0.88

3. A school measures the dimensions of its rectangular playing field to the nearest metre. The length was recorded as 350 m and the width as 200 m. Express the ranges in which the length and width lie using inequalities.

4. A boy's mass was measured to the nearest 0.1 kg. If his mass was recorded as 58.9 kg, illustrate on a number line the range within which it must lie.

5. An electronic clock is accurate to $\frac{1}{1000}$ of a second. The duration of a flash from a camera is timed at 0.004 seconds. Express the upper and lower bounds of the duration of the flash using inequalities.

6. The following numbers are rounded to the degree of accuracy shown in brackets. Express the lower and upper bounds of these numbers as an inequality.
   a) $x = 4.83$ (2 d.p.)
   b) $y = 5.05$ (2 d.p.)
   c) $z = 10.0$ (1 d.p.)
   d) $p = -100.00$ (2 d.p.)

Student Assessment 2

1. The following numbers are rounded to the nearest 100. Illustrate on a number line the range in which they must lie.
   a) 500
   b) 7000
   c) 0
   d) −32 000

2. The following numbers are expressed correct to three significant figures. Represent the limits of each number using inequalities.
   a) 254
   b) 40.5
   c) 0.410
   d) 100

3. The dimensions of a rectangular courtyard are given to the nearest 0.5 m. The length is recorded as 20.5 m and the width as 10.0 m. Represent the limits of these dimensions using inequalities.

4. The circumference c of a tree is measured to the nearest 2 mm. If its circumference is measured as 245.6 cm, illustrate on a number line the range within which it must lie.

5. The time it takes Earth to rotate around the Sun is given as 365.25 days to five significant figures. What are the upper and lower bounds of this time?

6. The following numbers are rounded to the degree of accuracy shown in brackets. Express the lower and upper bounds of these numbers as an inequality.

a) 10.90 (2 d.p.)
b) 3.00 (2 d.p.)
c) 0.5 (1 d.p.)
d) −175.00 (2 d.p.)

10 RATIO, PROPORTION AND MEASURES OF RATE

Demonstrate an understanding of the elementary ideas and notation of ratio, direct and inverse proportion and common measures of rate; divide a quantity in a given ratio.

NB: All diagrams are not drawn to scale.

Direct proportion

Workers in a pottery factory are paid according to how many plates they produce. The wage paid to them is said to be in **direct proportion** to the number of plates made. As the number of plates made increases so does their wage. Other workers are paid for the number of hours worked. For them the wage paid is in **direct proportion** to the number of hours worked. There are two main methods for solving problems involving direct proportion: the ratio method and the unitary method.

Worked example A bottling machine fills 500 bottles in 15 minutes. How many bottles will it fill in $1\frac{1}{2}$ hours?

Note: The time units must be the same, so for either method the $1\frac{1}{2}$ hours must be changed to 90 minutes.

The ratio method

Let x be the number of bottles filled. Then:

$$\frac{x}{90} = \frac{500}{15}$$

$$\text{so } x = \frac{500 \times 90}{15} = 3000$$

3000 bottles are filled in $1\frac{1}{2}$ hours.

The unitary method

In 15 minutes 500 bottles are filled.
Therefore in 1 minute $\frac{500}{15}$ bottles are filled.
So in 90 minutes $90 \times \frac{500}{15}$ bottles are filled.
In $1\frac{1}{2}$ hours, 3000 bottles are filled.

Exercise 10.1

Use either the ratio method or the unitary method to solve the problems below.

1. A machine prints four books in 10 minutes. How many will it print in 2 hours?
2. A farmer plants five apple trees in 25 minutes. If he continues to work at a constant rate, how long will it take him to plant 200 trees?

3. A television set uses 3 units of electricity in 2 hours. How many units will it use in 7 hours? Give your answer to the nearest unit.

4. A bricklayer lays 1500 bricks in an 8-hour day. Assuming he continues to work at the same rate, calculate:
a) how many bricks he would expect to lay in a five-day week,
b) how long to the nearest hour it would take him to lay 10 000 bricks.

5. A machine used to paint white lines on a road uses 250 litres of paint for each 8 km of road marked. Calculate:
a) how many litres of paint would be needed for 200 km of road,
b) what length of road could be marked with 4000 litres of paint.

6. An aircraft is cruising at 720 km/h and covers 1000 km. How far would it travel in the same period of time if the speed increased to 800 km/h?

7. A production line travelling at 2 m/s labels 150 tins. In the same period of time how many will it label at:
a) 6 m/s b) 1 m/s c) 1.6 m/s?

If the information is given in the form of a ratio, the method of solution is the same.

Worked example Tin and copper are mixed in the ratio 8 : 3. How much tin is needed to mix with 36 g of copper?

The ratio method

Let x grams be the mass of tin needed.

$$\frac{x}{36} = \frac{8}{3}$$

$$\text{Therefore } x = \frac{8 \times 36}{3}$$
$$= 96$$

So 96 g of tin is needed.

The unitary method

3 g of copper mixes with 8 g of tin.
1 g of copper mixes with $\frac{8}{3}$g of tin.
So 36 g of copper mixes with $36 \times \frac{8}{3}$g of tin.
Therefore 36 g of copper mixes with 96 g of tin.

Exercise 10.2

Use either the ratio method or unitary method to solve the problems below.

1. A production line produces 8 cars in 3 hours.
   a) Calculate how many it will produce in 48 hours.
   b) Calculate how long it will take to produce 1000 cars.

2. A machine produces six golf balls in fifteen seconds. Calculate how many are produced in:
   a) 5 minutes   b) 1 hour   c) 1 day.

3. A cassette recorder uses 0.75 units of electricity in 90 minutes. Calculate:
   a) how many units it will use in 8 hours,
   b) how long it will operate for 15 units of electricity.

4. A combine harvester takes 2 hours to harvest a 3 hectare field. If it works at a constant rate, calculate:
   a) how many hectares it will harvest in 15 hours,
   b) how long it will take to harvest a 54 hectare field.

5. A road-surfacing machine can re-surface 8 m of road in 40 seconds. Calculate how long it will take to re-surface 18 km of road, at the same rate.

6. A sailing yacht is travelling at 1.5 km/h and covers 12 km. If its speed increased to 2.5 km/h, how far would it expect to travel in the same period of time?

7. A plate-making machine produces 3 dozen plates in 8 minutes.
   a) How many plates are produced in one hour?
   b) How long would it take to produce 20 gross of plates?

Exercise 10.3

1. Sand and gravel are mixed in the ratio 5 : 3 to form ballast.
   a) How much gravel is mixed with 750 kg of sand?
   b) How much sand is mixed with 750 kg of gravel?

2. A recipe uses 150 g butter, 500 g flour, 50 g sugar and 100 g currants to make 18 small cakes.
   a) How much of each ingredient will be needed to make 6 dozen cakes?
   b) How many whole cakes could be made with 1 kg of butter?

3. A paint mix uses red and white paint in a ratio of 1 : 12.
   a) How much white paint will be needed to mix with 1.4 litres of red paint?
   b) If a total of 15.5 litres of paint is mixed, calculate the amount of white paint and the amount of red paint used. Give your answers to the nearest 0.1 litre.

4. A tulip farmer sells sacks of mixed bulbs to local people. The bulbs develop into two different colours of tulips, red and yellow. The colours are packaged in a ratio of 8 : 5, respectively.

a) If a sack contains 200 red bulbs, calculate the number of yellow bulbs.

b) If a sack contains 351 bulbs in total, how many of each colour would you expect to find?

c) One sack is packaged with a bulb mixture in the ratio 7 : 5 by mistake. If the sack contains 624 bulbs, how many more yellow bulbs would you expect to have compared with a normal sack of 624 bulbs?

5. A pure fruit juice is made by mixing the juices of oranges and mangoes in the ratio of 9 : 2.

a) If 189 litres of orange juice are used, calculate the number of litres of mango juice needed.

b) If 605 litres of the juice are made, calculate the number of litres of orange juice and mango juice used.

Divide a quantity in a given ratio

Worked examples

a) Divide 20 m in the ratio 3 : 2.

The ratio method

3 : 2 gives 5 parts.
$\frac{3}{5} \times 20\text{ m} = 12\text{ m}$
$\frac{2}{5} \times 20\text{ m} = 8\text{ m}$
20 m divided in the ratio 3 : 2 is 12 m : 8 m.

The unitary method

3 : 2 gives 5 parts.
5 parts is equivalent to 20 m.
1 part is equivalent to $\frac{20}{5}$ m.
Therefore 3 parts is $3 \times \frac{20}{5}$ m; that is 12 m.
Therefore 2 parts is $2 \times \frac{20}{5}$ m; that is 8 m.

b) A factory produces cars in red, blue, white and green in the ratio 7 : 5 : 3 : 1. Out of a production of 48 000 cars how many are white?

$7 + 5 + 3 + 1$ gives a total of 16 parts.
Therefore the total number of white cars $= \frac{3}{16} \times 48\,000 = 9000$.

Exercise 10.4

1. Divide 150 in the ratio 2 : 3.

2. Divide 72 in the ratio 2 : 3 : 4.

3. Divide 5 kg in the ratio 13 : 7.

4. Divide 45 minutes in the ratio 2 : 3.

5. Divide 1 hour in the ratio 1 : 5.

6. $\frac{7}{8}$ of a can of coke is water, the rest is syrup. What is the ratio of water to syrup?

7. $\frac{5}{9}$ of a litre carton of orange is pure orange juice, the rest is water. How many millilitres of each are in the carton?

8. 55% of students in a school are boys.
   a) What is the ratio of boys to girls?
   b) How many boys and how many girls are there if the school has 800 students?

9. A piece of wood is cut in the ratio 2 : 3. What fraction of the length is the longer piece?

10. If the original piece of wood in question 9 is 80 cm long, how long is the shorter piece?

11. A gas pipe is 7 km long. A valve is positioned in such a way that it divides the length of the pipe in the ratio 4 : 3. Calculate the distance of the valve from each end of the pipe.

12. The size of the angles of a quadrilateral are in the ratio 1 : 2 : 3 : 3. Calculate the size of each angle.

13. The angles of a triangle are in the ratio 3 : 5 : 4. Calculate the size of each angle.

14. A millionaire leaves 1.4 million dollars in his will to be shared between his three children in the ratio of their ages. If they are 24, 28 and 32 years old, calculate to the nearest dollar the amount they will each receive.

15. A small company makes a profit of £8000. This is divided between the directors in the ratio of their initial investments. If Alex put £20 000 into the firm, Maria £35 000 and Ahmet £25 000, calculate the amount of the profit they will each receive.

Inverse proportion

Sometimes an increase in one quantity causes a decrease in another quantity. For example, if fruit is to be picked by hand, the more people there are picking the fruit, the less time it will take. The time taken is said to be **inversely proportional** to the number of people picking the fruit.

Worked examples

a) If 8 people can pick the apples from the trees in 6 days, how long will it take 12 people?

8 people take 6 days.
1 person will take 6×8 days.
Therefore 12 people will take $\frac{6 \times 8}{12}$ days, i.e. 4 days.

b) A cyclist averages a speed of 27 km/h for 4 hours. At what average speed would she need to cycle to cover the same distance in 3 hours?

Completing it in 1 hour would require cycling at 27×4 km/h.

Completing it in 3 hours requires cycling at

$\frac{27 \times 4}{3}$ km/h, that is, 36 km/h.

Exercise 10.5

1. A teacher shares sweets among 8 students so that they get 6 each. How many sweets would they each have been given had there been 12 students?

2. The table below represents the relationship between the speed and the time taken for a train to travel between two stations.

Speed (km/h)	60			120	90	50	10
Time (h)	2	3	4				

Copy and complete the table.

3. A school can buy 150 books costing £12 each. If the price is reduced by 20%, how many more books could be bought?

4. Six people can dig a trench in 8 hours.

a) How long would it take:

i) 4 people ii) 12 people iii) 1 person?

b) How many people would it take to dig the trench in:

i) 3 hours ii) 16 hours iii) 1 hour?

5. Chairs in a hall are arranged in 35 rows of 18.

a) How many rows would there be with 21 chairs to a row?

b) How many chairs would there be in each row if there were 15 rows?

6. A train travelling at 100 km/h takes 4 hours for a journey. How long would it take a train travelling at 60 km/h?

7. A worker in a sugar factory packs 24 cardboard boxes with 15 bags of sugar in each. If he had boxes which held 18 bags of sugar each, how many fewer boxes would be needed?

8. A swimming pool is filled in 30 hours by two identical pumps. How much quicker would it be filled if five similar pumps were used instead?

Student Assessment 1

1. A ruler 30 cm long is broken into two parts in the ratio 8 : 7. How long are the two parts?

2. A recipe needs 400 g of flour to make 8 cakes. How much flour would be needed in order to make two dozen cakes?

3. To make 6 jam tarts, 120 g of jam is needed. How much jam is needed to make 10 tarts?

4. The scale of a map is 1 : 25 000.
   a) Two villages are 8 cm apart on the map. How far apart are they in real life? Give your answer in kilometres.
   b) The distance from a village to the edge of a lake is 12 km in real life. How far apart would they be on the map? Give your answer in centimetres.

5. A motorbike uses petrol and oil mixed in the ratio 13 : 2.
   a) How much of each is there in 30 litres of mixture?
   b) How much petrol would be mixed with 500 ml of oil?

6. a) A model car is a $\frac{1}{40}$ scale model. Express this as a ratio.
   b) If the length of the real car is 5.5 m, what is the length of the model car?

7. An aunt gives a brother and sister £2000 to be divided in the ratio of their ages. If the girl is 13 years old and the boy 12 years old, how much will each get?

8. The angles of a triangle are in the ratio 2 : 5 : 8. Find the size of each of the angles.

9. A photocopying machine is capable of making 50 copies each minute.
   a) If four identical copiers are used simultaneously how long would it take to make a total of 50 copies?
   b) How many of these copiers would be needed to make 6000 copies in 15 minutes?

10. It takes 16 hours for three bricklayers to build a wall. Calculate how long it would take for eight bricklayers to build a similar wall.

Student Assessment 2

1. A piece of wood is cut in the ratio 3 : 7.
   a) What fraction of the whole is the longer piece?
   b) If the wood is 1.5 m long, how long is the shorter piece?

2. A recipe for two people requires $\frac{1}{4}$ kg of rice to 150 g of meat.
   a) How much meat would be needed for five people?
   b) How much rice would there be in 1 kg of the final dish?

3. The scale of a map is 1 : 10 000.
  a) Two rivers are 4.5 cm apart on the map, how far apart are they in real life? Give your answer in metres.
  b) Two towns are 8 km apart in real life. How far apart are they on the map? Give your answer in centimetres.

4. a) A model train is a $\frac{1}{25}$ scale model. Express this as a ratio.
  b) If the length of the model engine is 7 cm, what is the true length of the engine?

5. Divide 3 tonnes in the ratio 2 : 5 : 13.

6. The ratio of the angles of a quadrilateral is 2 : 3 : 3 : 4. Calculate the size of each of the angles.

7. The ratio of the interior angles of a pentagon is 2 : 3 : 4 : 4 : 5. Calculate the size of the largest angle.

8. A large swimming pool takes 36 hours to fill using three identical pumps.
  a) How long would it take to fill using eight identical pumps?
  b) If the pool needs to be filled in 9 hours, how many of these pumps will be needed?

9. The first triangle is an enlargement of the second. Calculate the size of the missing sides and angles.

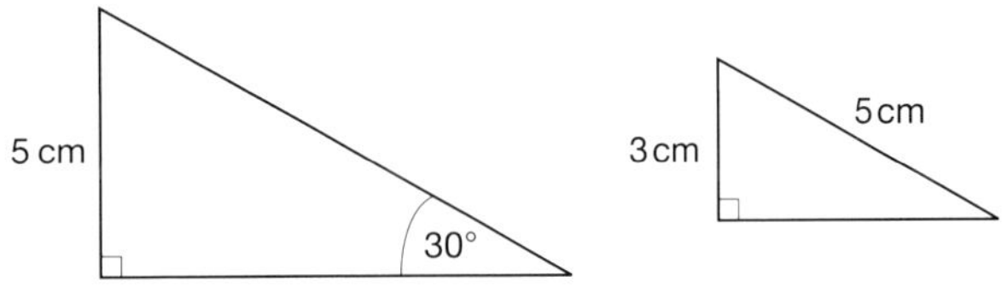

10. A tap issuing water at a rate of 1.2 litres per minute fills a container in 4 minutes.
  a) How long would it take to fill the same container if the rate was decreased to 1 litre per minute? Give your answer in minutes and seconds.
  b) If the container is to be filled in 3 minutes, calculate the rate at which the water should flow.

11 PERCENTAGES

Calculate a given percentage of a quantity; express one quantity as a percentage of another; calculate percentage increase or decrease.

You should already be familiar with the percentage equivalent of simple fractions and decimals as outlined in the table below.

Fraction	Decimal	Percentage
$\frac{1}{2}$	0.5	50%
$\frac{1}{4}$	0.25	25%
$\frac{3}{4}$	0.75	75%
$\frac{1}{8}$	0.125	12.5%
$\frac{3}{8}$	0.375	37.5%
$\frac{5}{8}$	0.625	62.5%
$\frac{7}{8}$	0.875	87.5%
$\frac{1}{10}$	0.1	10%
$\frac{2}{10}$ or $\frac{1}{5}$	0.2	20%
$\frac{3}{10}$	0.3	30%
$\frac{4}{10}$ or $\frac{2}{5}$	0.4	40%
$\frac{6}{10}$ or $\frac{3}{5}$	0.6	60%
$\frac{7}{10}$	0.7	70%
$\frac{8}{10}$ or $\frac{4}{5}$	0.8	80%
$\frac{9}{10}$	0.9	90%

Simple percentages

Worked examples

a) Of 100 sheep in a field, 88 are ewes.

i) What percentage of the sheep are ewes?
88 out of 100 are ewes
= 88%

ii) What percentage are not ewes?
12 out of 100
= 12%

b) A gymnast scored marks out of 10 from five judges.
They were: 8.0, 8.2, 7.9, 8.3 and 7.6.
Express these marks as percentages.

$\frac{8.0}{10} = \frac{80}{100} = 80\%$ $\frac{8.2}{10} = \frac{82}{100} = 82\%$ $\frac{7.9}{10} = \frac{79}{100} = 79\%$

$\frac{8.3}{10} = \frac{83}{100} = 83\%$ $\frac{7.6}{10} = \frac{76}{100} = 76\%$

c) Convert the following percentages into fractions and decimals:

i) 27% ii) 5%

$$\frac{27}{100} = 0.27 \qquad \frac{5}{100} = 0.05$$

Exercise 11.1

1. In a survey of 100 cars, 47 were white, 23 were blue and 30 were red. Express each of these numbers as a percentage of the total.

2. $\frac{7}{10}$ of the surface of the Earth is water. Express this as a percentage.

3. There are 200 birds in a flock. 120 of them are female. What percentage of the flock are:
a) female? b) male?

4. Write these percentages as fractions of 100:
a) 73% b) 28%
c) 10% d) 25%

5. Write these fractions as percentages:
a) $\frac{27}{100}$ b) $\frac{3}{10}$
c) $\frac{7}{50}$ d) $\frac{1}{4}$

6. Convert the following percentages to decimals:
a) 39% b) 47% c) 83%
d) 7% e) 2% f) 20%

7. Convert the following decimals to percentages:
a) 0.31 b) 0.67 c) 0.09
d) 0.05 e) 0.2 f) 0.75

Calculating a percentage of a quantity

Worked examples

a) Find 25% of 300 m.

25% can be written as 0.25.
0.25×300 m = 75 m.

b) Find 35% of 280 m.

35% can be written as 0.35.
0.35×280 m = 98 m.

Exercise 11.2

1. Write the percentage equivalent of the following fractions:
a) $\frac{1}{4}$ b) $\frac{2}{3}$ c) $\frac{5}{8}$
d) $1\frac{4}{5}$ e) $4\frac{9}{10}$ f) $3\frac{7}{8}$

2. Write the decimal equivalent of the following:
a) $\frac{3}{4}$ b) 80% c) $\frac{1}{5}$
d) 7% e) $1\frac{7}{8}$ f) $\frac{1}{6}$

3. Evaluate the following:

a) 25% of 80	b) 80% of 125	c) 62.5% of 80
d) 30% of 120	e) 90% of 5	f) 25% of 30

4. Evaluate the following:

a) 17% of 50	b) 50% of 17	c) 65% of 80
d) 80% of 65	e) 7% of 250	f) 250% of 7

5. In a class of 30 students, 20% have black hair, 10% have blonde hair and 70% have brown hair. Calculate the number of students with
   a) black hair,
   b) blonde hair,
   c) brown hair.

6. A survey conducted among 120 schoolchildren looked at which type of meat they preferred. 55% said they preferred beef, 20% said they preferred chicken, 15% preferred lamb and 10% pork. Calculate the number of children in each category.

7. A survey was carried out in a school to see what nationality its students were. Of the 220 students in the school, 65% were English, 20% were Pakistani, 5% were Greek and 10% belonged to other nationalities. Calculate the number of students of each nationality.

8. A shopkeeper keeps a record of the number of items he sells in one day. Of the 150 items he sold, 46% were newspapers, 24% were pens, 12% were books whilst the remaining 18% were other assorted items. Calculate the number of each item he sold.

Expressing one quantity as a percentage of another

To express one quantity as a percentage of another, first write the first quantity as a fraction of the second and then multiply by 100.

Worked example

In an examination a girl obtains 69 marks out of 75. Express this result as a percentage.

$$\frac{69}{75} \times 100\% = 92\%$$

Exercise 11.3

1. For each of the following, express the first quantity as a percentage of the second.

a) 24 out of 50	b) 46 out of 125
c) 7 out of 20	d) 45 out of 90
e) 9 out of 20	f) 16 out of 40
g) 13 out of 39	h) 20 out of 35

2. A hockey team plays 42 matches. It wins 21, draws 14 and loses the rest. Express each of these results as a percentage of the total number of games played.

3. Four candidates stood in an election:

A received 24 500 votes
B received 18 200 votes
C received 16 300 votes
D received 12 000 votes

Express each of these results as a percentage of the total votes cast.

4. A car manufacturer produces 155 000 cars a year. The cars are available for sale in six different colours. The numbers sold of each colour were:

Red	55 000
Blue	48 000
White	27 500
Silver	10 200
Green	9300
Black	5000

Express each of these as a percentage of the total number of cars produced. Give your answers to 1 d.p.

Percentage increases and decreases

Worked examples

a) A garage increases the price of a truck by 12%. If the original price was £14 500, calculate its new price.
Note: the original price represents 100%, therefore the increase can be represented as 112%.

New price = 112% of £14 500
= $1.12 \times$ £14 500
= £16 240

b) A doctor in Thailand has a salary of 18 000 baht per month. If her salary increases by 8%, calculate:
i) the amount extra she receives per month,
ii) her new monthly salary.

i) Increase = 8% of 18 000 baht
= $0.08 \times$ 18 000 baht = 1440 baht
ii) New salary = old salary + increase
= 18 000 + 1440 baht per month
= 19 440 baht per month

c) A shop is having a sale. It sells a set of tools costing \$130 at a 15% discount. Calculate the sale price of the tools.
Note: the old price represents 100%, therefore the new price can be represented as $(100 - 15)\% = 85\%$.

85% of \$130 = $0.85 \times$ \$130
= \$110.50

Exercise 11.4

1. Increase the following by the given percentage:
   a) 150 by 25% b) 230 by 40% c) 7000 by 2%
   d) 70 by 250% e) 80 by 12.5% f) 75 by 62%

2. Decrease the following by the given percentage:
   a) 120 by 25% b) 40 by 5% c) 90 by 90%
   d) 1000 by 10% e) 80 by 37.5% f) 75 by 42%

3. In the following questions, the first number is increased to become the second number. Calculate the percentage increase in each case.
   a) $50 \rightarrow 60$ b) $75 \rightarrow 135$ c) $40 \rightarrow 84$
   d) $30 \rightarrow 31.5$ e) $18 \rightarrow 33.3$ f) $4 \rightarrow 13$

4. In the following questions, the first number is decreased to become the second number. Calculate the percentage decrease in each case.
   a) $50 \rightarrow 25$ b) $80 \rightarrow 56$ c) $150 \rightarrow 142.5$
   d) $3 \rightarrow 0$ e) $550 \rightarrow 352$ f) $20 \rightarrow 19$

5. A farmer increases the yield on his farm by 15%. If his previous yield was 6500 tonnes, what is his present yield?

6. The cost of a computer in Singapore is reduced by 12.5% in a sale. If the computer was priced at $7800, what is its price in the sale?

7. A winter coat is priced at £100. In the sale its price is reduced by 25%.
   a) Calculate the sale price of the coat.
   b) After the sale, its price is increased by 25% again. Calculate the coat's price after the sale.

8. A farmer takes 250 chickens to be sold at a market. In the first hour he sells 8% of his chickens. In the second hour he sells 10% of those that were left.
   a) How many chickens has he sold in total?
   b) What percentage of the original number did he manage to sell in the two hours?

9. The number of fish on a fish farm increases each month by approximately 10%. If there were originally 350 fish, calculate to the nearest 100 how many fish there would be after 12 months.

Student Assessment 1

1. Copy the table below and fill in the missing values:

Fraction	Decimal	Percentage
$\frac{3}{4}$		
	0.8	
$\frac{5}{8}$		
	1.5	

2. Find 40% of 1600 m.

3. A shop increases the price of a television set by 8%. If the original price was £320, what is the new price?

4. A car loses 55% of its value after four years. If it cost \$22 500 when new, what is its value after the four years?

5. Express the first quantity as a percentage of the second.
a) 40 cm, 2 m b) 25 mins, 1 hour c) 450 g, 2 kg
d) 3 m, 3.5 m e) 70 kg, 1 tonne f) 75 cl, 2.5 litres

6. A house is bought for 75 000 rand, then resold for 87 000 rand. Calculate the percentage profit.

7. A pair of shoes is priced at £45. During a sale the price is reduced by 20%.
a) Calculate the sale price of the shoes.
b) What is the percentage increase in the price if after the sale it is once again restored to £45?

8. The population of a town increases by 5% each year. If in 1997 the population was 86 000, in which year is the population expected to exceed 100 000 for the first time?

Student Assessment 2

1. Copy the table below and fill in the missing values:

Fraction	Decimal	Percentage
	0.25	
$\frac{3}{5}$		
		$62\frac{1}{2}\%$
$2\frac{1}{4}$		

2. Find 30% of 2500 m.

3. In a sale a shop reduces its prices by 12.5%. What is the sale price of a desk previously costing 2400 Hong Kong dollars?

4. In the last six years the value of a house has increased by 35%. If it cost £72 000 six years ago, what is its value now?

5. Express the first quantity as a percentage of the second.
   a) 35 mins, 2 hours b) 650 g, 3 kg
   c) 5 m, 4 m d) 15 s, 3 mins
   e) 600 kg, 3 tonnes f) 35 cl, 3.5 litres

6. Shares in a company are bought for $600. After a year, the same shares are sold for $550. Calculate the percentage depreciation.

7. In a sale the price of a jacket originally costing 850 francs is reduced by 200 francs. Any item not sold by the last day of the sale is reduced by a further 50%. If the jacket is sold on the last day of the sale:
   a) calculate the price it is finally sold for,
   b) calculate the overall percentage reduction in price.

8. Each day the population of a type of insect increases by approximately 10%. How many days will it take for the population to double?

12 USE OF AN ELECTRONIC CALCULATOR

Use an electronic calculator efficiently; apply appropriate checks of accuracy.

There are many different types of calculator available today. These include basic calculators, scientific calculators and the latest graphical calculators. However, these are all useless unless you make use of their potential. The following chapter is aimed at familiarising you with some of the basic operations.

The four basic operations

Worked examples

a) Using a calculator, work out the answer to the following:

$12.3 + 14.9 =$

[1] [2] [.] [3] [+] [1] [4] [.] [9] [=] 27.2

b) Using a calculator, work out the answer to the following:

$16.3 \times 10.8 =$

[1] [6] [.] [3] [×] [1] [0] [.] [8] [=] 176.04

c) Using a calculator, work out the answer to the following:

$4.1 \times -3.3 =$

[4] [.] [1] [×] [3] [.] [3] [+/−] [=] −13.53

Exercise 12.1

1. Using a calculator, work out the answers to the following:

a) $9.7 + 15.3$ b) $13.6 + 9.08$
c) $12.9 + 4.92$ d) $115.0 + 6.24$
e) $86.13 + 48.2$ f) $108.9 + 47.2$

2. Using a calculator, work out the answers to the following:

a) $15.2 - 2.9$ b) $12.4 - 0.5$
c) $19.06 - 20.3$ d) $4.32 - 4.33$
e) $-9.1 - 21.2$ f) $-6.3 - 2.1$
g) $-28 - -15$ h) $-2.41 - -2.41$

3. Using a calculator, work out the answers to the following:

a) 9.2×8.7 b) 14.6×8.1
c) $4.1 \times 3.7 \times 6$ d) $9.3 \div 3.1$
e) 14.2×-3 f) $15.5 \div -5$
g) -2.2×-2.2 h) $-20 \div -4.5$

The order of operations

When carrying out calculations, care must be taken to ensure that they are carried out in the correct order.

Worked examples

a) Use a scientific calculator to work out the answer to the following:

$2 + 3 \times 4 =$

[2] [+] [3] [×] [4] [=] 14

b) Use a scientific calculator to work out the answer to the following:

$(2 + 3) \times 4 =$

[(] [2] [+] [3] [)] [×] [4] [=] 20

The reason why different answers are obtained is because, by convention, the operations have different priorities. These are as follows:

1. brackets
2. mutiplication/division
3. addition/subtraction

Therefore in ***Worked example* a)** 3×4 is evaluated first, and then the 2 is added, whilst in ***Worked example* b)** $(2 + 3)$ is evaluated first, followed by multiplication by 4.

Exercise 12.2

In the following questions, evaluate the answers:

i) in your head,
ii) using a scientific calculator.

1. a) $8 \times 3 + 2$ b) $4 \div 2 + 8$
c) $12 \times 4 - 6$ d) $4 + 6 \times 2$
e) $10 - 6 \div 3$ f) $6 - 3 \times 4$

2. a) $7 \times 2 + 3 \times 2$ b) $12 \div 3 + 6 \times 5$
c) $9 + 3 \times 8 - 1$ d) $36 - 9 \div 3 - 2$
e) $14 \times 2 - 16 \div 2$ f) $4 + 3 \times 7 - 6 \div 3$

3. a) $(4 + 5) \times 3$ b) $8 \times (12 - 4)$
c) $3 \times (8 + 3) - 3$ d) $(4 + 11) \div (7 - 2)$
e) $4 \times 3 \times (7 + 5)$ f) $24 \div 3 \div (10 - 5)$

Exercise 12.3

In each of the following questions:

i) Copy the calculation and put in any brackets which are needed to make it correct.
ii) Check your answer using a scientific calculator.

1. a) $6 \times 2 + 1 = 18$ b) $1 + 3 \times 5 = 16$
c) $8 + 6 \div 2 = 7$ d) $9 + 2 \times 4 = 44$
e) $9 \div 3 \times 4 + 1 = 13$ f) $3 + 2 \times 4 - 1 = 15$

2. a) $12 \div 4 - 2 + 6 = 7$ b) $12 \div 4 - 2 + 6 = 12$
c) $12 \div 4 - 2 + 6 = -5$ d) $12 \div 4 - 2 + 6 = 1.5$
e) $4 + 5 \times 6 - 1 = 33$ f) $4 + 5 \times 6 - 1 = 29$
g) $4 + 5 \times 6 - 1 = 53$ h) $4 + 5 \times 6 - 1 = 45$

It is important to use brackets when dealing with more complex calculations.

Worked examples

a) Evaluate the following using a scientific calculator:

$$\frac{12+9}{10-3} =$$

[( ] [1] [2] [+] [9] [ )] [÷] [( ] [1] [0] [−] [3] [ )] [=] 3

b) Evaluate the following using a scientific calculator:

$$\frac{20+12}{4^2} =$$

[( ] [2] [0] [+] [1] [2] [ )] [÷] [4] [x^2] [=] 2

c) Evaluate the following using a scientific calculator:

$$\frac{90+38}{4^3} =$$

[( ] [9] [0] [+] [3] [8] [ )] [÷] [4] [x^y] [3] [=] 2

Note: different types of calculator have different 'to the power of' buttons.

Exercise 12.4

Using a scientific calculator, evaluate the following:

1. a) $\dfrac{9+3}{6}$ b) $\dfrac{30-6}{5+3}$

c) $\dfrac{40+9}{12-5}$ d) $\dfrac{15 \times 2}{7+8} + 2$

e) $\dfrac{100+21}{11} + 4 \times 3$ f) $\dfrac{7 + 2 \times 4}{7-2} - 3$

2. a) $\dfrac{4^2 - 6}{2+8}$ b) $\dfrac{3^2 + 4^2}{5}$

c) $\dfrac{6^3 - 4^2}{4 \times 25}$ d) $\dfrac{3^3 \times 4^4}{12^2} + 2$

e) $\dfrac{3 + 3^3}{5} + \dfrac{4^2 - 2^3}{8}$ f) $\dfrac{(6+3) \times 4}{2^3} - 2 \times 3$

Student Assessment 1

1. Using a calculator, work out the answers to the following:

a) $6.9 + 18.2$ b) $12.2 - 4.9$
c) $-5.1 + 10$ d) $4.8 - -8.2$
e) 3.3×-4.2 f) $-15 \div -4$

2. Evaluate the following:

a) $6 \times 8 - 4$ b) $3 + 5 \times 2$
c) $3 \times 3 + 4 \times 4$ d) $3 + 3 \times 4 + 4$
e) $(5 + 2) \times 7$ f) $18 \div 2 \div (5 - 2)$

3. Copy the following, if necessary putting in brackets to make the statement correct:

a) $7 - 4 \times 2 = 6$ b) $12 + 3 \times 3 + 4 = 33$
c) $5 + 5 \times 6 - 4 = 20$ d) $5 + 5 \times 6 - 4 = 56$

4. Evaluate the following using a calculator:

a) $\dfrac{2^4 - 3^2}{2}$ b) $\dfrac{(8 - 3) \times 3}{5} + 7$

Student Assessment 2

1. Using a calculator, work out the answers to the following:

a) $7.1 + 8.02$ b) $2.2 - 5.8$
c) $-6.1 + 4$ d) $4.2 - -5.2$
e) -3.6×4.1 f) $-18 \div -2.5$

2. Evaluate the following:

a) $3 \times 9 - 7$ b) $12 + 6 \div 2$
c) $3 + 4 \div 2 \times 4$ d) $6 + 3 \times 4 - 5$
e) $(5 + 2) \div 7$ f) $14 \times 2 \div (9 - 2)$

3. Copy the following, if necessary putting in brackets to make the statement correct:

a) $7 - 5 \times 3 = 6$ b) $16 + 4 \times 2 + 4 = 40$
c) $4 + 5 \times 6 - 1 = 45$ d) $1 + 5 \times 6 - 6 = 30$

4. Using a calculator, evaluate the following:

a) $\dfrac{3^3 - 4^2}{2}$ b) $\dfrac{(15 - 3) \div 3}{2} + 7$

13 MEASURES

Use current units of mass, length, area, volume and capacity in practical situations, and express quantities in terms of larger or smaller units.

NB: All diagrams are not drawn to scale.

The metric system uses a variety of units for length, mass and capacity.

- The common units of length are: kilometre (km), metre (m), centimetre (cm) and millimetre (mm).
- The common units of mass are: tonne (t), kilogram (kg), gram (g) and milligram (mg).
- The common units of capacity are: litre (L or l) and millilitre (mL or ml).

Note: 'centi' comes from the Latin *centum* meaning hundred (a centimetre is one hundredth of a metre);
'milli' comes from the Latin *mille* meaning thousand (a millimetre is one thousandth of a metre);
'kilo' comes from the Greek *Khilloi* meaning thousand (a kilometre is one thousand metres).

It may be useful to have some practical experience of estimating lengths, volumes and capacities before starting the following exercises.

Exercise 13.1

Copy and complete the sentences below:

1. a) There are ... centimetres in one metre.
b) There are ... millimetres in one metre.
c) One metre is one ... of a kilometre.
d) There are ... kilograms in one tonne.
e) There are ... grams in one kilogram.
f) One milligram is one ... of a gram.
g) One thousand kilograms is one ....
h) One thousandth of a gram is one ....
i) One thousand millilitres is one ....
j) One thousandth of a litre is one ....

2. Which of the units below would be used to measure the following?

mm, cm, m, km, mg, g, kg, t, ml, litres

a) your height
b) the length of your finger
c) the mass of a shoe
d) the amount of liquid in a cup
e) the height of a van
f) the mass of a ship
g) the capacity of a swimming pool
h) the length of a highway
i) the mass of an elephant
j) the capacity of the petrol tank of a car

3. Use a ruler to draw lines of the following lengths:
a) 6 cm
b) 18 cm
c) 41 mm
d) 8.7 cm
e) 67 mm

4. Draw four lines in your exercise book and label them A, B, C and D.
a) Estimate their lengths in mm.
b) Measure them to the nearest mm.

5. Copy the questions below and put in the correct unit:
a) A tree in the school grounds is 28 . . . tall.
b) The distance to the nearest big city is 45 . . . .
c) The depth of a lake is 18 . . . .
d) A woman weighs about 60 . . . .
e) The capacity of a bowl is 5 . . . .
f) The distance Ahmet can run in 10 seconds is about 70 . . . .
g) The mass of my car is about 1.2 . . . .
h) Ayse walks about 1700 . . . to school.
i) A melon has a mass of 650 . . . .
j) The amount of blood in your body is 5 . . . .

Converting from one unit to another

Length

1 km = 1000 m

Therefore 1 m $= \frac{1}{1000}$ km

1 m = 1000 mm

Therefore 1 mm $= \frac{1}{1000}$ m

1 m = 100 cm

Therefore 1 cm $= \frac{1}{100}$ m

1 cm = 10 mm

Therefore 1 mm $= \frac{1}{10}$ cm

Worked examples

a) Change 5.8 km into m.
Since 1 km = 1000 m,
5.8 km is 5.8 × 1000 m

5.8 km = 5800 m

b) Change 4700 mm to m.
Since 1 m is 1000 mm,
4700 mm is 4700 ÷ 1000 m

4700 mm = 4.7 m

c) Convert 2.3 km into cm.
2.3 km is 2.3×1000 m = 2300 m
2300 m is 2300×100 cm

2.3 km = 230 000 cm

Exercise 13.2

1. Put in the missing unit to make the following statements correct:

a) 3 cm = 30 . . .
b) 25 . . . = 2.5 cm
c) 3200 m = 3.2 . . .
d) 7.5 km = 7500 . . .
e) 300 . . . = 30 cm
f) 6000 mm = 6 . . .
g) 3.2 m = 3200 . . .
h) 4.2 . . . = 4200 mm
i) 1 million mm = 1 . . .
j) 2.5 km = 2500 . . .

2. Convert the following to millimetres:

a) 2 cm
b) 8.5 cm
c) 23 cm
d) 1.2 m
e) 0.83 m
f) 0.05 m
g) 62.5 cm
h) 0.087 m
i) 0.004 m
j) 2 m

3. Convert the following to metres:

a) 3 km
b) 4700 mm
c) 560 cm
d) 6.4 km
e) 0.8 km
f) 96 cm
g) 62.5 cm
h) 0.087 km
i) 0.004 km
j) 12 mm

4. Convert the following to kilometres:

a) 5000 m
b) 6300 m
c) 1150 m
d) 2535 m
e) 250 000 m
f) 500 m
g) 70 m
h) 8 m
i) 1 million m
j) 700 million m

Mass

1 tonne is 1000 kg

Therefore 1 kg = $\frac{1}{1000}$ tonne

1 kilogram is 1000 g

Therefore 1 g = $\frac{1}{1000}$ kg

1 g is 1000 mg

Therefore 1 mg = $\frac{1}{1000}$ g

Worked examples

a) Convert 8300 kg to tonnes.
Since 1000 kg = 1 tonne, 8300 kg is $8300 \div 1000$ tonnes

8300 kg = 8.3 tonnes

b) Convert 2.5 g to mg.

Since 1 g is 1000 mg, 2.5 g is 2.5×1000 mg

2.5 g = 2500 mg

Exercise 13.3

1. Convert the following:
   a) 3.8 g to mg
   b) 28 500 kg to tonnes
   c) 4.28 tonnes to kg
   d) 320 mg to g
   e) 0.5 tonnes to kg

2. One item has a mass of 630 g, another item has a mass of 720 g. Express the total mass in kg.

3. a) Express the total of the following in kg:
   1.2 tonne, 760 kg, 0.93 tonne, 640 kg
   b) Express the total of the following in g:
   460 mg, 1.3 g, 1260 mg, 0.75 g
   c) A cat weighs 2800 g and a dog weighs 6.5 kg. What is the total weight in kg of the two animals?
   d) In one bag of shopping Imran has items of total mass 1350 g. In another bag there are items of total mass 3.8 kg. What is the mass in kg of both bags of shopping?
   e) What is the total weight in kg of the fruit listed below?
   apples 3.8 kg, pears 1400 g, bananas 0.5 kg, oranges 7500 g, grapes 0.8 kg

Capacity

1 litre is 1000 millilitres

Therefore 1 ml $= \frac{1}{1000}$ litre

Exercise 13.4

1. Convert the following to litres:
   a) 8400 ml
   b) 650 ml
   c) 87 500 ml
   d) 50 ml
   e) 2500 ml

2. Convert the following to ml:
   a) 3.2 litres
   b) 0.75 litre
   c) 0.087 litre
   d) 8 litres
   e) 0.008 litre
   f) 0.3 litre

3. Calculate the following and give the totals in ml:
   a) 3 litres + 1500 ml
   b) 0.88 litre + 650 ml
   c) 0.75 litre + 6300 ml
   d) 450 ml + 0.55 litre

4. Calculate the following and give the total in litres:
   a) 0.75 litre + 450 ml
   b) 850 ml + 490 ml
   c) 0.6 litre + 0.8 litre
   d) 80 ml + 620 ml + 0.7 litre

Perimeter, area and volume

The perimeter of a shape is the distance around the outside of the shape. It is a length and is measured in km, m, cm, mm, etc.

The area of a shape is the amount of surface it covers. Area is measured in square units: km^2, m^2, cm^2, mm^2, etc.

The volume of an object is the amount of space that it occupies. Volume is measured in cubic units: km^3 (rarely), m^3, cm^3, mm^3, etc.

Exercise 13.5

1. Find the perimeter of the following shapes:

a)

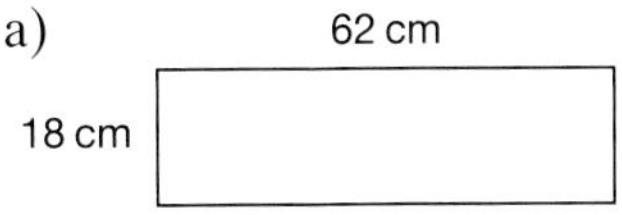

b)

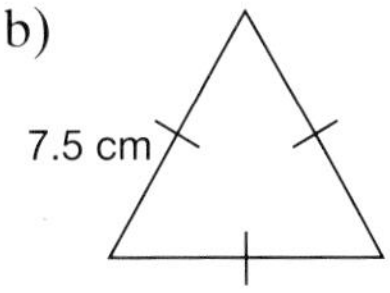

c)

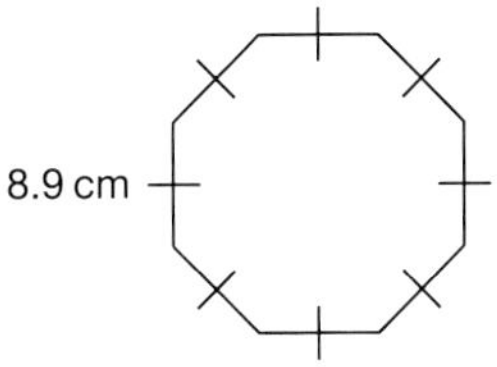

d)

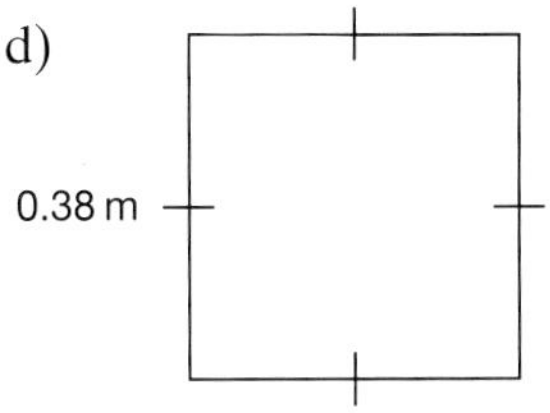

2. Find the perimeter of the following shapes in metres (m). Then find their area in square metres (m^2):

a)

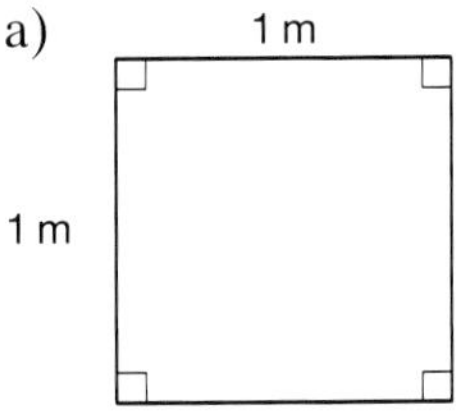

b)

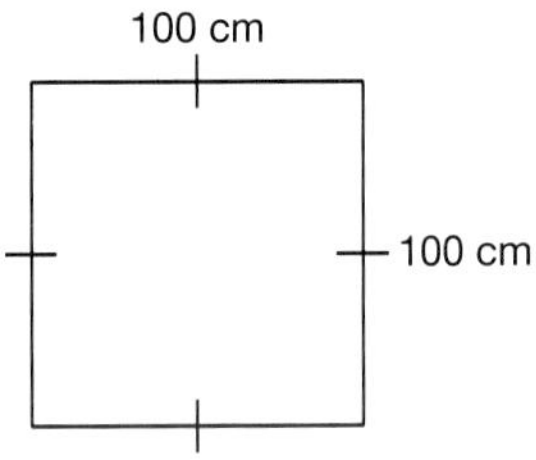

c)

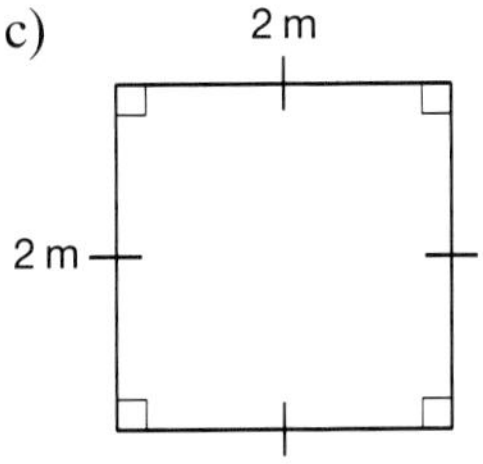

d)

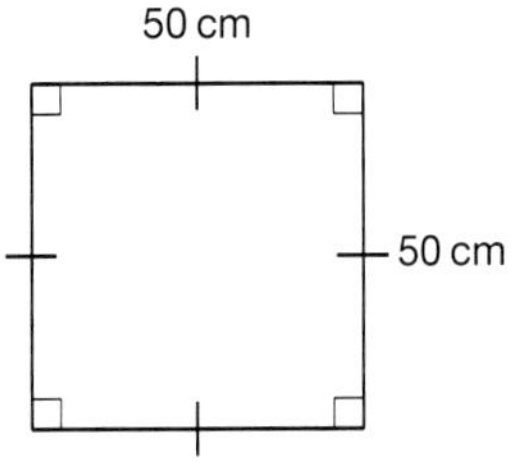

e)

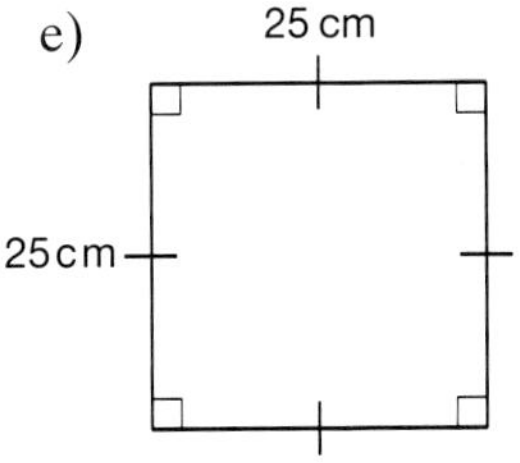

f)

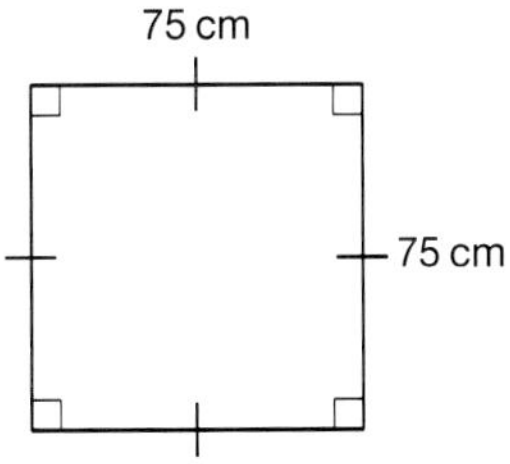

g)

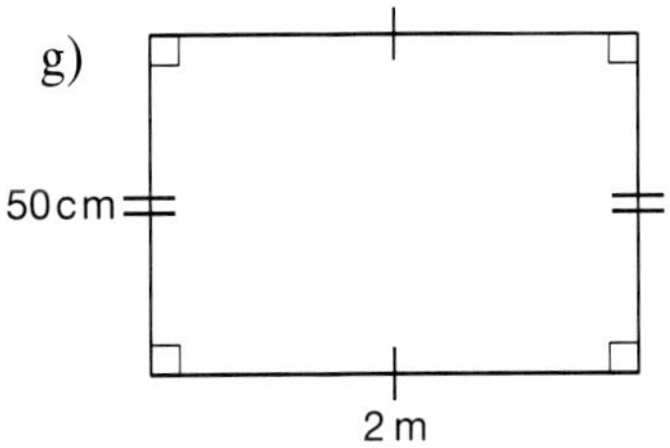

h)

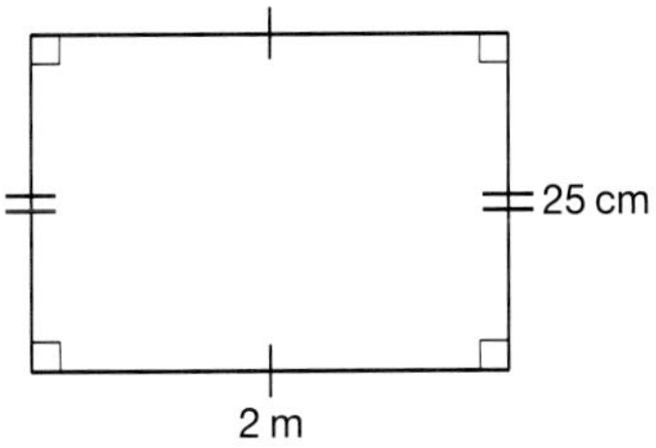

3. Find the perimeter of the shapes in question 2 in centimetres (cm). Then find their area in square centimetres (cm^2).

4. Find the volume of each of the cubes drawn below in cubic metres (m^3):

a)

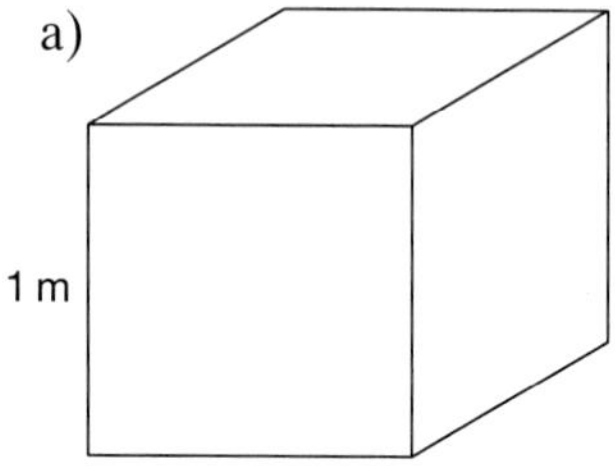

b)

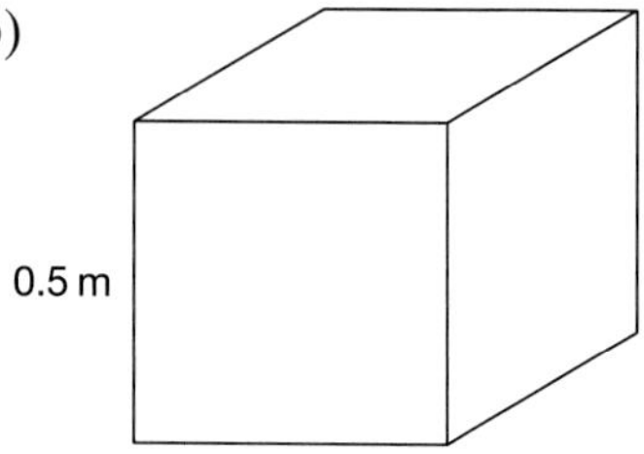

c)

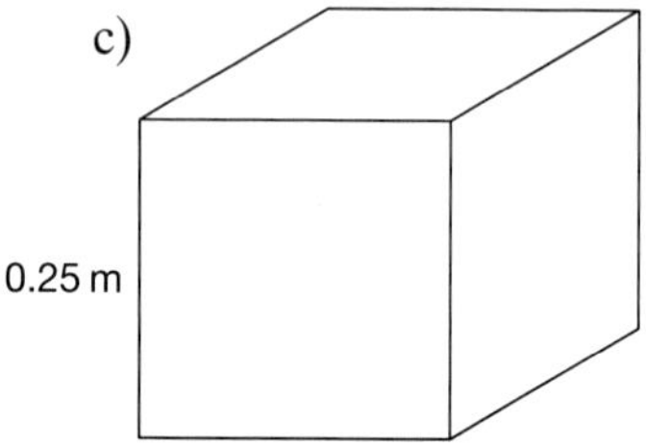

d)

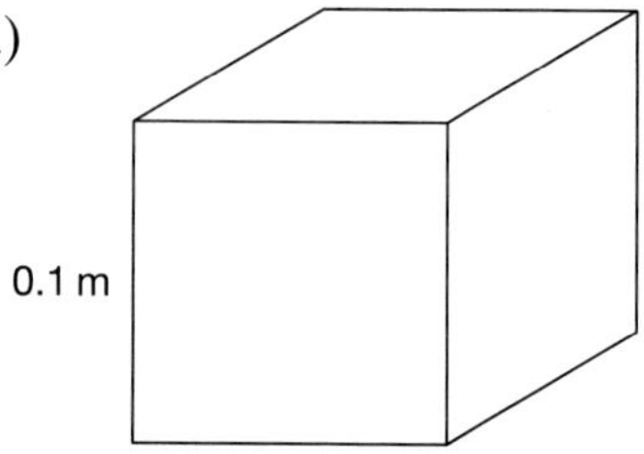

5. Find the volume of each of the cuboids drawn below in cubic metres (m^3):

a)

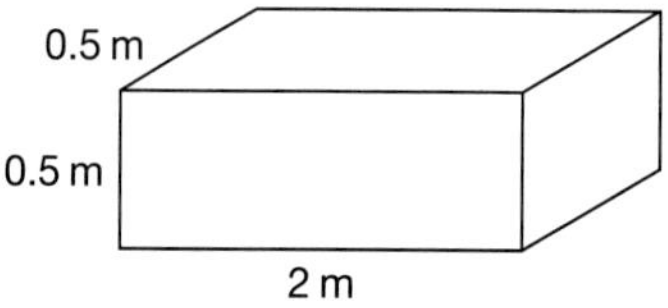

b)

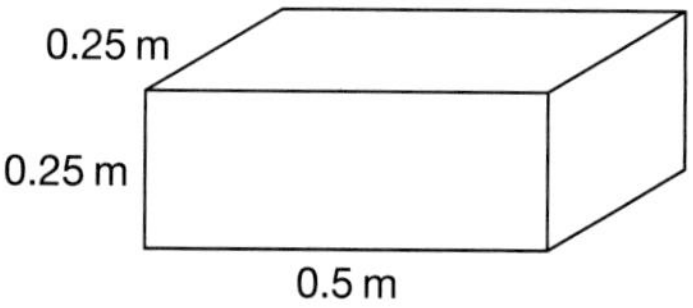

c)

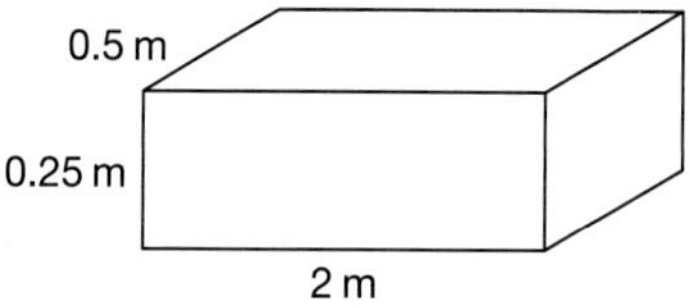

d)

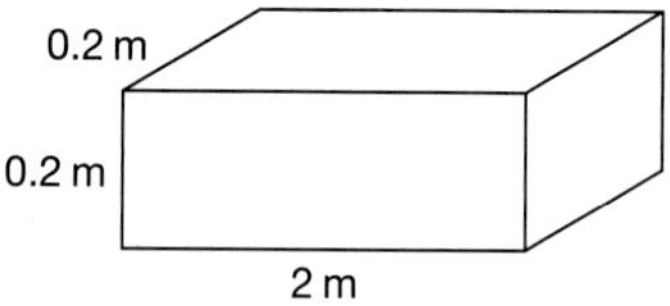

Student Assessment 1

1. Convert the following lengths into the units indicated:
   a) 2.6 cm to mm b) 62.5 cm to mm
   c) 0.88 m to cm d) 0.007 m to mm
   e) 4800 mm to m f) 7.81 km to m
   g) 6800 m to km h) 0.875 km to m
   i) 2 m to mm j) 0.085 m to mm

2. Convert the following masses into the units indicated:
   a) 4.2 g to mg b) 750 mg to g
   c) 3940 g to kg d) 4.1 kg to g
   e) 0.72 tonnes to kg f) 4100 kg to tonnes
   g) 6 280 000 mg to kg h) 0.83 tonnes to g
   i) 47 million kg to tonnes j) 1 kg to mg

3. Add the following masses (answer in kg):

   3.1 tonnes, 4860 kg, 0.37 tonnes

4. Convert the following liquid measures into the units indicated:
   a) 1800 ml to litres b) 3.2 litres to ml
   c) 0.083 litre to ml d) 250 000 ml to litres

5. Find the perimeter in cm of each of the shapes below:

   a)
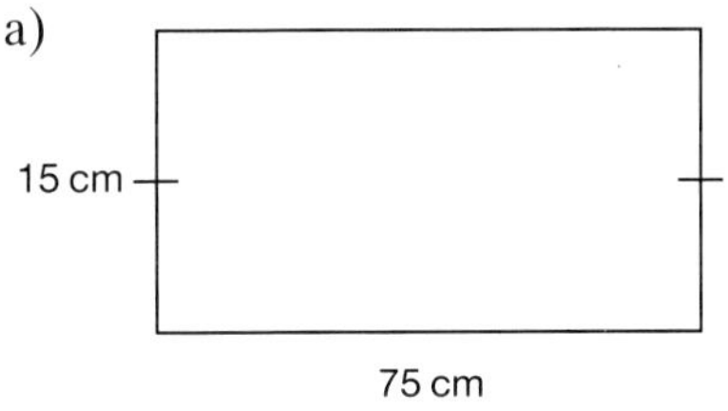

   b)
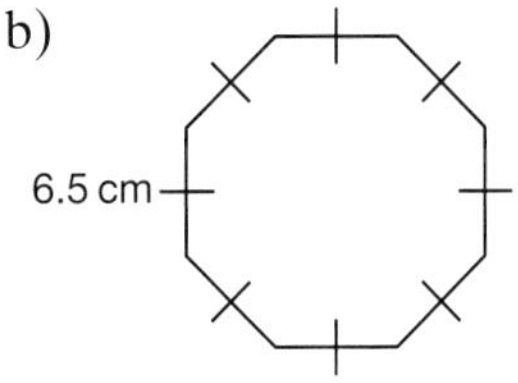

   c)
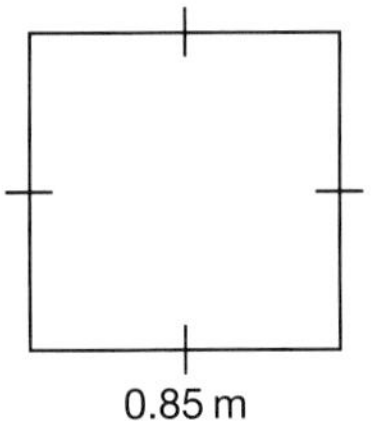

   d)
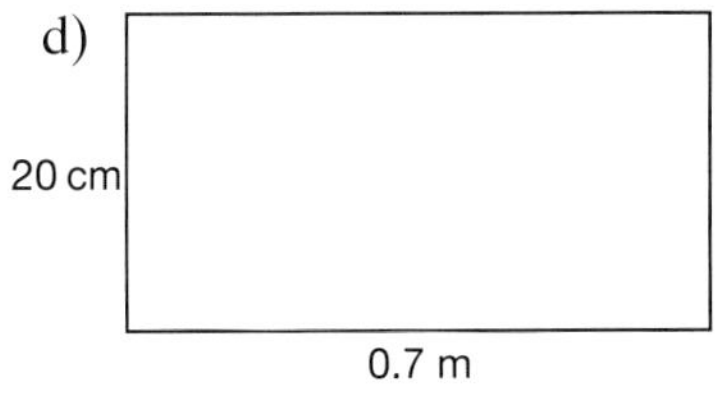

6. Find the area in cm^2 of each of the rectangles below:

   a)
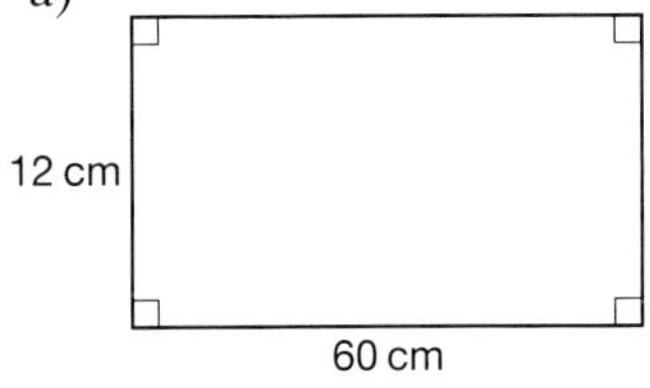

   b)
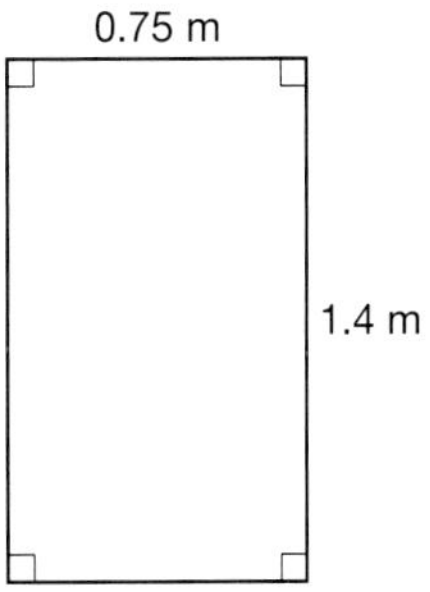

7. Find the volume in cm^3 of each of the cuboids below:

a)

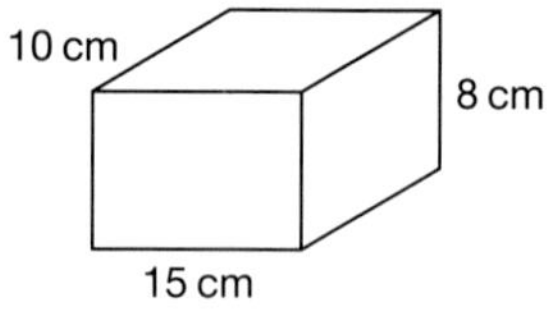

b)

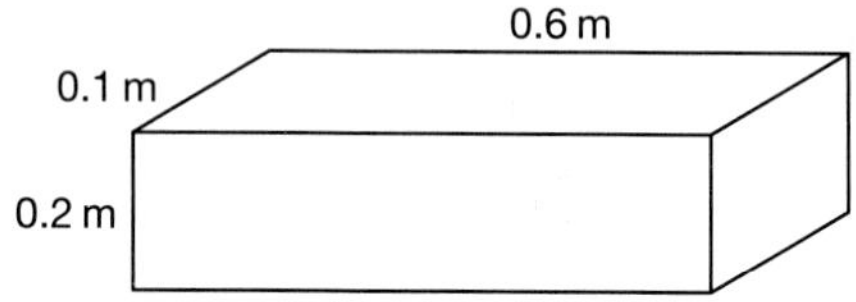

Student Assessment 2

1. Convert the following lengths into the units indicated:

a) 4.7 cm to mm
b) 0.003 m to mm
c) 3100 mm to cm
d) 6.4 km to m
e) 49 000 m to km
f) 4 m to mm
g) 0.4 cm to mm
h) 0.034 m to mm
i) 460 mm to cm
j) 50 000 m to km

2. Convert the following masses into the units indicated:

a) 3.6 mg to g
b) 550 mg to g
c) 6500 g to kg
d) 6.7 kg to g
e) 0.37 tonnes to kg
f) 1510 kg to tonnes
g) 380 000 kg to tonnes
h) 0.077 kg to g
i) 6 million mg to kg
j) 2 kg to mg

3. Subtract 1570 kg from 2 tonnes.

4. Convert the following measures of capacity to the units indicated:

a) 3400 ml to litres
b) 6.7 litres to ml
c) 0.73 litre to ml
d) 300 000 ml to litres.

5. Find the perimeter in cm of each of the shapes below:

a)

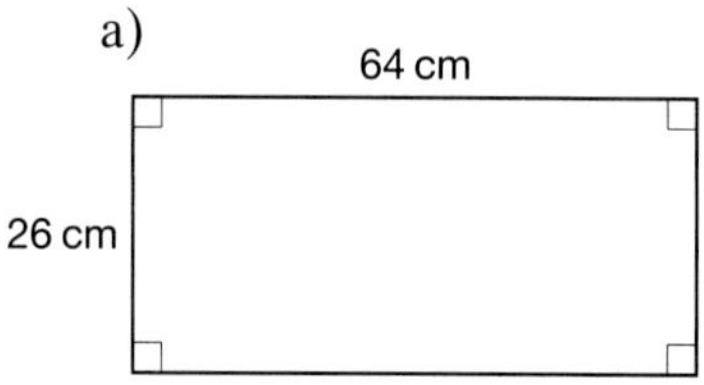

b)

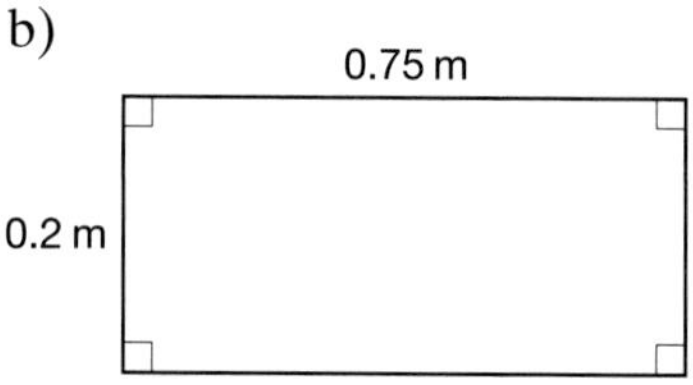

c)

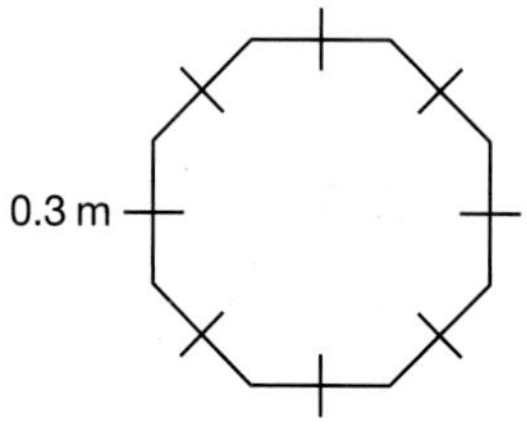

6. Find the area in cm^2 of each of the rectangles below:

a)

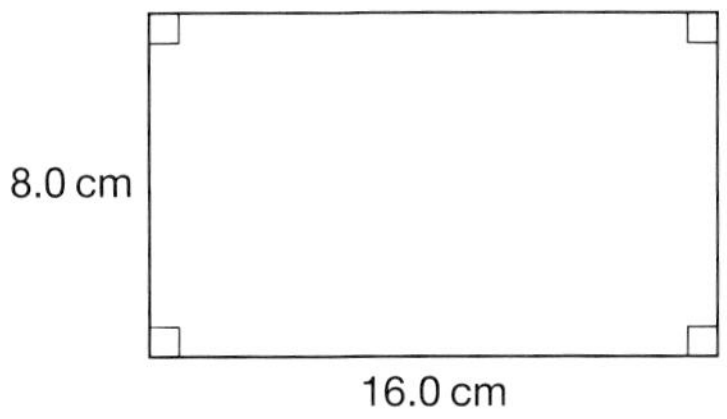

b)

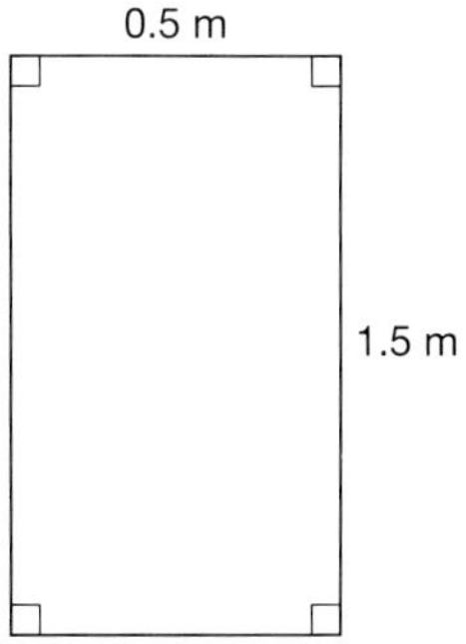

7. Find the volume in m^3 of each of the cuboids below:

a)

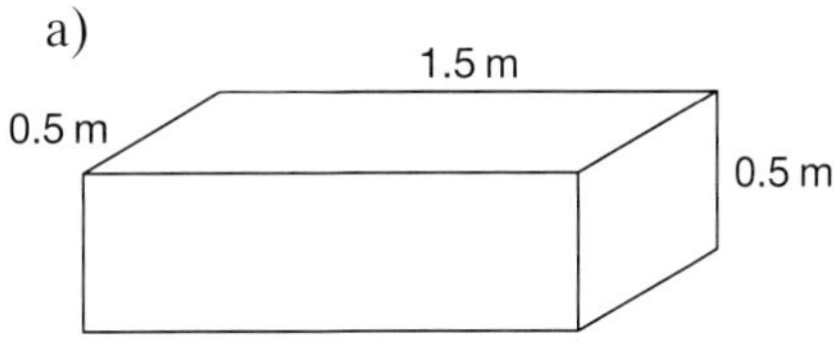

b)

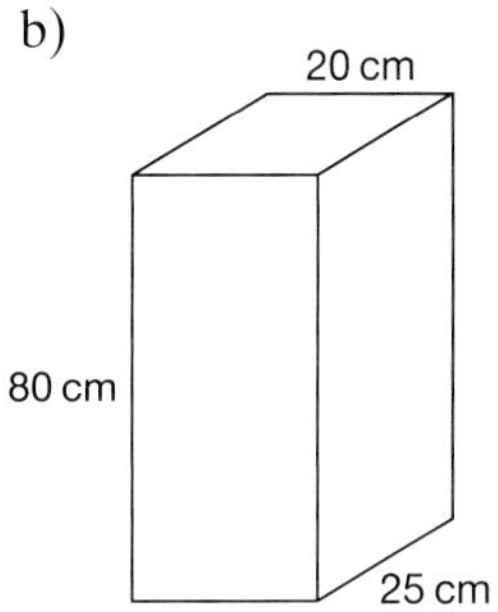

14 TIME

Calculate times in terms of the 24-hour clock; read clocks, dials and timetables.

Times are usually given in terms of the 12-hour clock. We tend to say, 'I get up at seven o'clock in the morning, play football at half past two in the afternoon, and go to bed before eleven o'clock'.

These times can be written as 7 a.m., 2.30 p.m. and 11 p.m.

In order to save confusion, most timetables are written using the 24-hour clock.

7 a.m. is written as 07.00
3 p.m. is written as 15.00
6.45 p.m. is written as 18.45

To change p.m. times to 24-hour clock times, add 12 hours.

To change 24-hour clock times later than 12.00 noon to 12-hour clock times, subtract 12 hours.

Exercise 14.1

1. Change these times into those on the 24-hour clock:

a) 2.30 p.m. b) 9 p.m. c) 8.45 a.m. d) 6 a.m.
e) midday f) 10.55 p.m. g) 7.30 a.m. h) 7.30 p.m.
i) 1 a.m. j) midnight

2. Change these times into those on the 24-hour clock:

a) A quarter past seven in the morning
b) Eight o'clock at night
c) Ten past nine in the morning
d) A quarter to nine in the morning
e) A quarter to three in the afternoon
f) Twenty to eight in the evening

3. These times are written for the 24-hour clock. Rewrite them using a.m. and p.m.

a) 07.20 b) 09.00 c) 14.30 d) 18.25
e) 23.40 f) 01.15 g) 00.05 h) 11.35
i) 17.50 j) 23.59 k) 04.10 l) 05.45

4. A journey to work takes a woman three quarters of an hour. If she catches the bus at the following times, when does she arrive?

a) 07.20 b) 07.55 c) 08.20 d) 08.45

5. The same woman catches buses home at the times shown below. The journey takes 55 minutes. If she catches the bus at the following times, when does she arrive?

a) 17.25 b) 17.50 c) 18.05 d) 18.20

6. A boy cycles to school each day. His journey takes 70 minutes. When will he arrive if he leaves home at:

a) 07.15 b) 08.25 c) 08.40 d) 08.55?

7. The train into the city from a village takes 1 hour and 40 minutes. Copy and complete the train timetable below.

Depart	**Arrive**
06.15	
	08.10
09.25	
	12.00
13.18	
	16.28
18.54	
	21.05

8. The same journey by bus takes 2 hours and 5 minutes. Copy and complete the bus timetable below.

Depart	**Arrive**
06.00	
	08.50
08.55	
	11.14
13.48	
	16.22
21.25	
	00.10

9. A coach runs from Cambridge to the airports at Stansted, Gatwick and Heathrow. The time taken for the journey remains constant. Copy and complete the timetables below for outward and return journeys.

Cambridge	04.00	08.35	12.50	19.45	21.10
Stansted	05.15				
Gatwick	06.50				
Heathrow	07.35				
Heathrow	06.25	09.40	14.35	18.10	22.15
Gatwick	08.12				
Stansted	10.03				
Cambridge	11.00				

10. British Airways aircraft fly twice a day from London to Johannesburg. The flight time is 11 hours and 20 minutes. Copy and complete the timetable below.

	London	Jo'burg	London	Jo'burg
Sunday	06.15		14.20	
Monday		18.43		05.25
Tuesday	07.20		15.13	
Wednesday		19.12		07.30
Thursday	06.10		16.27	
Friday		17.25		08.15
Saturday	09.55		18.50	

11. The flight time from London to Kuala Lumpur is 13 hours and 45 minutes. Copy and complete the timetable below.

	London	Kuala Lumpur	London	Kuala Lumpur	London	Kuala Lumpur
Sunday	08.28		14.00		18.30	
Monday		22.00		03.15		09.50
Tuesday	09.15		15.25		17.55	
Wednesday		21.35		04.00		08.22
Thursday	07.00		13.45		18.40	
Friday		00.10		04.45		07.38
Saturday	10.12		14.20		19.08	

Worked example

A train covers the 480 km journey from Paris to Lyon at an average speed of 100 km/h. If the train leaves Paris at 08.35, when does it arrive in Lyon?

Time taken = distance/speed
Paris to Lyon = 480/100 hours, that is, 4.8 hours.
4.8 hours is 4 hours and (0.8×60 minutes), that is, 4 hours and 48 minutes.
Departure 08.35; arrival 08.35 + 04.48

Arrival time is 13.23.

Exercise 14.2

Find the time in hours and minutes for the following journeys of the given distance at the average speed stated:

1. a) 240 km at 60 km/h   b) 340 km at 40 km/h
c) 270 km at 80 km/h   d) 100 km at 60 km/h
e) 70 km at 30 km/h   f) 560 km at 90 km/h
g) 230 km at 100 km/h   h) 70 km at 50 km/h
i) 4500 km at 750 km/h   j) 6000 km at 800 km/h

2. Grand Prix racing cars cover a 120 km race at the following average speeds. How long do the first five cars take to complete the race? Answer in minutes and seconds.

First 240 km/h	Second 220 km/h	Third 210 km/h
Fourth 205 km/h	Fifth 200 km/h	

3. A train covers the 1500 km distance from Amsterdam to Barcelona at an average speed of 100 km/h. If the train leaves Amsterdam at 09.30, when does it arrive in Barcelona?

4. A plane takes off at 16.25 for the 3200 km journey from Moscow to Athens. If the plane flies at an average speed of 600 km/h, when will it land in Athens?

5. A plane leaves London for Boston, a distance of 5200 km, at 09.45. The plane travels at an average speed of 800 km/h. If Boston time is five hours behind British time, what is the time in Boston when the aircraft lands?

Student Assessment 1

1. Change the times below into those on the 24-hour clock:
   a) 4.35 a.m.
   b) 6.30 p.m.
   c) a quarter to eight in the morning
   d) half past seven in the evening

2. The times below are written for the 24-hour clock. Rewrite them using a.m. and p.m.
   a) 08.45 b) 18.35 c) 21.12 d) 00.15

3. A journey to school takes a girl 25 minutes. What time does she arrive if she leaves home at the following times?
   a) 07.45 b) 08.15 c) 08.38

4. A bus service visits the towns on the timetable below. Copy the timetable and fill in the missing times, given the following information:

   The journey from:

   Alphaville to Betatown takes 37 minutes
   Betatown to Gammatown takes 18 minutes
   Gammatown to Deltaville takes 42 minutes.

Alphaville	07.50		
Betatown		11.38	
Gammatown			16.48
Deltaville			

5. Find the times for the following journeys of given distance at the average speed stated. Give your answers in hours and minutes.
   a) 250 km at 50 km/h
   b) 375 km at 100 km/h
   c) 80 km at 60 km/h
   d) 200 km at 120 km/h
   e) 70 km at 30 km/h
   f) 300 km at 80 km/h

Student Assessment 2

1. Change the times below to those on the 24-hour clock:
   a) 5.20 a.m.
   b) 8.15 p.m.
   c) ten to nine in the morning
   d) half past eleven at night

2. The times below are written for the 24-hour clock. Rewrite them using a.m. and p.m.
   a) 07.15 b) 16.43 c) 19.30 d) 00.35

3. A journey to school takes a boy 22 minutes. When does he arrive if he leaves home at the following times?
   a) 07.48 b) 08.17 c) 08.38

4. A train stops at the following stations. Copy the timetable and fill in the times, given the following information:

The journey from:

Apple to Peach is 1 hr 38 minutes
Peach to Pear is 2 hrs 4 minutes
Pear to Plum is 1 hr 53 minutes.

Apple	10.14		
Peach		17.20	
Pear			23.15
Plum			

5. Find the time for the following journeys of given distance at the average speed stated. Give your answers in hours and minutes.

a) 350 km at 70 km/h
b) 425 km at 100 km/h
c) 160 km at 60 km/h
d) 450 km at 120 km/h
e) 600 km at 160 km/h

15 MONEY

Calculate using money and convert from one currency to another.

In 2002, 1 euro could be exchanged for 1.60 Australian dollars ($A).
A graph to enable conversion from euro to dollars and dollars to euro can be seen below.

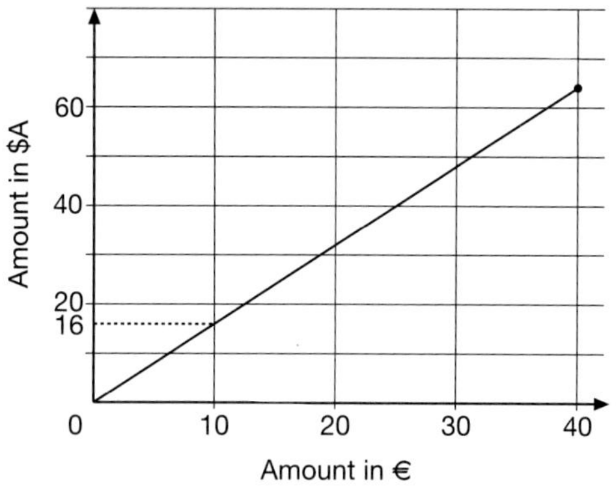

Exercise 15.1

1. Use the conversion graph above to convert the following to Australian dollars:
   a) €20 b) €30 c) €5 d) €25
   e) €35 f) €15

2. Use the graph above to convert the following to euro:
   a) $20 b) $30 c) $40 d) $35
   e) $25 f) $48

3. 1 euro could be exchanged for 50 Indian rupees. Draw a conversion graph. Use an appropriate scale with the horizontal scale up to €100. Use your graph to convert the following to rupees:
   a) €10 b) €40 c) €50 d) €90
   e) €1000 f) €720

4. Use your graph from Q.3 to convert the following to euro:
   a) 100 rupees b) 720 rupees c) 630 rupees
   d) 450 rupees e) 5400 rupees f) 2700 rupees

The table (right) shows the exchange rate for €1 into various currencies.

Draw conversion graphs for the exchange rates shown to answer Q5–10.

Zimbabwe	50 Zimbabwe dollars
South Africa	9.5 rand
Turkey	140 000 Turkish lira
Japan	120 yen
Malta	0.4 lira
U.S.A	0.95 dollars

5. How many Zimbabwe dollars would you receive for the following?
a) €20 b) €50 c) €75
d) €30 e) €250

6. How many euro would a visitor to the UK receive for the following number of South African rand?
a) 200 b) 550 c) 670 d) 10 000

7. In the grand Bazaar in Istanbul, a visitor sees three carpets priced at 120 million Turkish lira, 400 million Turkish lira and 880 million Turkish lira. Draw and use a conversion graph to find the prices in euro.

8. €1 can be exchanged for US$ 0.95.
€1 can also be exchanged for 120 yen.
Draw a conversion graph for US dollars to Japanese yen, and answer the questions below:
a) How many yen would you receive for:
i) $300 ii) $750 iii) $1000
b) How many US dollars would you receive for:
i) 5000 yen ii) 8500 yen iii) 1 million yen

9. Use the currency table on page 96 to draw a conversion graph for Maltese lira to Zimbabwe dollars. Use the graph to find the number of dollars you would receive for:
a) 50 lira b) 200 lira c) 120 lira d) 600 lira

10. Use the currency table on page 96 to draw a conversion graph for Turkish lira to Zimbabwe dollars. Use the graph to find the number of lira you would receive for:
a) 500 dollars b) 700 dollars
c) 1200 dollars

Student Assessment 1

A visitor from Hong Kong receives 8 Pakistan rupees for each Hong Kong dollar.

Draw a conversion graph for Hong Kong dollars and Pakistan rupees, and use it to answer the questions below.

1. How many Pakistan rupees would you get for:
   a) HK$30 b) HK$240 c) HK$5000?
2. How many Hong Kong dollars would you get for:
   a) 120 rupees b) 450 rupees c) 1 thousand rupees?

Below is a currency conversion table showing the amount of foreign currency received for 1 euro. Draw the appropriate conversion graphs to enable you to answer Q.3–5.

New Zealand	1.95 dollars
Brazil	2.5 reals
India	50 Indian rupees

3. Convert the following number of New Zealand dollars into euro:
   a) 320 dollars b) 72 dollars c) 960 dollars
4. Convert the following number of German marks into New Zealand dollars:
   a) 100 euro b) 35 euro c) 450 euro
5. Convert the following number of Brazilian reals into Indian rupees:
   a) 60 reals b) 300 reals c) 100 reals

Student Assessment 2

1. 2.7 Australian dollars can be exchanged for €1.60. Draw a conversion graph to find the number of Australian dollars you would get for:
   a) €50 b) €30 c) €70
2. Use your graph from Q.1 to find the number of euro you could receive for:
   a) $A54 b) $A81 c) $A100

Below is a currency conversion table showing the amount of foreign currency received for €1. Draw the appropriate conversion graphs to answer Q.3–5:

Namibia	9.4 Namibian dollars
Canada	1.5 dollars
Hong Kong	7.4 dollars

3. Convert the following number of Canadian dollars into Hong Kong dollars:

a) €$60 b) €$1200 c) €$150

4. Convert the following number of Hong Kong dollars into Namibian dollars:

a) HK$110 b) HK$660 c) HK$1000

5. Convert the following number of Namibian dollars into Canadian dollars:

a) NAB$330 b) NAB$500 c) NAB$75

16 PERSONAL AND HOUSEHOLD FINANCE

Problems involving earnings, simple interest, compound interest, discount, and profit and loss.

Net pay is what is left after deductions such as tax, insurance and pension contributions are taken from **gross earnings**. That is, Net pay = Gross pay − Deductions

A **bonus** is an extra payment sometimes added to an employee's basic pay.

In many companies there is a fixed number of hours that an employee is expected to work. Any work done in excess of this **basic week** is paid at a higher rate, referred to as **overtime**. Overtime may be 1.5 times basic pay, called **time and a half**, or twice basic pay, called **double time**.

Exercise 16.1

1. Copy the table below and find the net pay for the following employees:

	Gross pay (£)	Deductions (£)	Net pay (£)
a) A Ahmet	162.00	23.50	
b) B Martinez	205.50	41.36	
c) C Stein	188.25	33.43	
d) D Wong	225.18	60.12	

2. Copy and complete the table below for the following employees:

	Basic pay (£)	Overtime (£)	Bonus (£)	Gross pay (£)
a) P Small	144	62	23	
b) B Smith	152		31	208
c) A Chang		38	12	173
d) U Zafer	115	43		213
e) M Said	128	36	18	

3. Copy and complete the table below for the following employees:

	Gross pay (£)	Tax (£)	Pension (£)	Net pay (£)
a) A Hafar	203	54	18	
b) K Zyeb		65	23	218
c) H Such	345		41	232
d) K Donald	185	23		147

4. Find the basic pay in each of the cases below. Copy and complete the table.

	No. of hours worked	Basic rate per hour (£)	Gross pay (£)
a)	40	3.15	
b)	44	4.88	
c)	38	5.02	
d)	35	8.30	
e)	48	7.25	

5. Copy and complete the table below, which shows basic pay and overtime at time and a half.

	Basic hours worked	Rate per hour (£)	Basic pay (£)	Overtime hours worked	Overtime pay (£)	Total gross pay (£)
a)	40	3.60		8		
b)	35		203.00	4		
c)	38	4.15		6		
d)		6.10	256.20	5		
e)	44	5.25		4		
f)		4.87	180.19	3		
g)	36	6.68		6		
h)	45	7.10	319.50	7		

6. In Q.5, deductions amount to 32% of the total gross pay. Calculate the net pay for each employee.

Piece work is a method of payment where an employee is paid for the number of articles made, not for time taken.

Exercise 16.2

1. Four children help to pick grapes in a vineyard in Spain. They are paid 5.50 euro for each basket of grapes. Copy and complete the table below.

	Mon	Tue	Wed	Thur	Fri	Total	Gross pay
Pepe	4	5	7	6	6		
Felicia	3	4	4	5	5		
Delores	5	6	6	5	6		
Juan	3	4	6	6	6		

2. Five women work in a pottery factory in Portugal, making plates. They are paid 5 euro for every dozen plates made. Copy and complete the following table, which shows the number of plates that each woman produces.

	Mon	Tue	Wed	Thur	Fri	Total	Gross pay
a) Maria	240	360	288	192	180		
b) Bea	168	192	312	180	168		
c) Joanna	288	156	192	204	180		
d) Bianca	228	144	108	180	120		
e) Selina	192	204	156	228	144		

3. A group of five women work at home making clothes. The patterns and material are provided by the company, and they are paid in South African rand for each article produced at the rates shown below.

Jacket 25 rand   Trousers 11 rand

Shirt 13 rand   Dress 12 rand

The five women make the number of articles of clothing shown in the table below. Find each woman's gross pay. If the deductions amount to 15% of gross earnings, calculate each woman's net pay.

	Jackets	Shirts	Trousers	Dresses
Mrs Smith	3	12	7	0
Mrs Jones	8	5	2	9
Ms White	0	14	12	2
Miss Green	6	8	3	12
Mrs Brown	4	9	16	5

4. A Greek school organises a sponsored walk. Below is a list of how far pupils walked, the amount they were sponsored per mile, and the total each raised.

a) Copy and complete the table.

Distance walked (km)	Amount per km in euro	Total raised
10	0.80	
	0.65	9.10
18	0.38	
	0.72	7.31
12		7.92
	1.20	15.60
15	1	
	0.88	15.84
18		10.44
17		16.15

b) How much was raised in total?
c) This total was divided between three children's charities in the ratio of 2 : 3 : 5. How much did each charity receive?

Interest can be defined as money added by a bank to sums deposited by customers. The money deposited is called the **principal**. The **percentage interest** is the given rate and the money is left for a fixed period of time.

A formula can be obtained for **simple interest**:

$$SI = \frac{Ptr}{100}$$

where SI = simple interest, i.e. the interest paid
P = the principal
t = time in years
r = rate percent

Worked example Find the simple interest earned on £250 deposited for 6 years at 8% p.a.

$$SI = \frac{Ptr}{100}$$
$$SI = \frac{250 \times 6 \times 8}{100}$$
$$SI = 120$$

So the interest paid is £120.

Exercise 16.3

All rates of interest given here are annual rates.

1. Find the simple interest paid in the following cases:

a) Principal £300	rate 6%	time 4 years
b) Principal £750	rate 8%	time 7 years
c) Principal £425	rate 6%	time 4 years
d) Principal £2800	rate 4.5%	time 2 years
e) Principal £6500	rate 6.25%	time 8 years
f) Principal £880	rate 6%	time 7 years

Worked example How long will it take for a sum of £250 invested at 8% to earn interest of £80?

$$SI = \frac{Ptr}{100}$$
$$80 = \frac{250 \times t \times 8}{100}$$
$$80 = 20t$$
$$4 = t$$

It will take 4 years.

2. Calculate how long it will take for the following amounts of interest to be earned at the given rate.

a) P = £500	r = 6%	SI = £150
b) P = £5800	r = 4%	SI = £96

c) $P = £4000$ $r = 7.5\%$ $SI = £1500$
d) $P = £2800$ $r = 8.5\%$ $SI = £1904$
e) $P = £900$ $r = 4.5\%$ $SI = £243$
f) $P = £400$ $r = 9\%$ $SI = £252$

Worked example What rate per year must be paid for a principal of £750 to earn interest of £180 in 4 years?

$$SI = \frac{Ptr}{100}$$

$$180 = \frac{750 \times 4 \times r}{100}$$

$$180 = 30r$$

$$6 = r$$

The rate must be 6% per year.

3. Calculate the rate of interest per year which will earn the given amount of interest:
a) Principal £400 time 4 years interest £112
b) Principal £800 time 7 years interest £224
c) Principal £2000 time 3 years interest £210
d) Principal £1500 time 6 years interest £675
e) Principal £850 time 5 years interest £340
f) Principal £1250 time 2 years interest £275

Worked example Find the principal which will earn interest of £120 in 6 years at 4%.

$$SI = \frac{Ptr}{100}$$

$$120 = \frac{P \times 6 \times 4}{100}$$

$$120 = \frac{24P}{100}$$

$$12\,000 = 24P$$

$$500 = P$$

So the principal is £500.

4. Calculate the principal which will earn the interest below in the number of years and rate given:
a) SI = £80 time = 4 years rate = 5%
b) SI = £36 time = 3 years rate = 6%
c) SI = £340 time = 5 years rate = 8%
d) SI = £540 time = 6 years rate = 7.5%
e) SI = £540 time = 3 years rate = 4.5%
f) SI = £348 time = 4 years rate = 7.25%

5. What rate of interest is paid on a deposit of £2000 which earns £400 interest in 5 years?

6. How long will it take a principal of £350 to earn £56 interest at 8% per year?

7. A principal of £480 earns £108 interest in 5 years. What rate of interest was being paid?
8. A principal of £750 becomes a total of £1320 in 8 years. What rate of interest was being paid?
9. £1500 is invested for 6 years at 3.5% per year. What is the interest earned?
10. £500 is invested for 11 years and becomes £830 in total. What rate of interest was being paid?

Compound interest

Compound interest means that not only is interest paid on the principal amount but interest is paid on the interest. It is compounded (or added to).

This sounds complicated but the example below will make it clear.

A builder is going to build six houses on a plot of land in Spain. He borrows 500 000 euro at 10% and will pay off the loan in full after three years.

At the end of the first year he will owe
€500 000 + 10% of €500 000
i.e. 550 000 euro

At the end of the second year he will owe
€550 000 + 10% of €550 000
i.e. €550 000 + €55 000
= 605 000 euro

At the end of the third year he will owe
€605 000 + 10% of €605 000
i.e. €605 000 + €60 500
= 665 500 euro

His interest will be 665 500 – 500 000 euro
i.e. 165 500 euro

At simple interest he would pay 50 000 euro a year
i.e. 150 000 euro in total.

The extra 15 500 euro was the compound interest.

11. A shipping company borrows £70 million at 5% compound interest to build a new cruise ship. If it repays the debt after 3 years, how much interest will the company pay?
12. A woman borrows 100 000 euro to improve her house. She borrows the money at 15% interest and repays it in full after 3 years. What interest will she pay?
13. A man owes £5000 on his credit cards. The annual percentage rate (APR) is 20%. If he lets the debt grow how much will he owe in 4 years?

14. A school increases its intake by 10%. If it starts with 1000 students, how many will it have at the beginning of the fourth year of expansion?

15. 8 million tonnes of fish were caught in the North Sea in 2005. If the catch is reduced by 20% each year for 4 years, what amount can be caught at the end of this time?

It is possible to calculate the time taken for a debt to grow using compound interest.

Worked example How many years will it take for a debt to double at 27% compound interest?

After 1 year the debt will be D(1 + 27%) or 1.27D.
After 2 years the debt will be D(1.27 × 1.27) or 1.61D.
After 3 years the debt will be D(1.27 × 1.27 × 1.27) or 2.05D.

So the debt will have doubled after 3 years.

Note that the amount of the debt does not affect this calculation. Note also that if the debt were reducing it would take the same number of years to halve.

16. How many years would it take a debt to double at a compound interest rate of 42%?

17. How many years would it take a debt to double at a compound interest rate of 15%?

18. A car loses value at 27% compound interest each year. How many years will it take to halve in value?

19. The price of a house increases by 20% each year compound increase. How many years before it doubles in price?

20. A boat loses value at a rate of 15% compound per year. How many years before its value has halved?

Profit and loss

Foodstuffs and manufactured goods are produced at a cost, known as the **cost price**, and sold at the **selling price**. If the selling price is greater than the cost price, a profit is made.

Worked example A market trader buys oranges in boxes of 12 dozen for £14.40 per box. He buys three boxes and sells all the oranges for 12p each. What is his profit or loss?

Cost price: 3 × £14.40 = £43.20
Selling price: 3 × 144 × 12p = £51.84
In this case he makes a profit of £51.84 − £43.20
His profit is £8.64.

A second way of solving this problem would be:
£14.40 for a box of 144 oranges is 10p each.
So cost price of each orange is 10p, and selling price of each orange is 12p. The profit is 2p per orange.
So 3 boxes would give a profit of $3 \times 144 \times 2$p.
That is, £8.64.

Exercise 16.4

1. A market trader buys peaches in boxes of 120. He buys 4 boxes at a cost price of £13.20 per box. He sells 425 peaches at 12p each – the rest are ruined. How much profit or loss does he make?
2. A shopkeeper buys 72 bars of chocolate for £5.76. What is his profit if he sells them for 12p each?
3. A holiday company charters an aircraft to fly to Malta at a cost of £22 000. It then sells 195 seats at £185 each. Calculate the profit made per seat if the plane has 200 seats.
4. A car is priced at £7200. The car dealer allows a customer to pay a one-third deposit and 12 payments of £420 per month. How much extra does it cost the customer?
5. At an auction a company sells 150 television sets for an average of £65 each. The production cost was £10 000. How much loss did the company make?
6. A market trader sells tools and small electrical goods. Find his profit or loss at the end of a day in which he sells each of the following:
   a) 15 torches: cost price £2 each, selling price £2.30 each
   b) 60 plugs: cost price £10 a dozen, selling price £1.10 each
   c) 200 video tapes: cost price £9 for 10, selling price £1.30 each
   d) 5 personal stereos: cost price £82, selling price £19 each
   e) 96 batteries costing £1 for 6, selling price 59p for 3
   f) 3 clock radios costing £65, sold for £14 each

Percentage profit and loss

Most profits or losses are expressed as a percentage.

Profit or loss, divided by cost price, multiplied by 100 = % profit or loss.

Worked example

A woman buys a car for £7500 and sells it two years later for £4500. Calculate her loss over two years as a percentage of the cost price.

Cost Price = £7500 Selling Price = £4500 Loss = £3000

$$\text{Loss \%} = \frac{3000}{7500} \times 100 = 40$$

Her loss is 40%.

When something becomes worth less over a period of time, it is said to **depreciate**.

Exercise 16.5

1. Find the depreciation of the following cars as a percentage of the cost price. (C.P. = Cost Price, S.P. = Selling Price)

a) VW	C.P. £4500	S.P. £4005
b) Rover	C.P. £9200	S.P. £6900
c) Mercedes	C.P. £11 000	S.P. £5500
d) Toyota	C.P. £4350	S.P. £3480
e) Fiat	C.P. £6850	S.P. £4795
f) Ford	C.P. £7800	S.P. £2600

2. A company manufactures electrical items for the kitchen. Find the percentage profit on each of the following:

a) Cooker	C.P. £240	S.P. £300
b) Fridge	C.P. £50	S.P. £65
c) Freezer	C.P. £80	S.P. £96
d) Microwave	C.P. £120	S.P. £180
e) Washing machine	C.P. £260	S.P. £340
f) Dryer	C.P. £70	S.P. £91

3. A developer builds a number of different kinds of house on a village site. Given the cost prices and the selling prices below, which type of house gives the developer the largest percentage profit?

	Cost price (£)	Selling price (£)
Type A	40 000	52 000
Type B	65 000	75 000
Type C	81 000	108 000
Type D	110 000	144 000
Type E	132 000	196 000

4. Students in a school organise a disco. The disco company charges £350 hire charge. The students sell 280 tickets at £2.25. What is the percentage profit?

5. A shop sells secondhand sailing yachts. Calculate the percentage profit on each of the following:

	Cost price (£)	Selling price (£)
Mirror	400	546
Wayfarer	1100	1540
Laser	900	1305
Fireball	1250	1720

Student Assessment 1

1. A boy worked 3 hours a day each weekday for £3.75 per hour. What was his 4-weekly gross payment?

2. A woman works at home making curtains. In one week she makes 4 pairs of long curtains and 13 pairs of short curtains. What is her gross pay if she receives £2.10 for each long curtain, and £1.85 for each short curtain?

3. Calculate the missing numbers from the simple interest table below:

Principal (£)	Rate (%)	Time (years)	Interest (£)
200	9	3	(a)
350	7	(b)	98
520	(c)	5	169
(d)	3.75	6	189

4. A car cost £7200 new and sold for £5400 after two years. What was the percentage average annual depreciation?

5. A farmer sold eight cows at market at an average sale price of £48 each. If his total costs for rearing all the animals was £432, what was his percentage loss on each animal?

Student Assessment 2

1. A girl works in a shop on Saturdays for 8.5 hours. She is paid £3.60 per hour. What is her gross pay for 4 weeks' work?

2. A potter makes cups and saucers in a factory. He is paid £1.44 per dozen cups and £1.20 per dozen saucers. What is his gross pay if he makes 9 dozen cups and 11 dozen saucers in one day?

3. Calculate the missing numbers from the simple interest table below:

Principal (£)	Rate (%)	Time (years)	Interest (£)
300	6	4	(a)
250	(b)	3	60
480	5	(c)	96
650	(d)	8	390
(e)	3.75	4	187.50

4. A family house was bought for £48 000 twelve years ago. It is now valued at £120 000. What is the average annual increase in the value of the house?

5. An electrician bought five broken washing machines for £550. He repaired them and sold them for £143 each. What was his percentage profit?

ICT SECTION

Number Activity 1

The step patterns below follow a rule.

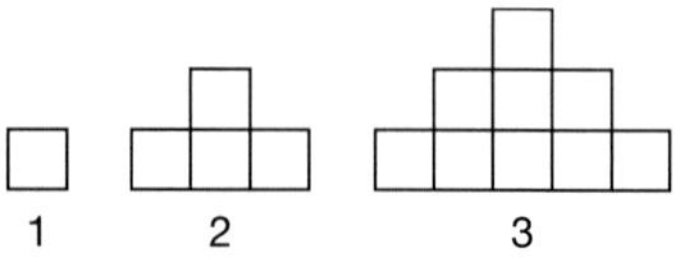

1. On squared paper, draw the next two patterns in this sequence.

2. Count the number of squares used in each of the first five patterns. Enter the results into a table on a spreadsheet, similar to the one shown below.

	A	B
1	Pattern	Number of squares
2	1	
3	2	
4	3	
5	4	
6	5	
7	10	
8	20	
9	50	

3. The number of squares needed for each pattern follows a rule. Describe the rule.

4. By writing a formula in cell B7 and copying it down to B9, use the spreadsheet to generate the results for the 10th, 20th and 50th patterns.

5. Repeat Q.1–4 for the following patterns:

a)

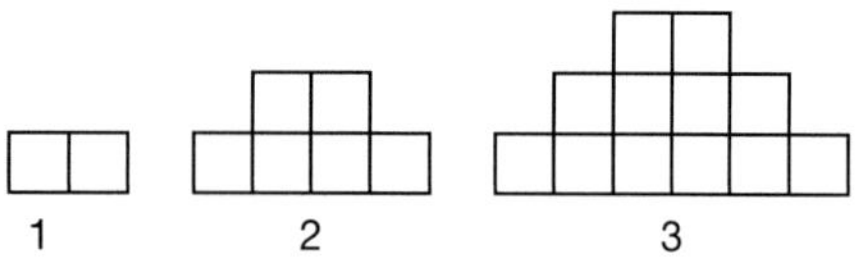

b)

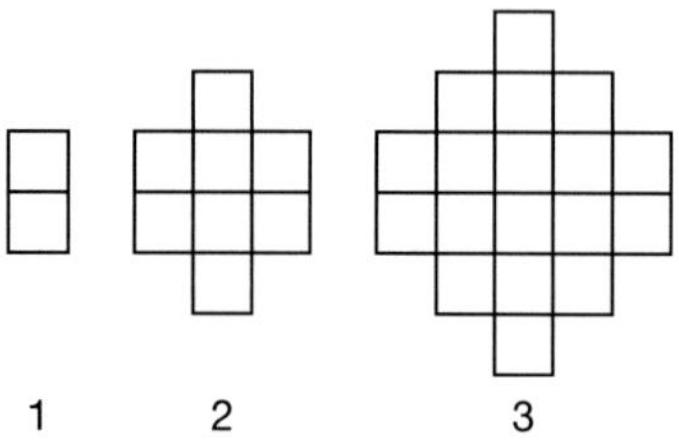

Number Activity 2

In this activity, you will be using both the internet and a spreadsheet in order to produce currency conversions.

1. Log on to the internet and search for a website that shows the exchange rates between different currencies.
2. Compare your own currency with another currency of your choice. Write down the exchange rate, e.g. £1 = €1.40.
3. Using a spreadsheet construct a currency converter. An example is shown below:

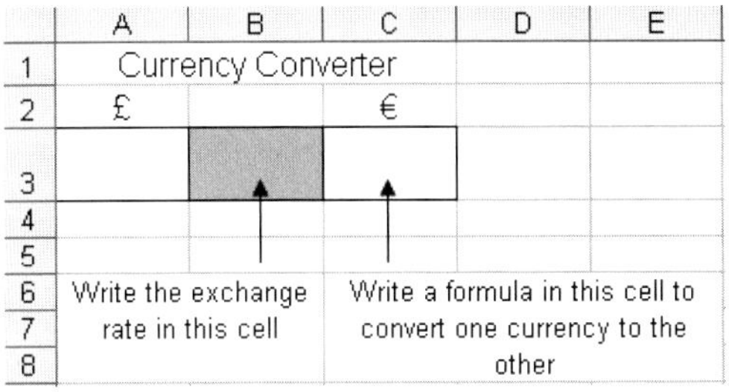

4. By entering different amounts of your own currency, use the currency converter to calculate the correct conversion. Record your answers in a table.
5. Repeat steps 1–4 for five different currencies of your choice.

REVIEWS FOR CHAPTERS 1–16

Review 1

1. List the prime factors of 120 in index form.
2. Draw a square of side 3.2 units. Use it to show how to calculate $(3.2)^2$.
3. London has a noon temperature of 3 °C on the 1st January. Boston is 15 °C colder. What is the temperature in Boston?
4. Change the following fractions to decimals:
   a) $\frac{3}{5}$ b) $\frac{3}{4}$ c) $\frac{3}{8}$
5. Write the following lengths in order of magnitude, starting with the smallest:

   236 mm 18 cm 0.05 m $\frac{1}{4}$m
6. Write 85 million in standard form.
7. Copy these equivalent fractions and fill in the blanks:
   a) $\frac{15}{35} = \frac{}{7}$ b) $\frac{9}{21} = \frac{}{63}$
8. 1 mile is 1760 yards. Calculate the number of yards in 15 miles.
9. Calculate the perimeter of a regular hexagon of side 5.2 cm.
10. Find the area and perimeter of a square of side 1.1 cm.

Review 2

1. Which of the following is an irrational number?

   1.3, $3\sqrt{8}$, $\sqrt{7}$, 0.675, 0.33, 4
2. The following numbers are expressed correct to two significant figures.
   Representing each number by the letter x, express the range in which each must lie, using an inequality:
   a) 180 b) 0.75
3. A machine prints 7 books in 50 minutes. How many books will be printed in 5 hours?
4. Find 35% of 560.
5. Change the following times to those shown on a 24-hour clock:
   a) a quarter to eight in the morning b) 8.30 p.m.
6. £1 sterling can be exchanged for 9.5 South African rand. Draw a conversion graph and use it to convert the following:
   a) £6 into rand b) 60 rand into pounds

7. Express the following in metres:
a) 160 cm b) 2.8 km c) 12 cm

8. Calculate the simple interest on £650 for 3 years at 7.5%.

9. Calculate the perimeter of an equilateral triangle of side 13.5 cm.

10. Calculate the volume of a shoe box which is a cuboid of sides 25 cm, 12 cm and 18 cm.

Review 3

1. What is the HCF of 72, 144 and 108?

2. Draw a table of $y = x^2$ for value of x from 0 to +8. Use your table to draw a graph of $y = x^2$. Use the graph to estimate (to 1 d.p.):
a) $(3.2)^2$ b) $(7.5)^2$

3. The temperature in Moscow on New Year's Day is −28 °C. The temperature in Sydney on New Year's Day is 28 °C. What is the difference in temperature between the two cities?

4. Change the following decimals to fractions. Give each fraction in its simplest form.
a) 0.25 b) 0.625 c) 1.05

5. Represent each of the following inequalities on a number line, where x is a real number:
a) $x < 3$ b) $x \geqslant 4$ c) $1 < x < 4$
d) $3 > x > -1$

6. Write 0.000 007 83 in standard form.

7. Evaluate $3\frac{2}{5} - 1\frac{1}{2}$.

8. Round the following numbers to the number of significant figures shown in brackets:
a) 68.3 (1 s.f.) b) 478 700 (2 s.f.) c) 645.380 (1 s.f.)
d) 645.380 (3 s.f.)

9. Calculate the perimeter of a regular octagon of side 17.5 cm.

10. Calculate the area and perimeter of a square of side 7.5 cm.

Review 4

1. Calculate the next two terms in the following sequences:
a) 108, 96, 84, 72, ...
b) 1, 2, 3, 5, 8, 13, 21, ...

2. The following numbers are expressed to the nearest whole number. Illustrate on a number line the range in which each must lie:
a) 6 b) −4 c) 100

3. A bricklayer lays 1200 bricks in a 9-hour day. Assuming he works at the same rate, how many bricks will he lay in 45 working hours?

4. Increase 450 euro by 8%.

5. A bus journey from Cambridge to Bourn takes 1 hour and 45 minutes. What time will a bus leaving Cambridge at 08.50 arrive in Bourn?

6. 15 South African rand can be exchanged for 11 Hong Kong dollars. Draw a conversion graph and use it to convert the following:
   a) 200 rand into dollars   b) 100 HK dollars into rand

7. Express the following masses in kg:
   a) 28 000 g   b) 2400 g   c) 150 g

8. How long will it take a sum of 5000 euro invested at 4% to earn simple interest of 800 euro?

9. Calculate the perimeter of a regular pentagon of side 3.8 cm.

10. Calculate the volume of a cube of side 1.5 cm.

Review 5

1. What is the LCM of 7, 8, and 12?

2. Calculate the following square roots without using a calculator:
   a) $\sqrt{49}$ b) $\sqrt{0.81}$ c) $\sqrt{0.04}$

3. A helicopter flying at 450 m above sea level drops a sonar device onto the ocean floor. If the ocean is 1680 m deep at this point, how far above the sonar device is the helicopter?

4. Work out 345×17.

5. A month has at least 28 days, but not more than 31 days. Illustrate this information using inequalities.

6. Write the answer to the calculation of 200 000 multiplied by 6 500 000 in standard form.

7. Evaluate $6\frac{3}{7} \div 1\frac{1}{14}$.

8. Estimate the answer to the following: $\dfrac{3.8 \times 25.9}{1.6 \times 6.8}$

9. Calculate the perimeter of a decagon of side 4.3 cm.

10. Calculate the volume of a cube of side 1.1 cm.

Review 6

1. State whether the following are rational numbers or not:
   a) 1.3   b) π c) $\sqrt{0.04}$ d) $\sqrt{7}$ e) $\sqrt{0.85}$

2. A girl's mass was measured to the nearest 0.1 kg. If her mass was 48.2 kg, illustrate on a number line the range within which it must lie.

3. A paint mix uses red and white paint in a ratio of 3 : 10. How much red paint should be mixed with 7.5 litres of white paint?

4. A television set priced at £280 is reduced by 30% in a sale. What is the new price?

5. Express 'a quarter to four in the afternoon' as it would appear on a 24-hour clock.

6. 80 Bangladeshi taka can be exchanged for 1.4 US dollars. Draw a conversion graph and use it to convert the following:
   a) 1250 taka into dollars b) 50 dollars into taka

7. Express the following in litres:
   a) 8500 ml b) 400 ml c) 15 500 ml

8. What rate of simple interest per year must be paid on a principal of 375 euro, in order to earn 90 euro in 4 years?

9. Calculate the area and perimeter of a square of side 6.5 cm.

10. Calculate the volume of a cuboid 12 cm long, 8 cm wide and 8 cm deep.

Review 7

1. Evaluate the following:
   a) $(-6) + (-4)$ b) $(-6) \times (-4)$ c) $(-54) \div (+9)$

2. Without using a calculator, find the value of:
   a) $\sqrt{\frac{36}{49}}$ b) $\sqrt{1\frac{19}{81}}$ c) $\sqrt{6\frac{1}{4}}$

3. The library of Celsius in Ephesus was built in 380 BC. How old is this building?

4. Work out $4840 \div 23$, giving your answer to 1 d.p.

5. Write the following fractions in order of magnitude, starting with the smallest:

   $\frac{1}{2}$ $\frac{2}{3}$ $\frac{4}{7}$ $\frac{3}{5}$ $\frac{5}{9}$

6. A star is 4 light years away from Earth. If the speed of light is 3×10^5 km/s, calculate the distance the star is away from the Earth. Give your answer in kilometres in standard form.

7. Evaluate $1\frac{1}{2} \times 3\frac{3}{4} \div \frac{7}{8}$.

8. A rectangle's dimensions are given as 8.3 m by 58.2 m. Estimate its area.

9. Calculate the perimeter of a regular heptagon of side 6.3 cm.

10. Calculate the volume of a cube of side 2.4 cm.

Review 8

1. For each of the sequences shown below give an expression for the nth term:
   a) 7, 13, 19, 25, 31, ...
   b) 0, 5, 10, 15, 20, ...

2. The numbers below are rounded to the degree of accuracy shown in brackets. Express the lower and upper bounds of these numbers as an inequality.
   a) $x = 4.55$ (2 d.p.) b) $y = -10.0$ (1 d.p.)

3. Divide 28 m in the ratio 3 : 4.

4. Find 75% of 640.

5. How long will it take a train travelling at an average speed of 120 km/h to travel 300 km?

6. 1 euro can be converted into 48 Indian rupees. Draw a conversion graph and use it to convert the following:
   a) €25 into rupees
   b) 680 rupees into euro

7. Express the following lengths in metres:
   a) 650 cm b) 0.85 cm c) 15 cm

8. Calculate the simple interest on £340 for 4 years at 3.75%.

9. Calculate the perimeter and area of a rectangular field 110 m by 75 m.

10. Calculate the volume of a cube of side 3.5 cm.

Review 9

1. If $\frac{x}{y} = -7$, copy and complete the table below:

x	+21	+14	+7	1	−0.28	−7	−10.5
y							

2. Use long multiplication to calculate:
   a) $(2.5)^2$ b) $(0.9)^2$ c) $(1.01)^2$

3. Marcus Flavius died in AD 38 at the age of 73. In what year was he born?

4. Give the following fractions as percentages:
   a) $\frac{4}{5}$ b) $\frac{7}{8}$ c) $2\frac{1}{4}$

5. Insert one of the symbols =, > or < into the space to make the following statements correct:
   a) $7 \times 9 \ldots 9 \times 7$ b) $82 \ldots 42 + 42$
   c) 30 cm ... 0.4 m d) $3^4 \ldots 9^2$

6. Write the answer to the following calculation in standard form:

$$(5.8 \times 10^7) + (8.6 \times 10^6)$$

7. Evaluate $3\frac{1}{4} + 1\frac{2}{3} - 3\frac{7}{8}$.

8. If 1 km is $\frac{5}{8}$ of 1 mile, calculate the number of kilometres in 75 miles.

9. Divide 45 in the ratio 4 : 5.

10. Calculate $\frac{4}{5}$ of 280.

Review 10

1. For each of the sequences given below, explain the pattern in words:
   a) −7, −4, −1, +2, +5
   b) 1, 1, 2, 3, 5, 8, 13

2. The following numbers are rounded to the nearest 100. Illustrate on a number line the range in which each must lie.
   a) 600 b) −1000

3. Divide 60 m in the ratio 2 : 3 : 5.

4. A tax of 15% is added to a garage bill of $180. What is the total bill?

5. A train travels for $4\frac{1}{4}$ hours at an average speed of 120 km/h. How far does it travel?

6. 550 000 Turkish lira can be exchanged for 1.6 US dollars. Draw a conversion graph and use it to convert the following:
   a) 12 dollars into lira b) 2 million lira into dollars

7. Express the following in grams:
   a) 0.085 kg b) 0.003 kg c) 0.5 kg

8. Find the principal which will earn interest of 90 euros in four years at 8% simple interest.

9. Evaluate $1\frac{3}{5} \div \frac{5}{16}$.

10. Change the following fractions to percentages:
    a) $\frac{3}{5}$ b) $\frac{3}{4}$ c) $\frac{3}{8}$

Review 11

1. List the LCM and HCF of the following sets of numbers
   a) 4, 6, 12 b) 13, 9, 39

2. Without using a calculator, find:
   a) $\sqrt{64}$ b) $\sqrt{225}$ c) $\sqrt{0.0081}$

3. The temperature at 6 p.m. is 3 °C. If the temperature falls by 12 °C in the next twelve hours, then rises by 1.5 °C per hour for the next three hours, what will the temperature be at 9 a.m.?

4. Copy and complete the table below:

Fraction	Decimal	Percentage
$\frac{1}{8}$		
	0.375	
		140%

5. Illustrate each of the following inequalities on a number line where y is a real number:
   a) $y > 4$
   b) $-4 < y < 5$
   c) $0 \leqslant y \leqslant 4$
   d) $-2 \leqslant y < 2$

6. Jupiter is about 780 million km from the sun. How many seconds will it take light travelling at 3×10^5 km/s to travel from the Sun to Jupiter?

7. Evaluate $(1\frac{1}{2} + 3\frac{3}{8}) \div (2\frac{1}{4} - \frac{5}{8})$.

8. A cuboid's dimensions are given as 14.9 cm by 12.3 cm by 4.5 cm. Estimate its volume.

9. Evaluate $4\frac{1}{2} \div \frac{2}{3} \times \frac{5}{6}$.

10. Calculate the simple interest earned on £480 in 7 years at 5%.

Review 12

1. Explain in words the difference between a rational number and an irrational number.

2. The following numbers are expressed correct to three significant figures. Represent the limits of each number using inequalities.
   a) 266
   b) 30.7

3. It takes 12 hours for five bricklayers to build a wall. Calculate how long it would take 8 bricklayers to build a similar wall.

4. Express 750 kg as a percentage of 2.5 tonnes.

5. A train covers 600 km in $4\frac{1}{2}$ hours. Calculate its average speed.

6. 1 US$ converts to 60 Pakistan rupees. Draw a conversion graph and use it to convert the following:
   a) 35 dollars into rupees
   b) 3000 rupees into dollars

7. Express the following volumes in millilitres:
a) 2.5 litres b) 0.03 litre c) 0.0075 litre

8. Calculate the simple interest on £550 for 3 years at 7.8%.

9. Calculate the area and perimeter of a right-angled triangle with sides 5 cm, 12 cm and 13 cm.

10. Write $7\frac{3}{4}$ million in standard form.

ALGEBRA

17 INDICES

Use and interpret positive, negative and zero indices.

The index refers to the power to which a number is raised. In the example 5^3, the number 5 is raised to the power 3. The 3 is known as the **index**. Indices is the plural of index.

Worked examples

a) $5^3 = 5 \times 5 \times 5$
$= 125$

b) $7^4 = 7 \times 7 \times 7 \times 7$
$= 2401$

c) $3^1 = 3$

Laws of indices

When working with numbers involving indices there are three basic laws which can be applied. These are:

(1) $a^m \times a^n = a^{m+n}$

(2) $a^m \div a^n$ or $\dfrac{a^m}{a^n} = a^{m-n}$

(3) $(a^m)^n = a^{mn}$

Positive indices

Worked examples

a) Simplify $4^3 \times 4^2$.

$$4^3 \times 4^2 = 4^{(3+2)} = 4^5$$

b) Simplify $2^5 \div 2^3$.

$$2^5 \div 2^3 = 2^{(5-3)} = 2^2$$

c) Evaluate $3^3 \times 3^4$.

$$3^3 \times 3^4 = 3^{(3+4)} = 3^7 = 2187$$

d) Evaluate $(4^2)^3$.

$$(4^2)^3 = 4^{(2\times3)} = 4^6 = 4096$$

Exercise 17.1

1. Using indices, simplify the following expressions:

a) $3 \times 3 \times 3$
b) $2 \times 2 \times 2 \times 2 \times 2$
c) 4×4
d) $6 \times 6 \times 6 \times 6$
e) $8 \times 8 \times 8 \times 8 \times 8 \times 8$
f) 5

2. Simplify the following using indices:

a) $2 \times 2 \times 2 \times 3 \times 3$
b) $4 \times 4 \times 4 \times 4 \times 4 \times 5 \times 5$
c) $3 \times 3 \times 4 \times 4 \times 4 \times 5 \times 5$
d) $2 \times 7 \times 7 \times 7 \times 7$
e) $1 \times 1 \times 6 \times 6$
f) $3 \times 3 \times 3 \times 4 \times 4 \times 6 \times 6 \times 6 \times 6 \times 6$

3. Write out the following in full:

a) 4^2 b) 5^7
c) 3^5 d) $4^3 \times 6^3$
e) $7^2 \times 2^7$ f) $3^2 \times 4^3 \times 2^4$

4. Without a calculator, work out the value of the following:

a) 2^5 b) 3^4
c) 8^2 d) 6^3
e) 10^6 f) 4^4
g) $2^3 \times 3^2$ h) $10^3 \times 5^3$

Exercise 17.2

1. Simplify the following using indices:

a) $3^2 \times 3^4$ b) $8^5 \times 8^2$
c) $5^2 \times 5^4 \times 5^3$ d) $4^3 \times 4^5 \times 4^2$
e) $2^1 \times 2^3$ f) $6^2 \times 3^2 \times 3^3 \times 6^4$
g) $4^5 \times 4^3 \times 5^5 \times 5^4 \times 6^2$ h) $2^4 \times 5^7 \times 5^3 \times 6^2 \times 6^6$

2. Simplify the following:

a) $4^6 \div 4^2$ b) $5^7 \div 5^4$
c) $2^5 \div 2^4$ d) $6^5 \div 6^2$
e) $\frac{6^5}{6^2}$ f) $\frac{8^6}{8^5}$
g) $\frac{4^8}{4^5}$ h) $\frac{3^9}{3^2}$

3. Simplify the following:

a) $(5^2)^2$ b) $(4^3)^4$
c) $(10^2)^5$ d) $(3^3)^5$
e) $(6^2)^4$ f) $(8^2)^3$

4. Simplify the following:

a) $\frac{2^2 \times 2^4}{2^3}$ b) $\frac{3^4 \times 3^2}{3^5}$
c) $\frac{5^6 \times 5^7}{5^2 \times 5^8}$ d) $\frac{(4^2)^5 \times 4^2}{4^7}$
e) $\frac{4^4 \times 2^5 \times 4^2}{4^3 \times 2^3}$ f) $\frac{6^3 \times 6^3 \times 8^5 \times 8^6}{8^6 \times 6^2}$
g) $\frac{(5^5)^2 \times (4^4)^3}{5^8 \times 4^9}$ h) $\frac{(6^3)^4 \times 6^3 \times 4^9}{6^8 \times (4^2)^4}$

5. Simplify the following:

a) $c^5 \times c^3$

b) $m^4 \div m^2$

c) $(b^3)^5 \div b^6$

d) $\dfrac{m^4n^9}{mn^3}$

e) $\dfrac{6a^6b^4}{3a^2b^3}$

f) $\dfrac{12x^5y^7}{4x^2y^5}$

g) $\dfrac{4u^3v^6}{8u^2v^3}$

h) $\dfrac{3x^6y^5z^3}{9x^4y^2z}$

6. Simplify the following:

a) $4a^2 \times 3a^3$

b) $2a^2b \times 4a^3b^2$

c) $(2p^2)^3$

d) $(4m^2n^3)^2$

e) $(5p^2)^2 \times (2p^3)^3$

f) $(4m^2n^2) \times (2mn^3)^3$

g) $\dfrac{(6x^2y^4)^2 \times (2xy)^3}{12x^6y^8}$

h) $(ab)^d \times (ab)^e$

The zero index

The zero index indicates that a number is raised to the power 0. Any number raised to the power 0 is equal to 1. This can be explained by applying the laws of indices.

$a^m \div a^n = a^{m-n}$ therefore $\dfrac{a^m}{a^m} = a^{m-m}$

$= a^0$

However, $\dfrac{a^m}{a^m} = 1$

Therefore $a^0 = 1$

Negative indices

A negative index indicates that a number is being raised to a negative power: e.g. 4^{-3}.

Another law of indices states that $a^{-m} = \dfrac{1}{a^m}$. This can be proved as follows.

$$a^{-m} = a^{0-m}$$

$$= \frac{a^0}{a^m} \text{ (from the second law of indices)}$$

$$= \frac{1}{a^m}$$

Therefore $a^{-m} = \dfrac{1}{a^m}$

Exercise 17.3

Without using a calculator, evaluate the following:

1. a) $2^3 \times 2^0$ b) $5^2 \div 6^0$
c) $5^2 \times 5^{-2}$ d) $6^3 \times 6^{-3}$
e) $(4^0)^2$ f) $4^0 \div 2^2$

2. a) 4^{-1} b) 3^{-2}
c) 6×10^{-2} d) 5×10^{-3}
e) 100×10^{-2} f) 10^{-3}

3. a) 9×3^{-2} b) 16×2^{-3}
c) 64×2^{-4} d) 4×2^{-3}
e) 36×6^{-3} f) 100×10^{-1}

4. a) $\frac{3}{2^{-2}}$ b) $\frac{4}{2^{-3}}$
c) $\frac{9}{5^{-2}}$ d) $\frac{5}{4^{-2}}$
e) $\frac{7^{-3}}{7^{-4}}$ f) $\frac{8^{-6}}{8^{-8}}$

Student Assessment 1

1. Using indices, simplify the following:
a) $2 \times 2 \times 2 \times 5 \times 5$
b) $2 \times 2 \times 3 \times 3 \times 3 \times 3 \times 3$

2. Write the following out in full:
a) 4^3
b) 6^4

3. Work out the value of the following without using a calculator:
a) $2^3 \times 10^2$
b) $1^4 \times 3^3$

4. Simplify the following using indices:
a) $3^4 \times 3^3$
b) $6^3 \times 6^2 \times 3^4 \times 3^5$
c) $\dfrac{4^5}{2^3}$
d) $\dfrac{(6^2)^3}{6^5}$
e) $\dfrac{3^5 \times 4^2}{3^3 \times 4^0}$
f) $\dfrac{4^{-2} \times 2^6}{2^2}$

5. Without using a calculator, evaluate the following:
a) $2^4 \times 2^{-2}$
b) $\dfrac{3^5}{3^3}$
c) $\dfrac{5^{-5}}{5^{-6}}$
d) $\dfrac{2^5 \times 4^{-3}}{2^{-1}}$

Student Assessment 2

1. Using indices, simplify the following:
a) $3 \times 2 \times 2 \times 3 \times 27$
b) $2 \times 2 \times 4 \times 4 \times 4 \times 2 \times 32$

2. Write the following out in full:
a) 6^5
b) 2^{-5}

3. Work out the value of the following without using a calculator:
a) $3^3 \times 10^3$
b) $1^{-4} \times 5^3$

4. Simplify the following using indices:
a) $2^4 \times 2^3$
b) $7^5 \times 7^2 \times 3^4 \times 3^8$
c) $\dfrac{4^8}{2^{10}}$
d) $\dfrac{(3^3)^4}{27^3}$
e) $\dfrac{7^6 \times 4^2}{4^3 \times 7^6}$
f) $\dfrac{8^{-2} \times 2^6}{2^{-2}}$

5. Without using a calculator, evaluate the following:
a) $5^2 \times 5^{-1}$
b) $\dfrac{4^5}{4^3}$
c) $\dfrac{7^{-5}}{7^{-7}}$
d) $\dfrac{3^{-5} \times 4^2}{3^{-6}}$

18 ALGEBRAIC REPRESENTATION AND MANIPULATION

Use letters to express generalised numbers and express basic arithmetic processes algebraically; substitute numbers for words and letters in formulae; transform simple formulae; manipulate directed numbers; use brackets and extract common factors.

Expanding brackets

When removing brackets, every term inside the bracket must be multiplied by whatever is outside the bracket.

Worked examples

a) $3(x + 4)$
$= 3x + 12$

b) $5x(2y + 3)$
$= 10xy + 15x$

c) $2a(3a + 2b - 3c)$
$= 6a^2 + 4ab - 6ac$

d) $-4p(2p - q + r^2)$
$= -8p^2 + 4pq - 4pr^2$

e) $-2x^2\left(x + 3y - \frac{1}{x}\right)$
$= -2x^3 - 6x^2y + 2x$

f) $\frac{-2}{x}\left(-x + 4y + \frac{1}{x}\right)$
$= 2 - \frac{8y}{x} - \frac{2}{x^2}$

Exercise 18.1

Expand the following:

1. a) $2(a + 3)$ b) $4(b + 7)$
c) $5(2c + 8)$ d) $7(3d + 9)$
e) $9(8e - 7)$ f) $6(4f - 3)$

2. a) $3a(a + 2b)$ b) $4b(2a + 3b)$
c) $2c(a + b + c)$ d) $3d(2b + 3c + 4d)$
e) $e(3c - 3d - e)$ f) $f(3d - e - 2f)$

3. a) $2(2a^2 + 3b^2)$ b) $4(3a^2 + 4b^2)$
c) $-3(2c + 3d)$ d) $-(2c + 3d)$
e) $-4(c^2 - 2d^2 + 3e^2)$ f) $-5(2e - 3f^2)$

4. a) $2a(a + b)$ b) $3b(a - b)$
c) $4c(b^2 - c^2)$ d) $3d^2(a^2 - 2b^2 + c^2)$
e) $-3e^2(4d - e)$ f) $-2f(2d - 3e^2 - 2f)$

Exercise 18.2

Expand the following:

1. a) $4(x - 3)$ b) $5(2p - 4)$
c) $-6(7x - 4y)$ d) $3(2a - 3b - 4c)$
e) $-7(2m - 3n)$ f) $-2(8x - 3y)$

2. a) $3x(x - 3y)$ b) $a(a + b + c)$
c) $4m(2m - n)$ d) $-5a(3a - 4b)$
e) $-4x(-x + y)$ f) $-8p(-3p + q)$

3. a) $-(2x^2 - 3y^2)$ b) $-(-a + b)$
c) $-(-7p + 2q)$ d) $\frac{1}{2}(6x - 8y + 4z)$
e) $\frac{3}{4}(4x - 2y)$ f) $\frac{1}{5}x(10x - 15y)$

4. a) $3r(4r^2 - 5s + 2t)$ b) $a^2(a + b + c)$
c) $3a^2(2a - 3b)$ d) $pq(p + q - pq)$
e) $m^2(m - n + nm)$ f) $a^3(a^3 + a^2b)$

Exercise 18.3

Expand and simplify the following:

1. a) $2a + 2(3a + 2)$ b) $4(3b - 2) - 5b$
c) $6(2c - 1) - 7c$ d) $-4(d + 2) + 5d$
e) $-3e + (e - 1)$ f) $5f - (2f - 7)$

2. a) $2(a + 1) + 3(b + 2)$ b) $4(a + 5) - 3(2b + 6)$
c) $3(c - 1) + 2(c - 2)$ d) $4(d - 1) - 3(d - 2)$
e) $-2(e - 3) - (e - 1)$ f) $2(3f - 3) + 4(1 - 2f)$

3. a) $2a(a + 3) + 2b(b - 1)$
b) $3a(a - 4) - 2b(b - 3)$
c) $2a(a + b + c) - 2b(a + b - c)$
d) $a^2(c^2 + d^2) - c^2(a^2 + d^2)$
e) $a(b + c) - b(a - c)$
f) $a(2d + 3e) - 2e(a - c)$

Exercise 18.4

Expand and simplify the following:

1. a) $3a - 2(2a + 4)$ b) $8x - 4(x + 5)$
c) $3(p - 4) - 4$ d) $7(3m - 2n) + 8n$
e) $6x - 3(2x - 1)$ f) $5p - 3p(p + 2)$

2. a) $7m(m + 4) + m^2 + 2$ b) $3(x - 4) + 2(4 - x)$
c) $6(p + 3) - 4(p - 1)$ d) $5(m - 8) - 4(m - 7)$
e) $3a(a + 2) - 2(a^2 - 1)$ f) $7a(b - 2c) - c(2a - 3)$

3. a) $\frac{1}{2}(6x + 4) + \frac{1}{3}(3x + 6)$
b) $\frac{1}{4}(2x + 6y) + \frac{3}{4}(6x - 4y)$
c) $\frac{1}{8}(6x - 12y) + \frac{1}{2}(3x - 2y)$
d) $\frac{1}{5}(15x + 10y) + \frac{3}{10}(5x - 5y)$
e) $\frac{2}{3}(6x - 9y) + \frac{1}{3}(9x + 6y)$
f) $\frac{x}{7}(14x - 21y) - \frac{x}{2}(4x - 6y)$

Simple factorising

When factorising, the largest possible factor is removed from each of the terms and placed outside the brackets.

Worked examples Factorise the following expressions:

a) $10x + 15$
$= 5(2x + 3)$

b) $8p - 6q + 10r$
$= 2(4p - 3q + 5r)$

c) $-2q - 6p + 12$
$= 2(-q - 3p + 6)$

d) $2a^2 + 3ab - 5ac$
$= a(2a + 3b - 5c)$

e) $6ax - 12ay - 18a^2$
$= 6a(x - 2y - 3a)$

f) $3b + 9ba - 6bd$
$= 3b(1 + 3a - 2d)$

Exercise 18.5

Factorise the following:

1. a) $4x - 6$ b) $18 - 12p$
c) $6y - 3$ d) $4a + 6b$
e) $3p - 3q$ f) $8m + 12n + 16r$

2. a) $3ab + 4ac - 5ad$ b) $8pq + 6pr - 4ps$
c) $a^2 - ab$ d) $4x^2 - 6xy$
e) $abc + abd + fab$ f) $3m^2 + 9m$

3. a) $3pqr - 9pqs$ b) $5m^2 - 10mn$
c) $8x^2y - 4xy^2$ d) $2a^2b^2 - 3b^2c^2$
e) $12p - 36$ f) $42x - 54$

4. a) $18 + 12y$ b) $14a - 21b$
c) $11x + 11xy$ d) $4s - 16t + 20r$
e) $5pq - 10qr + 15qs$ f) $4xy + 8y^2$

5. a) $m^2 + mn$ b) $3p^2 - 6pq$
c) $pqr + qrs$ d) $ab + a^2b + ab^2$
e) $3p^3 - 4p^4$ f) $7b^3c + b^2c^2$

6. a) $m^3 - m^2n + mn^2$ b) $4r^3 - 6r^2 + 8r^2s$
c) $56x^2y - 28xy^2$ d) $72m^2n + 36mn^2 - 18m^2n^2$

7. a) $3a^2 - 2ab + 4ac$ b) $2ab - 3b^2 + 4bc$
c) $2a^2c - 4b^2c + 6bc^2$ d) $39cd^2 + 52c^2d$

8. a) $12ac - 8ac^2 + 4a^2c$ b) $34a^2b - 51ab^2$
c) $33ac^2 + 121c^3 - 11b^2c^2$ d) $38c^3d^2 - 57c^2d^3 + 95c^2d^2$

9. a) $15\frac{a}{c} - 25\frac{b}{c} + 10\frac{d}{c}$ b) $46\frac{a}{c^2} - 23\frac{b}{c^2}$
c) $\frac{1}{2a} - \frac{1}{4a}$ d) $\frac{3}{5d} - \frac{1}{10d} + \frac{4}{15d}$

10. a) $\frac{5}{a^2} - \frac{3}{a}$ b) $\frac{6}{b^2} - \frac{3}{b}$
c) $\frac{2}{3a} - \frac{3}{3a^2}$ d) $\frac{3}{5d^2} - \frac{4}{5d}$

Substitution

Worked examples Evaluate the expressions below if $a = 3$, $b = 4$ and $c = -5$:

a) $2a + 3b - c$
$= 6 + 12 + 5$
$= 23$

b) $3a - 4b + 2c$
$= 9 - 16 - 10$
$= -17$

c) $-2a + b - 3c$
$= -6 + 8 + 15$
$= 17$

d) $a^2 + b^2 + c^2$
$= 9 + 16 + 25$
$= 50$

e) $3a(2b - 3c)$
$= 9(8 + 15)$
$= 9 \times 23$
$= 207$

f) $-2c(-a + 2b)$
$= 10(-3 + 8)$
$= 10 \times 5$
$= 50$

Exercise 18.6

Evaluate the following expression if:

$a = 2 \quad b = 3 \quad c = 5$

1. a) $3a + 2b$ b) $4a - 3b$
c) $a - b - c$ d) $3a - 2b + c$

2. a) $-b(a + b)$ b) $-2c(a - b)$
c) $-3a(a - 3c)$ d) $-4b(b - c)$

3. a) $a^2 + b^2$ b) $b^2 + c^2$
c) $2a^2 - 3b^2$ d) $3c^2 - 2b^2$

4. a) $-a^2$ b) $(-a)^2$
c) $-b^3$ d) $(-b)^3$

5. a) $-c^3$ b) $(-c)^3$
c) $(-ac)^2$ d) $-(ac)^2$

Exercise 18.7

Evaluate the following expressions if $p = 4$, $q = -2$, $r = 3$ and $s = -5$:

1. a) $2p + 4q$ b) $5r - 3s$
c) $3q - 4s$ d) $6p - 8q + 4s$
e) $3r - 3p + 5q$ f) $-p - q + r + s$

2. a) $2p - 3q - 4r + s$ b) $3s - 4p + r + q$
c) $p^2 + q^2$ d) $r^2 - s^2$
e) $p(q - r + s)$ f) $r(2p - 3q)$

3. a) $2s(3p - 2q)$ b) $pq + rs$
c) $2pr - 3rq$ d) $q^3 - r^2$
e) $s^3 - p^3$ f) $r^4 - q^5$

4. a) $-2pqr$ b) $-2p(q + r)$
c) $-2rq + r$ d) $(p + q)(r - s)$
e) $(p + s)(r - q)$ f) $(r + q)(p - s)$

5. a) $(2p + 3q)(p - q)$ b) $(q + r)(q - r)$
c) $q^2 - r^2$ d) $p^2 - r^2$
e) $(p + r)(p - r)$ f) $(-s + p)q^2$

Transformation of formulae

In the formula $a = 2b + c$, 'a' is the subject. In order to make either b or c the subject, the formula has to be rearranged.

Worked examples Rearrange the following formulae to make the **bold** letter the subject:

a) $a = 2b + \mathbf{c}$
$a - 2b = \mathbf{c}$

b) $2r + \mathbf{p} = q$
$\mathbf{p} = q - 2r$

c) $ab = \mathbf{c}d$
$\frac{ab}{d} = \mathbf{c}$

d) $\frac{a}{b} = \frac{c}{\mathbf{d}}$
$a\mathbf{d} = cb$
$\mathbf{d} = \frac{cb}{a}$

Exercise 18.8

In the following questions, make the **bold** letter the subject of the formula:

1. a) $a + \mathbf{b} = c$ b) $\mathbf{b} + 2c = d$ c) $2b + \mathbf{c} = 4a$
d) $3d + \mathbf{b} = 2a$

2. a) $a\mathbf{b} = c$ b) $a\mathbf{c} = bd$ c) $\mathbf{a}b = c + 3$
d) $a\mathbf{c} = b - 4$

3. a) $m + \mathbf{n} = r$ b) $\mathbf{m} + n = p$ c) $2m + \mathbf{n} = 3p$
d) $3x = 2p + \mathbf{q}$ e) $\mathbf{a}b = cd$ f) $ab = c\mathbf{d}$

4. a) $3\mathbf{x}y = 4m$ b) $7pq = 5\mathbf{r}$ c) $3\mathbf{x} = c$
d) $3\mathbf{x} + 7 = y$ e) $5\mathbf{y} - 9 = 3r$ f) $5y - 9 = 3\mathbf{x}$

5. a) $6\mathbf{b} = 2a - 5$ b) $6b = 2\mathbf{a} - 5$ c) $3x - 7y = 4\mathbf{z}$
d) $3\mathbf{x} - 7y = 4z$ e) $3x - 7\mathbf{y} = 4z$ f) $2\mathbf{p}r - q = 8$

6. a) $\frac{\mathbf{p}}{4} = r$ b) $\frac{4}{\mathbf{p}} = 3r$ c) $\frac{1}{5}n = 2\mathbf{p}$
d) $\frac{1}{5}\mathbf{n} = 2p$ e) $\mathbf{p}(q + r) = 2t$ f) $p(\mathbf{q} + r) = 2t$

7. a) $3m - n = \mathbf{r}t(p + q)$ b) $3m - n = r\mathbf{t}(p + q)$
c) $3\mathbf{m} - n = rt(p + q)$ d) $3m - \mathbf{n} = rt(p + q)$
e) $3m - n = rt(\mathbf{p} + q)$ f) $3m - n = rt(p + \mathbf{q})$

8. a) $\frac{ab}{c} = \mathbf{d}e$ b) $\frac{\mathbf{a}b}{c} = de$ c) $\frac{ab}{\mathbf{c}} = de$
d) $\frac{\mathbf{a} + b}{c} = d$ e) $\frac{a}{c} + \mathbf{b} = d$ f) $\frac{a}{\mathbf{c}} + b = d$

Student Assessment 1

1. Expand the following:
   a) $3(a + 4)$ b) $7(4b - 2)$
   c) $3c(c + 4d)$ d) $5d(3c - 2d)$
   e) $-6(2e - 3f)$ f) $-(2f - 3g)$

2. Expand the following and simplify where possible:
   a) $3a + 3(a + 4)$ b) $2(3b - 2) - 5b$
   c) $-3c - (c - 1)$ d) $2(d - 1) + 3(d + 1)$
   e) $-2e(e + 3) + 4(e^2 + 2)$
   f) $2f(d + e + f) - 2e(d + e + f)$

3. Factorise the following:
   a) $3a + 12$ b) $15b^2 + 25b$
   c) $4cf - 6df + 8ef$ d) $4d^3 - 2d^4$

4. If $a = 3$, $b = 4$ and $c = 5$, evaluate the following:
   a) $a + b + c$ b) $4a - 3b$
   c) $a^2 + b^2 + c^2$ d) $(a + c)(a - c)$

5. Rearrange the following formulae to make the **bold** letter the subject:
   a) $\boldsymbol{a} + b = c$ b) $c = \boldsymbol{b} - d$
   c) $a\boldsymbol{c} = bd$ d) $3\boldsymbol{d} - e = 2c$
   e) $\boldsymbol{e}(f - g) = 3a$ f) $e(\boldsymbol{f} - g) = 3a$

Student Assessment 2

1. Expand the following:
   a) $4(a + 2)$ b) $5(2b - 3)$
   c) $2c(c + 2d)$ d) $3d(2c - 4d)$
   e) $-5(3e - f)$ f) $-(-f + 2g)$

2. Expand the following and simplify where possible:
   a) $2a + 5(a + 2)$ b) $3(2b - 3) - b$
   c) $-4c - (4 - 2c)$ d) $3(d + 2) - 2(d + 4)$
   e) $-e(2e + 3) + 3(2 + e^2)$ f) $f(d - e - f) - e(e + f)$

3. Factorise the following:
   a) $7a + 14$ b) $26b^2 + 39b$
   c) $3cf - 6df + 9gf$ d) $5d^2 - 10d^3$

4. If $a = 2$, $b = 3$ and $c = 5$, evaluate the following:
   a) $a - b - c$ b) $2b - c$
   c) $a^2 - b^2 + c^2$ d) $(a + c)^2$

5. Rearrange the following formulae to make the **bold** letter the subject:
   a) $\boldsymbol{a} - b = c$ b) $2c = \boldsymbol{b} - 3d$
   c) $ad = b\boldsymbol{c}$ d) $e = 5\boldsymbol{d} - 3c$
   e) $4a = \boldsymbol{e}(f + g)$ f) $4a = e(\boldsymbol{f} + g)$

Student Assessment 3

1. Expand the following and simplify where possible:

a) $5(2a - 6b + 3c)$ b) $3x(5x - 9)$

c) $-5y(3xy + y^2)$ d) $3x^2(5xy + 3y^2 - x^3)$

e) $5p - 3(2p - 4)$

f) $4m(2m - 3) + 2(3m^2 - m)$

g) $\frac{1}{3}(6x - 9) + \frac{1}{4}(8x + 24)$

h) $\frac{m}{4}(6m - 8) + \frac{m}{2}(10m - 2)$

2. Factorise the following:

a) $12a - 4b$ b) $x^2 - 4xy$

c) $8p^3 - 4p^2q$ d) $24xy - 16x^2y + 8xy^2$

3. If $x = 2$, $y = -3$ and $z = 4$, evaluate the following:

a) $2x + 3y - 4z$ b) $10x + 2y^2 - 3z$

c) $z^2 - y^3$ d) $(x + y)(y - z)$

e) $z^2 - x^2$ f) $(z + x)(z - x)$

4. Rearrange the following formulae to make the **bold** letter the subject:

a) $x = 3p + \boldsymbol{q}$ b) $3m - 5\boldsymbol{n} = 8r$ c) $2m = \frac{3\boldsymbol{y}}{t}$

d) $x(\boldsymbol{w} + y) = 2y$ e) $\frac{xy}{2\boldsymbol{p}} = \frac{rs}{t}$ f) $\frac{\boldsymbol{x} + y}{w} = m + n$

Student Assessment 4

1. Expand the following and simplify where possible:

a) $3(2x - 3y + 5z)$ b) $4p(2m - 7)$

c) $-4m(2mn - n^2)$ d) $4p^2(5pq - 2q^2 - 2p)$

e) $4x - 2(3x + 1)$ f) $4x(3x - 2) + 2(5x^2 - 3x)$

g) $\frac{1}{5}(15x - 10) - \frac{1}{3}(9x - 12)$ h) $\frac{x}{2}(4x - 6) + \frac{x}{4}(2x + 8)$

2. Factorise the following:

a) $16p - 8q$ b) $p^2 - 6pq$

c) $5p^2q - 10pq^2$ d) $9pq - 6p^2q + 12q^2p$

3. If $a = 4$, $b = 3$ and $c = -2$, evaluate the following:

a) $3a - 2b + 3c$ b) $5a - 3b^2$

c) $a^2 + b^2 + c^2$ d) $(a + b)(a - b)$

e) $a^2 - b^2$ f) $b^3 - c^3$

4. Rearrange the following formulae to make the **bold** letter the subject:

a) $p = 4m + \boldsymbol{n}$ b) $4x - 3\boldsymbol{y} = 5z$

c) $2x = \frac{3\boldsymbol{y}}{5p}$ d) $m(x + \boldsymbol{y}) = 3w$

e) $\frac{pq}{4\boldsymbol{r}} = \frac{mn}{t}$ f) $\frac{p + \boldsymbol{q}}{r} = m - n$

19 EQUATIONS

Solve simple linear equations in one unknown; solve simultaneous linear equations in two unknowns.

NB All diagrams are not drawn to scale.

An equation is formed when the value of an unknown quantity is needed.

Simple linear equations

Worked examples Solve the following linear equations:

a)
$$\begin{aligned} 3x + 8 &= 14 \\ 3x &= 6 \\ x &= 2 \end{aligned}$$

b)
$$\begin{aligned} 12 &= 20 + 2x \\ -8 &= 2x \\ -4 &= x \end{aligned}$$

c)
$$\begin{aligned} 3(p + 4) &= 21 \\ 3p + 12 &= 21 \\ 3p &= 9 \\ p &= 3 \end{aligned}$$

d)
$$\begin{aligned} 4(x - 5) &= 7(2x - 5) \\ 4x - 20 &= 14x - 35 \\ 4x + 15 &= 14x \\ 15 &= 10x \\ 1.5 &= x \end{aligned}$$

Exercise 19.1

Solve the following linear equations:

1. a) $5a - 2 = 18$ b) $7b + 3 = 17$
c) $9c - 12 = 60$ d) $6d + 8 = 56$
e) $4e - 7 = 33$ f) $12f + 4 = 76$

2. a) $4a = 3a + 7$ b) $8b = 7b - 9$
c) $7c + 5 = 8c$ d) $5d - 8 = 6d$

3. a) $3a - 4 = 2a + 7$ b) $5b + 3 = 4b - 9$
c) $8c - 9 = 7c + 4$ d) $3d - 7 = 2d - 4$

4. a) $6a - 3 = 4a + 7$ b) $5b - 9 = 2b + 6$
c) $7c - 8 = 3c + 4$ d) $11d - 10 = 6d - 15$

5. a) $\frac{a}{4} = 3$ b) $\frac{1}{4}b = 2$
c) $\frac{c}{5} = 2$ d) $\frac{1}{5}d = 3$
e) $4 = \frac{e}{3}$ f) $-2 = \frac{1}{8}f$

6. a) $\frac{a}{3} + 1 = 4$ b) $\frac{b}{5} + 2 = 6$
c) $8 = 2 + \frac{c}{3}$ d) $-4 = 3 + \frac{d}{5}$
e) $9 = 5 + \frac{2e}{3}$ f) $-7 = \frac{3f}{2} - 1$

7. a) $\frac{2a}{3} = 3$ b) $5 = \frac{3b}{2}$
c) $\frac{4c}{5} = 2$ d) $7 = \frac{5d}{8}$
e) $1 + \frac{3e}{8} = -5$ f) $2 = \frac{5f}{7} - 8$

8. a) $\frac{a+3}{2} = 4$ b) $\frac{b+5}{3} = 2$

c) $5 = \frac{c-2}{3}$ d) $2 = \frac{d-5}{3}$

e) $3 = \frac{2e-1}{5}$ f) $6 = \frac{4e-2}{5}$

9. a) $3(a+1) = 9$ b) $5(b-2) = 25$

c) $8 = 2(c-3)$ d) $14 = 4(3-d)$

e) $21 = 3(5-e)$ f) $36 = 9(5-2f)$

10. a) $\frac{a+2}{3} = \frac{a-3}{2}$ b) $\frac{b-1}{4} = \frac{b+5}{3}$

c) $\frac{2-c}{5} = \frac{7-c}{4}$ d) $\frac{8+d}{7} = \frac{7+d}{6}$

e) $\frac{3-e}{4} = \frac{5-e}{2}$ f) $\frac{10+f}{3} = \frac{5-f}{2}$

Exercise 19.2

Solve the following linear equations:

1. a) $3x = 2x - 4$ b) $5y = 3y + 10$

c) $2y - 5 = 3y$ d) $p - 8 = 3p$

e) $3y - 8 = 2y$ f) $7x + 11 = 5x$

2. a) $3x - 9 = 4$ b) $4 = 3x - 11$

c) $6x - 15 = 3x + 3$ d) $4y + 5 = 3y - 3$

e) $8y - 31 = 13 - 3y$ f) $4m + 2 = 5m - 8$

3. a) $7m - 1 = 5m + 1$ b) $5p - 3 = 3 + 3p$

c) $12 - 2k = 16 + 2k$ d) $6x + 9 = 3x - 54$

e) $8 - 3x = 18 - 8x$ f) $2 - y = y - 4$

4. a) $\frac{x}{2} = 3$ b $\frac{1}{2}y = 7$

c) $\frac{x}{4} = 1$ d) $\frac{1}{4}m = 3$

5. a) $\frac{x}{3} - 1 = 4$ b) $\frac{x}{5} + 2 = 1$

c) $\frac{2}{3}x = 5$ d) $\frac{3}{4}x = 6$

e) $\frac{1}{5}x = \frac{1}{2}$ f) $\frac{2x}{5} = 4$

6. a) $\frac{x+1}{2} = 3$ b) $4 = \frac{x-2}{3}$

c) $\frac{x-10}{3} = 4$ d) $8 = \frac{5x-1}{3}$

e) $\frac{2(x-5)}{3} = 2$ f) $\frac{3(x-2)}{4} = 4x - 8$

7. a) $6 = \frac{2(y-1)}{3}$ b) $2(x+1) = 3(x-5)$

c) $5(x-4) = 3(x+2)$ d) $\frac{3+y}{2} = \frac{y+1}{4}$

e) $\frac{7+2x}{3} = \frac{9x-1}{7}$ f) $\frac{2x+3}{4} = \frac{4x-2}{6}$

Constructing equations

In many cases, when dealing with the practical applications of mathematics, equations need to be constructed first before they can be solved. Often the information is either given within the context of a problem or in a diagram.

Worked examples

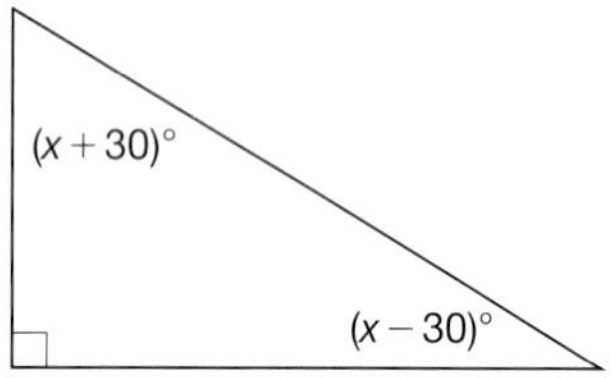

a) Find the size of each of the angles in the triangle (left) by constructing an equation and solving it to find the value of x.

The sum of the angles of a triangle is 180°.

$$\begin{aligned}(x+30)+(x-30)+90 &= 180\\ 2x+90 &= 180\\ 2x &= 90\\ x &= 45\end{aligned}$$

The three angles are therefore: 90°, $x + 30 = 75°$, $x - 30 = 15°$.
Check: 90° + 75° + 15° = 180°.

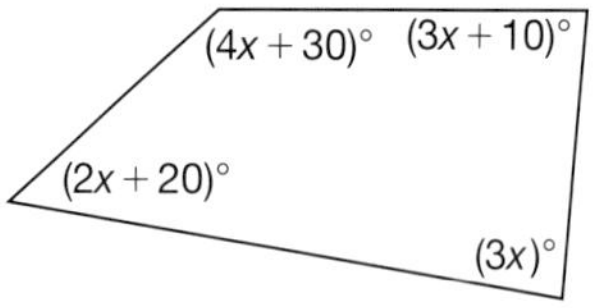

b) Find the size of each of the angles in the quadrilateral (left) by constructing an equation and solving it to find the value of x.

The sum of the angles of a quadrilateral is 360°.

$$\begin{aligned}4x+30+3x+10+3x+2x+20 &= 360\\ 12x+60 &= 360\\ 12x &= 300\\ x &= 25\end{aligned}$$

The angles are:

$$\begin{aligned}4x+30 &= (4\times 25)+30 = 130°\\ 3x+10 &= (3\times 25)+10 = 85°\\ 3x &= 3\times 25 = 75°\\ 2x+20 &= (2\times 25)+20 = \underline{70°}\\ \text{Total} &= 360°\end{aligned}$$

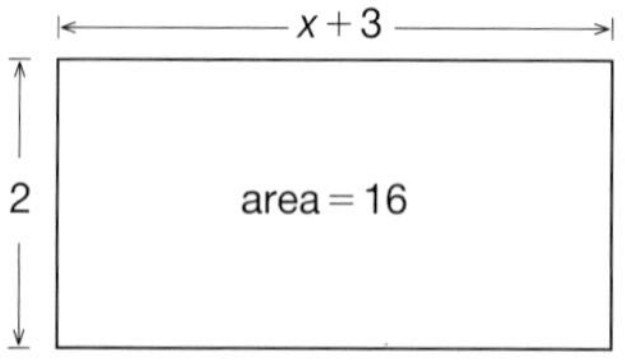

c) Construct an equation and solve it to find the value of x in the diagram (left).

Area of rectangle = base × height

$$\begin{aligned}2(x+3) &= 16\\ 2x+6 &= 16\\ 2x &= 10\\ x &= 5\end{aligned}$$

Exercise 19.3

In Q.1–3:

i) construct an equation in terms of x,
ii) solve the equation,
iii) calculate the value of each of the angles,
iv) check your answers.

1. a)

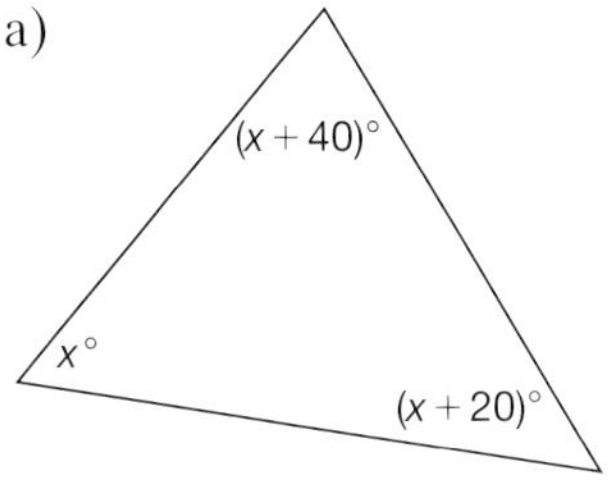

b)

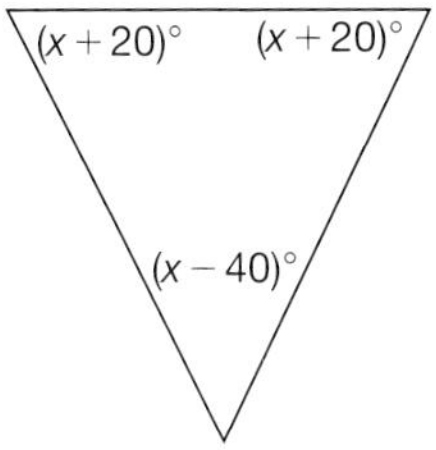

c)

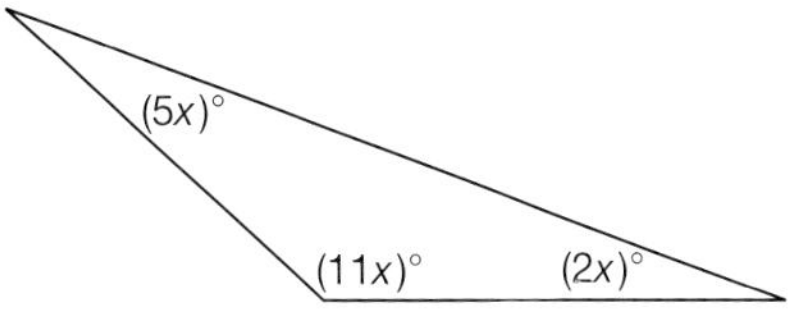

d)

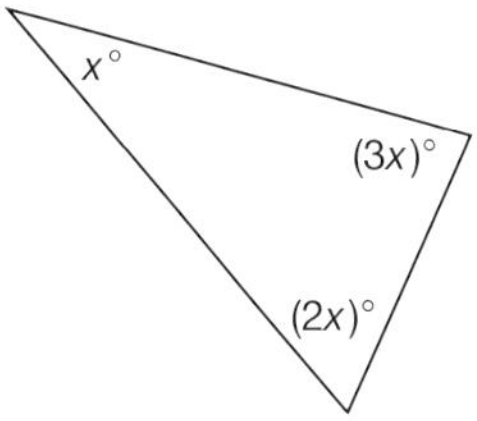

e)

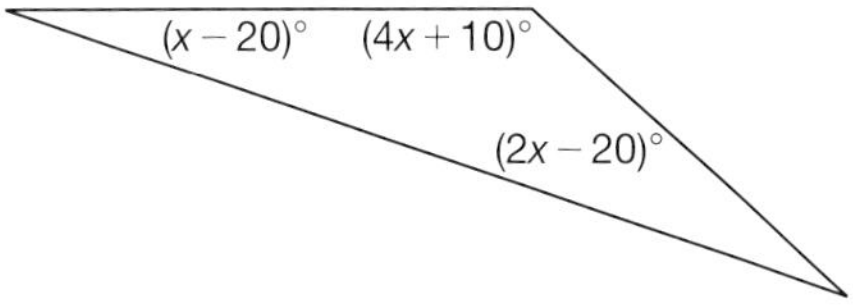

f)

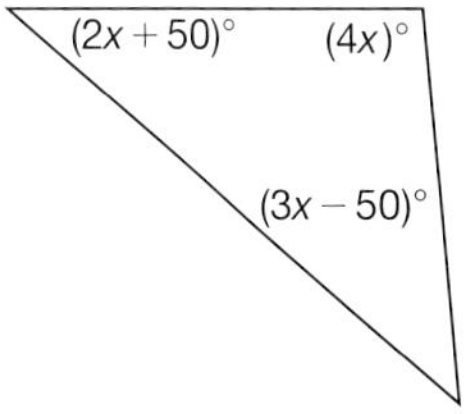

2. a)

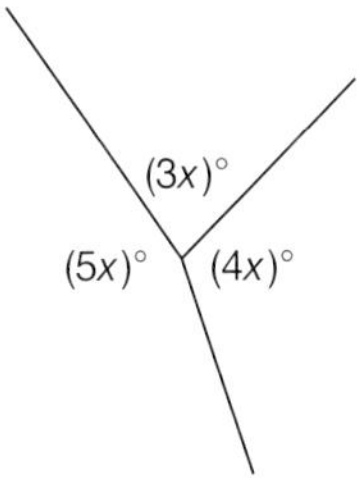

b)

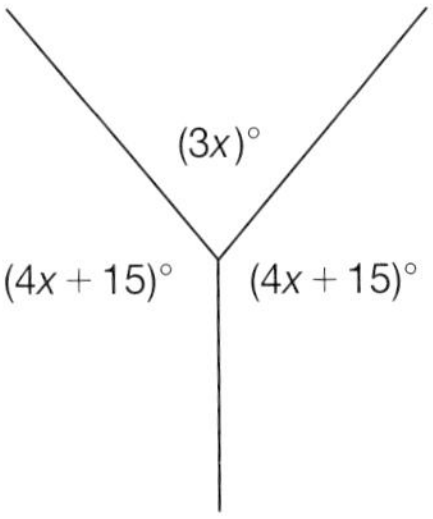

c)

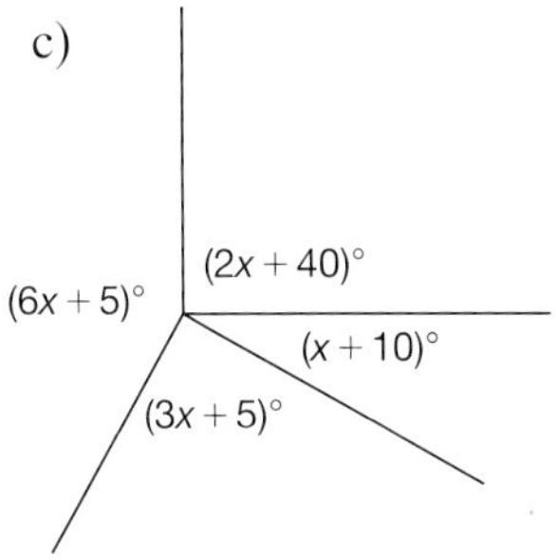

d)

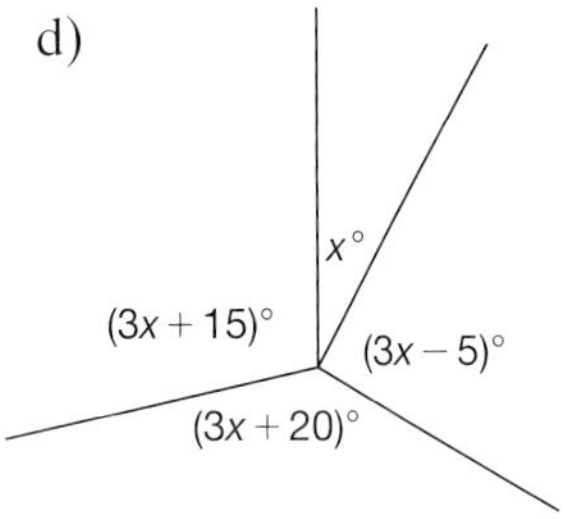

3. a)

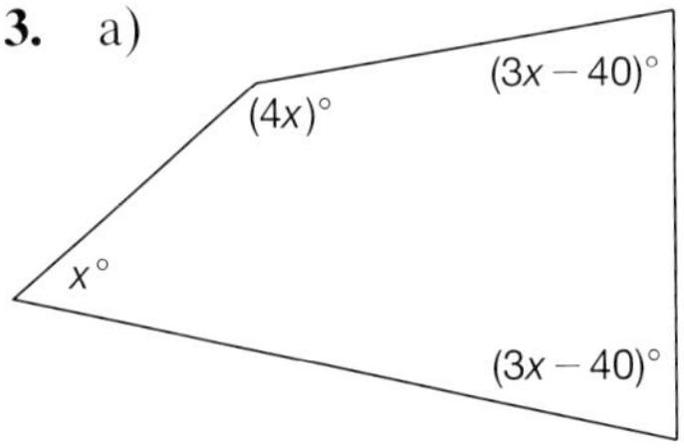

b)

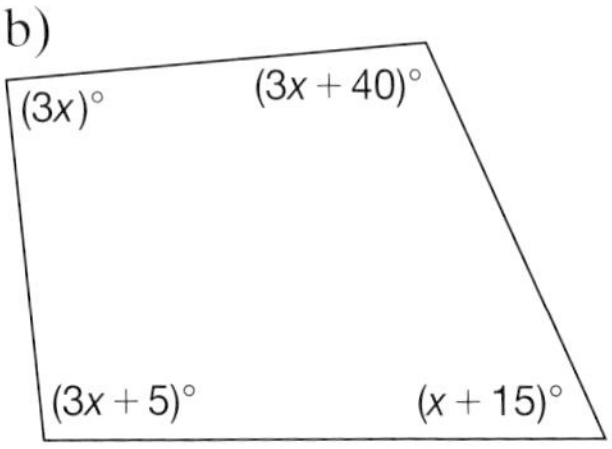

c)

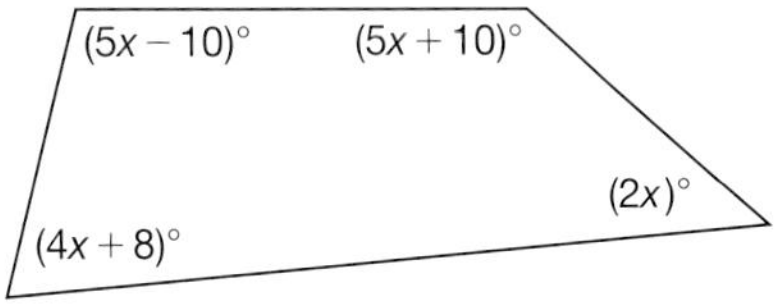

d)

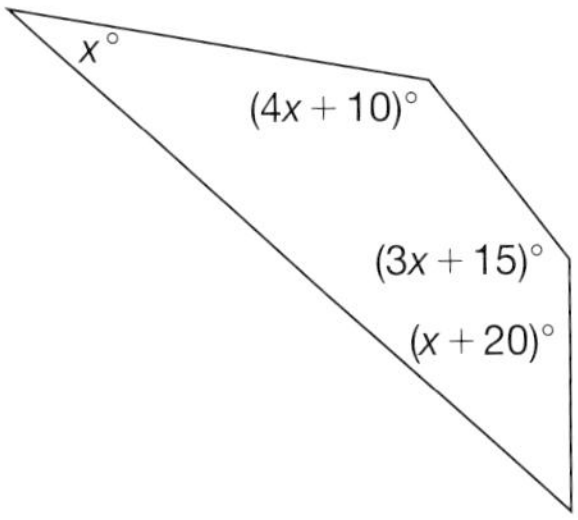

e)

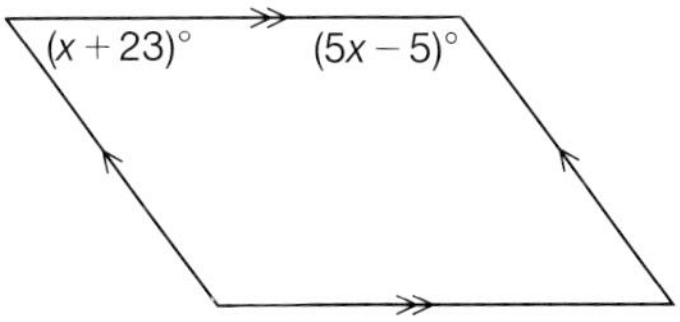

f)

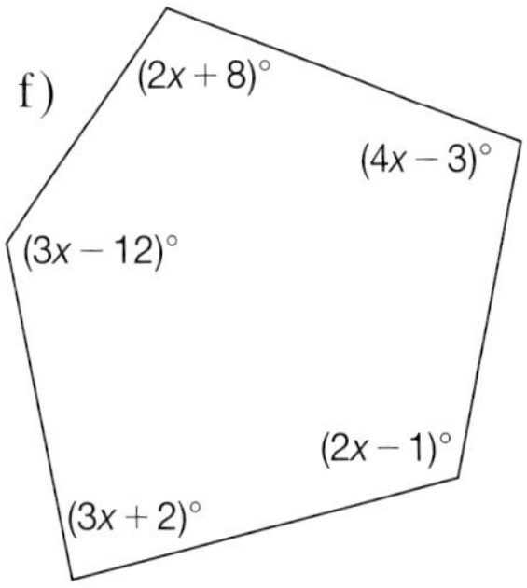

4. By constructing an equation and solving it, find the value of x in each of these isosceles triangles:

a)
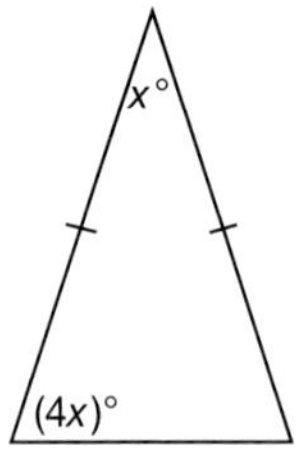

b)
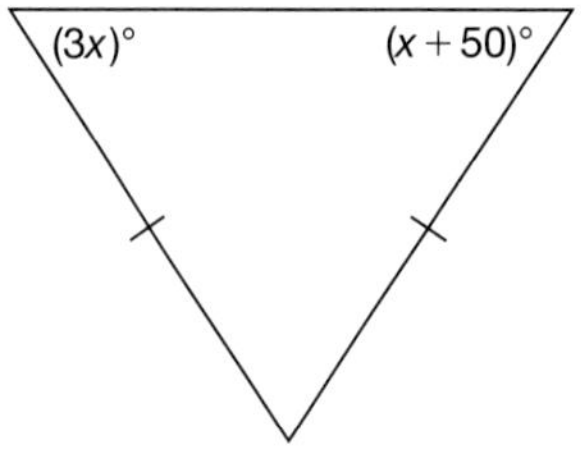

c)
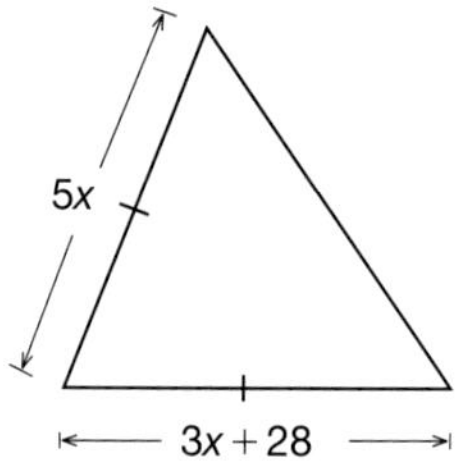

d)
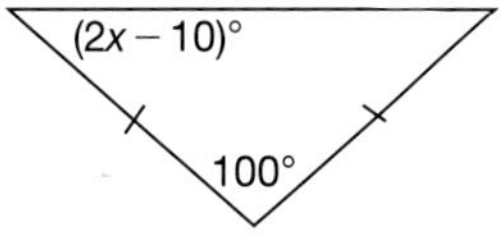

e)
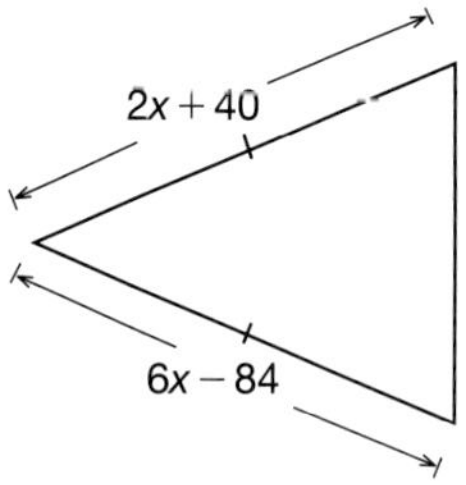

f)
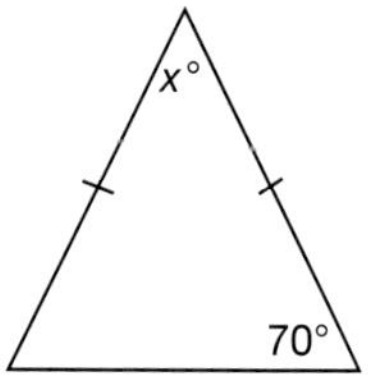

5. Using angle properties, calculate the value of x in each of these questions:

a)
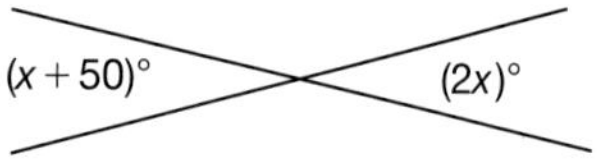

b)
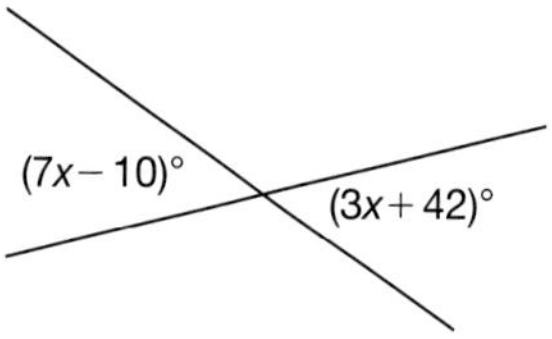

c)
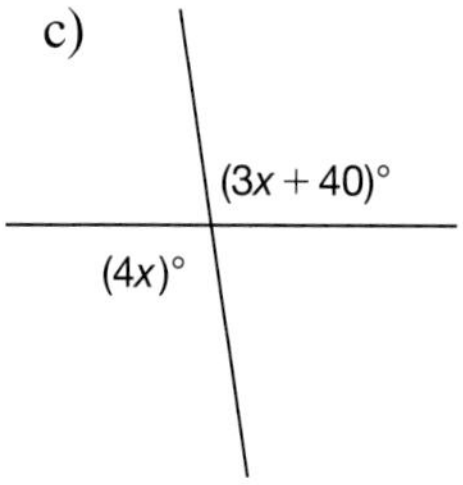

d)
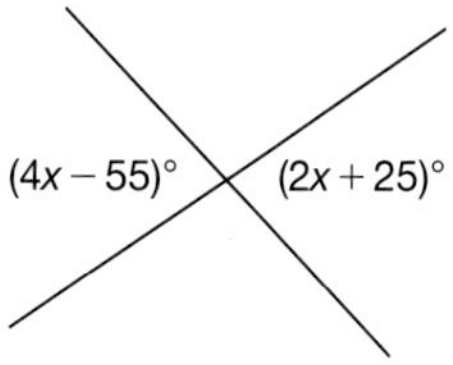

6. Calculate the value of x:

a)
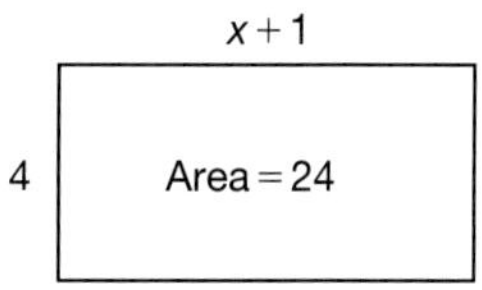

b)
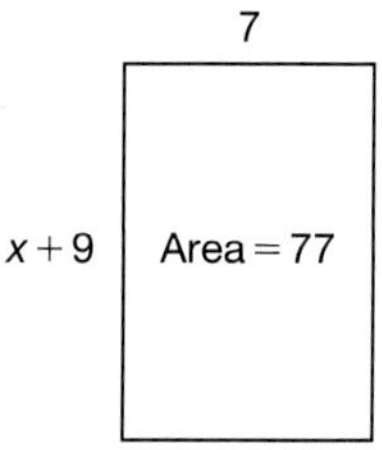

c)
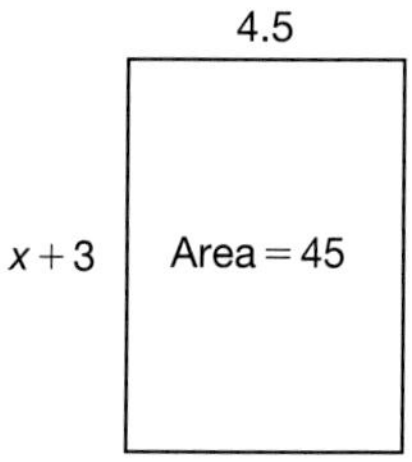

d)
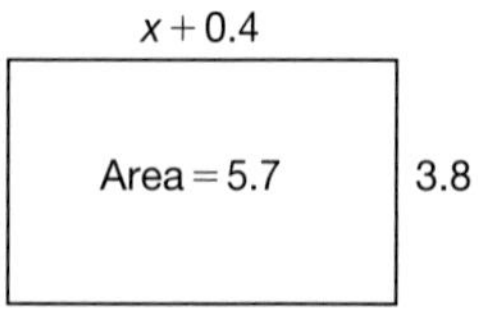

e)
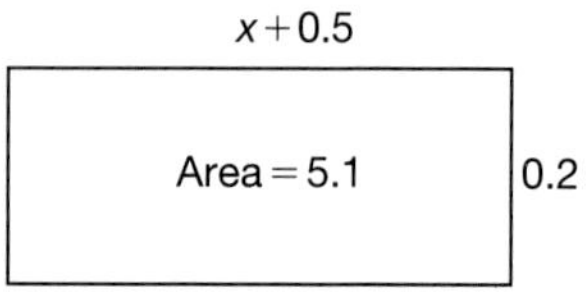

f)
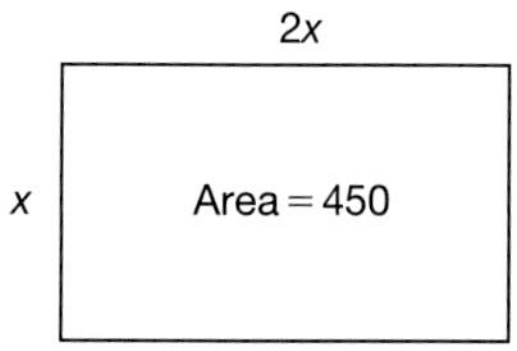

Sometimes the question is put into words and an equation has to be formed from those words.

Worked examples

If I multiply a number by 4 and then add 6 the total is 26. What is the number?

Let the number be n

Then
$$\begin{aligned} 4n + 6 &= 26 \\ 4n &= 26 - 6 \\ 4n &= 20 \\ n &= 5 \end{aligned}$$

So the number is 5.

Find the number in each case by forming an equation.

7. a) If I multiply a number by 7 and add 1 the total is 22.
b) If I multiply a number by 9 and add 7 the total is 70.

8. a) If I multiply a number by 8 and add 12 the total is 92.
b) If I add 2 to a number and then multiply by 3 the answer is 18.

9. a) If I add 12 to a number and then multiply by 5 the answer is 100.
b) If I add 6 to a number and divide it by 4 the answer is 5.

10. a) If I add 3 to a number and multiply by 4 the answer is the same as multiplying by 6 and adding 8.
b) If I add 1 to a number and then multiply by 3 it is the same as multiplying by 5 and substracting 3.

Simultaneous equations

When the values of two unknowns are needed, two equations need to be formed and solved. The process of solving two equations and finding a common solution is known as solving equations simultaneously.

The two most common ways of solving simultaneous equations algebraically are by **elimination** and by **substitution**.

By elimination

The aim of this method is to eliminate one of the unknowns by either adding or subtracting the two equations.

Worked examples

Solve the following simultaneous equations by finding the values of x and y which satisfy both equations.

a) $3x + y = 9$ (1)
$5x - y = 7$ (2)

By adding equations (1) + (2) we eliminate the variable y:

$$\begin{aligned} 8x &= 16 \\ x &= 2 \end{aligned}$$

To find the value of y we substitute $x = 2$ into either equation (1) or (2).
Substituting $x = 2$ into equation (1):

$$3x + y = 9$$
$$6 + y = 9$$
$$y = 3$$

To check that the solution is correct, the values of x and y are substituted into equation (2). If they are correct then the left-hand side of the equation will equal the right-hand side.

$$5x - y = 7$$
$$10 - 3 = 7$$
$$7 = 7$$

b) $4x + y = 23$ (1)
$x + y = 8$ (2)

By subtracting the equations, i.e. (1) − (2), we eliminate the variable y:

$$3x = 15$$
$$x = 5$$

By substituting $x = 5$ into equation (2), y can be calculated:

$$x + y = 8$$
$$5 + y = 8$$
$$y = 3$$

Check by substituting both values into equation (1):

$$4x + y = 23$$
$$20 + 3 = 23$$
$$23 = 23$$

By substitution

The same equations can also be solved by the method known as **substitution**.

Worked examples

a) $3x + y = 9$ (1)
$5x - y = 7$ (2)

Equation (2) can be rearranged to give: $y = 5x - 7$
This can now be substituted into equation (1):

$$3x + (5x - 7) = 9$$
$$3x + 5x - 7 = 9$$
$$8x - 7 = 9$$
$$8x = 16$$
$$x = 2$$

To find the value of y, $x = 2$ is substituted into either equation (1) or (2) as before, giving $y = 3$.

b) $4x + y = 23$ (1)
$x + y = 8$ (2)

Equation (2) can be rearranged to give $y = 8 - x$.
This can be substituted into equation (1):

$$\begin{aligned} 4x + (8 - x) &= 23 \\ 4x + 8 - x &= 23 \\ 3x + 8 &= 23 \\ 3x &= 15 \\ x &= 5 \end{aligned}$$

y can be found as before, giving the result $y = 3$.

Exercise 19.4

Solve the following simultaneous equations either by elimination or by substitution:

1. a) $x + y = 6$, $x - y = 2$ b) $x + y = 11$, $x - y - 1 = 0$ c) $x + y = 5$, $x - y = 7$
d) $2x + y = 12$, $2x - y = 8$ e) $3x + y = 17$, $3x - y = 13$ f) $5x + y = 29$, $5x - y = 11$

2. a) $3x + 2y = 13$, $4x = 2y + 8$ b) $6x + 5y = 62$, $4x - 5y = 8$ c) $x + 2y = 3$, $8x - 2y = 6$
d) $9x + 3y = 24$, $x - 3y = -14$ e) $7x - y = -3$, $4x + y = 14$ f) $3x = 5y + 14$, $6x + 5y = 58$

3. a) $2x + y = 14$, $x + y = 9$ b) $5x + 3y = 29$, $x + 3y = 13$ c) $4x + 2y = 50$, $x + 2y = 20$
d) $x + y = 10$, $3x = -y + 22$ e) $2x + 5y = 28$, $4x + 5y = 36$ f) $x + 6y = -2$, $3x + 6y = 18$

4. a) $x - y = 1$, $2x - y = 6$ b) $3x - 2y = 8$, $2x - 2y = 4$ c) $7x - 3y = 26$, $2x - 3y = 1$
d) $x = y + 7$, $3x - y = 17$ e) $8x - 2y = -2$, $3x - 2y = -7$ f) $4x - y = -9$, $7x - y = -18$

5. a) $x + y = -7$, $x - y = -3$ b) $2x + 3y = -18$, $2x = 3y + 6$
c) $5x - 3y = 9$, $2x + 3y = 19$ d) $7x + 4y = 42$, $9x - 4y = -10$
e) $4x - 4y = 0$, $8x + 4y = 12$ f) $x - 3y = -25$, $5x - 3y = -17$

6. a) $2x + 3y = 13$, $2x - 4y + 8 = 0$ b) $2x + 4y = 50$, $2x + y = 20$
c) $x + y = 10$, $3y = 22 - x$ d) $5x + 2y = 28$, $5x + 4y = 36$
e) $2x - 8y = 2$, $2x - 3y = 7$ f) $x - 4y = 9$, $x - 7y = 18$

7. a) $-4x = 4y$
$4x - 8y = 12$

b) $3x = 19 + 2y$
$-3x + 5y = 5$

c) $3x + 2y = 12$
$-3x + 9y = -12$

d) $3x + 5y = 29$
$3x + y = 13$

e) $-5x + 3y = 14$
$5x + 6y = 58$

f) $-2x + 8y = 6$
$2x = 3 - y$

If neither x nor y can be eliminated by simply adding or subtracting the two equations then it is necessary to multiply one or both of the equations. The equations are multiplied by a number in order to make the coefficients of x (or y) numerically equal.

Worked examples

a) $3x + 2y = 22$ (1)
$x + y = 9$ (2)

To eliminate y, equation (2) is multiplied by 2:

$3x + 2y = 22$ (1)
$2x + 2y = 18$ (3)

By subtracting (3) from (1), the variable y is eliminated:

$x = 4$

Substituting $x = 4$ into equation (2), we have:

$x + y = 9$
$4 + y = 9$
$y = 5$

Check by substituting both values into equation (1):

$3x + 2y = 22$
$12 + 10 = 22$
$22 = 22$

b) $5x - 3y = 1$ (1)
$3x + 4y = 18$ (2)

To eliminate the variable y, equation (1) is multiplied by 4, and equation (2) is multiplied by 3.

$20x - 12y = 4$ (3)
$9x + 12y = 54$ (4)

By adding equations (3) and (4), the variable y is eliminated:

$29x = 58$
$x = 2$

Substituting $x = 2$ into equation (2) gives:

$3x + 4y = 18$
$6 + 4y = 18$
$4y = 12$
$y = 3$

Check by substituting both values into equation (1):

$$5x - 3y = 1$$
$$10 - 9 = 1$$
$$1 = 1$$

Exercise 19.5

Solve the following simultaneous equations either by elimination or by substitution.

1. a) $2a + b = 5$, $3a - 2b = 4$
b) $3b + 2c = 18$, $2b - c = 5$
c) $4c - d = 18$, $2c + 2d = 14$
d) $d + 5e = 17$, $2d - 8e = -2$
e) $3e - f = 5$, $e + 2f = 11$
f) $f + 3g = 5$, $2f - g = 3$

2. a) $a + 2b = 8$, $3a - 5b = 9$
b) $4b - 3c = 17$, $b + 5c = 10$
c) $6c - 4d = -2$, $5c + d = 7$
d) $5d + e = 18$, $2d + 3e = 15$
e) $e + 2f = 14$, $3e - f = 7$
f) $7f - 5g = 9$, $f + g = 3$

3. a) $3a - 2b = -5$, $a + 5b = -4$
b) $b + 2c = 3$, $3b - 5c = -13$
c) $c - d = 4$, $3c + 4d = 5$
d) $2d + 3e = 2$, $3d - e = -8$
e) $e - 2f = -7$, $3e + 3f = -3$
f) $f + g = -2$, $3f - 4g = 1$

4. a) $2x + y = 7$, $3x + 2y = 12$
b) $5x + 4y = 21$, $x + 2y = 9$
c) $x + y = 7$, $3x + 4y = 23$
d) $2x - 3y = -3$, $3x + 2y = 15$
e) $4x = 4y + 8$, $x + 3y = 10$
f) $x + 5y = 11$, $2x - 2y = 10$

5. a) $x + y = 5$, $3x - 2y + 5 = 0$
b) $2x - 2y = 6$, $x - 5y = -5$
c) $2x + 3y = 15$, $2y = 15 - 3x$
d) $x - 6y = 0$, $3x - 3y = 15$
e) $2x - 5y = -11$, $3x + 4y = 18$
f) $x + y = 5$, $2x - 2y = -2$

6. a) $3y = 9 + 2x$, $3x + 2y = 6$
b) $x + 4y = 13$, $3x - 3y = 9$
c) $2x = 3y - 19$, $3x + 2y = 17$
d) $2x - 5y = -8$, $-3x - 2y = -26$
e) $5x - 2y = 0$, $2x + 5y = 29$
f) $8y = 3 - x$, $3x - 2y = 9$

7. a) $4x + 2y = 5$
$3x + 6y = 6$
b) $4x + y = 14$
$6x - 3y = 3$
c) $10x - y = -2$
$-15x + 3y = 9$
d) $-2y = 0.5 - 2x$
$6x + 3y = 6$
e) $x + 3y = 6$
$2x - 9y = 7$
f) $5x - 3y = -0.5$
$3x + 2y = 3.5$

Exercise 19.6

1. The sum of two numbers is 17 and their difference is 3. Find the two numbers by forming two equations and solving them simultaneously.

2. The difference between two numbers is 7. If their sum is 25, find the two numbers by forming two equations and solving them simultaneously.

3. Find the values of x and y.

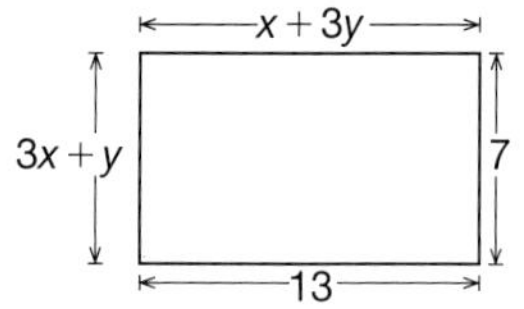

4. Find the values of x and y.

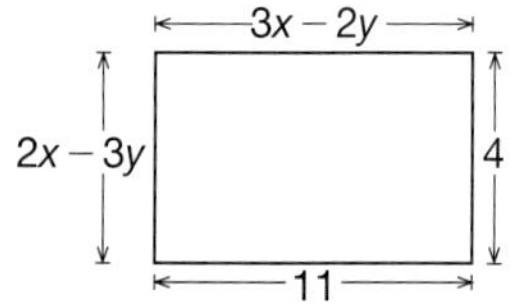

5. A man's age is three times his son's age. Ten years ago he was five times his son's age. By forming two equations and solving them simultaneously, find both of their ages.

6. A grandfather is ten times older than his granddaughter. He is also 54 years older than her. How old is each of them?

Student Assessment 1

Solve the following equations:

1. a) $a + 9 = 15$ b) $3b + 7 = -14$
c) $3 - 5c = 18$ d) $4 - 7d = -24$

2. a) $5a + 7 = 4a - 3$ b) $8 - 3b = 4 - 2b$
c) $6 - 3c = c + 8$ d) $4d - 3 = d + 9$

3. a) $\frac{a}{5} = 2$ b) $\frac{b}{7} = 3$
c) $4 = c - 2$ d) $6 = \frac{1}{3}d$

4. a) $\frac{a}{2} + 1 = 5$ b) $\frac{b}{3} - 2 = 2$
c) $7 = \frac{c}{3} - 1$ d) $1 = \frac{1}{3}d - 2$

5. a) $\frac{a - 2}{3} = \frac{a + 2}{2}$ b) $\frac{b + 5}{4} = \frac{2 + b}{3}$
c) $4(c - 5) = 3(c + 1)$ d) $6(2 + 3d) = 5(4d - 2)$

Solve the following simultaneous equations:

6. a) $a + 2b = 4$
$3a + b = 7$
b) $b - 2c = -2$
$3b + c = 15$
c) $2c - 3d = -5$
$4c + d = -3$
d) $4d + 5e = 0$
$d + e = -1$

Student Assessment 2

Solve the following equations:

1. a) $a + 5 = 7$ b) $4b - 3 = 13$
c) $2 - 3c = 11$ d) $9 - 6d = 39$

2. a) $a - 4 = 3a + 4$ b) $3 - b = 4b - 7$
c) $3c + 1 = 8 - 4c$ d) $6d - 9 = 3 - 4d$

3. a) $\frac{a}{7} = 3$ b) $\frac{b}{4} = -2$
c) $4 = \frac{c}{-3}$ d) $-5 = \frac{d}{-3}$

4. a) $\frac{1}{2}a - 2 = 3$ b) $\frac{1}{3}b = \frac{2}{3}b - 5$
c) $\frac{1}{5}c = 2 - \frac{4}{5}c$ d) $1 = \frac{3}{7}d + 2$

5. a) $\frac{a - 4}{4} = \frac{2 + a}{3}$ b) $\frac{1 + b}{5} = \frac{2 + b}{3}$
c) $2(c - 2) = 2(5c - 3)$ d) $4(d + 1) - 7(3 - d) = 5$

Solve the following simultaneous equations:

6. a) $a + 4b = 9$
$4a + b = 21$
b) $b + 2c = -3$
$4b - 3c = 10$
c) $3c + 4d = 8$
$c + 6d = 5$
d) $4d - 3e = 1$
$d + e = -5$

Student Assessment 3

Solve the following equations:

1. a) $x + 7 = 16$ b) $2x - 9 = 13$
c) $8 - 4x = 24$ d) $5 - 3x = -13$

2. a) $7 - m = 4 + m$ b) $5m - 3 = 3m + 11$
c) $6m - 1 = 9m - 13$ d) $18 - 3p = 6 + p$

3. a) $\dfrac{x}{-5} = 2$ b) $4 = \frac{1}{3}x$
c) $\dfrac{x + 2}{3} = 4$ d) $\dfrac{2x - 5}{7} = \dfrac{5}{2}$

4. a) $\frac{2}{3}(x - 4) = 8$ b) $4(x - 3) = 7(x + 2)$
c) $4 = \frac{2}{7}(3x + 8)$ d) $\frac{3}{4}(x - 1) = \frac{5}{8}(2x - 4)$

Solve the following simultaneous equations:

5. a) $2x + 3y = 16$, $2x - 3y = 4$ b) $4x + 2y = 22$, $-2x + 2y = 2$
c) $x + y = 9$, $2x + 4y = 26$ d) $2x - 3y = 7$, $-3x + 4y = -11$

Student Assessment 4

Solve the following equations:

1. a) $y + 9 = 3$ b) $3x - 5 = 13$
c) $12 - 5p = -8$ d) $2.5y + 1.5 = 7.5$

2. a) $5 - p = 4 + p$ b) $8m - 9 = 5m + 3$
c) $11p - 4 = 9p + 15$ d) $27 - 5r = r - 3$

3. a) $\dfrac{p}{-2} = -3$ b) $6 = \frac{2}{5}x$
c) $\dfrac{m - 7}{5} = 3$ d) $\dfrac{4t - 3}{3} = 7$

4. a) $\frac{2}{5}(t - 1) = 3$ b) $5(3 - m) = 4(m - 6)$
c) $5 = \frac{2}{3}(x - 1)$ d) $\frac{4}{5}(t - 2) = \frac{1}{4}(2t + 8)$

Solve the following simultaneous equations:

5. a) $x + y = 11$, $x - y = 3$ b) $5p - 3q = -1$, $-2p - 3q = -8$
c) $3x + 5y = 26$, $x - y = 6$ d) $2m - 3n = -9$, $3m + 2n = 19$

20 GRAPHS IN PRACTICAL SITUATIONS

Interpret and use graphs in practical situations including conversion graphs, distance–time graphs and travel graphs.

Conversion graphs

A straight line graph can be used to convert one set of units to another. Examples include converting from one currency to another, converting distance in miles to kilometres and converting temperature from degrees Celsius to degrees Fahrenheit.

Worked example The graph below converts US dollars into Chinese yuan based on an exchange rate of $1 = 8.80 yuan.

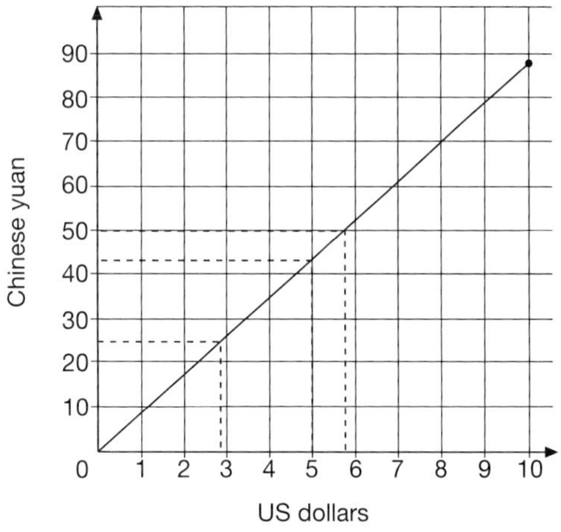

i) Using the graph estimate the number of yuan equivalent to $5.

A line is drawn up from $5 until it reaches the plotted line, then across to the *y*-axis.
From the graph it can be seen that $5 ≈ 44 yuan.
(≈ is the symbol for 'is approximately equal to')

ii) Using the graph, what would be the cost in dollars of a drink costing 25 yuan?

A line is drawn across from 25 yuan until it reaches the plotted line, then down to the *x*-axis.
From the graph it can be seen that the cost of the drink ≈ $2.80.

iii) If a meal costs 200 yuan, use the graph to estimate its cost in US dollars.

The graph does not go up to 200 yuan, therefore a factor of 200 needs to be used, e.g. 50 yuan.
From the graph 50 yuan ≈ \$5.70, therefore it can be deduced that 200 yuan ≈ \$22.80 (i.e. 4 × \$5.70).

Exercise 20.1

1. Given that 80 km = 50 miles, draw a conversion graph up to 100 km. Using your graph, estimate:
   a) how many miles is 50 km,
   b) how many kilometres is 80 miles,
   c) the speed in miles per hour (mph) equivalent to 100 km/h,
   d) the speed in km/h equivalent to 40 mph.

2. You can roughly convert temperature in degrees Celsius to degrees Fahrenheit by doubling the degrees Celsius and adding 30.
   Draw a conversion graph up to 50 °C. Use your graph to estimate the following:
   a) the temperature in °F equivalent to 25 °C,
   b) the temperature in °C equivalent to 100 °F,
   c) the temperature in °F equivalent to 0 °C,
   d) the temperature in °C equivalent to 200 °F.

3. Given that 0 °C = 32 °F and 50 °C = 122 °F, on the graph you drew for question 2, draw a true conversion graph.
   i) Use the true graph to calculate the conversions in Q.2.
   ii) Where would you say the rough conversion is most useful?

4. Long-distance calls from New York to Harare are priced at 85 cents/min off peak and \$1.20/min at peak times.
   a) Draw, on the same axes, conversion graphs for the two different rates.
   b) From your graph estimate the cost of an 8-minute call made off peak.
   c) Estimate the cost of the same call made at peak rate.
   d) A call is to be made from a telephone box. If the caller has only \$4 to spend, estimate how much more time he can talk for if he rings at off peak instead of at peak times.

5. A maths exam is marked out of 120. Draw a conversion graph and use it to change the following marks to percentages.
   a) 80
   b) 110
   c) 54
   d) 72

Speed, distance and time

Students may already be familiar with the following formulae:

distance = speed × time

Rearranging the formula gives:

$$\text{time} = \frac{\text{distance}}{\text{speed}} \text{ and } \text{speed} = \frac{\text{distance}}{\text{time}}$$

Where the speed is not constant:

$$\text{average speed} = \frac{\text{total distance}}{\text{total time}}$$

Exercise 20.2

1. Find the average speed of an object moving:
   a) 30 m in 5 s b) 48 m in 12 s
   c) 78 km in 2 h d) 50 km in 2.5 h
   e) 400 km in 2 h 30 min f) 110 km in 2 h 12 min

2. How far will an object travel during:
   a) 10 s at 40 m/s b) 7 s at 26 m/s
   c) 3 hours at 70 km/h d) 4 h 15 min at 60 km/h
   e) 10 min at 60 km/h f) 1 h 6 min at 20 m/s?

3. How long will it take to travel:
   a) 50 m at 10 m/s b) 1 km at 20 m/s
   c) 2 km at 30 km/h d) 5 km at 70 m/s
   e) 200 cm at 0.4 m/s f) 1 km at 15 km/h?

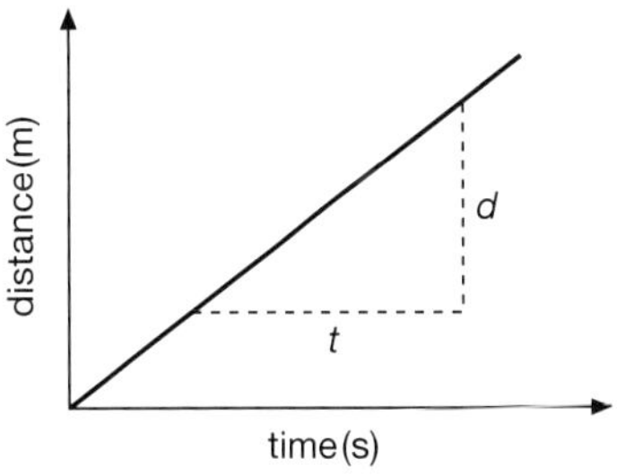

The graph of an object travelling at a constant speed is a straight line as shown (left).

$$\text{Gradient} = \frac{d}{t}$$

The units of the gradient are m/s, hence the gradient of a distance–time graph represents the speed at which the object is travelling.

Worked example

The graph (left) represents an object travelling at constant speed.

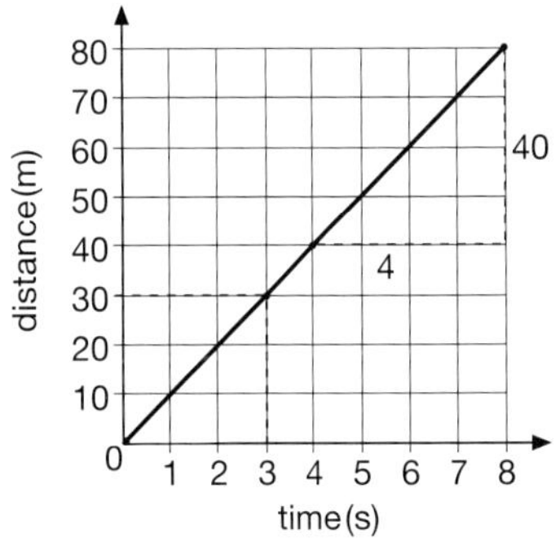

i) From the graph calculate how long it took to cover a distance of 30 m.

   The time taken to travel 30 m is 3 seconds.

ii) Calculate the gradient of the graph.

   Taking two points on the line, gradient $= \frac{40}{4} = 10$.

iii) Calculate the speed at which the object was travelling.

   Gradient of a distance–time graph = speed.
   Therefore the speed is 10 m/s.

Exercise 20.3

1. Draw a distance–time graph for the first 10 seconds of an object travelling at 6 m/s.

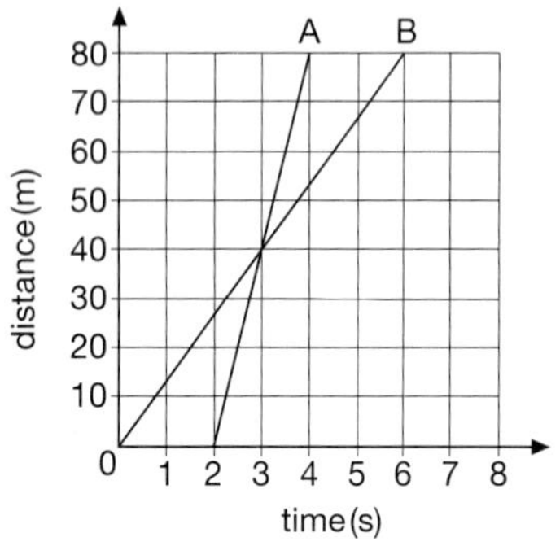

2. Draw a distance–time graph for the first 10 seconds of an object travelling at 5 m/s. Use your graph to estimate:

   a) the time taken to travel 25 m,
   b) how far the object travels in 3.5 seconds.

3. Two objects A and B set off from the same point and move in the same straight line. B sets off first, whilst A sets off 2 seconds later. Using the distance–time graph (left) estimate:
   a) the speed of each of the objects,
   b) how far apart the objects would be 20 seconds after the start.

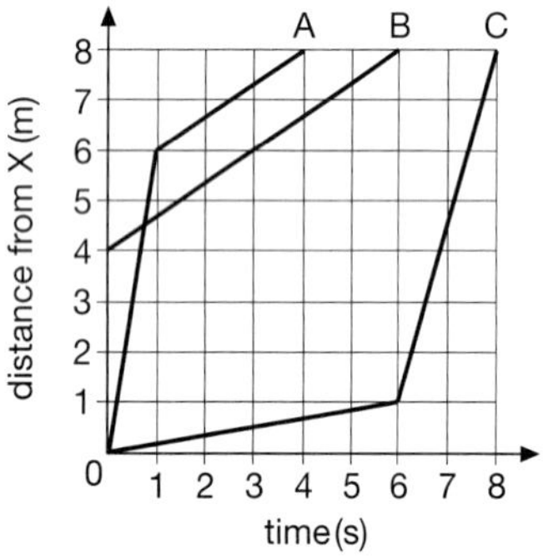

4. Three objects A, B and C move in the same straight line away from a point X. Both A and C change their speed during the journey, whilst B travels at a constant speed throughout. From the distance–time graph (left) estimate:

   a) the speed of object B,
   b) the two speeds of object A,
   c) the average speed of object C,
   d) how far object C is from X, 3 seconds from the start,
   e) how far apart objects A and C are 4 seconds from the start.

Travel graphs

The graphs of two or more journeys can be shown on the same axes. The shape of the graph gives a clear picture of the movement of each of the objects.

Worked example

Car X and Car Y both reach point B 100 km from A at 11 a.m.

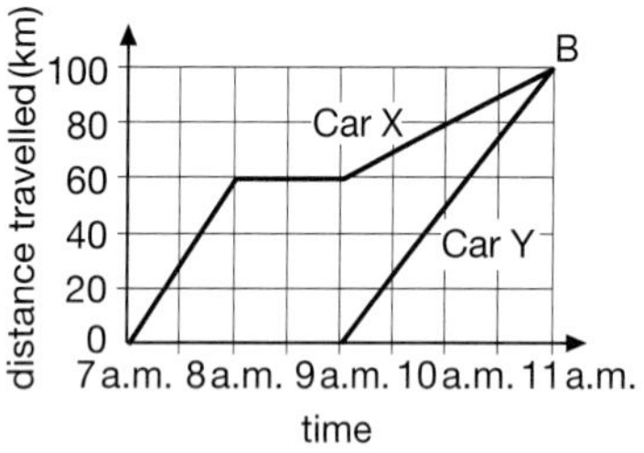

i) Calculate the speed of Car X between 7 a.m. and 8 a.m.

$$\text{speed} = \frac{\text{distance}}{\text{time}}$$

$$= \frac{60}{1}\ \text{km/h}$$

$$= 60\ \text{km/h}$$

ii) Calculate the speed of Car Y between 9 a.m. and 11 a.m.

$$\text{speed} = \frac{100}{2}\ \text{km/h}$$

$$= 50\ \text{km/h}$$

iii) Explain what is happening to Car X between 8 a.m. and 9 a.m.

No distance has been travelled, therefore Car X is stationary.

Exercise 20.4

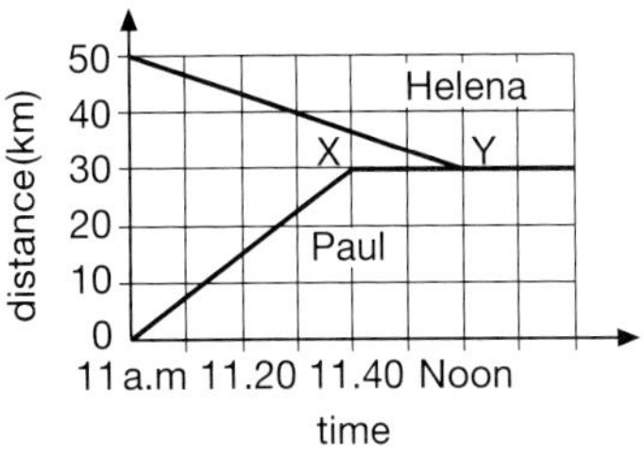

1. Two friends Paul and Helena arrange to meet for lunch at noon. They live 50 km apart and the restaurant is 30 km from Paul's home. The travel graph (left) illustrates their journeys.
   a) What is Paul's average speed between 11 a.m. and 11.40 a.m.?
   b) What is Helena's average speed between 11 a.m. and noon?
   c) What does the line XY represent?

2. A car travels at a speed of 60 km/h for 1 hour. It stops for 30 minutes, then continues at a constant speed of 80 km/h for a further 1.5 hours. Draw a distance–time graph for this journey.

3. A girl cycles for 1.5 hours at 10 km/h. She stops for an hour, then travels for a further 15 km in 1 hour. Draw a distance–time graph of the girl's journey.

4. Two friends leave their houses at 4 p.m. The houses are 4 km apart and the friends travel towards each other on the same road. Fyodor walks at 7 km/h and Yin at 5 km/h.
   a) On the same axes, draw a distance–time graph of their journeys.
   b) From your graph estimate the time at which they meet.
   c) Estimate the distance from Fyodor's house to the point where they meet.

5. A train leaves a station P at 6 p.m. and travels to station Q 150 km away. It travels at a steady speed of 75 km/h. At 6.10 p.m. another train leaves Q for P at a steady speed of 100 km/h.
   a) On the same axes draw a distance–time graph to show both journeys.
   b) From the graph estimate the time at which the trains pass each other.
   c) At what distance from station Q do both trains pass each other?
   d) Which train arrives at its destination first?

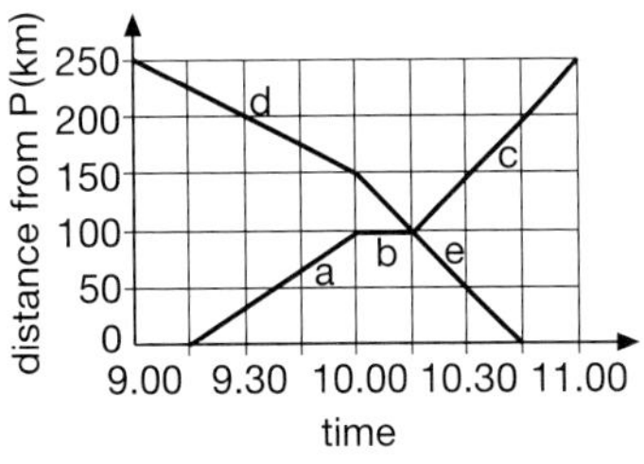

6. A train sets off from town P at 9.15 a.m. and heads towards town Q 250 km away. Its journey is split into the three stages, a, b and c. At 9.00 a.m. a second train left town Q heading for town P. Its journey was split into the two stages, d and e. Using the graph (left), calculate the following:
   a) the speed of the first train during stages a, b and c,
   b) the speed of the second train during stages d and e.

Student Assessment 1

1. In 2002, 1 euro had an exchange rate of 9.5 South African rand and 1.7 Singapore dollars.
   a) Draw a conversion graph for rand to Singapore dollars up to 80 rand.
   b) Estimate from your graph how many dollars you would get for 50 rand.
   c) Estimate from your graph how many rand you would get for 11.5 dollars.

2. A South African taxi driver has a fixed charge of 20 rand and then charges 6 rand per km.
   a) Draw a conversion graph to enable you to estimate the cost of the following taxi rides:
      i) 5 km,
      ii) 8.5 km.
   b) If a trip cost 80 rand, estimate from your graph the distance travelled.

3. An electricity account can work in two ways:
   - account A which involves a fixed charge of \$5 and then a rate of 7c per unit,
   - account B which involves no fixed charge but a rate of 9.5c per unit.

   a) On the same axes draw a graph up to 400 units for each type of account, converting units used to cost.
   b) Use your graph to advise a customer on which account to use.

4. A car travels at 60 km/h for 1 hour. The driver then takes a 30-minute break. After her break, she continues at 80 km/h for 90 minutes.
   a) Draw a distance–time graph for her journey.
   b) Calculate the total distance travelled.

5. Two trains depart at the same time from cities M and N, which are 200 km apart. One train travels from M to N, the other from N to M. The train departing from M travels a distance of 60 km in the first hour, 120 km in the next 1.5 hours and then the rest of the journey at 40 km/h. The train departing from N travels the whole distance at a speed of 100 km/h. Assuming all speeds are constant:
   a) draw a travel graph to show both journeys,
   b) estimate how far from city M the trains are when they pass each other,
   c) estimate how long after the start of the journey the trains pass each other.

Student Assessment 2

1. Absolute zero, 0 degrees Kelvin (K), is equivalent to −273 °C. 0 °C is equivalent to 273 K. Draw a conversion graph which will convert K into °C. Use your graph to estimate the following:
   a) the temperature in K equivalent to −40 °C,
   b) the temperature in °C equivalent to 100 K.

2. A German plumber has a call-out charge of 70 euro and then charges a rate of 50 euro per hour.
   a) Draw a conversion graph and estimate the cost of the following:
      i) a job lasting $4\frac{1}{2}$ hours,
      ii) a job lasting $6\frac{3}{4}$ hours.
   b) If a job cost 245 euro, estimate from your graph how long it took to complete.

3. A boy lives 3.5 km from his school. He walks home at a constant speed of 9 km/h for the first 10 minutes. He then stops and talks to his friends for 5 minutes. He finally runs the rest of his journey home at a constant speed of 12 km/h.
   a) Illustrate this information on a distance–time graph.
   b) Use your graph to estimate the total time it took the boy to get home that day.

4. Below are four distance–time graphs A, B, C and D. Two of them are not possible.
   a) Which two graphs are impossible?
   b) Explain why the two you have chosen are not possible.

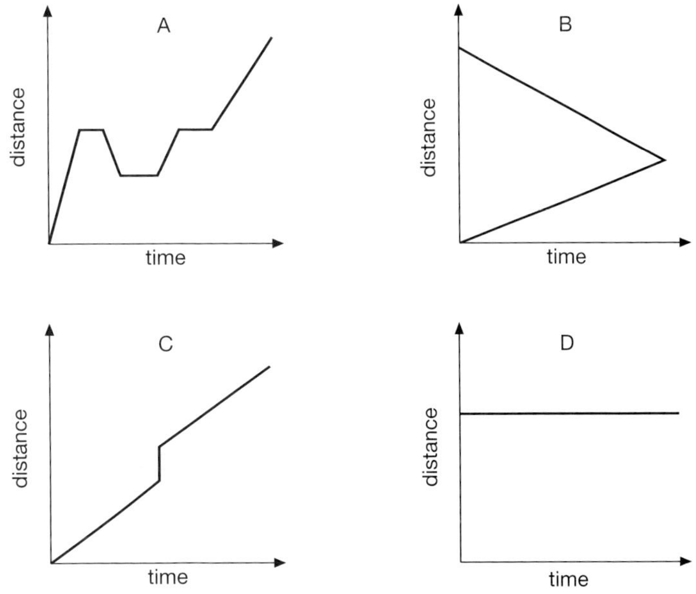

21 STRAIGHT LINE GRAPHS AND PARABOLAS

Construct tables of values for functions of the form $ax + b$; $\pm x^2 + ax + b$, a/x, where x can be positive or negative but not zero and a and b are integers; draw and interpret such graphs; find the gradient of a straight line graph; calculate the gradient of a straight line from the coordinates of two points on it; interpret and obtain the equation of a straight line graph in the form $y = mx + c$; determine the equation of a line parallel to a given line; solve linear and quadratic equations approximately by graphical methods.

Linear Function

Lines are made of an infinite number of points. This section deals with those points which form a straight line. Each point on a straight line, if plotted on a pair of axes, will have particular coordinates. The relationship between the coordinates of the points on a straight line indicates the characteristics of that straight line.

Gradient

The **gradient** of a straight line refers to its 'steepness' or 'slope'. The gradient of a straight line is **constant**, i.e. it does not change. The gradient can be calculated by considering the coordinates of any two points (x_1, y_1), (x_2, y_2) on the line.

It is calculated using the following formula:

$$\text{gradient} = \frac{\text{vertical distance between the two points}}{\text{horizontal distance between the two points}}$$

By considering the x and y coordinates of the two points this can be rewritten as:

$$\text{gradient} = \frac{y_2 - y_1}{x_2 - x_1}$$

Worked examples

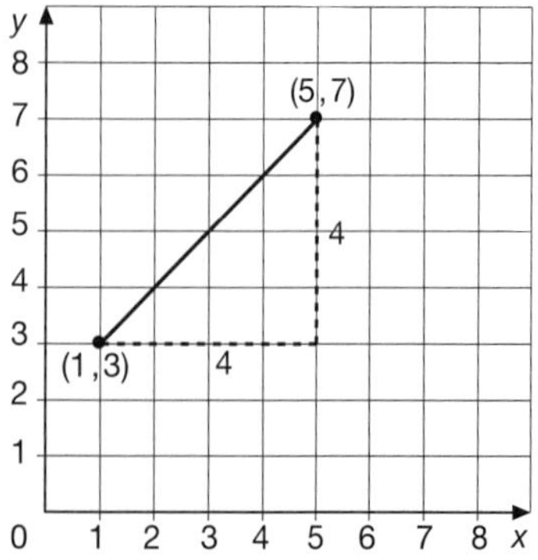

a) The coordinates of two points on a straight line are (1, 3) and (5, 7). Plot the two points on a pair of axes and calculate the gradient of the line joining them.

$$\text{gradient} = \frac{7 - 3}{5 - 1} = \frac{4}{4}$$
$$= 1$$

Note: It does not matter which point we choose to be (x_1, y_1) or (x_2, y_2) as the gradient will be the same. In the example above, choosing the other way round:

$$\text{gradient} = \frac{3 - 7}{1 - 5} = \frac{-4}{-4}$$
$$= 1$$

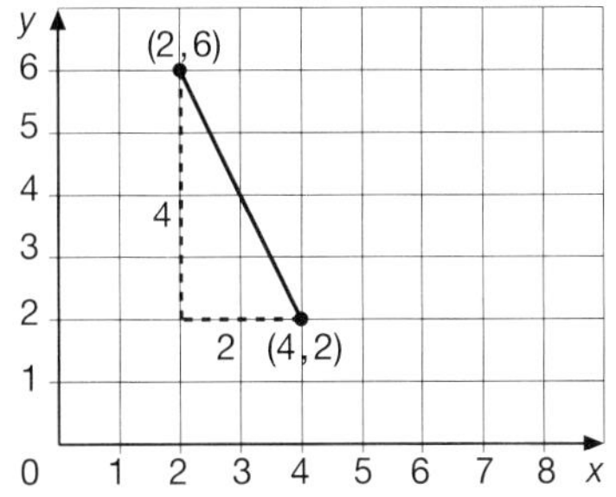

b) The coordinates of two points on a straight line are (2, 6) and (4, 2). Plot the two points on a pair of axes and calculate the gradient of the line joining them.

$$\text{gradient} = \frac{6-2}{2-4} = \frac{4}{-2} = -2$$

To check whether or not the sign of the gradient is correct the following guideline is useful:

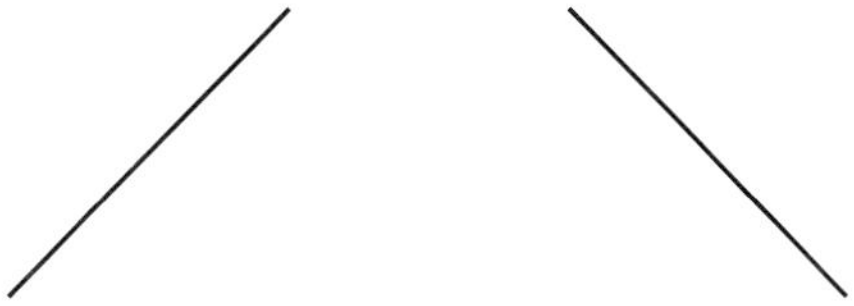

A line sloping this way will have a positive gradient

A line sloping this way will have a negative gradient

Exercise 21.1

1. With the aid of axes if necessary, calculate the gradient of the line joining the following pairs of points:

a) (5, 6) (1, 2)
b) (6, 4) (3, 1)
c) (1, 4) (5, 8)
d) (0, 0) (4, 8)
e) (2, 1) (4, 7)
f) (0, 7) (−3, 1)
g) (−3, −3) (−1, 5)
h) (4, 2) (−4, −2)
i) (−3, 5) (4, 5)
j) (2, 0) (2, 6)
k) (−4, 3) (4, 5)
l) (3, 6) (−3, −3)

2. With the aid of axes if necessary, calculate the gradient of the line joining the following pairs of points:

a) (1, 4) (4, 1)
b) (3, 6) (7, 2)
c) (2, 6) (6, −2)
d) (1, 2) (9, −2)
e) (0, 3) (−3, 6)
f) (−3, −5) (−5, −1)
g) (−2, 6) (2, 0)
h) (2, −3) (8, 1)
i) (6, 1) (−6, 4)
j) (−2, 2) (4, −4)
k) (−5, −3) (6, −3)
l) (3, 6) (5, −2)

Equations of straight lines

The coordinates of every point on a straight line all have a common relationship. This relationship when expressed algebraically as an equation in terms of x and/or y is known as the equation of the straight line.

Worked examples

a) By looking at the coordinates of some of the points on the line below, establish the equation of the straight line.

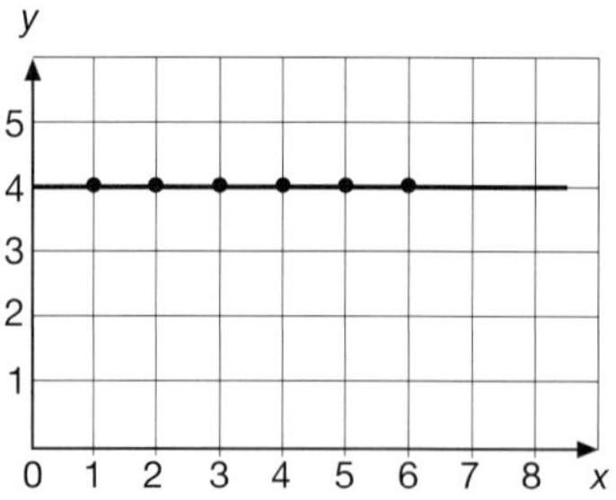

x	y
1	4
2	4
3	4
4	4
5	4
6	4

Some of the points on the line have been identified and their coordinates entered in a table above. By looking at the table it can be seen that the only rule all the points have in common is that $y = 4$.

Hence the equation of the straight line is $y = 4$.

b) By looking at the coordinates of some of the points on the line below, establish the equation of the straight line.

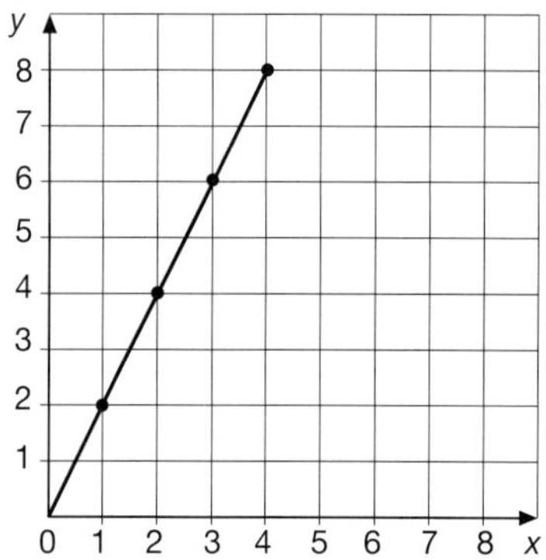

x	y
1	2
2	4
3	6
4	8

Once again by looking at the table it can be seen that the relationship between the x and y coordinates is that each y coordinate is twice the corresponding x coordinate.

Hence the equation of the straight line is $y = 2x$.

Exercise 21.2

In each of the following identify the coordinates of some of the points on the line and use these to find the equation of the straight line.

a)

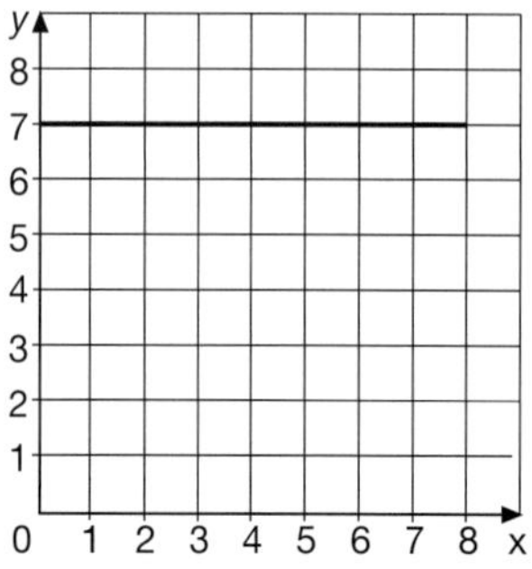

b)

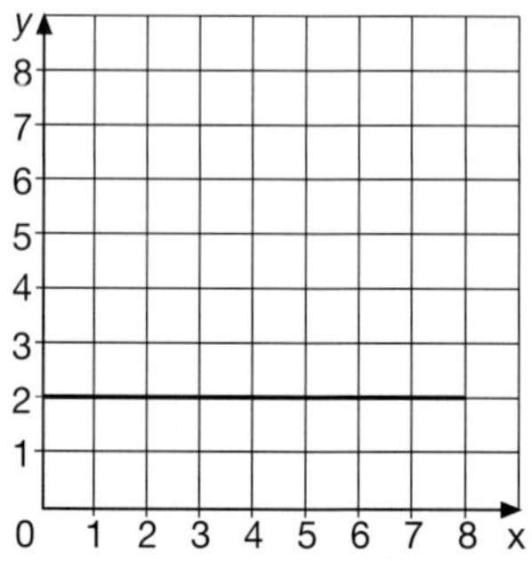

c)
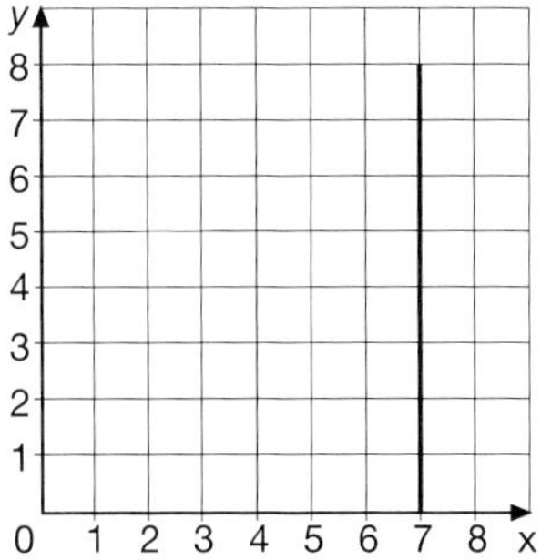

d)
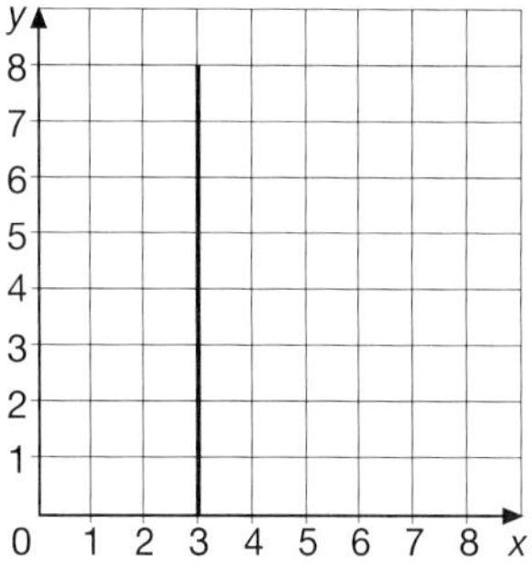

e)
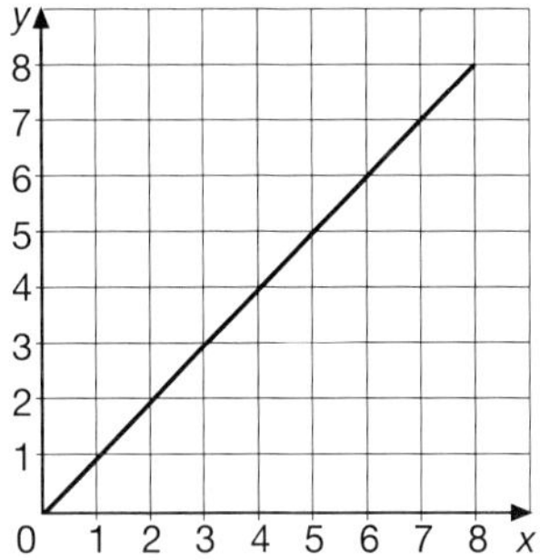

f)
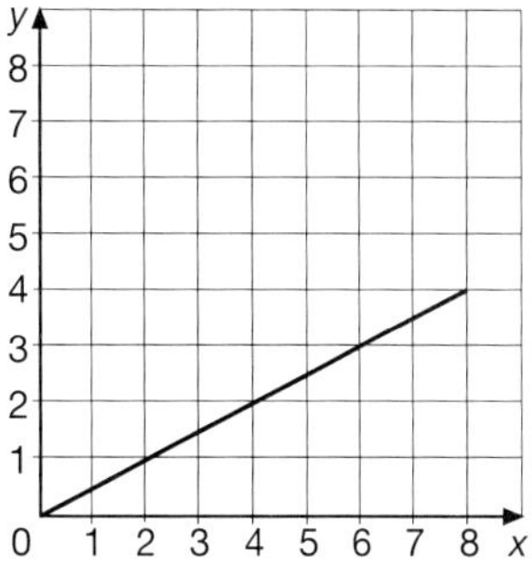

g)
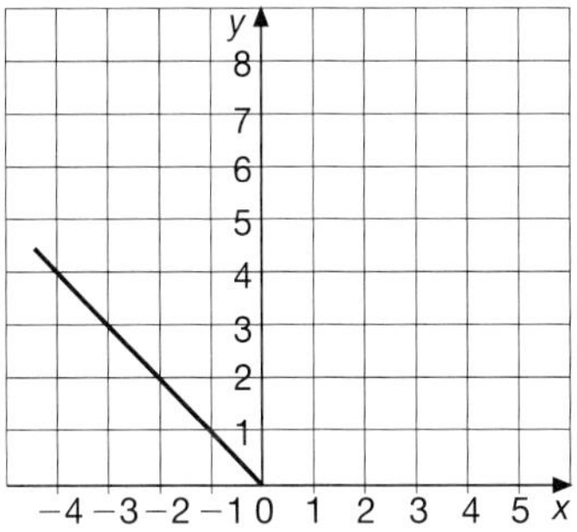

h)
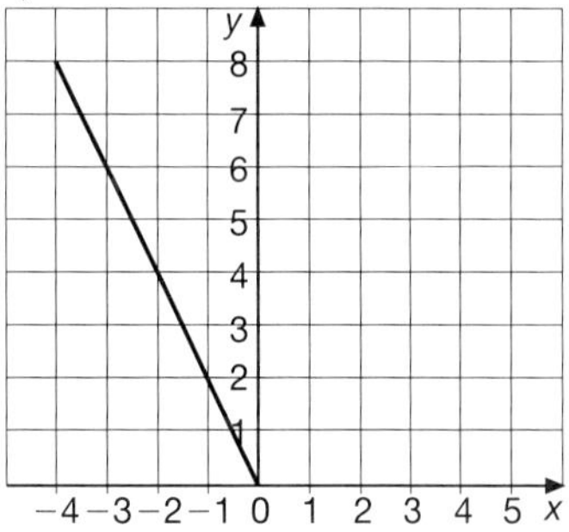

Exercise 21.3

1. In each of the following identify the coordinates of some of the points on the line and use these to find the equation of the straight line.

a)
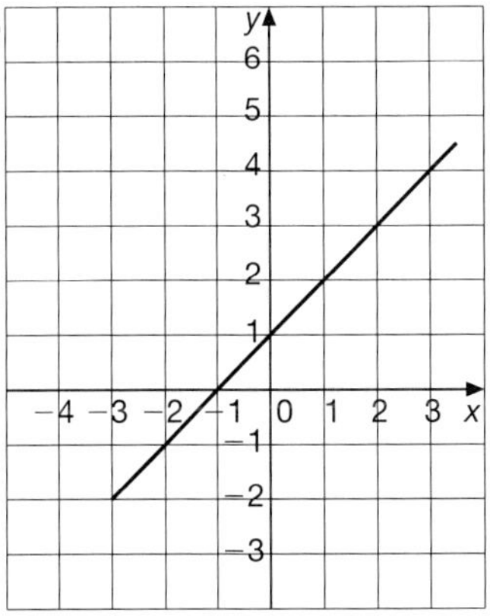

b)
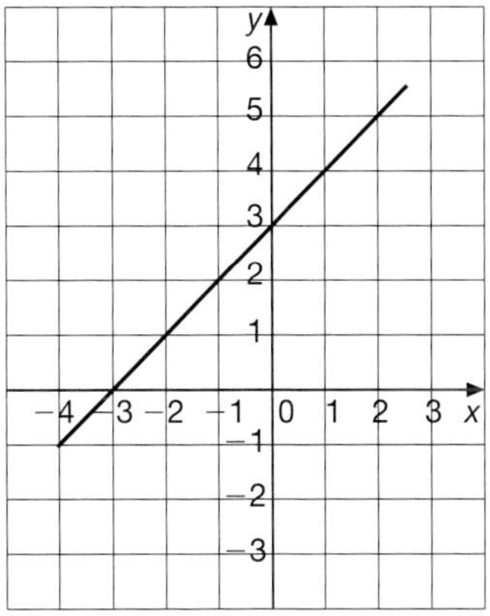

c)
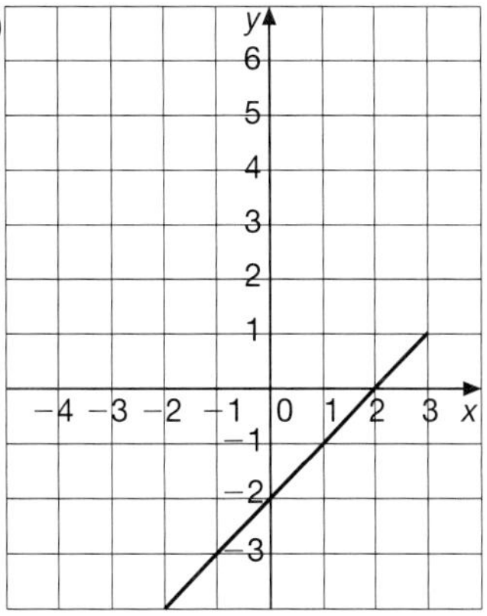

d)

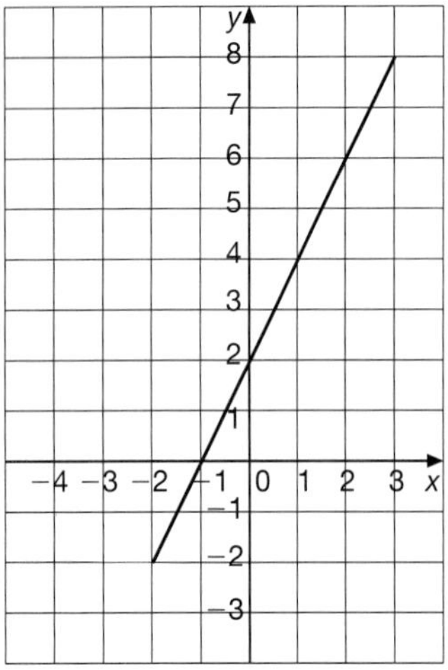

e)

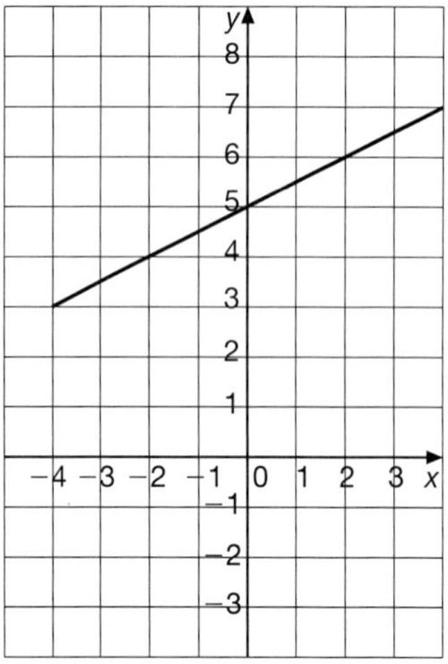

f)

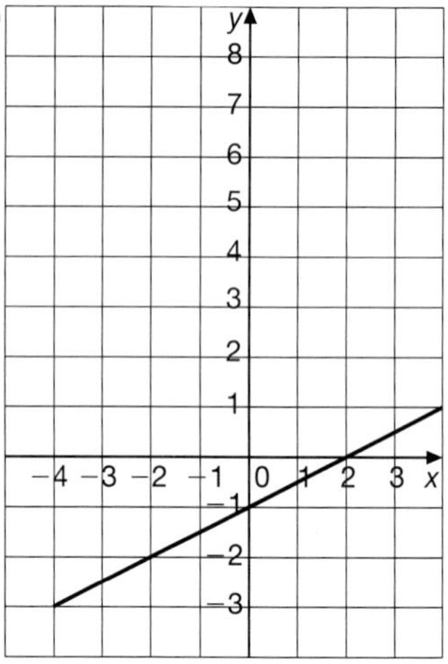

2. In each of the following identify the coordinates of some of the points on the line and use these to find the equation of the straight line.

a)

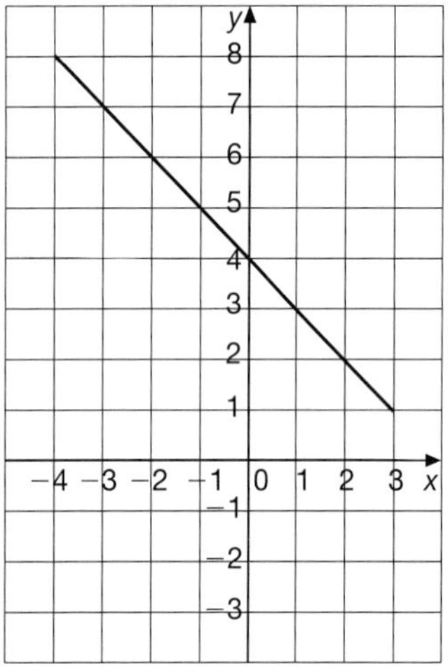

b)

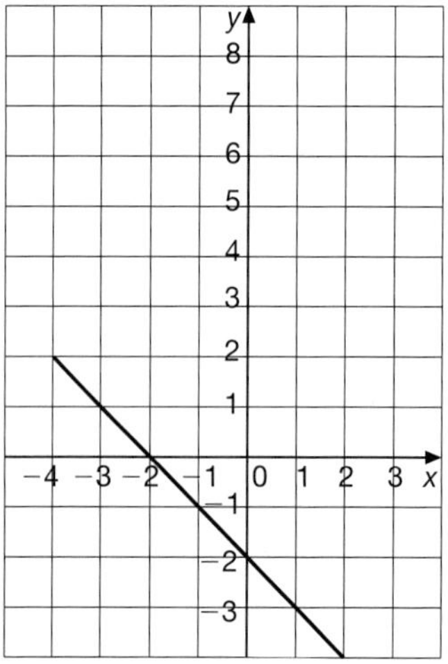

c)

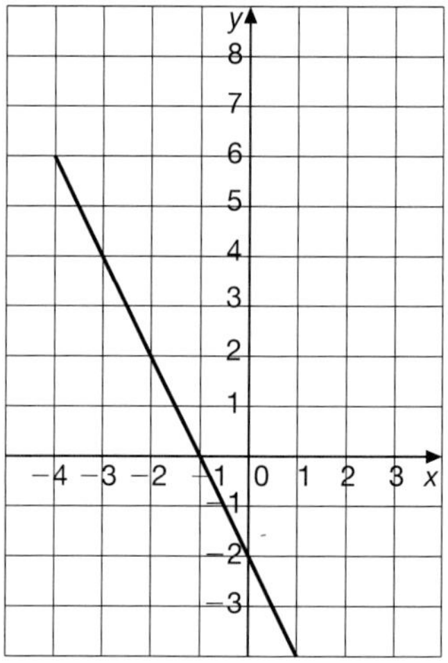

d)

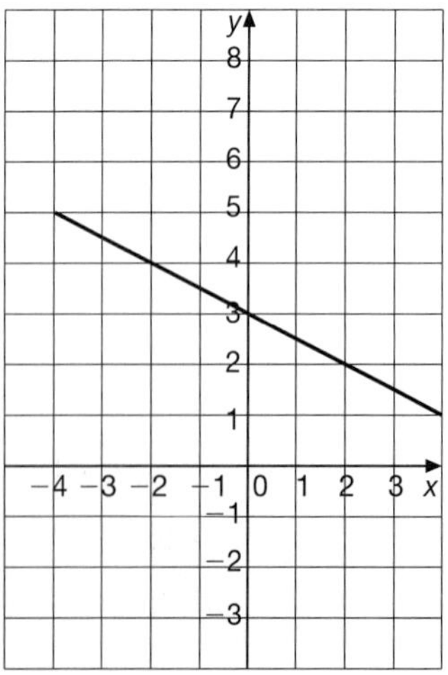

e)

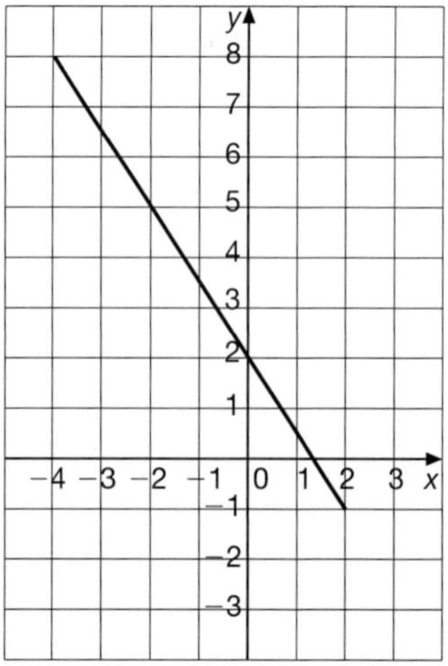

f)

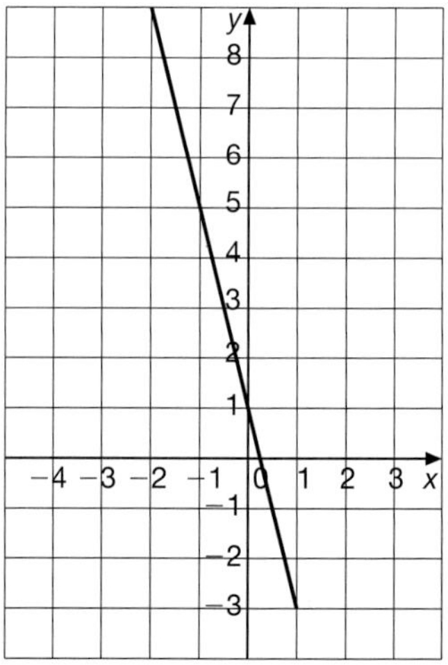

3. a) For each of the graphs in Q.1 and 2 calculate the gradient of the straight line.
b) What do you notice about the gradient of each line and its equation?
c) What do you notice about the equation of the straight line and where the line intersects the y-axis?

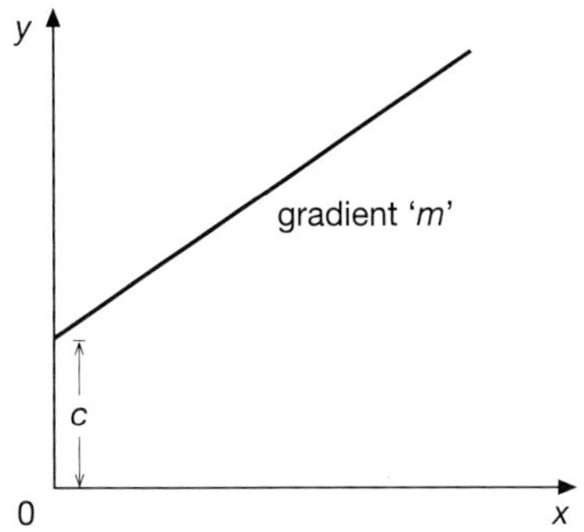

In general the equation of any straight line can be written in the form:

$$y = mx + c$$

where 'm' represents the gradient of the straight line and 'c' the intercept with the y-axis. This is shown in the diagram (left).

By looking at the equation of a straight line written in the form $y = mx + c$, it is therefore possible to deduce the line's gradient and intercept with the y-axis without having to draw it.

Worked examples

a) Calculate the gradient and y-intercept of the following straight lines:

i) $y = 3x - 2$ gradient $= 3$
y-intercept $= -2$

ii) $y = -2x + 6$ gradient $= -2$
y-intercept $= 6$

b) Calculate the gradient and y-intercept of the following straight lines:

i) $2y = 4x + 2$

This needs to be rearranged into **gradient-intercept** form (i.e. $y = mx + c$).

$y = 2x + 1$ gradient $= 2$
y-intercept $= 1$

ii) $y - 2x = -4$

Rearranging into gradient-intercept form, we have:

$y = 2x - 4$ gradient $= 2$
y-intercept $= -4$

iii) $-4y + 2x = 4$

Rearranging into gradient-intercept form, we have:

$y = \frac{1}{2}x - 1$ gradient $= \frac{1}{2}$
y-intercept $= -1$

iv) $\dfrac{y + 3}{4} = -x + 2$

Rearranging into gradient-intercept form, we have:

$y + 3 = -4x + 8$
$y = -4x + 5$ gradient $= -4$
y-intercept $= 5$

Exercise 21.4

For the following linear equations, calculate both the gradient and y-intercept in each case.

1. a) $y = 3x + 1$ b) $y = 3x + 3$ c) $y = 3x - 2$

   What can you say about the lines drawn of equations a) b) c)?

   d) $y = -2x + 4$ e) $y = -2x + 1$ f) $y = -2x - 3$

   What can you say about the lines drawn of equations d) e) f)?

   g) $y = 7 - x$ h) $y = 3 - x$ i) $y = -(3 + x)$

   What can you say about the lines drawn of equations g) h) i)?

2. a) $y = 2x + 1$ b) $y = 3x + 5$ c) $y = x - 2$
   d) $y = \frac{1}{2}x + 4$ e) $y = -3x + 6$ f) $y = -\frac{2}{3}x + 1$
   g) $y = -x$ h) $y = -x - 2$ i) $y = -(2x - 2)$

3. a) $y - 3x = 1$ b) $y + \frac{1}{2}x - 2 = 0$
   c) $y + 3 = -2x$ d) $y + 2x + 4 = 0$
   e) $y - \frac{1}{4}x - 6 = 0$ f) $-3x + y = 2$
   g) $2 + y = x$ h) $8x - 6 + y = 0$
   i) $-(3x + 1) + y = 0$

4. a) $2y = 4x - 6$ b) $2y = x + 8$
   c) $\frac{1}{2}y = x - 2$ d) $\frac{1}{4}y = -2x + 3$
   e) $3y - 6x = 0$ f) $\frac{1}{3}y + x = 1$
   g) $6y - 6 = 12x$ h) $4y - 8 + 2x = 0$
   i) $2y - (4x - 1) = 0$

The equation of a line through two points

The equation of a straight line can be deduced once the coordinates of two points on the line are known.

Worked example Calculate the equation of the straight line passing through the points $(-3, 3)$ and $(5, 5)$.

Plotting the two points gives:

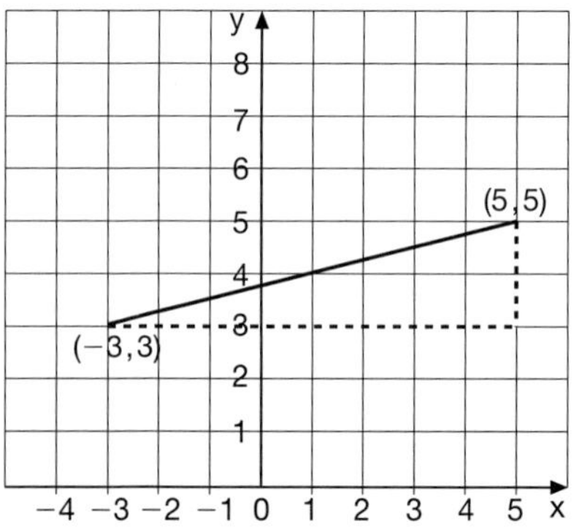

The equation of any straight line can be written in the general form $y = mx + c$. Here we have:

$$\text{gradient} = \frac{5-3}{5-(-3)} = \frac{2}{8}$$

$$\text{gradient} = \frac{1}{4}$$

The equation of the line now takes the form $y = \frac{1}{4}x + c$.

Since the line passes through the two given points, their coordinates must satisfy the equation. So to calculate the value of 'c' the x and y coordinates of one of the points are substituted into the equation. Substituting (5, 5) into the equation gives:

$$5 = \tfrac{1}{4} \times 5 + c$$
$$5 = \tfrac{5}{4} + c$$

Therefore $c = 5 - 1\frac{1}{4} = 3\frac{3}{4}$

The equation of the straight line passing through $(-3, 3)$ and $(5, 5)$ is:

$$y = \tfrac{1}{4}x + 3\tfrac{3}{4}$$

Exercise 21.5

Find the equation of the straight line which passes through each of the following pairs of points:

1. a) (1, 1) (4, 7)   b) (1, 4) (3, 10)
c) (1, 5) (2, 7)   d) (0, −4) (3, −1)
e) (1, 6) (2, 10)   f) (0, 4) (1, 3)
g) (3, −4) (10, −18)   h) (0, −1) (1, −4)
i) (0, 0) (10, 5)

2. a) (−5, 3) (2, 4)   b) (−3, −2) (4, 4)
c) (−7, −3) (−1, 6)   d) (2, 5) (1, −4)
e) (−3, 4) (5, 0)   f) (6, 4) (−7, 7)
g) (−5, 2) (6, 2)   h) (1, −3) (−2, 6)
i) (6, −4) (6, 6)

Drawing straight line graphs

To draw a straight line graph only two points need to be known. Once these have been plotted the line can be drawn between them and extended if necessary at both ends.

Worked examples

a) Plot the line $y = x + 3$.

To identify two points simply choose two values of x. Substitute these into the equation and calculate their corresponding y values.

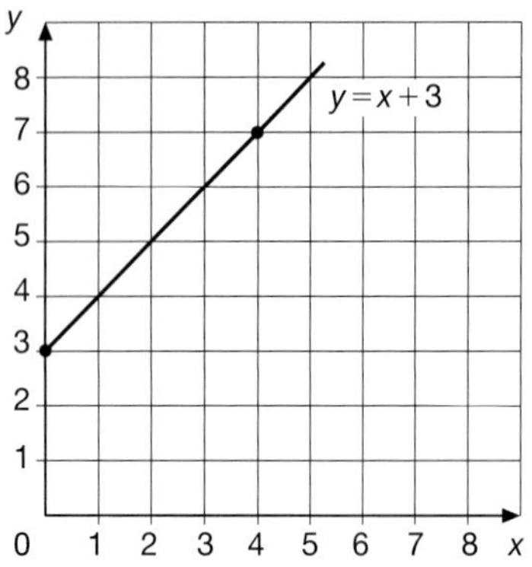

When $x = 0$, $y = 3$.
When $x = 4$, $y = 7$.

Therefore two of the points on the line are (0, 3) and (4, 7). The straight line $y = x + 3$ is plotted on the left.

b) Plot the line $y = -2x + 4$.

When $x = 2$, $y = 0$.
When $x = -1$, $y = 6$.

The coordinates of two points on the line are (2, 0) and (−1, 6).

Note that, in questions of this sort, it is often easier to rearrange the equation into gradient-intercept form first.

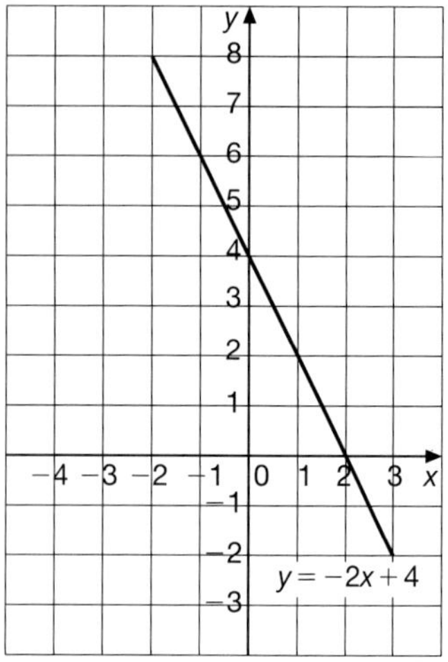

Exercise 21.6

1. Plot the following straight lines:
   a) $y = 2x + 4$ b) $y = 2x + 3$ c) $y = 2x - 1$
   d) $y = x - 4$ e) $y = x + 1$ f) $y = x + 3$
   g) $y = 1 - x$ h) $y = 3 - x$ i) $y = -(x + 2)$
2. Plot the following straight lines:
   a) $y = 2x + 3$ b) $y = x - 4$ c) $y = 3x - 2$
   d) $y = -2x$ e) $y = -x - 1$ f) $-y = x + 1$
   g) $-y = 3x - 3$ h) $2y = 4x - 2$ i) $y - 4 = 3x$
3. Plot the following straight lines:
   a) $-2x + y = 4$ b) $-4x + 2y = 12$
   c) $3y = 6x - 3$ d) $2x = x + 1$
   e) $3y - 6x = 9$ f) $2y + x = 8$
   g) $x + y + 2 = 0$ h) $3x + 2y - 4 = 0$
   i) $4 = 4y - 2x$

Quadratic functions

The general expression for a quadratic function takes the form $ax^2 + bx + c$ where a, b and c are constants. Some examples of quadratic functions are given below.

$$y = 2x^2 + 3x - 12 \qquad y = x^2 - 5x + 6 \qquad y = 3x^2 + 2x - 3$$

If a graph of a quadratic function is plotted, the smooth curve produced is called a **parabola**.

e.g. $y = x^2$

x	–4	–3	–2	–1	0	1	2	3	4
y	16	9	4	1	0	1	4	9	16

$y = -x^2$

x	–4	–3	–2	–1	0	1	2	3	4
y	–16	–9	–4	–1	0	–1	–4	–9	–16

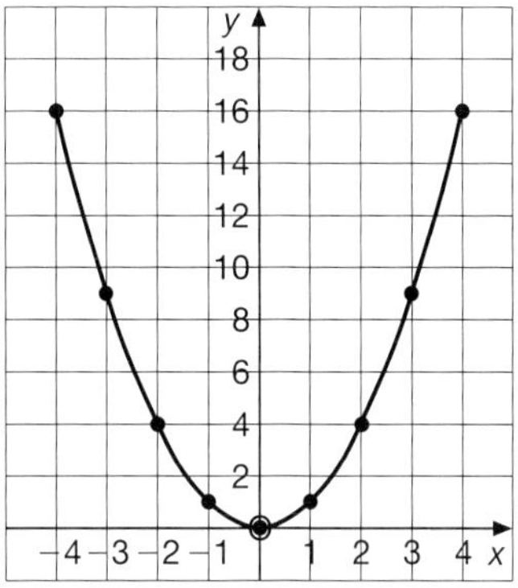

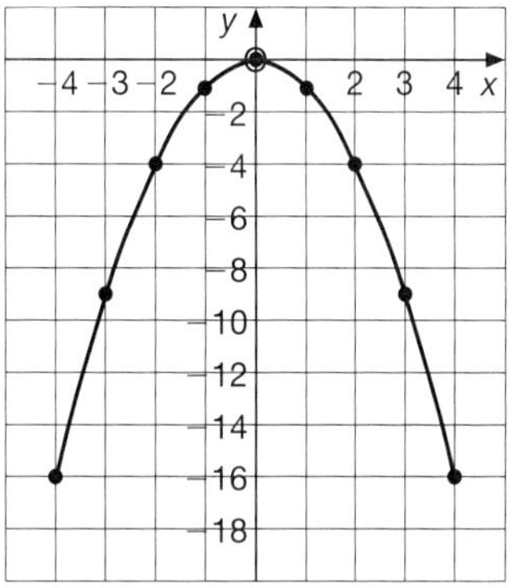

Worked examples **a)** Plot a graph of the function $y = x^2 - 5x + 6$ for $0 \leqslant x \leqslant 5$.

A table of values for x and y is given below:

x	0	1	2	3	4	5
y	6	2	0	0	2	6

This can then be plotted to give the graph:

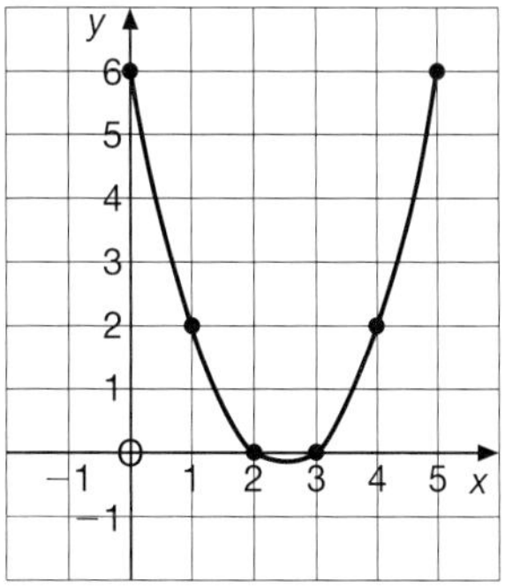

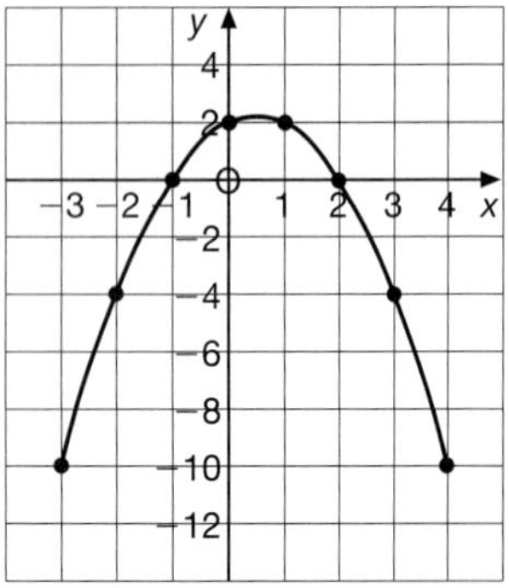

b) Plot a graph of the function $y = -x^2 + x + 2$ for $-3 \leqslant x \leqslant 4$.

Drawing up a table of values gives:

x	−3	−2	−1	0	1	2	3	4
y	−10	−4	0	2	2	0	−4	−10

The graph of the function is given on the left.

Exercise 21.7

For each of the following quadratic functions, construct a table of values and then draw the graph.

1. $y = x^2 + x - 2, \qquad -4 \leqslant x \leqslant 3$
2. $y = -x^2 + 2x + 3, \qquad -3 \leqslant x \leqslant 5$
3. $y = x^2 - 4x + 4, \qquad -1 \leqslant x \leqslant 5$
4. $y = -x^2 - 2x - 1, \qquad -4 \leqslant x \leqslant 2$
5. $y = x^2 - 2x - 15, \qquad -4 \leqslant x \leqslant 6$

Graphical solution of a quadratic equation

Worked example

i) Draw a graph of $y = x^2 - 4x + 3$ for $-2 \leqslant x \leqslant 5$.

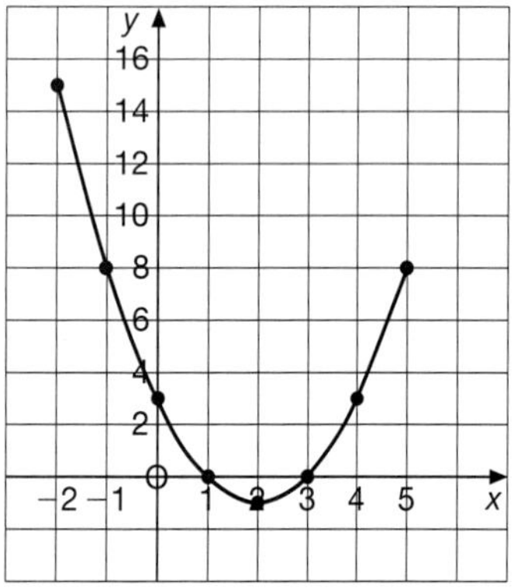

x	−2	−1	0	1	2	3	4	5
y	15	8	3	0	−1	0	3	8

ii) Use the graph to solve the equation $x^2 - 4x + 3 = 0$.

To solve the equation it is necessary to find the values of x when $y = 0$, i.e. where the graph crosses the x-axis. These points occur when $x = 1$ and $x = 3$, therefore these are the solutions.

Exercise 21.8

Solve each of the quadratic functions below by plotting a graph for the ranges of x stated.

1. $x^2 - x - 6 = 0, \qquad -4 \leqslant x \leqslant 4$
2. $-x^2 + 1 = 0, \qquad -4 \leqslant x \leqslant 4$
3. $x^2 - 6x + 9 = 0, \qquad 0 \leqslant x \leqslant 6$
4. $-x^2 - x + 12 = 0, \qquad -5 \leqslant x \leqslant 4$
5. $x^2 - 4x + 4 = 0, \qquad -2 \leqslant x \leqslant 6$

In the previous worked example, as $y = x^2 - 4x + 3$, a solution could be found to the equation $x^2 - 4x + 3 = 0$ by reading off where the graph crossed the x-axis. The graph can, however, also be used to solve other quadratic equations.

Worked example Use the graph of $y = x^2 - 4x + 3$ to solve the equation $x^2 - 4x + 1 = 0$.

$x^2 - 4x + 1 = 0$ can be rearranged to give:

$$x^2 - 4x + 3 = 2$$

Using the graph of $y = x^2 - 4x + 3$ and plotting the line $y = 2$ on the same graph gives the graph shown below.

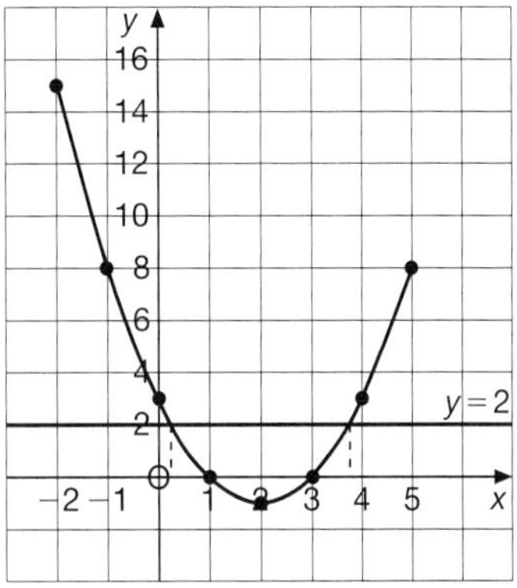

Where the curve and the line cross gives the solution to $x^2 - 4x + 3 = 2$ and hence also $x^2 - 4x + 1 = 0$.

Therefore the solutions to $x^2 - 4x + 1 = 0$ are $x \approx 0.3$ and 3.7.

Exercise 21.9

Using the graphs which you drew in Ex.21.8, solve the following quadratic equations. Show your method clearly.

1. $x^2 - x - 4 = 0$

2. $-x^2 - 1 = 0$

3. $x^2 - 6x + 8 = 0$

4. $-x^2 - x + 9 = 0$

5. $x^2 - 4x + 1 = 0$

The reciprocal function

Worked example Draw the graph of $y = \frac{2}{x}$ for $-4 \leqslant x \leqslant 4$.

x	−4	−3	−2	−1	0	1	2	3	4
y	−0.5	−0.7	−1	−2	–	2	1	0.7	0.5

This is a reciprocal function. If the graph is plotted, the curves produced are called a **hyperbola**.

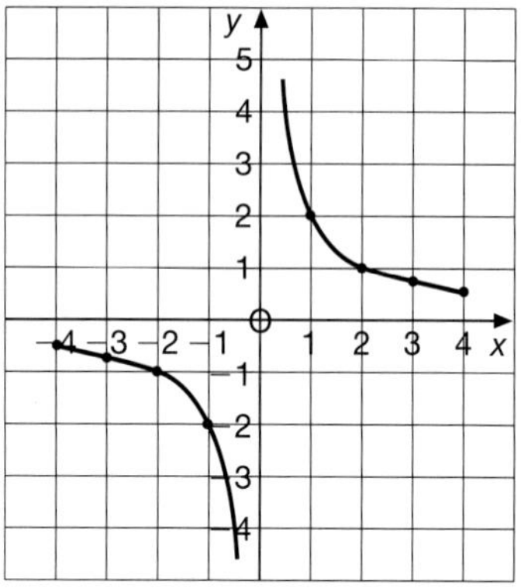

Exercise 21.10

1. Plot the graph of the function $y = \dfrac{1}{x}$ for $-4 \leqslant x \leqslant 4$.

2. Plot the graph of the function $y = \dfrac{3}{x}$ for $-4 \leqslant x \leqslant 4$.

3. Plot the graph of the function $y = \dfrac{5}{3x}$ for $-4 \leqslant x \leqslant 4$.

Student Assessment 1

1. Sketch the following lines on the same pair of axes, labelling each one clearly:
   a) $x = -2$ b) $y = 3$
   c) $y = 2x$ d) $y = -\frac{x}{2}$
2. Plot the graphs of the following linear equations:
   a) $y = x + 1$ b) $y = 3 - 3x$
3. Rearrange the following equations into the form $y = mx + c$, and then plot the graphs of the equations:
   a) $2x - y = 4$ b) $2y - 5x = 8$
4. Plot the graphs of the following functions between the given limits of x:
   a) $y = x^2 - 3$, $-4 \leqslant x \leqslant 4$
   b) $y = 3 - x^2$, $-4 \leqslant x \leqslant 4$
5. Plot the graph of $y = \frac{1}{x}$ for $-4 \leqslant x \leqslant 4$.

Student Assessment 2

1. Sketch the following lines on the same axes, labelling each one clearly:
   a) $x = 3$ b) $y = -2$
   c) $y = -3x$ d) $y = \frac{x}{4} + 4$
2. Plot the graphs of the following linear equations:
   a) $y = 2x + 3$ b) $y = 4 - x$
3. Rearrange the following equations into the form $y = mx + c$, and then plot the graphs of the equations:
   a) $2x - y = 3$ b) $-3x + 2y = 5$
4. Plot the following functions between the given limits of x:
   a) $y = x^2 - 5$, $-4 \leqslant x \leqslant 4$
   b) $y = 1 - x^2$, $-4 \leqslant x \leqslant 4$
5. Plot the graph of $y = \frac{1}{x + 2}$ for $-4 \leqslant x \leqslant 4$.

Student Assessment 3

1. Sketch the graphs of the following functions:
   a) $y = x^2$ b) $y = -x^2$
2. a) Copy and complete the table below for the function $y = x^2 + 8x + 15$.

x	−7	−6	−5	−4	−3	−2	−1	0	1	2
y		3				3				

   b) Plot a graph of the function.

3. Plot a graph of each of the functions below between the given limits of x.
   a) $y = -x^2 - 2x - 1, \quad -3 \leqslant x \leqslant 3$
   b) $y = x^2 + 2x - 7, \quad -5 \leqslant x \leqslant 3$

4. a) Plot the graph of the quadratic function $y = x^2 + 9x + 20$ for $-7 \leqslant x \leqslant -2$.
   b) Showing your method clearly, use your graph to solve the equation $x^2 = -9x - 14$.

5. a) Plot the graph of $y = \frac{1}{x}$ for $-4 \leqslant x \leqslant 4$.
   b) Showing your method clearly, use your graph to solve the equation $1 = -x^2 + 3x$.

Student Assessment 4

1. Sketch the graph of the function $y = \frac{1}{x}$.

2. a) Copy and complete the table below for the function $y = -x^2 - 7x - 12$.

x	−7	−6	−5	−4	−3	−2	−1	0	1	2
y		−6				−2				

   b) Plot a graph of the function.

3. Plot a graph of each of the functions below between the given limits of x.
   a) $y = x^2 - 3x - 10, \quad -3 \leqslant x \leqslant 6$
   b) $y = -x^2 - 4x - 4, \quad -5 \leqslant x \leqslant 1$

4. a) Plot the graph of the quadratic equation $y = -x^2 - x + 15$ for $-6 \leqslant x \leqslant 4$.
   b) Showing your method clearly, use your graph to solve the following equations:
   i) $10 = x^2 + x$ ii) $x^2 = -x + 5$

5. a) Plot the graph of $y = \frac{2}{x}$ for $-4 \leqslant x \leqslant 4$.
   b) Showing your method clearly, use your graph to solve the equation $x^2 + x = 2$.

ICT SECTION

Algebra Activity 1

You have seen that the solution of two simultaneous equations gives the coordinates of the point which satisfies both equations.

If the simultaneous equations were plotted, then the point at which the lines cross would correspond to the solution of the simultaneous equations.

e.g. Solving $x + y = 6$ and $x - y = 2$ produces the result $x = 4$ and $y = 2$, i.e. coordinates (4, 2).

A plot of both lines is shown below.
From the graph it can be seen that the point of intersection of the two lines occurs at (4, 2).

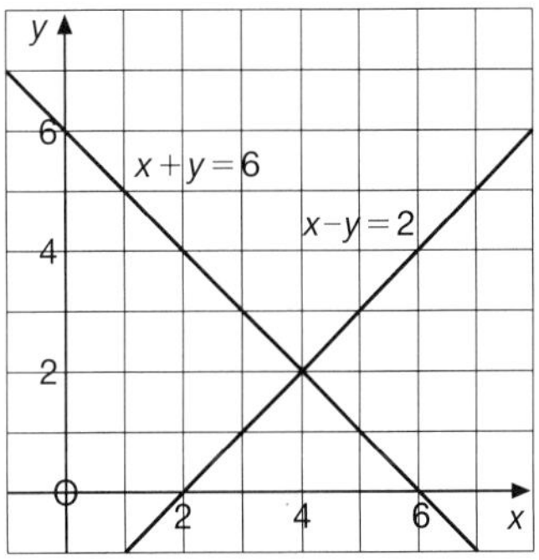

1. Using a graphing package, solve the following simultaneous equations graphically:

   a) $y = x$ and $x + y = 4$
   b) $y = 2x$ and $x + y = 3$
   c) $y = 2x$ and $y = 3$
   d) $y - x = 2$ and $y + \frac{1}{2}x = 5$
   e) $y + x = 5$ and $y + \frac{1}{2}x = 3$

2. Check your answers to Q.1 by solving the simultaneous equations algebraically.

Algebra Activity 2

In this activity you will be using a graphing package or graphical calculator to find the solution to quadratic and reciprocal functions.

You will know that if a quadratic equation is plotted, its solution is represented by where the graph intersects the x-axis.

e.g. When plotting $y = x^2 - 4x + 3$, the solution of $x^2 - 4x + 3 = 0$ occurs where the graph crosses the x-axis as shown in the graph below.
i.e. $x = 1$ and $x = 3$.

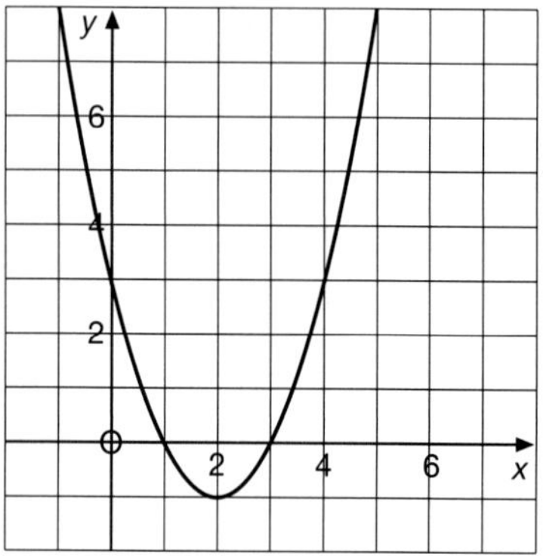

1. Using a graphing package or a graphical calculator solve the following equations graphically:

a) $x^2 + x - 2 = 0$
b) $x^2 - 7x + 6 = 0$
c) $x^2 + x - 12 = 0$
d) $2x^2 + 5x - 3 = 0$
e) $\frac{2}{x} - 2 = 0$
f) $\frac{2}{x} + 1 = 0$

REVIEWS FOR CHAPTERS 17–21

Review 13

1. Find the value of the following without using a calculator:
a) 2^3 b) 3^4 c) $2^4 \times 3^2$

2. Expand the following and simplify:
a) $3(2m - 6) + 2(3 - 4m)$ b) $2x(x - y) - 2y(y - x)$

3. Factorise the following:
a) $15ab - 10bc$ b) $6p^2q + 9pq^2 + 12pq$

4. If $x = 2$, $y = 3$ and $z = 5$, evaluate:
a) $3x - 4y + 2z$ b) $x^2 - y^2 - z^2$

5. Rearrange the following formulae to make the **bold** letter the subject:
a) $3c = 2b + \boldsymbol{a}$ b) $\dfrac{2\boldsymbol{b} + c}{3} = 4a$

6. Solve the following equations:
a) $3a + 4 = 19$ b) $\dfrac{b + 4}{3} = 1$

7. Solve the following simultaneous equations:
$a + 3b = 11$ $3a - 3b = -3$

8. Given that 50 miles ≈ 80 km, draw a conversion graph up to 100 miles. Use your graph to estimate:
a) how many km there are in 70 miles,
b) how many miles there are in 70 km.

9. Sketch the graphs of:
a) $x = 2$ b) $y = -3$

10. Sketch the graph of $y = x^2$ from -3 to $+3$. Use your graph to find $\sqrt{6.25}$.

Review 14

1. Find the value of the following without using a calculator:
a) 6^2 b) 5^3 c) $\dfrac{2^5}{4^2}$

2. Expand the following and simplify if possible:
a) $4(2a - 3b) + 5(2b - 3a)$ b) $a(2b - 4) + b(5 - 2a)$

3. Factorise the following:
a) $12x^2y - 48xy^2$ b) $3abc - 6b^2c + 9bc^2$

4. If $p = 2$, $c = -3$ and $d = -1$, evaluate:
a) $2p - c + 3d$ b) $p^2 + c^2 - d^2$

5. Rearrange the following formulae to make the **bold** letter the subject:
a) $2p = 3q - \boldsymbol{r}$ b) $x - y = \dfrac{(x - \boldsymbol{z})}{3}$

6. Solve the following equations:
a) $4a - 9 = -1$ b) $\frac{(b-5)}{2} = 2$

7. Solve the following simultaneous equations:

$$2a - 2b = 4$$
$$4a - b = 5$$

8. Given that 0 °C = 32 °F and 100 °C = 212 °F, draw a conversion graph between degrees Celsius and Fahrenheit. Use your graph to convert:
a) 50 °C to Fahrenheit b) 100 °F to Celsius

9. Sketch the graphs of:
a) $y = 2x$ b) $y = -x$

10. Sketch the graph of $y = -x^2$ from -3 to $+3$. Use your graph to find $-\sqrt{5.29}$

Review 15

1. Work out the value of the following without using a calculator:
a) 7^2 b) 2^6 c) $3^3 \times 2^3$

2. Expand the following and simplify if possible:
a) $5x(x - y + z)$ b) $3(a + 2b) - a(2b + 4)$

3. Factorise the following:
a) $39xy - 52yz$ b) $4pqr^2 + 8pq^2r - 16p^2qr$

4. If $a = 3$, $b = -3$ and $c = +6$, evaluate:
a) $a^3 + b^3$ b) $a^4 + abc - c^2$

5. Rearrange the following formulae to make the **bold** letter the subject:
a) $3b - c = 2\mathbf{d} - e$ b) $\frac{p}{\mathbf{q}} = rs$

6. Solve the following equations:
a) $8 = 6a - 1$ b) $3 = \frac{2c - 1}{4}$

7. Solve the following simultaneous equations:

$$a + b = -1$$
$$a - b = 5$$

8. A maths examination is marked out of 140. Draw a conversion graph to convert the following scores out of 140 into percentages:
a) 60 b) 100 c) 125

9. Sketch the graph of $y = \frac{1}{x}$ for $-3 \leq x \leq 3$.

10. a) Copy and complete the table below for the function $y = x^2 - 3$.

x	−3	−2	−1	0	1	2	3
y	6			−3			

b) Plot the graph of the function $y = x^2 - 3$ for $-3 \leqslant x \leqslant 3$.

Review 16

1. Work out the value of the following without using a calculator:

a) 9^2 b) 10^4 c) $\frac{9^2}{3^4}$

2. Factorise the following:

a) $17x^2 - 51xy$ b) $18abc - 27bc^2 + 36b^2c$

3. Expand the following and simplify if possible:

a) $3a(2b - 3) + 3b(3 - 2a)$ b) $\frac{1}{2}(6a + 4) - \frac{1}{3}(9a - 6)$

4. If $p = -2$, $q = -3$ and $r = -5$ evaluate:

a) $p^2 + q^2 + r^2$ b) $2pqr - 3qr$

5. Rearrange the following formulae to make the **bold** letter the subject:

a) $4a - 3b = c - 5\mathbf{d}$ b) $\frac{m}{2n} = \frac{r}{3\mathbf{s}}$

6. Solve the following equations:

a) $3 = 4a - 5$ b) $4 = \frac{5b - 3}{8}$

7. Solve the following simultaneous equations:

$$a + b = 9$$
$$2a - 2b = 2$$

8. 1 euro will convert to 2.8 Brazilian reals. Draw an appropriate conversion graph to convert the following:

a) €70 to reals b) 300 reals to euro

9. Sketch the graph of $y = -3x$.

10. a) Copy and complete the table below for the function $y = x^2 + x$.

x	−3	−2	−1	0	1	2	3
y	6				2		

b) Plot the graph of the function $y = x^2 + x$ for $-3 \leqslant x \leqslant 3$.

Review 17

1. Work out the value of the following without using a calculator:

a) $(\frac{1}{3})^2$ b) $(\frac{1}{2})^5$ c) $4^2 \times (\frac{1}{2})^3$

2. Factorise the following:

a) $24mn^2 - 36m^2n$ b) $34abc - 51bcd + 68bc^2$

3. Expand the following and simplify if possible:

a) $14a - 7(2a - 3)$ b) $-6b(2a - 3) + 3a(5b + 1)$

4. If $m = -1, n = 1$ and $t = -2$, evaluate:

a) $m^2 + n^2$ b) $2mn - nt + mt$

5. Rearrange the following formulae to make the **bold** letter the subject:

a) $\frac{a}{b} = -\frac{\mathbf{c}}{d}$ b) $\frac{\mathbf{a} + b}{c} = d$

6. Solve the following equations:

a) $\frac{m + 1}{3} = 5$ b) $\frac{3}{4}(2b - 1) = 2$

7. Solve the following simultaneous equations:

$$a + 3b = 7$$
$$2a + b = 9$$

8. An English examination is marked out of 80. Draw a conversion graph to convert the following marks out of 80 to percentages:

a) 50 b) 28 c) 72

9. Sketch the graphs of:

a) $x = -3$ b) $y = 2$

10. a) Copy and complete the table below for the function $y = x^2 + x - 2$.

x	−4	−3	−2	−1	0	1	2	3
y		4						10

b) Plot the graph of the function $y = x^2 + x - 2$ for $-4 \leqslant x \leqslant 3$. Use the graph to solve the quadratic equation $x^2 + x = 2$.

Review 18

1. Simplify the following using indices:
a) 3×3 b) $9^2 \div 3^3$

2. Expand and simplify the following:
a) $-2x(y - 3) + 3y(2 + x)$
b) $2a(a + b - c) - 2b(a - b - c) + 2c(a - b - c)$

3. Factorise the following:
a) $121x^2y - 77xy^2$ b) $45pqr - 30p^2q + 60pq^2$

4. If $p = 3$, $q = -3$ and $r = -1$, evaluate:
a) $p^2 + q^2$ b) $pqr - p^2r$

5. Rearrange the formulae to make the letter in **bold** the subject:
a) $\dfrac{m}{n} = \dfrac{p}{-\boldsymbol{q}}$ b) $\dfrac{(a - \boldsymbol{b})}{c} = d$

6. Solve the following equations:
a) $\dfrac{a + 4}{3} = 1$ b) $\frac{3}{5}(b + 1) = b$

7. Solve the following simultaneous equations:

$$m + n = 1$$
$$2m + 3n = -3$$

8. Absolute zero, 0 Kelvin (K) is equivalent to −273 °C. 0 °C is equivalent to 273 K. Draw a conversion graph to convert K into °C. Use your graph to convert:
a) −50 °C to K b) 200 K to °C

9. Sketch the graph of $y = \dfrac{2}{x}$.

10. a) Copy and complete the table below for the function $y = x^2 - 3x + 2$.

x	−2	−1	0	1	2	3	4
y	12						6

b) Plot the graph of the function $y = x^2 - 3x + 2$ for $-2 \leqslant x \leqslant 4$. Use your graph to solve the quadratic equation $x^2 - 3x + 2 = 0$.

SHAPE AND SPACE

22 GEOMETRICAL TERMS AND RELATIONSHIPS

Use the terms: line, parallel, right angle, acute, obtuse and reflex angles, perpendicular, similarity, congruence; use and interpret vocabulary of triangles, quadrilaterals, circles, polygons and simple solid figures, including nets.

NB: All diagrams are not drawn to scale.

Angles

Different types of angle have different names:

acute angles lie between 0° and 90°
right angles are exactly 90°
obtuse angles lie between 90° and 180°
reflex angles lie between 180° and 360°

Exercise 22.1

1. Draw and label one example of each of the following types of angle:

   right   acute   obtuse   reflex

2. Copy the angles below and write beneath each drawing the type of angle it shows:

a) 
b) 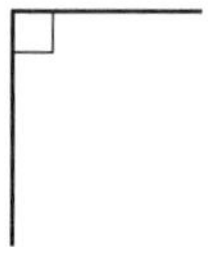
c) 
d) 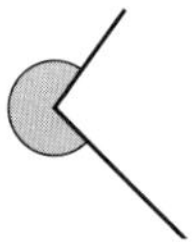
e) 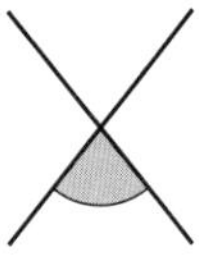
f) 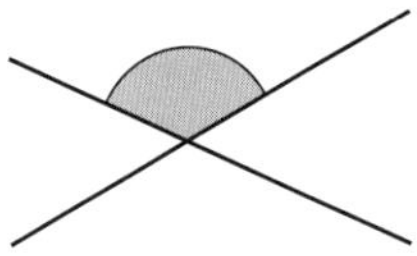
g) 
h) 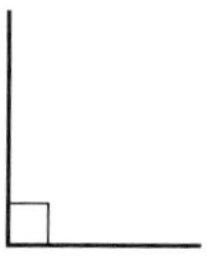

Two angles which add together to total 180° are called **supplementary** angles.
Two angles which add together to total 90° are called **complementary** angles.

3. State whether the following pairs of angles are supplementary, complementary or neither:

a) 70°, 20°	b) 90°, 90°	c) 40°, 50°	d) 80°, 30°
e) 15°, 75°	f) 145°, 35°	g) 133°, 57°	h) 33°, 67°
i) 45°, 45°	j) 140°, 40°		

4. Measure the angles in the diagram below:

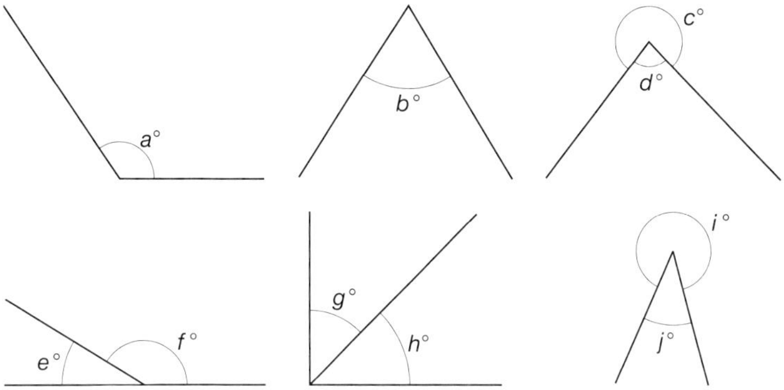

5. Use a protractor, ruler and pencil to draw the following angles accurately:

a) 45° b) 72° c) 90°
d) 142° e) 260° f) 318°

Angles can be named either by labelling them as shown in question 4, or by indicating the lines which meet to form the angle as shown below.

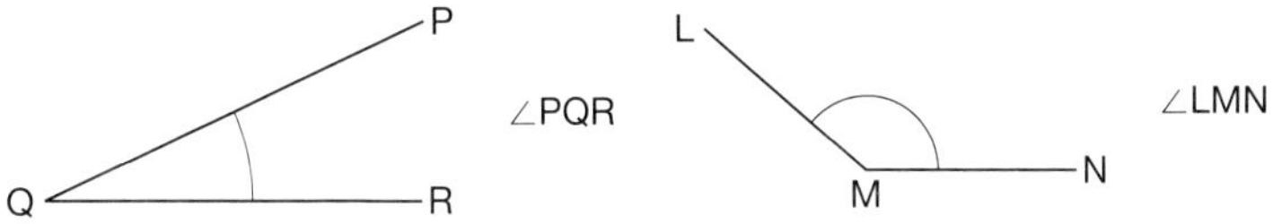

6. Sketch the shapes below. Name the angles marked with a single letter, in terms of the lines which meet to form the angles.

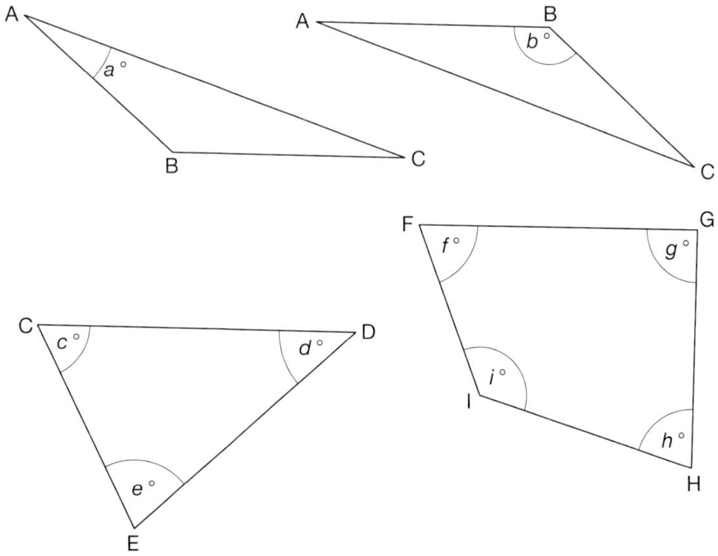

7. Measure accurately the angles marked in the sketches you made for Q.6.

To find the shortest distance between two points, you measure the length of the **straight line** which joins them.

Two lines which meet at right angles are **perpendicular** to each other.

So in the diagram below CD is **perpendicular** to AB, and AB is perpendicular to CD.

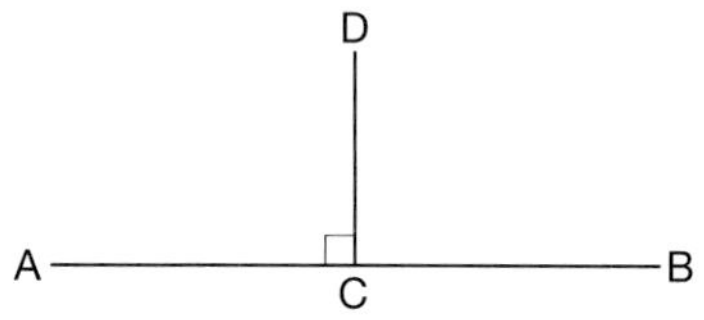

If the lines AD and BD are drawn to form a triangle, the line CD can be called the **height** or **altitude** of the triangle ABD.

8. In the diagrams below, state which pairs of lines are perpendicular to each other.

a)

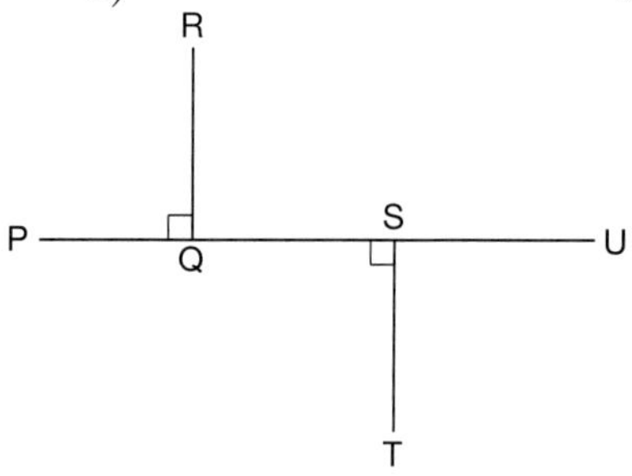

b)

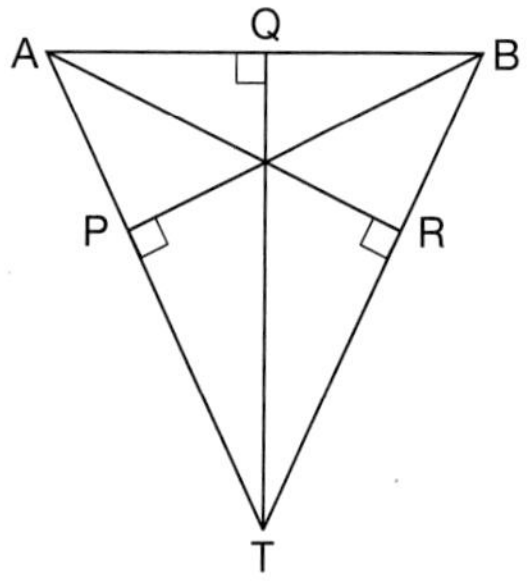

Parallel lines are straight lines which can be continued to infinity in either direction without meeting.

Railway lines are an example of parallel lines. Parallel lines are marked with arrows as shown:

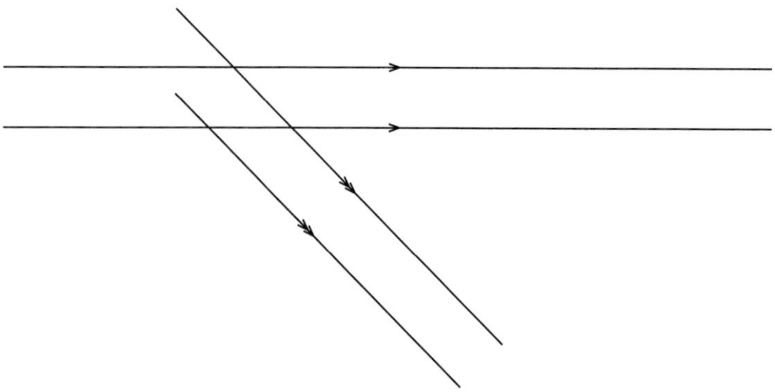

Triangles

Triangles can be described in terms of their sides or their angles, or both.

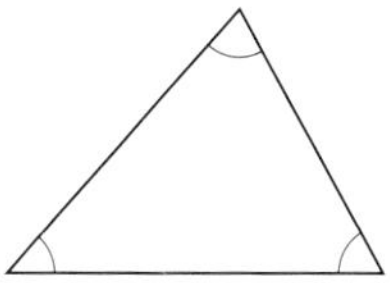

An **acute-angled** triangle has all its angles less than 90°.

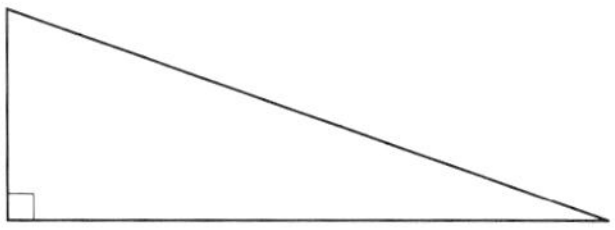

A **right-angled** triangle has an angle of 90°.

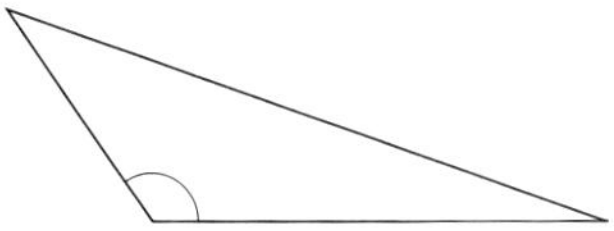

An **obtuse-angled** triangle has one angle greater than 90°.

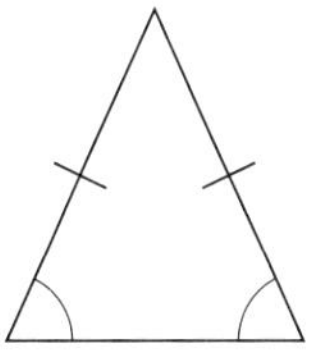

An **isosceles** triangle has two sides of equal length, and the angles opposite the equal sides are equal.

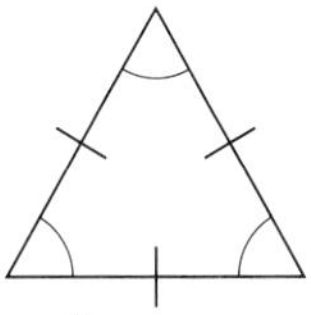

An **equilateral** triangle has three sides of equal length and three equal angles.

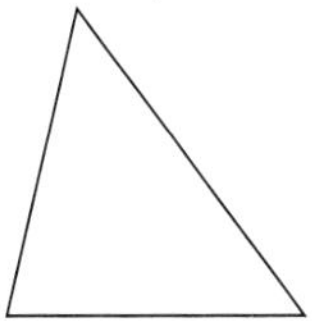

A **scalene** triangle has three sides of different lengths and all three angles are different.

Exercise 22.2

1. Describe the triangles below in two ways. The example on the right shows an **acute-angled isosceles** triangle.

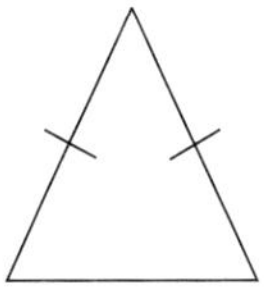

a)

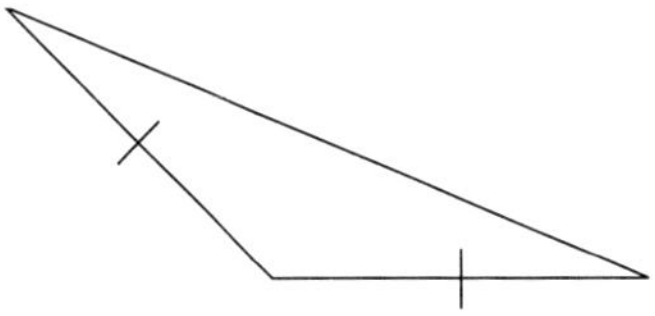

b)

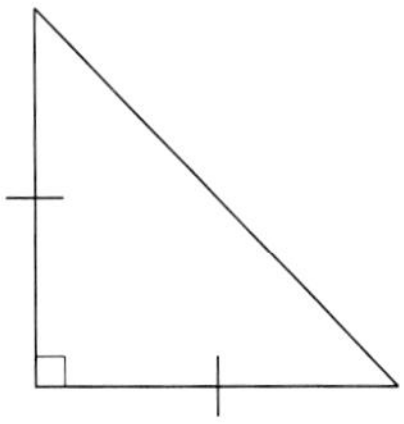

c)

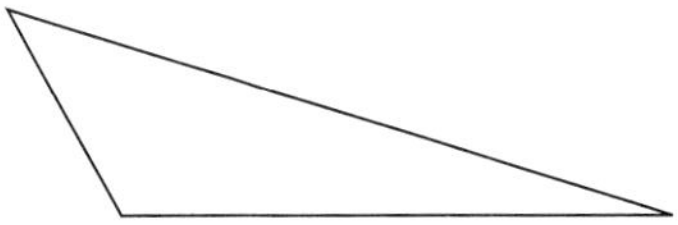

d)

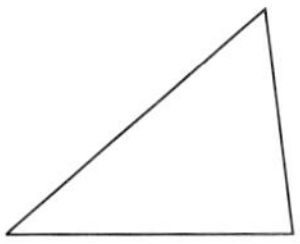

e)

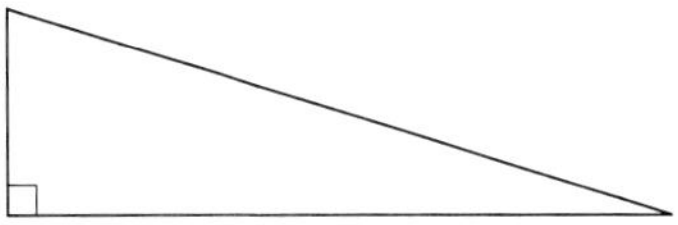

f)

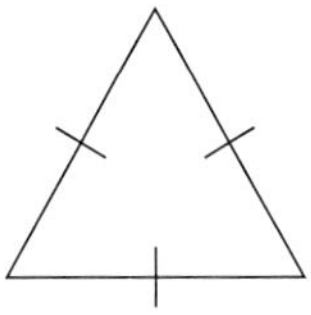

2. Draw the following triangles using a ruler and protractor:
a) an acute-angled isosceles triangle of sides 5 cm, 5 cm and 6 cm, and altitude 4 cm,
b) a right-angled scalene triangle of sides 6 cm, 8 cm and 10 cm,
c) an equilateral triangle of side 7.5 cm,
d) an obtuse-angled isosceles triangle of sides 13 cm, 13 cm and 24 cm, and altitude 5 cm.

Congruent triangles are **identical**. They have corresponding sides of the same length, and corresponding angles which are equal.

Triangles are congruent if any of the following can be proved:

- Three corresponding sides are equal (S, S, S);
- Two corresponding sides and the included angle are equal (S, A, S);
- Two angles and the corresponding side are equal (A, S, A);
- Each triangle has a right angle, and the hypotenuse and a corresponding side are equal in length.

3. Which of the triangles below are congruent?

a)
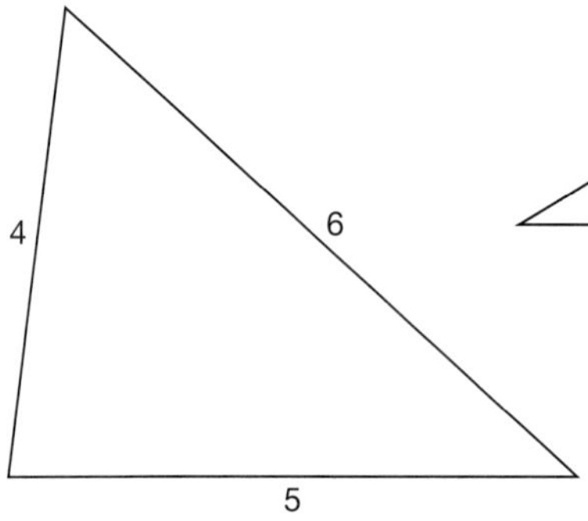

b)
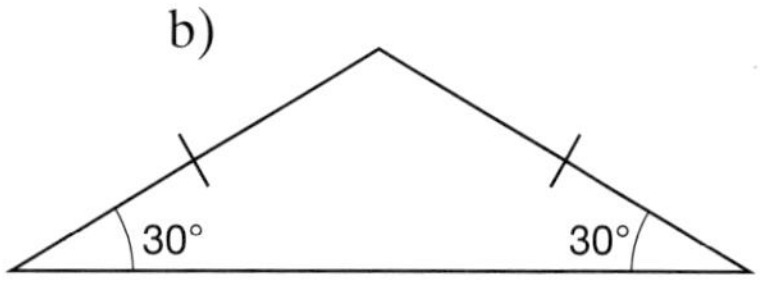

c)
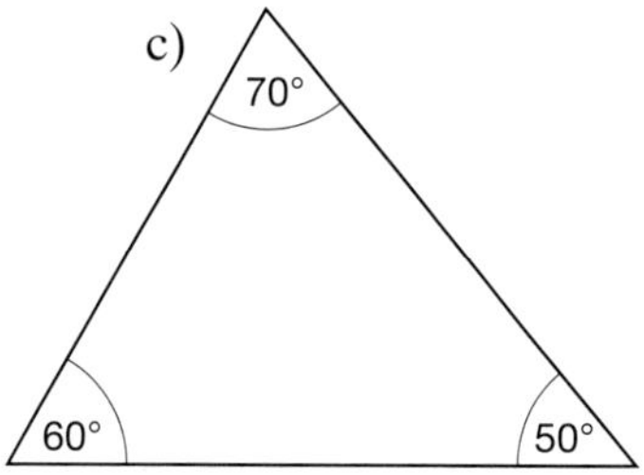

d)
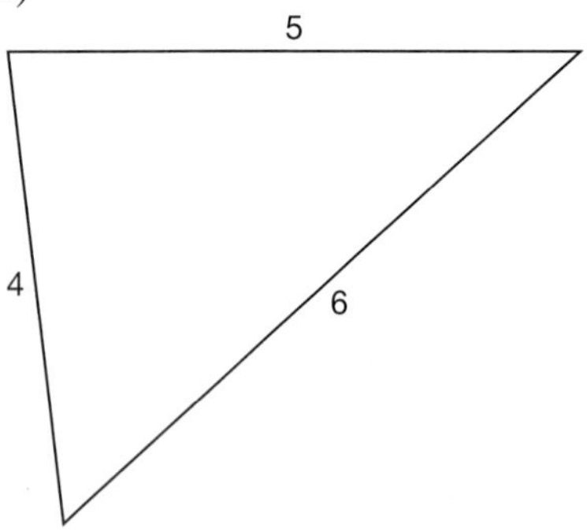

e)
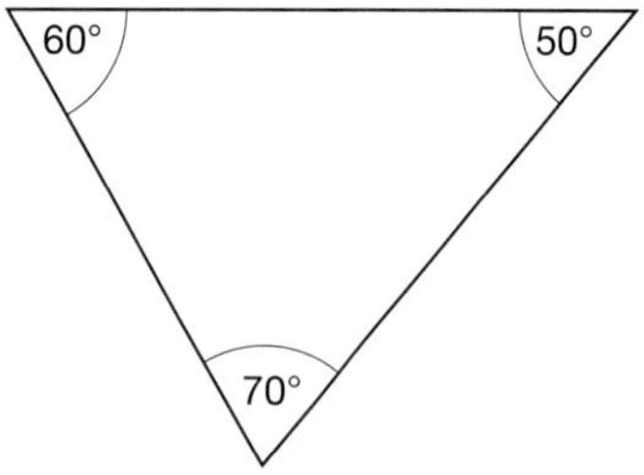

f)
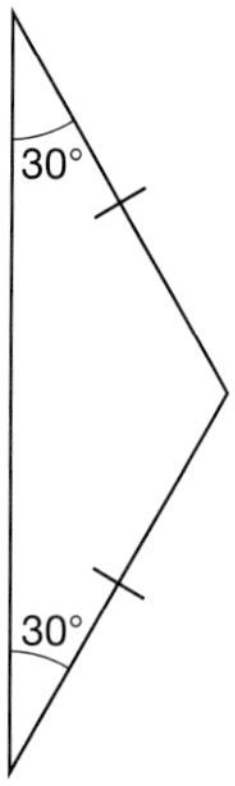

4. Make accurate drawings of pairs of triangles to illustrate the proofs of congruence given above.

5. In the diagrams below, identify pairs of congruent triangles. Give reasons for your answers.

a)

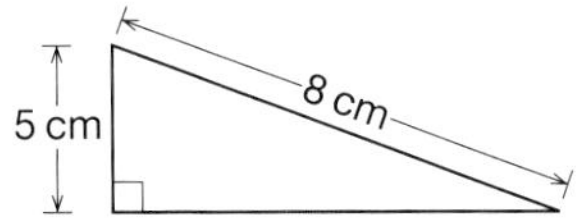

b)

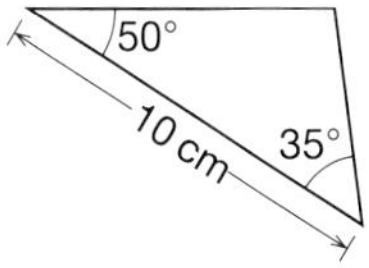

c)

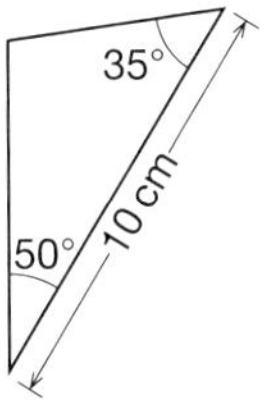

d)

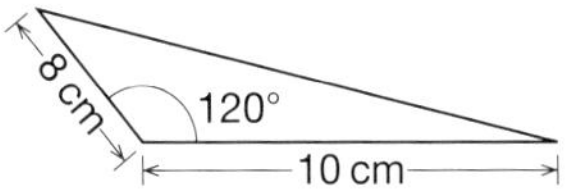

e)

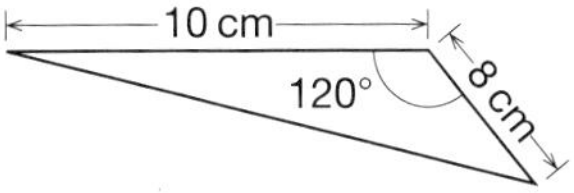

f)

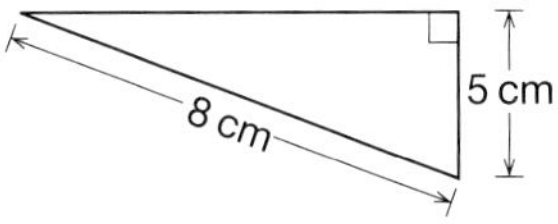

g)

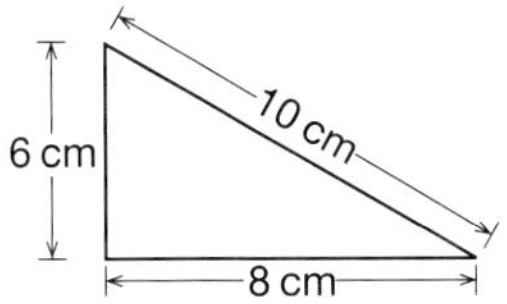

h)

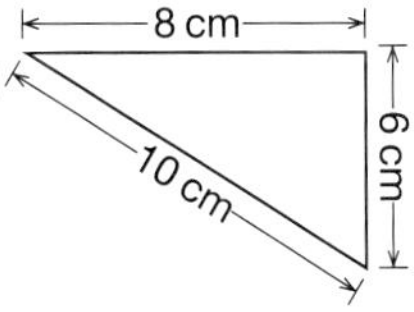

Vocabulary of the circle

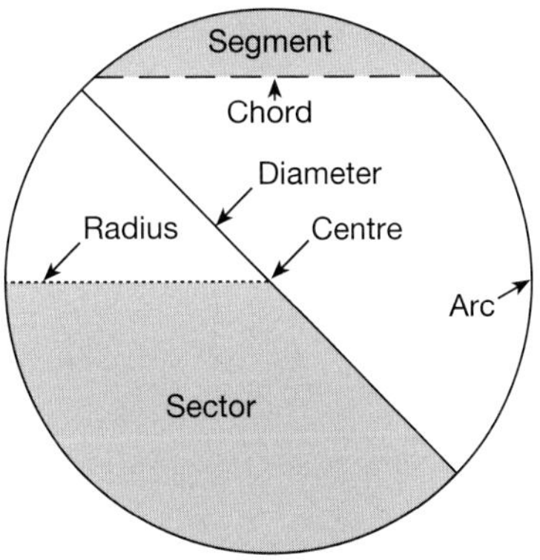

Quadrilaterals

- A **quadrilateral** is any four-sided, closed shape consisting only of straight lines.
- A **square** has four sides of equal length, and four right-angled corners.
- A **rectangle** has two pairs of sides of equal length, and four right-angled corners.
- A **parallelogram** has two pairs of sides of equal length, and two pairs of opposite angles which are equal.
- A **rhombus** is a parallelogram with all its sides of equal length.
- A **trapezium** is a quadrilateral with one pair of parallel sides.
- An **isosceles trapezium** has one pair of parallel sides and the other pair of sides are equal in length.
- A **kite** has two pairs of adjacent sides equal in length.

Exercise 22.3

1. Copy the diagrams and name each shape according to the definitions given above.

a)

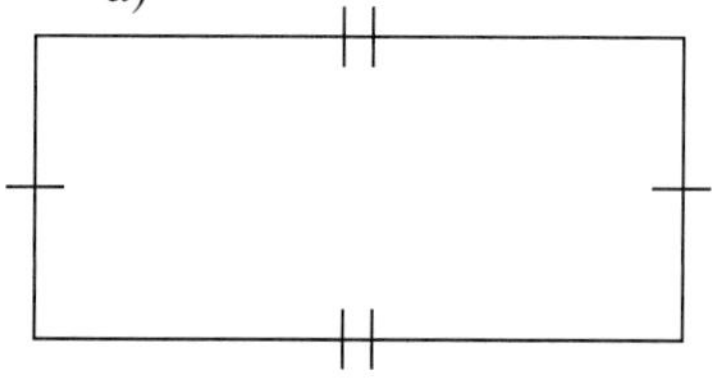

b)

c)

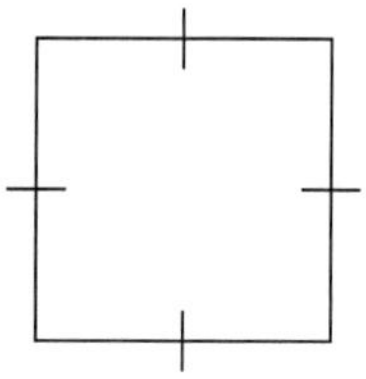

d)

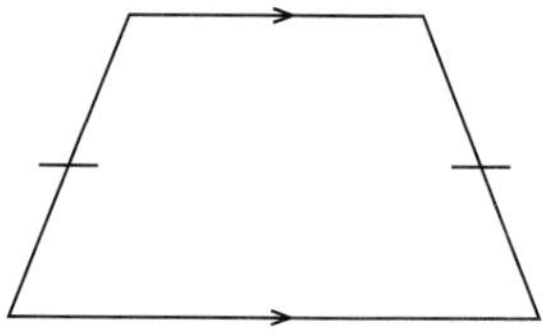

e)

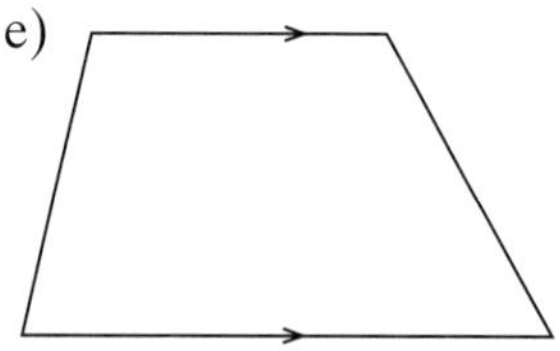

f)

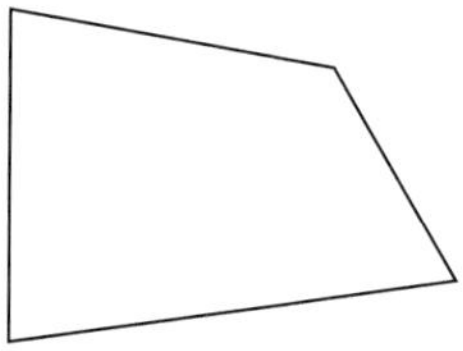

g)

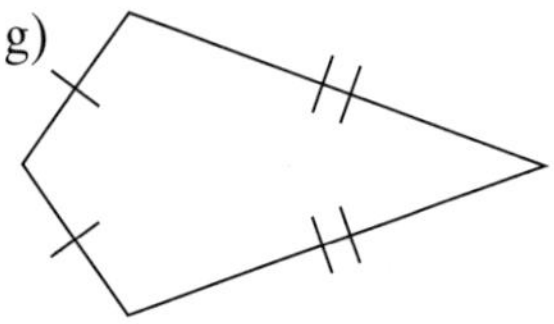

h)

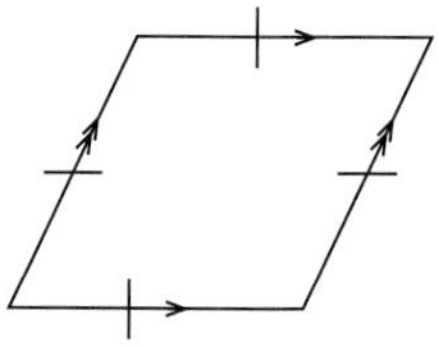

2. Copy and complete the following table. The first line has been started for you.

	Rectangle	Square	Parallelogram	Kite	Rhombus	Equilateral triangle
Opposite sides equal in length	Yes		Yes			
All sides equal in length						
All angles right angles						
Both pairs of opposite sides parallel						
Diagonals equal in length						
Diagonals intersect at right angles						
All angles equal						

Polygons

Any closed figure made up of straight lines is called a **polygon**.

If the sides are the same length and the interior angles are equal, the figure is called a **regular polygon**.

The names of the common polygons are:

3 sides	**tri**angle
4 sides	**quad**rilateral
5 sides	**penta**gon
6 sides	**hexa**gon
7 sides	**hepta**gon
8 sides	**octa**gon
10 sides	**deca**gon
12 sides	**dodeca**gon

Two polygons are said to be **similar** if

a) their angles are the same
b) corresponding sides are in proportion.

3. Draw a sketch of each of the shapes listed above.
4. Draw accurately a regular hexagon, a regular pentagon, and a regular octagon.

Nets

The diagram below is the **net** of a cube. It shows the faces of the cube opened out into a two-dimensional plan. The net of a three-dimensional shape can be folded up to make that shape.

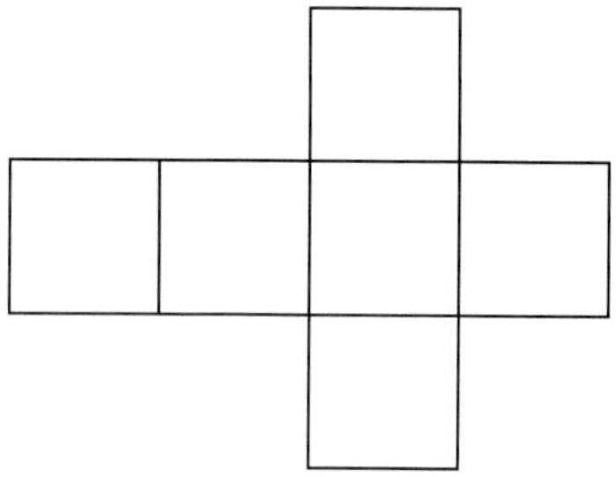

Exercise 22.4 Draw the following on squared paper:

1. Two other possible nets of a cube
2. The net of a cuboid (rectangular prism)
3. The net of a triangular prism
4. The net of a cylinder
5. The net of a square-based pyramid
6. The net of a tetrahedron

Student Assessment 1

1. Are the angles below acute, obtuse, reflex or right angles?
a) b) c) d)

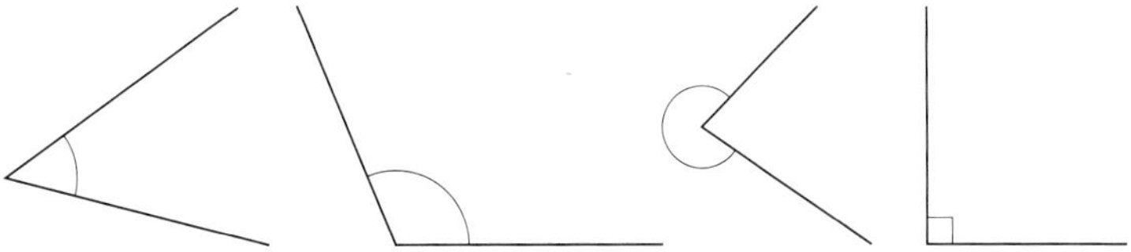

2. Use a ruler, pencil and protractor to draw angles of 135° and 72°.

3. Identify the types of triangles below in two ways (for example, obtuse-angled scalene triangle):
a) b)

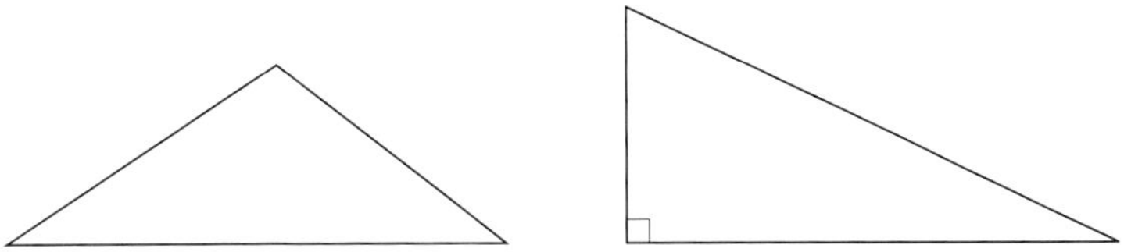

4. Draw two congruent isosceles triangles with base angles of 45°.

5. Draw a circle of radius 3 cm. Mark on it:
a) a diameter b) a chord c) a sector.

6. Draw a rhombus and write down three of its properties.

7. Using the triangle (below), explain fully why $\triangle ABC$ and $\triangle PBQ$ are similar.

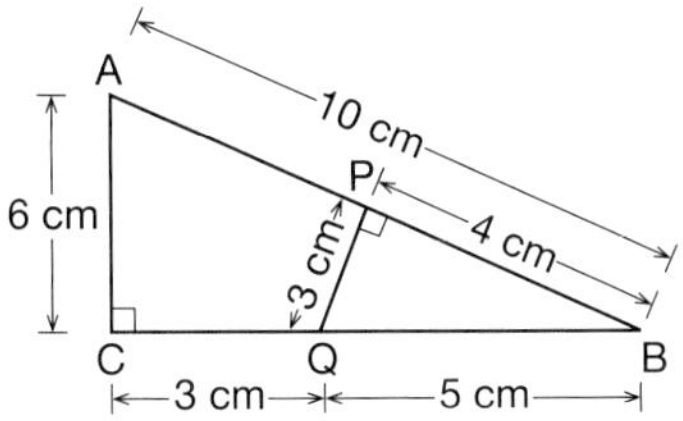

Student Assessment 2

1. Draw and label two pairs of intersecting parallel lines.
2. Draw an obtuse-angled isosceles triangle with base angles of 25°.
3. Make two statements about these two triangles:

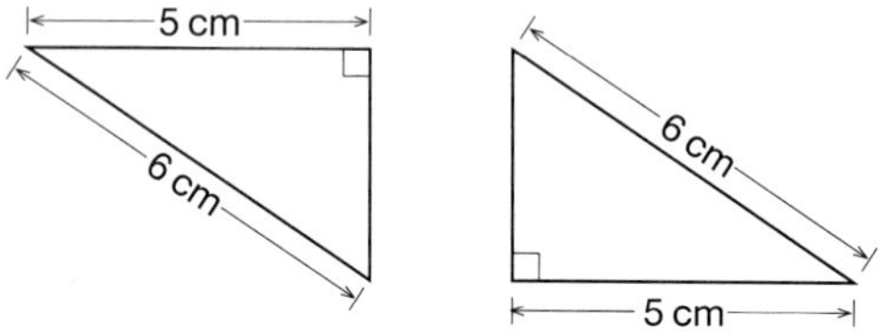

4. Draw a circle of radius 5 cm. Construct a regular hexagon within the circle.
5. Draw:
   a) a rhombus b) an isosceles triangle
   c) a trapezium d) a parallelogram.
6. On squared paper, draw the net of a triangular prism.
7. Which of the triangles below are similar?

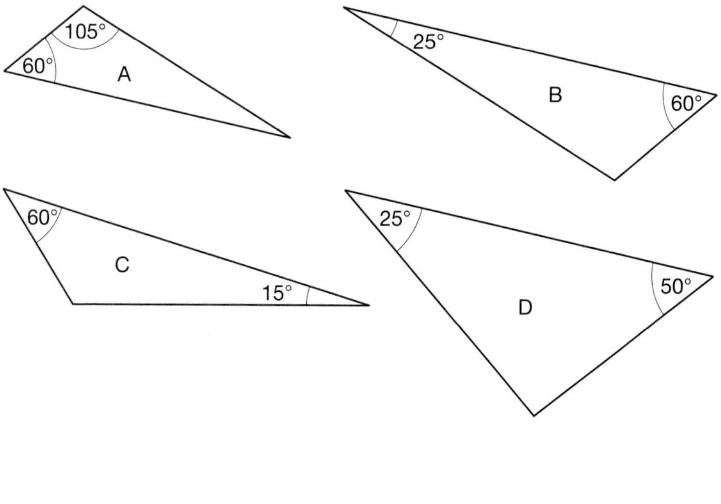

23 GEOMETRICAL CONSTRUCTIONS

Measure lines and angles; construct a triangle given the three sides using ruler and compasses only; construct other simple geometrical figures from given data using protractors and set squares as necessary; construct angle bisectors and perpendicular bisectors using straight edges and compasses only; read and make scale drawings.

Lines and angles

A straight line can be both drawn and measured accurately using a ruler.

Exercise 23.1

1. Using a ruler, measure the length of the following lines to the nearest mm:

   a)

   b)

   c)

   d)

   e)

   f)

2. Draw lines of the following lengths using a ruler:

   a) 3 cm b) 8 cm
   c) 4.6 cm d) 94 mm
   e) 38 mm f) 61 mm

An angle is a measure of turn. When drawn it can be measured using either a protractor or an angle measurer. The units of turn are degrees (°). Measuring with a protractor needs care, as there are two scales marked on it – an inner one and an outer one.

Worked examples **a)** Measure the angle drawn below:

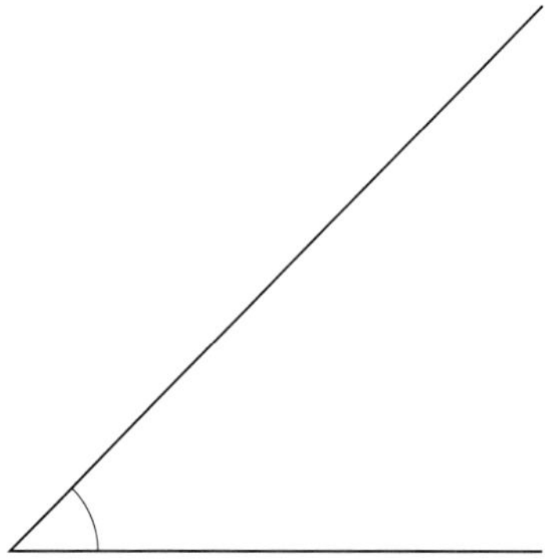

- Place the protractor over the angle so that the cross lies on the point where the two arms meet.
- Align the 0° with one of the arms:

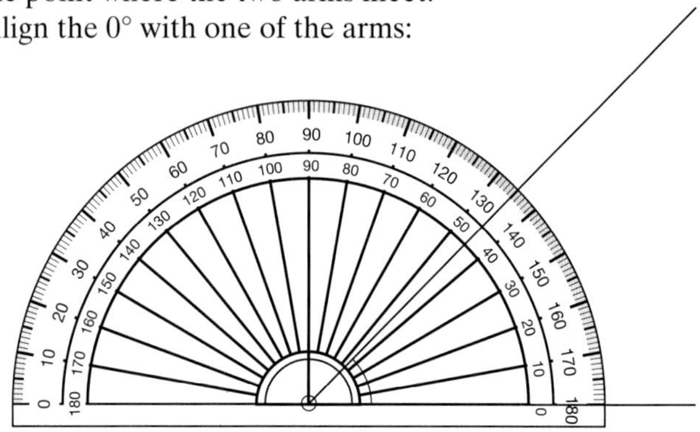

- Decide which scale is appropriate. In this case, it is the inner scale as it starts at 0°.
- Measure the angle using the inner scale.

The angle is 45°.

b) Draw an angle of 110°.

- Start by drawing a straight line.
- Place the protractor on the line so that the cross is on one of the end points of the line. Ensure that the line is aligned with the 0° on the protractor:

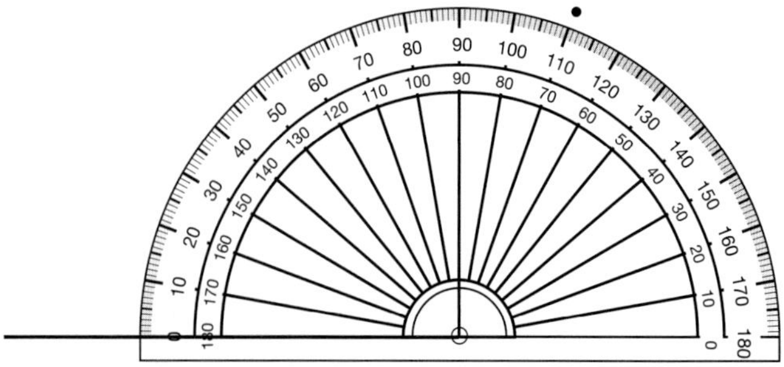

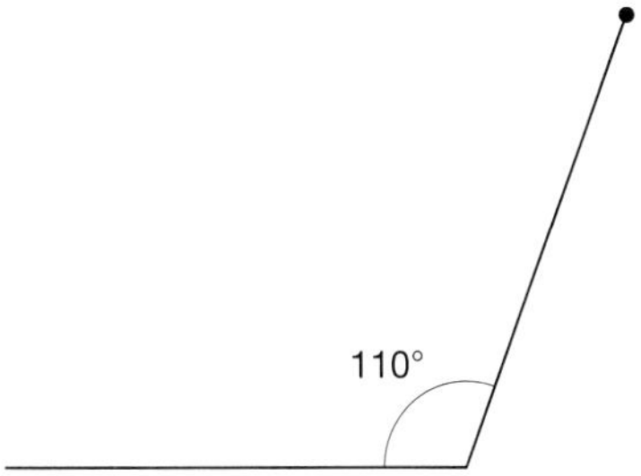

- Decide which scale to use. In this case, it is the outer scale as it starts at 0°.
- Mark where the protractor reads 110°.
- Join the mark made to the end point of the original line.

Exercise 23.2

1. Measure each of the following angles:

a)

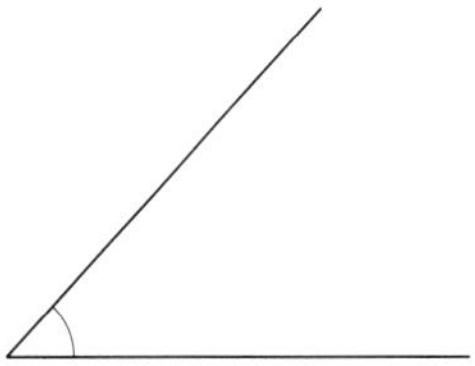

b)

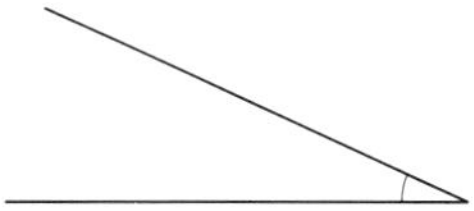

c)

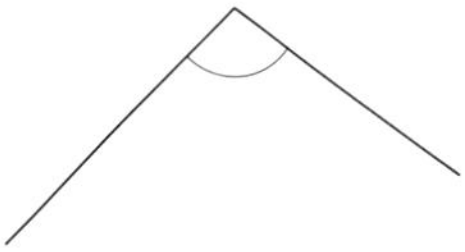

d)

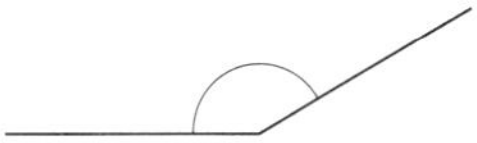

e)

f)

2. Measure each of the following angles:

a)

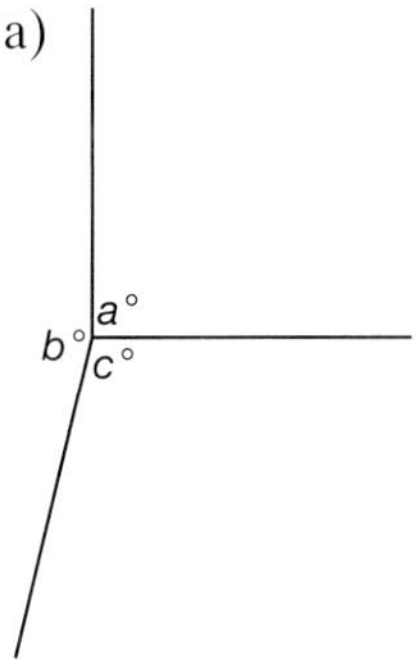

b)

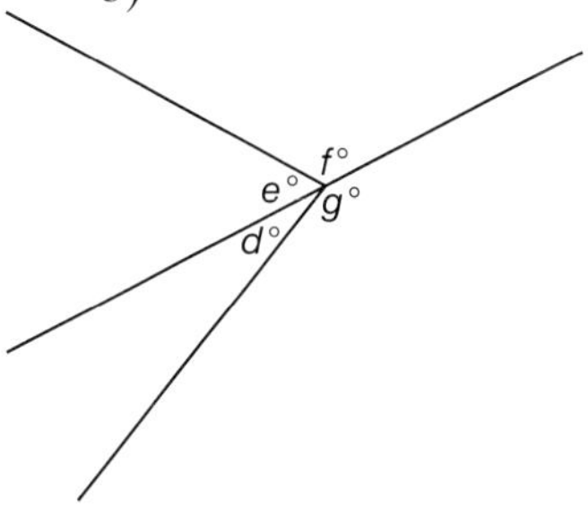

c)

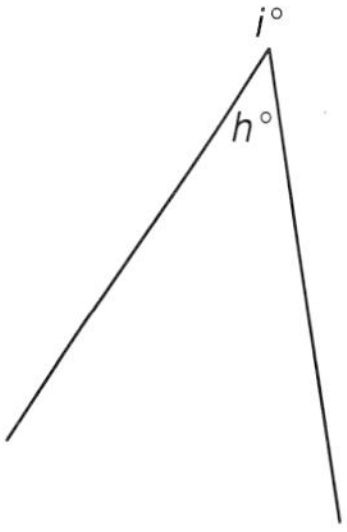

d)

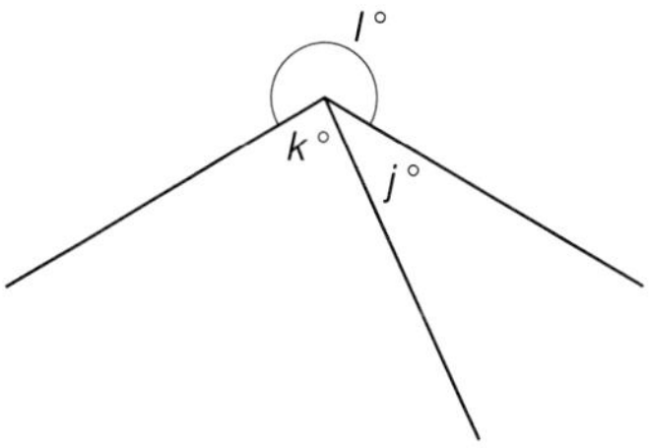

e)

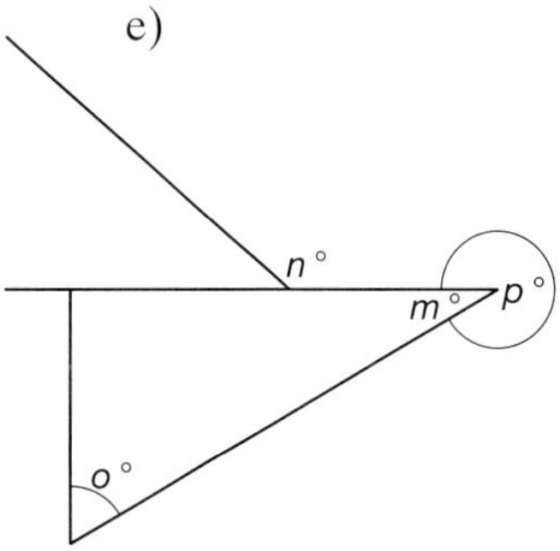

f)

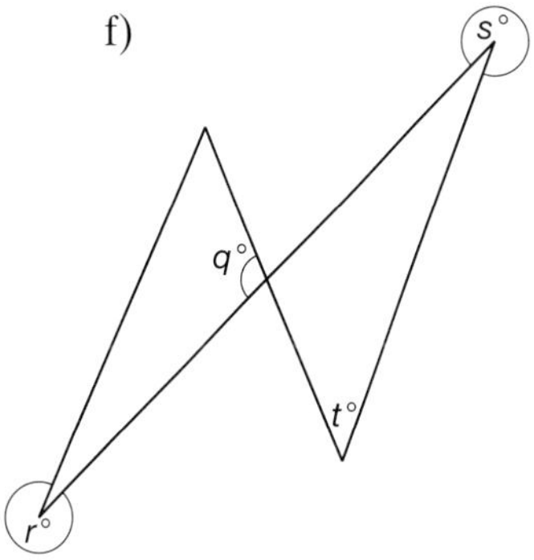

3. Draw angles of the following sizes:

a) 20°	b) 45°	c) 90°
d) 120°	e) 157°	f) 172°
g) 14°	h) 205°	i) 311°
j) 283°	k) 198°	l) 352°

Constructing triangles

Triangles can be drawn accurately by using a ruler and a pair of compasses. This is called **constructing** a triangle.

Worked example Construct the triangle ABC given that:

AB = 8 cm, BC = 6 cm and AC = 7 cm

■ Draw the line AB using a ruler:

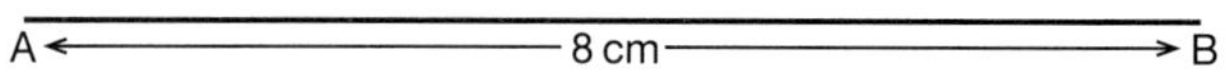

■ Open up a pair of compasses to 6 cm. Place the compass point on B and draw an arc:

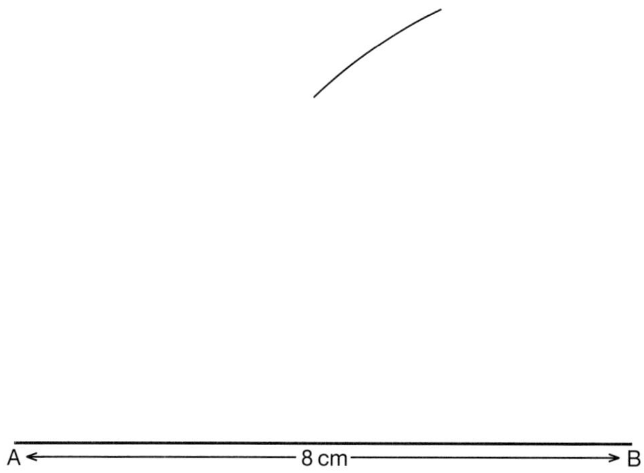

Note that every point on the arc is 6 cm away from B.

■ Open up the pair of compasses to 7 cm. Place the compass point on A and draw another arc, with centre A and radius 7 cm, ensuring that it intersects with the first arc. Every point on the second arc is 7 cm from A. Where the two arcs intersect is point C, as it is both 6 cm from B and 7 cm from A.

■ Join C to A and C to B:

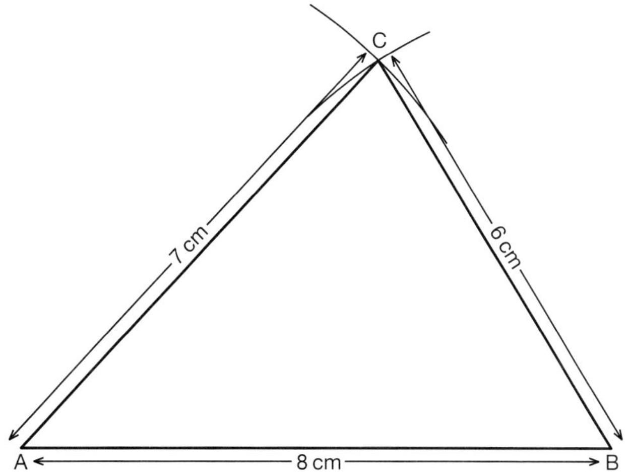

Exercise 23.3 Using only a ruler and a pair of compasses, construct the following triangles:

1. △ABC where AB = 10 cm, AC = 7 cm and BC = 9 cm
2. △LMN where LM = 4 cm, LN = 8 cm and MN = 5 cm
3. △PQR, an equilateral triangle of side length 7 cm
4. a) △ABC where AB = 8 cm, AC = 4 cm and BC = 3 cm
   b) Is this triangle possible? Explain your answer.

Constructing simple geometric figures

Exercise 23.4 1. Draw the following circles using a pair of compasses. In each case O is the centre.

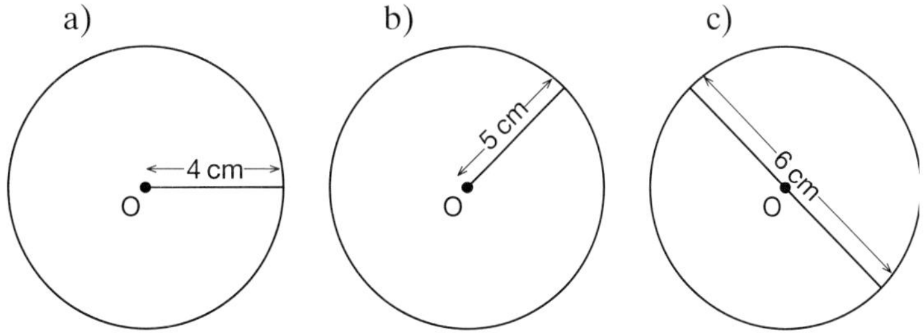

2. Using a pair of compasses, copy the following circle patterns:

a)

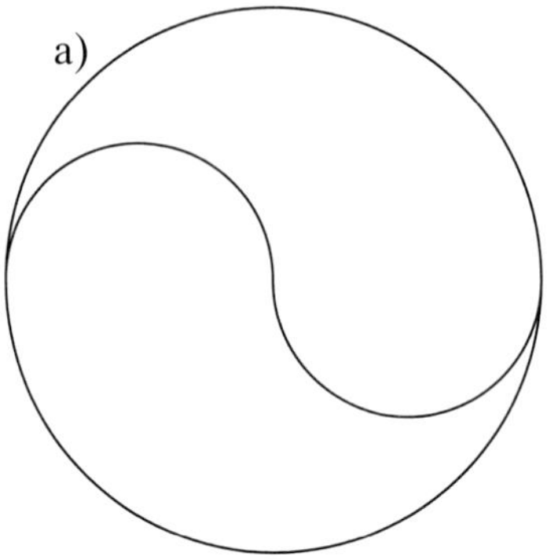

b)

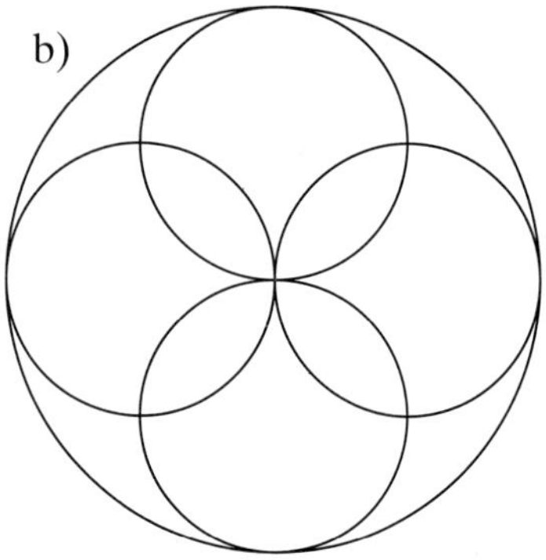

c)

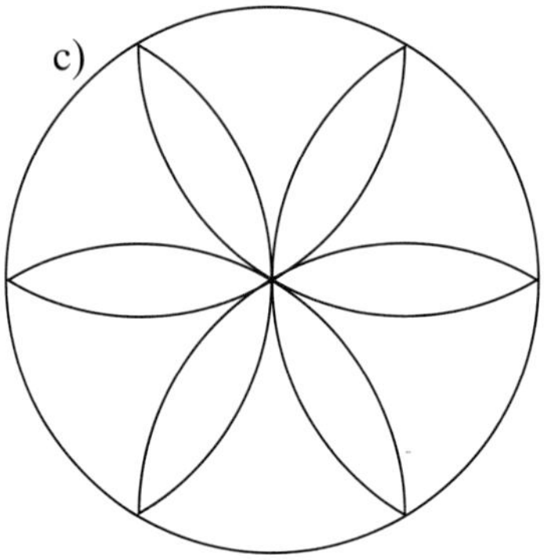

d)

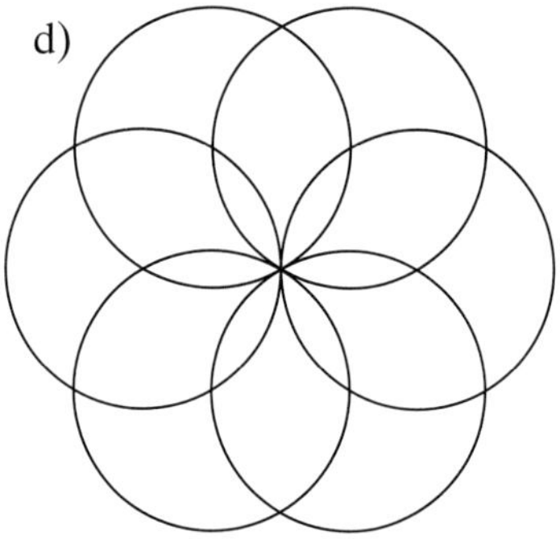

3. Draw some circle patterns of your own.

Worked example Construct a regular hexagon using a pair of compasses and a ruler.

- Open up a pair of compasses and draw a circle.
- Keeping the compasses the same radius, put the point on the circumference of the circle and draw an arc. Ensure that the arc intersects the circumference.
- Place the compass point on the point of intersection of the arc and the circumference, and draw another arc.
- Repeat the above procedures until you have drawn six arcs.

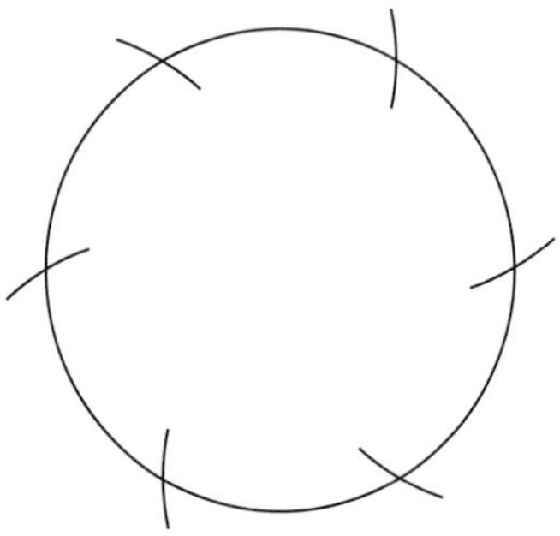

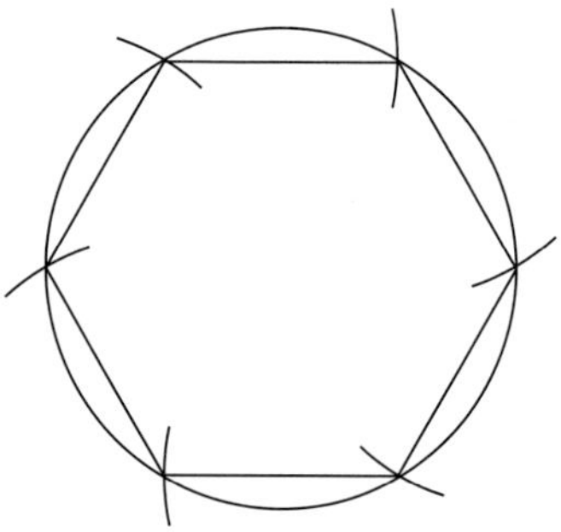

- Draw lines joining the six arcs.

4. Draw some patterns of your own involving regular hexagons.

Squares and rectangles can be constructed using only a ruler and a set square. The set square is used because it provides a right angle with which to draw lines perpendicular to each other.

Worked example Construct a square of side length 6 cm.

- Draw a line 6 cm long.
- Place a set square at the end of the line, ensuring that one of the perpendicular sides rests on the line drawn.
- Draw the perpendicular line 6 cm long.

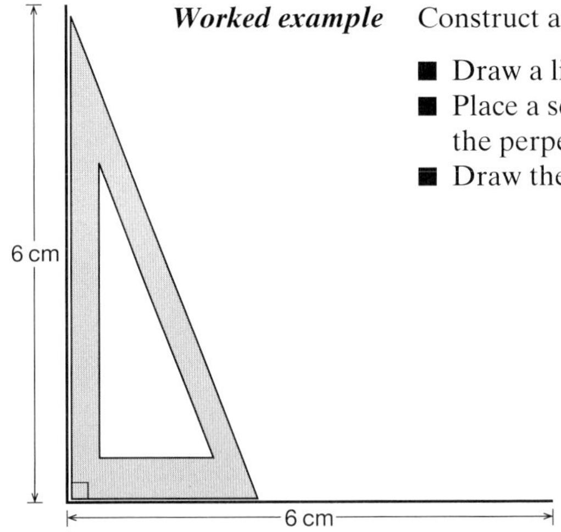

- Repeat this process for the remaining two sides.

Exercise 23.5

Use appropriate geometrical instruments to construct the following shapes:

1.

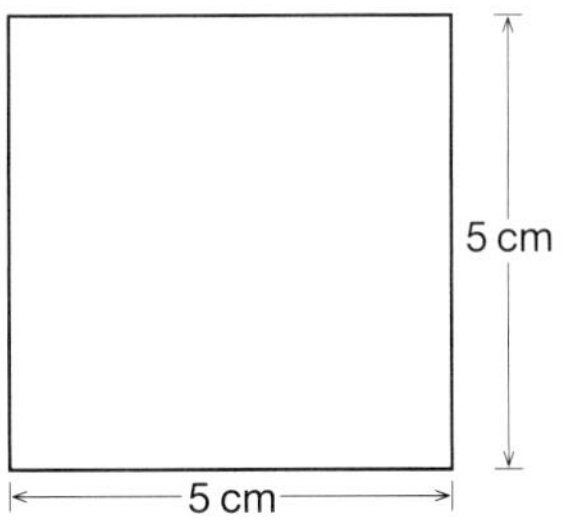

2.

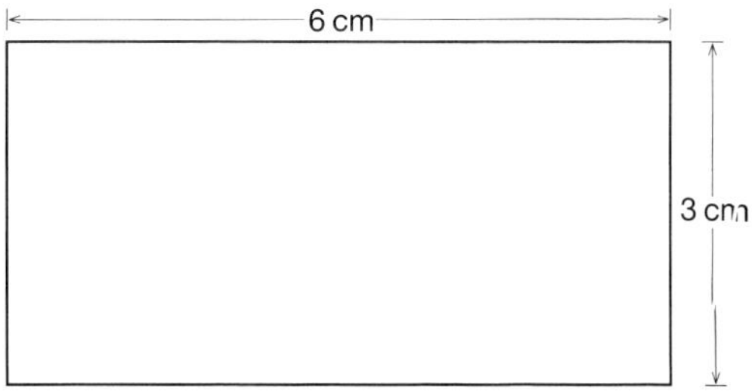

3.

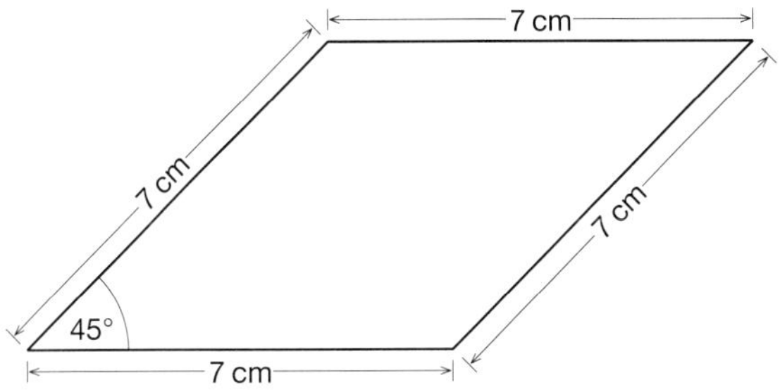

4.

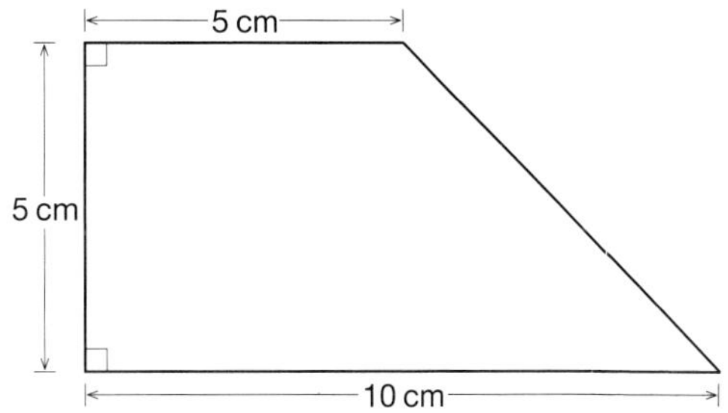

5.

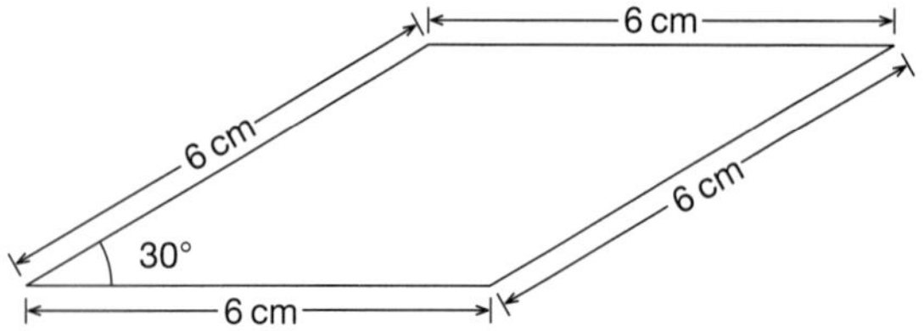

6.

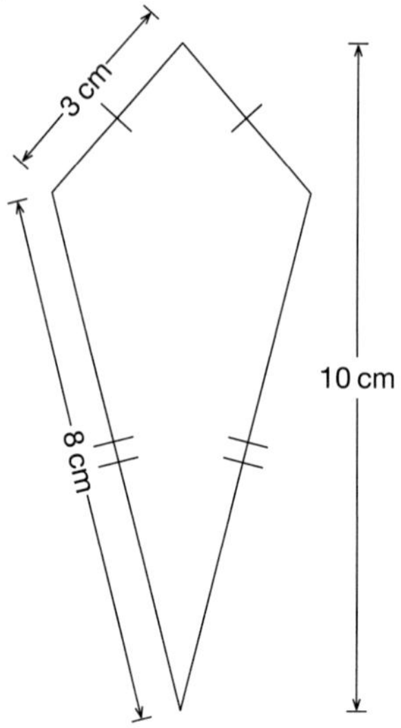

7.

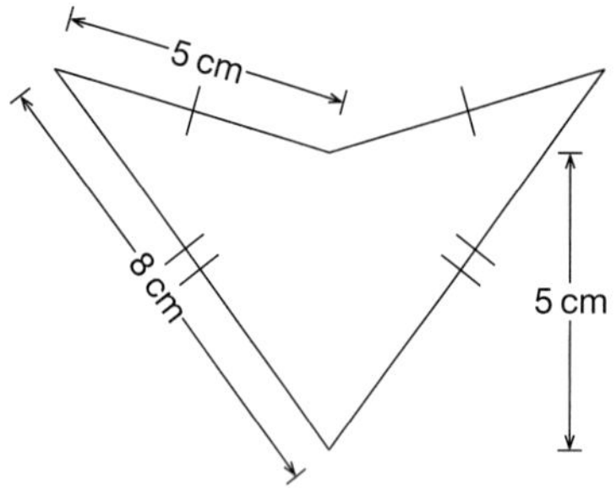

8.

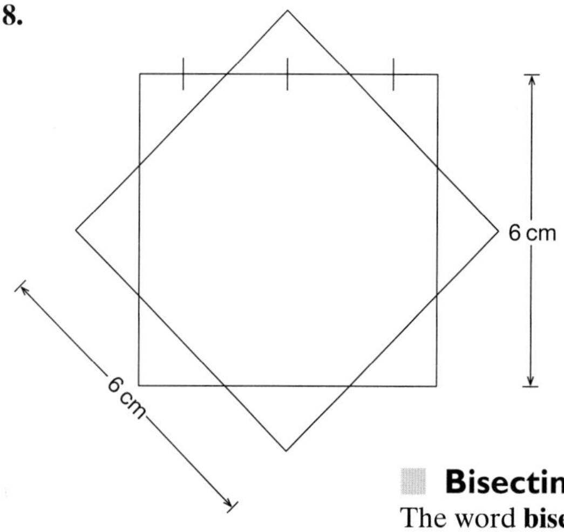

9.

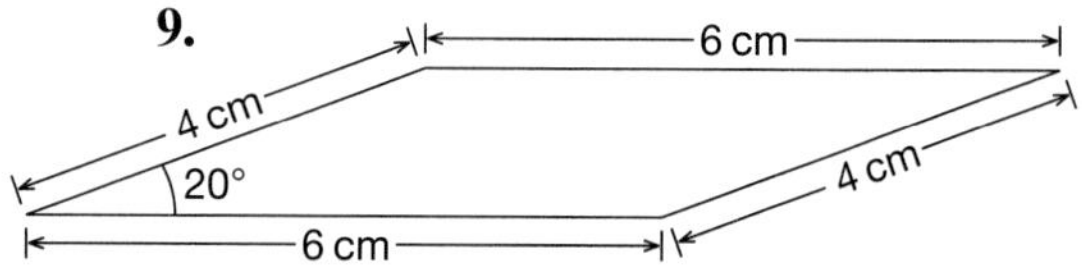

Bisecting lines and angles

The word **bisect** means 'to divide in half'. Therefore, to bisect an angle means to divide an angle in half. Similarly, to bisect a line means to divide a line in half. A **perpendicular bisector** to a line is another line which divides it in half and meets the original line at right angles.

To bisect either a line or an angle involves the use of a pair of compasses.

Worked examples **a)** A line AB is drawn below. Construct the perpendicular bisector to AB.

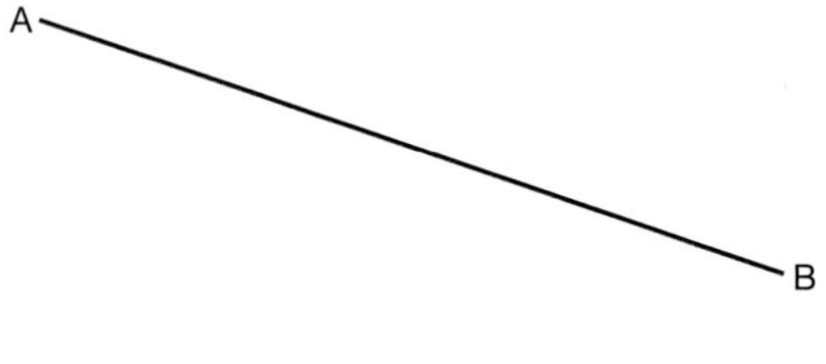

- Open a pair of compasses to more than half the distance AB.
- Place the compass point on A and draw arcs above and below AB.
- With the same radius, place the compass point on B and draw arcs above and below AB. Note that the two pairs of arcs should intersect (see diagram below).
- Draw a line through the two points where the arcs intersect:

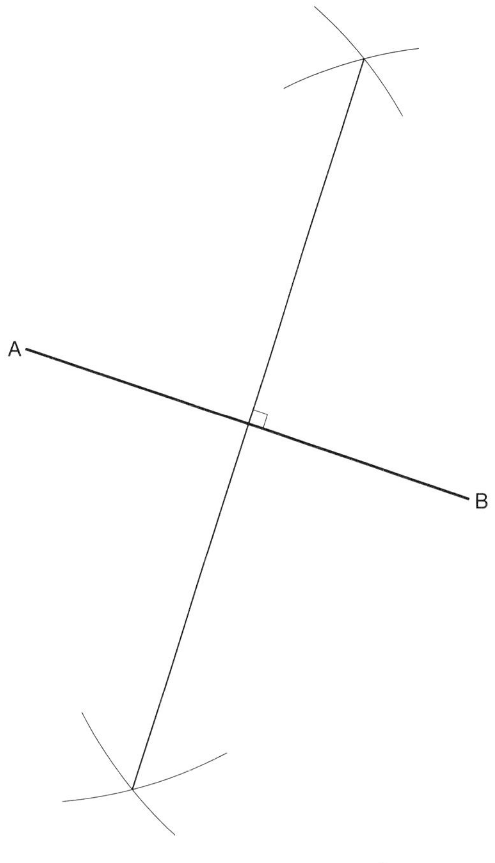

The line drawn is known as the perpendicular bisector of AB, as it divides AB in half and also meets it at right angles.

b) Using a pair of compasses, bisect the angle ABC below:

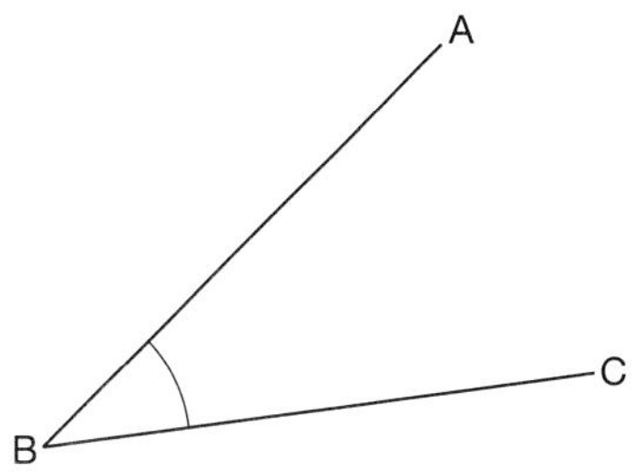

- Open a pair of compasses and place the point on B. Draw two arcs such that they intersect the arms of the angle:

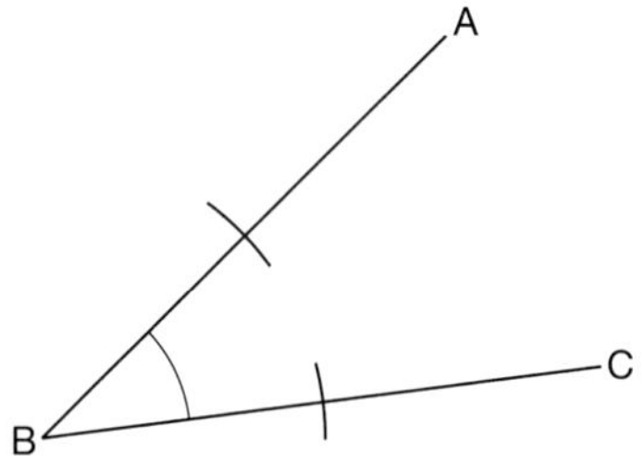

- Place the compasses in turn on the points of intersection, and draw another pair of arcs of the same radius. Ensure that they intersect.
- Draw a line through B and the point of intersection of the two arcs. This line bisects angle ABC.

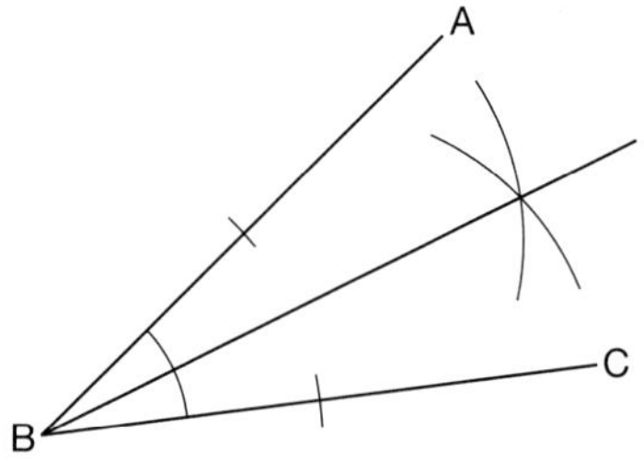

Exercise 23.6

1. Copy each of the lines drawn below on to plain paper, and construct the perpendicular bisectors.

a)

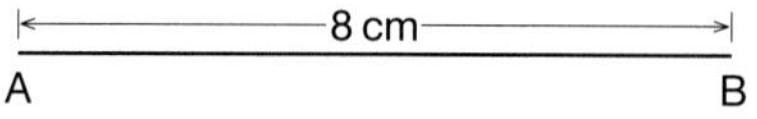

b)

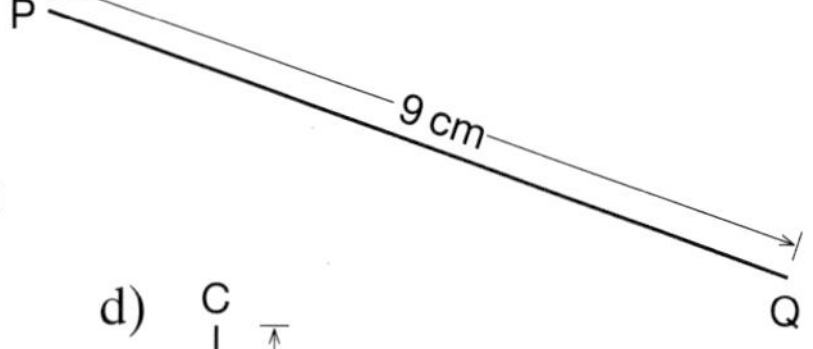

c)

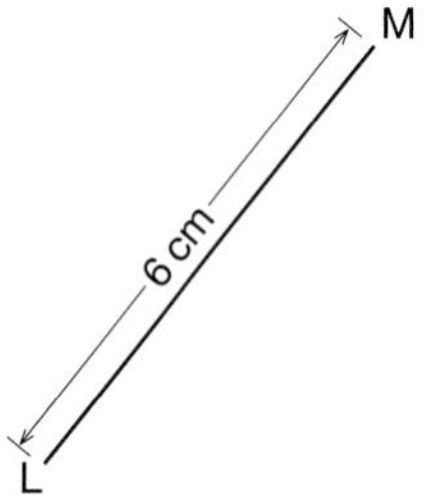

d)

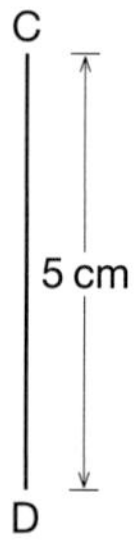

2. For each of the following:
i) draw the angle,
ii) bisect the angle using a pair of compasses.

a) 45° b) 70°
c) 130° d) 173°
e) 210° f) 312°

3. Copy the diagram below.

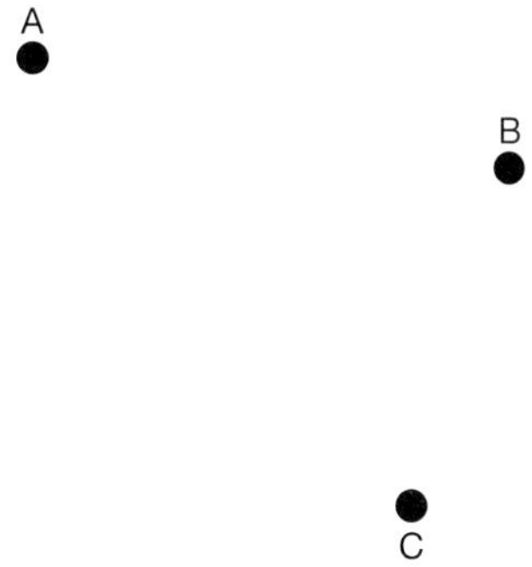

a) Construct the perpendicular bisector of AB.
b) Construct the perpendicular bisector of BC.
c) What can be said about the point of intersection of the two perpendicular bisectors?

4. Draw a triangle similar to the one shown below:

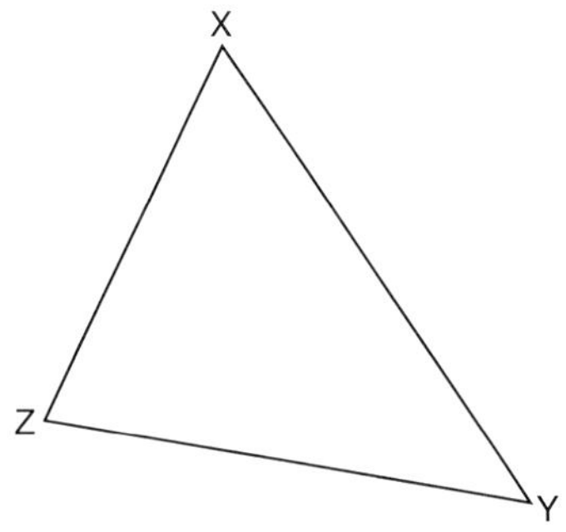

By construction, draw a circle to pass through points X, Y and Z. This is called the **circumcircle** of the triangle.

5. Draw a triangle similar to the one shown below:

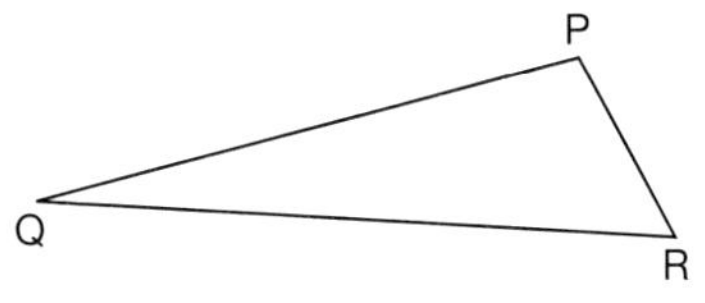

By construction, draw a circle to pass through points P, Q and R.

Scale drawings

Scale drawings are used when an accurate diagram, drawn in proportion, is needed. Common uses of scale drawings include maps and plans. The use of scale drawings involves understanding how to scale measurements.

Worked examples

a) A map is drawn to a scale of 1 : 10 000. If two objects are 1 cm apart on the map, how far apart are they in real life? Give your answer in metres.

A scale of 1 : 10 000, means that 1 cm on the map represents 10 000 cm in real life.

Therefore the distance $= 10\,000$ cm

$= 100$ m

b) A model boat is built to a scale of 1 : 50. If the length of the real boat is 12 m, calculate the length of the model boat in cm.

A scale of 1 : 50 means that 50 cm on the real boat is 1 cm on the model boat.

$12 \text{ m} = 1200 \text{ cm}$

Therefore the length of the model boat $= 1200 \div 50$ cm

$= 24$ cm

c) i) Construct to a scale of 1 : 1, a triangle ABC such that AB = 6 cm, AC = 5 cm and BC = 4 cm.

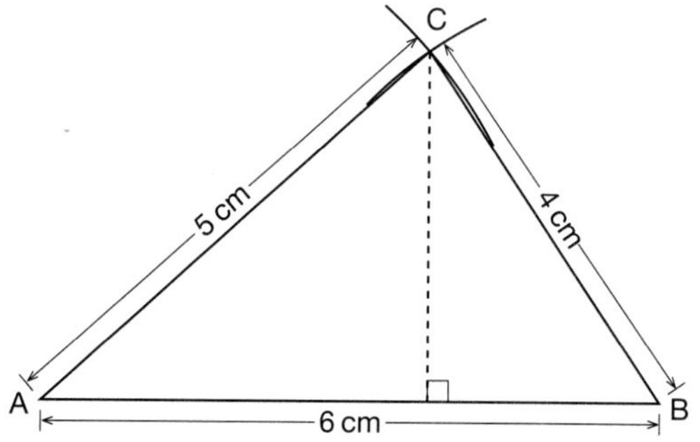

ii) Measure the perpendicular length of C from AB.
Perpendicular length is 3.3 cm.

iii) Calculate the area of the triangle.

$$\text{Area} = \frac{\text{base length} \times \text{perpendicular height}}{2}$$

$$\text{Area} = \frac{6 \times 3.3}{2} \text{ cm} = 9.9 \text{ cm}^2$$

Exercise 23.7

1. In the following questions, both the scale to which a map is drawn and the distance between two objects on the map are given.
   Find the real distance between the two objects, giving your answer in metres.

a) 1 : 10 000	3 cm	b) 1 : 10 000	2.5 cm
c) 1 : 20 000	1.5 cm	d) 1 : 8000	5.2 cm

2. In the following questions, both the scale to which a map is drawn and the true distance between two objects are given.
   Find the distance between the two objects on the map, giving your answer in cm.

a) 1 : 15 000	1.5 km	b) 1 : 50 000	4 km
c) 1 : 10 000	600 m	d) 1 : 25 000	1.7 km

3. A rectangular pool measures 20 m by 36 m as shown below:

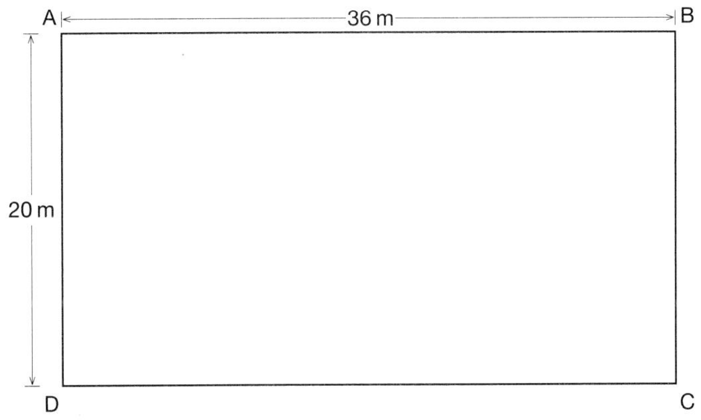

   a) Construct a scale drawing of the pool, using 1 cm for every 4 m.
   b) A boy swims across the pool in such a way that his path is the perpendicular bisector of BD. Show, by construction, the path that he takes.
   c) Work out the distance the boy swam.

4. A triangular enclosure is shown in the diagram below:

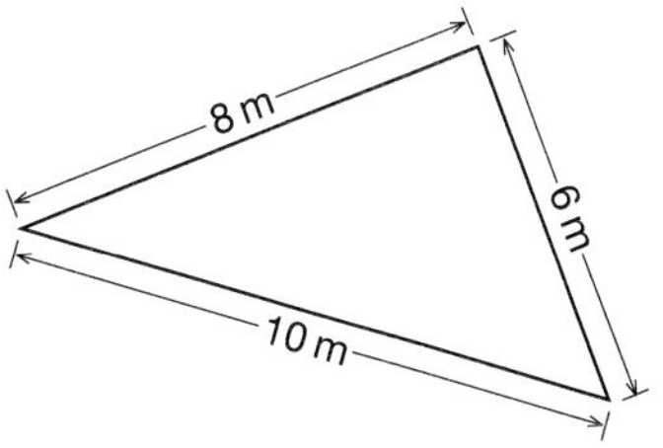

   a) Using a scale of 1 cm for each metre, construct a scale drawing of the enclosure.
   b) Calculate the true area of the enclosure.

5. Three radar stations A, B and C pick up a distress signal from a boat at sea.

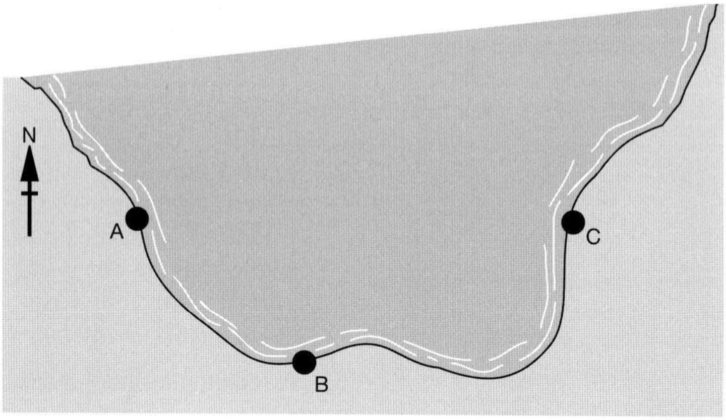

C is 24 km due East of A, AB = 12 km and BC = 18 km. The signal indicates that the boat is equidistant from all three radar stations.

a) By construction and using a scale of 1 cm for every 3 km, locate the position of the boat.
b) What is the boat's true distance from each radar station?

6. A plan view of a field is shown below:

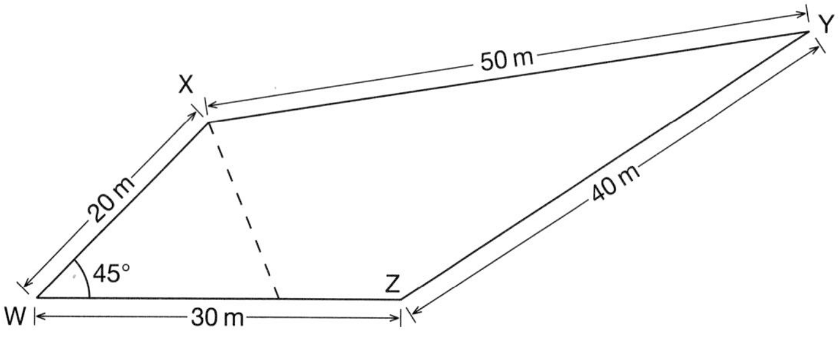

a) Using a scale of 1 cm for every 5 m, construct a scale drawing of the field.
b) A farmer divides the field by running a fence from X in such a way that it bisects ∠WXY. By construction, show the position of the fence on your diagram.
c) Work out the length of fencing used.

Student Assessment 1

1. a) Using a ruler, measure the length of the line below:

b) Draw a line 4.7 cm long.

2. a) Using a protractor, measure the angle below:

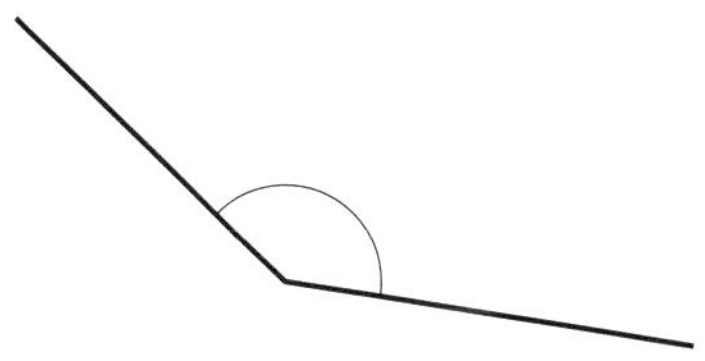

b) Draw an angle of 300°.

3. Construct $\triangle ABC$ such that AB = 8 cm, AC = 6 cm and BC = 12 cm.

4. Construct, using a pair of compasses, the following circle pattern:

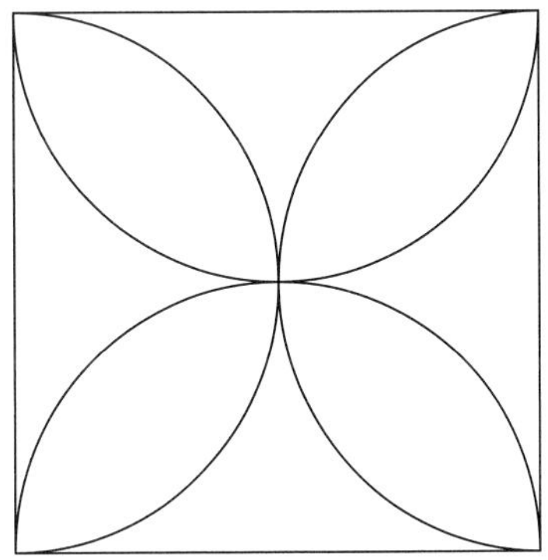

5. Three players, P, Q and R, are approaching a football. Their positions relative to each other are shown below:

The ball is equidistant from all three players. Copy the diagram and show, by construction, the position of the ball.

6. A plan of a living room is shown below:

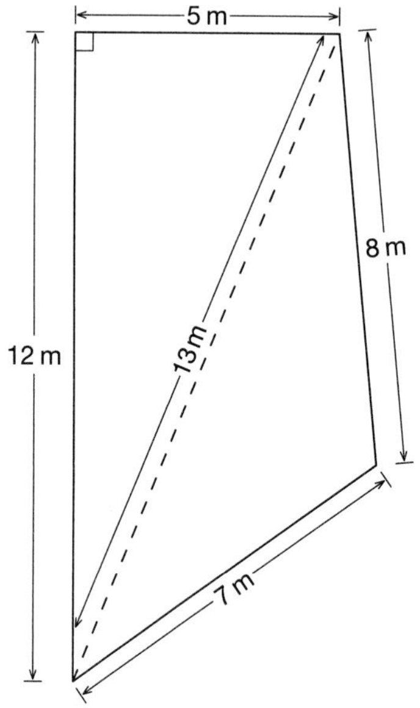

a) Using a pair of compasses, construct a scale drawing of the room using 1 cm for every metre.
b) Using a set square if necessary, calculate the total area of the actual living room.

Student Assessment 2

1. Measure each of the five angles of the pentagon below:

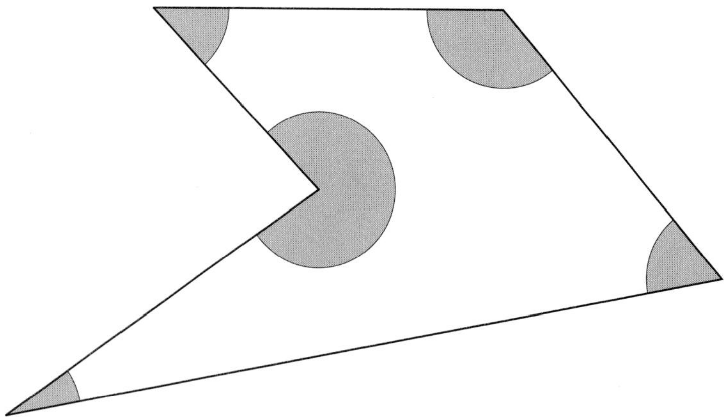

2. Draw, using a ruler and a protractor, a triangle with angles of 40°, 60° and 80°.

3. a) Draw an angle of 320°.
   b) Using a pair of compasses, bisect the angle.

4. Construct the circle pattern shown below:

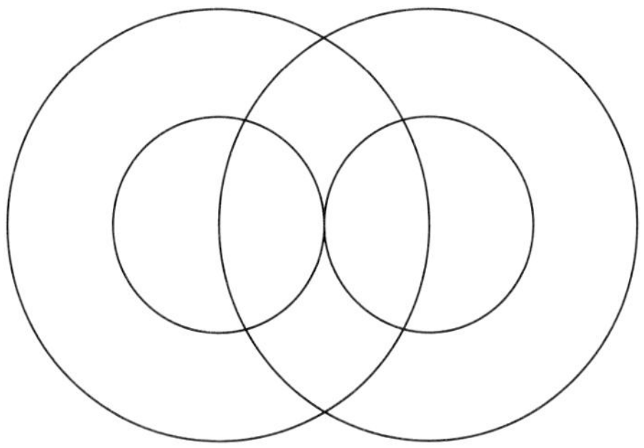

5. In the following questions, both the scale to which a map is drawn and the true distance between two objects are given. Find the distance between the two objects on the map, giving your answer in cm.
a) 1 : 20 000 4.4 km b) 1 : 50 000 12.2 km

6. a) Construct a regular hexagon with sides of length 3 cm.
b) Calculate its area, showing your method clearly.

24 ANGLE PROPERTIES

Calculate unknown angles using the following geometrical properties: angles at a point, angles formed within parallel lines, angle properties of triangles and quadrilaterals, angle properties of regular polygons, angle in a semi-circle, and angle between tangent and radius of a circle.

NB: All diagrams are not drawn to scale.

Angles at a point

One complete revolution is equivalent to a rotation of 360° about a point. Similarly, half a complete revolution is equivalent to a rotation of 180° about a point. These facts can be seen clearly by looking at either a circular angle measurer or a semi-circular protractor.

Worked examples

a) Calculate the size of the angle x in the diagram below:

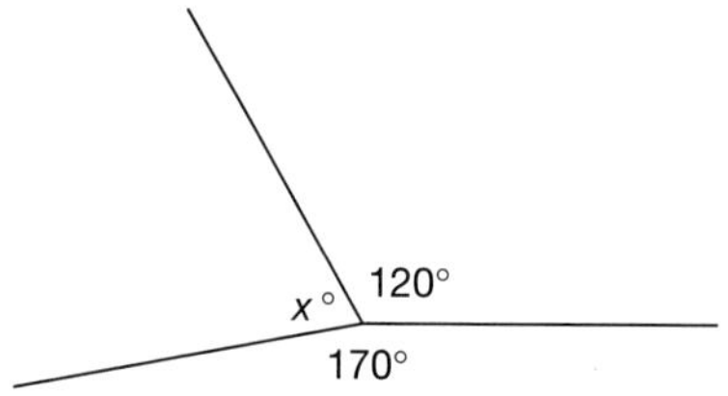

The sum of all the angles around a point is 360°. Therefore:

$$120 + 170 + x = 360$$
$$x = 360 - 120 - 170$$
$$x = 70$$

Therefore angle x is 70°.
Note that the size of the angle x is **calculated** and **not measured**.

b) Calculate the size of angle a in the diagram below:

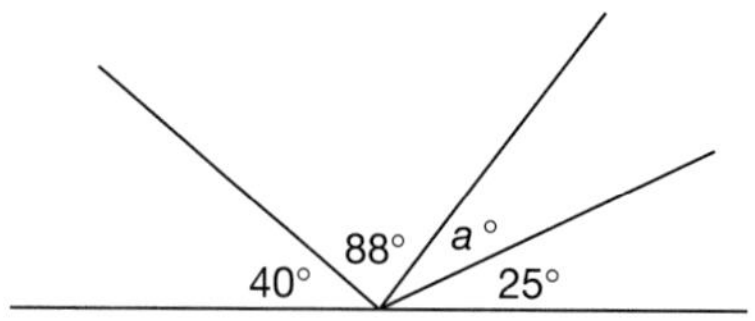

The sum of all the angles at a point on a straight line is 180°. Therefore:

$$40 + 88 + a + 25 = 180$$
$$a = 180 - 40 - 88 - 25$$
$$a = 27$$

Therefore angle a is 27°.

Exercise 24.1

1. In the following questions, calculate the size of angle x:

a)

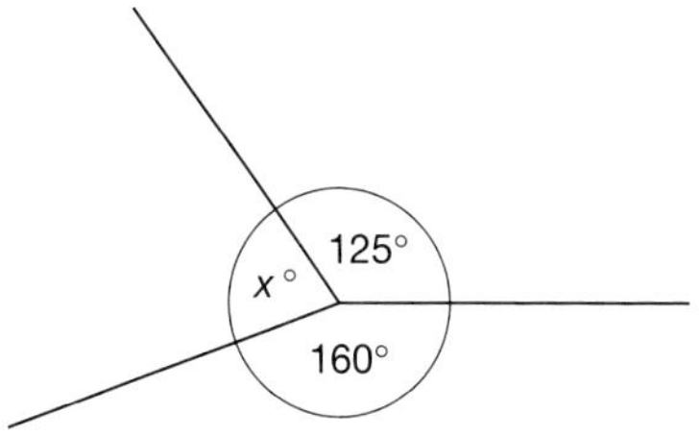

b)

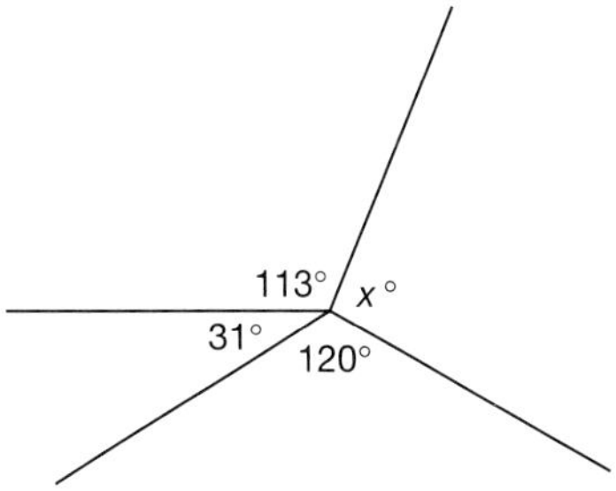

c)

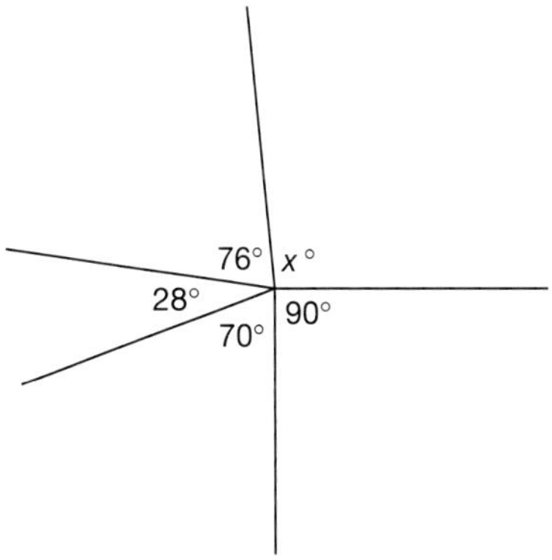

d)

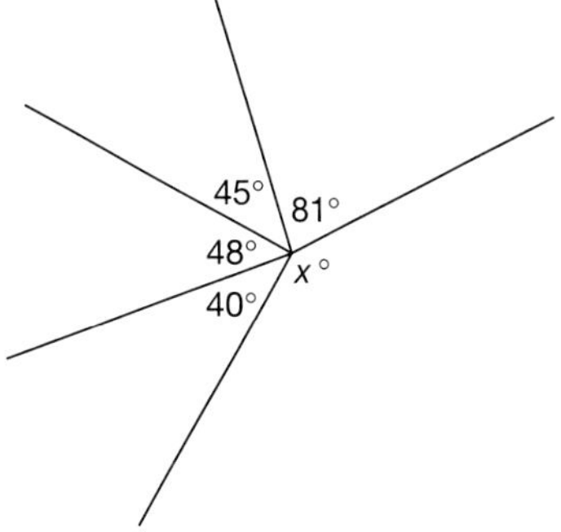

2. In the following questions, the angles lie about a point on a straight line. Calculate the size of angle *y* in each case:

a)

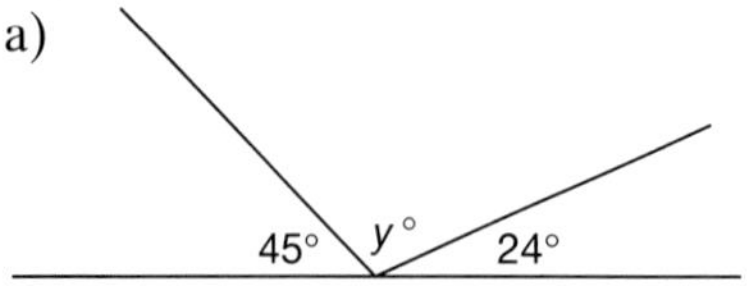

b)

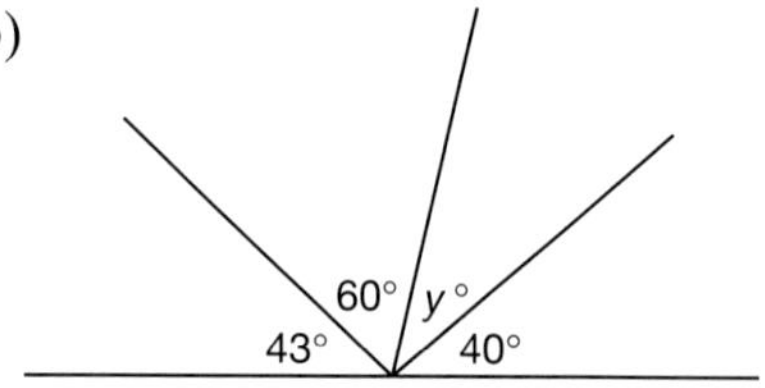

c)

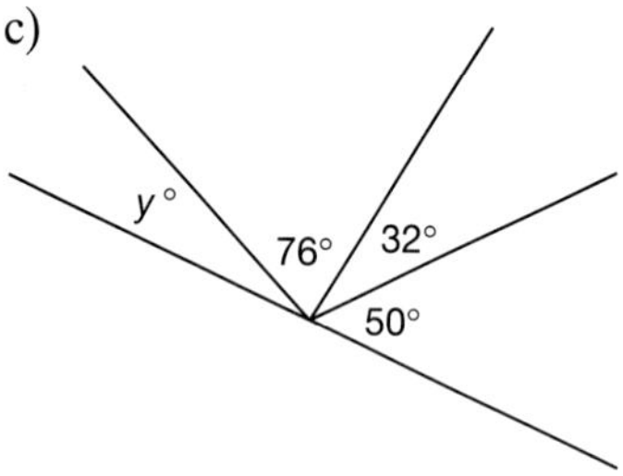

d)

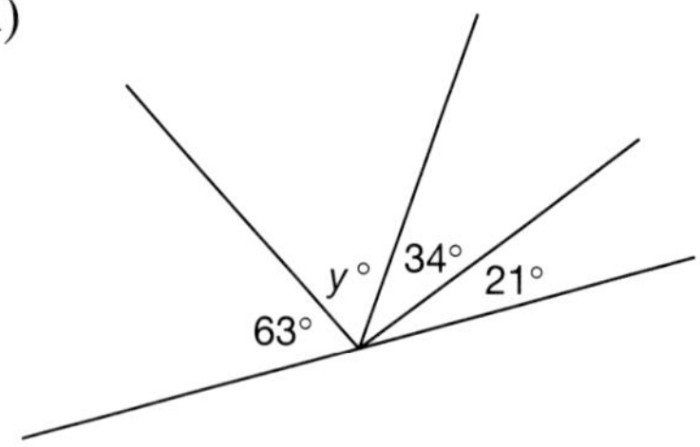

3. Calculate the size of angle *p* in each of the following:

a)

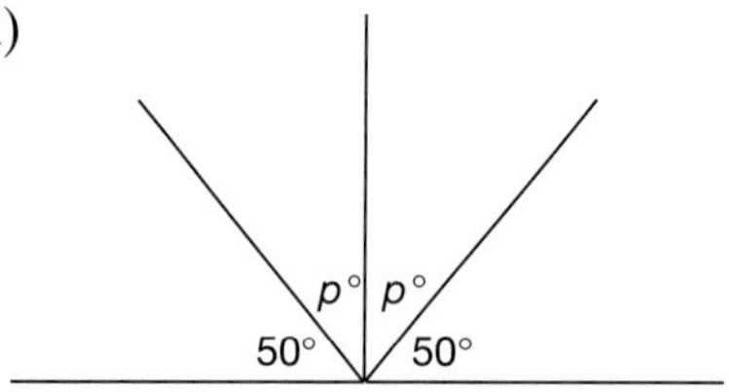

b)

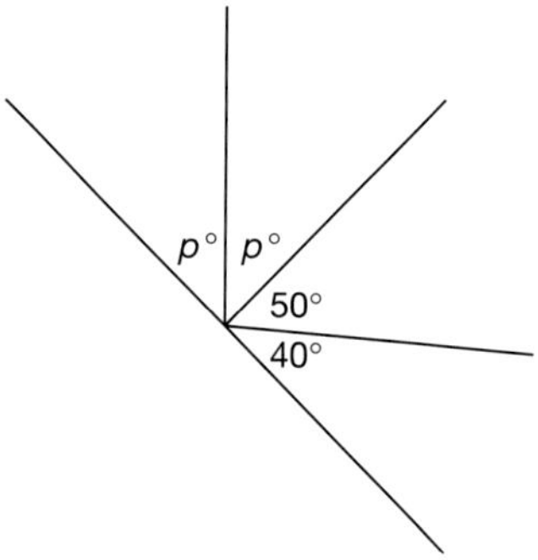

c)

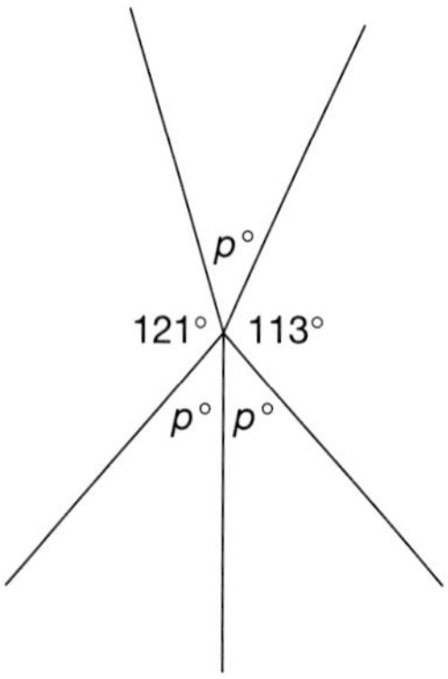

d)

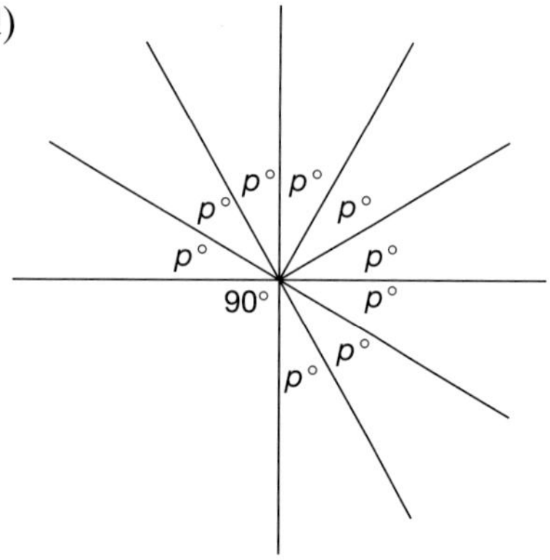

Angles formed within parallel lines

Exercise 24.2

1. Draw a similar diagram to the one shown below. Measure carefully each of the labelled angles and write them down.

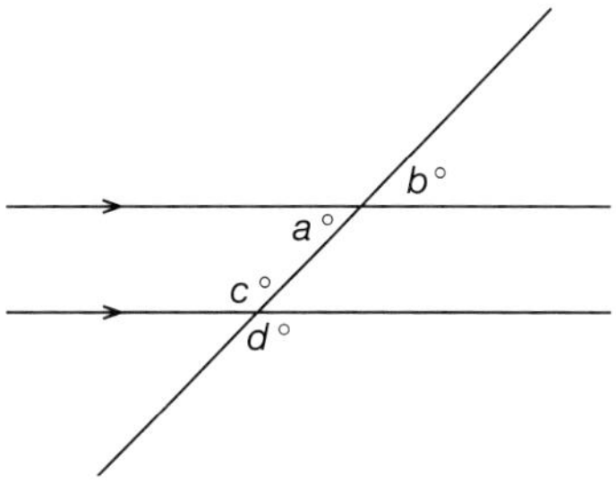

2. Draw a similar diagram to the one shown below. Measure carefully each of the labelled angles and write them down.

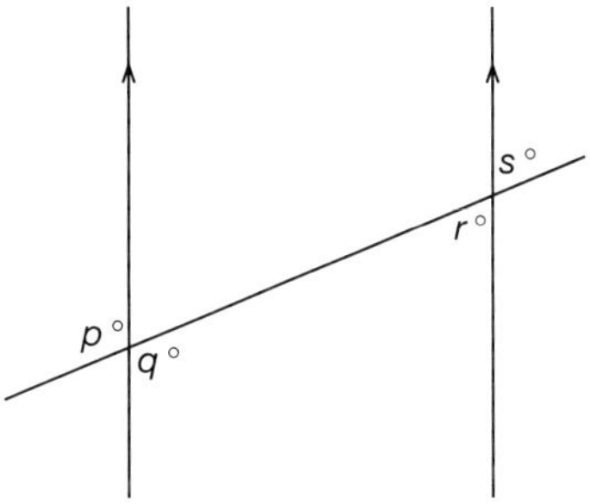

3. Draw a similar diagram to the one shown below. Measure carefully each of the labelled angles and write them down.

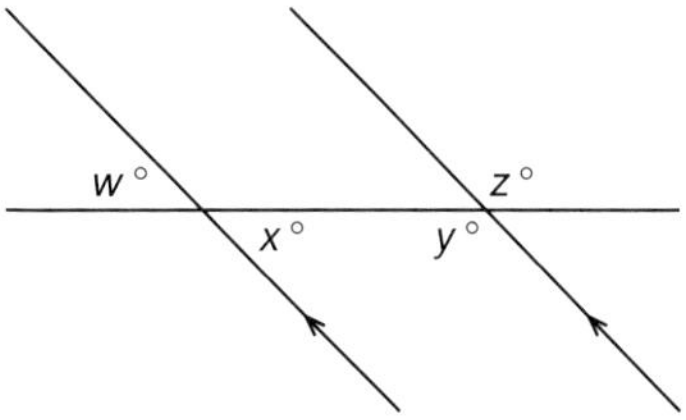

4. Write down what you have noticed about the angles you measured in Q.1–3.

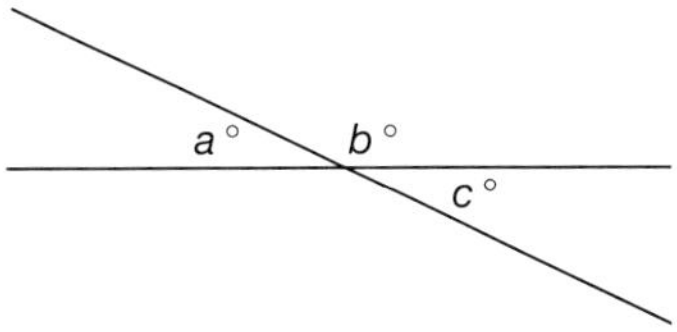

When two straight lines cross, it is found that the angles opposite each other are the same size. They are known as **vertically opposite angles**. By using the fact that angles at a point on a straight line add up to 180°, it can be shown why vertically opposite angles must always be equal in size.

$a + b = 180°$
$c + b = 180°$

Therefore, a is equal to c.

Exercise 24.3

1. Draw a similar diagram to the one shown below. Measure carefully each of the labelled angles and write them down.

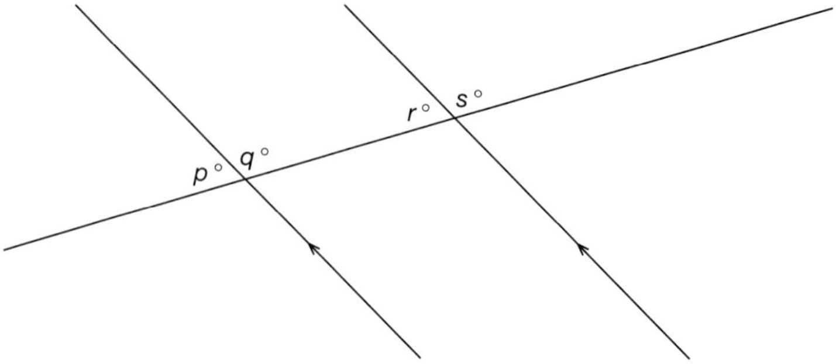

2. Draw a similar diagram to the one shown below. Measure carefully each of the labelled angles and write them down.

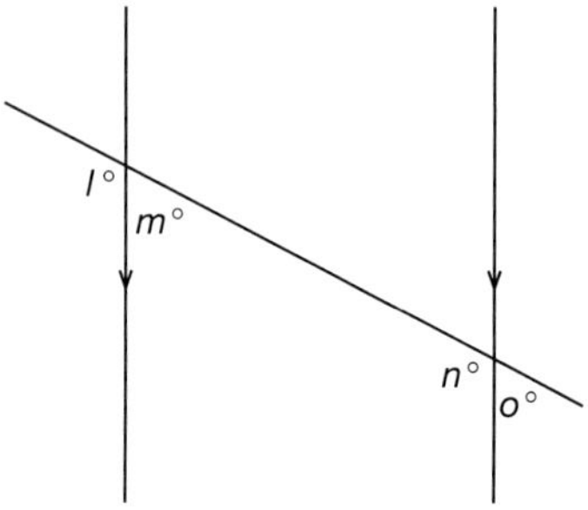

3. Draw a similar diagram to the one shown below. Measure carefully each of the labelled angles and write them down.

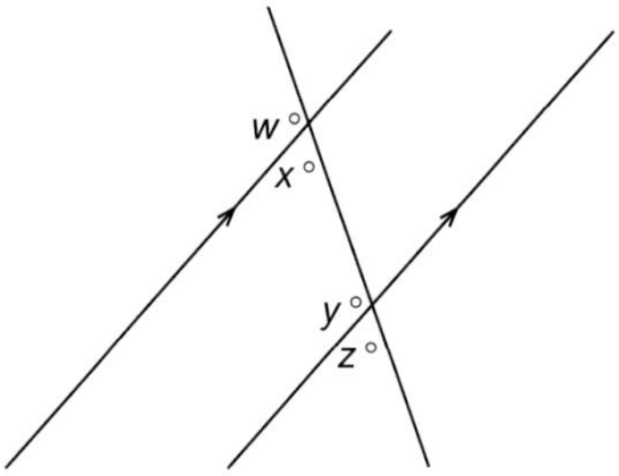

4. Write down what you have noticed about the angles you measured in Q.1–3.

When a line intersects two parallel lines, as in the diagram below, it is found that certain angles are the same size.

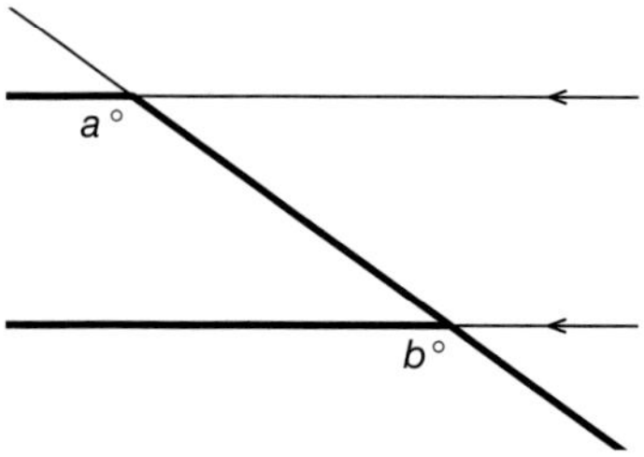

The angles a and b are equal and are known as **corresponding angles**. Corresponding angles can be found by looking for an 'F' formation in a diagram.

A line intersecting two parallel lines also produces another pair of equal angles, known as **alternate angles**. These can be shown to be equal by using the fact that both vertically opposite and corresponding angles are equal.

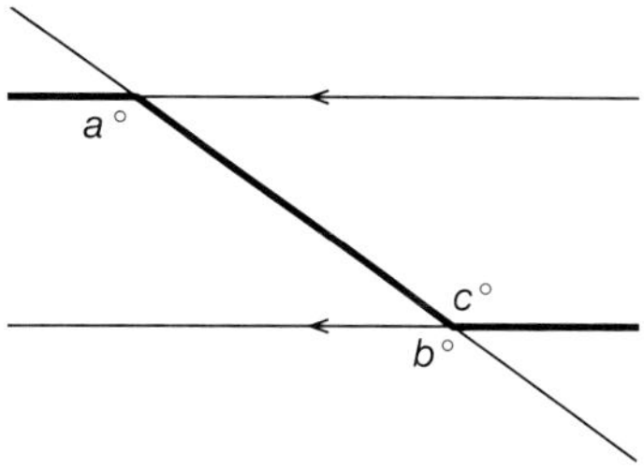

In the diagram above, $a = b$ (corresponding angles). But $b = c$ (vertically opposite). It can therefore be deduced that $a = c$.

Angles a and c are alternate angles. These can be found by looking for a 'Z' formation in a diagram.

Exercise 24.4

In each of the following questions, some of the angles are given. Deduce, giving your reasons, the size of the other labelled angles.

1.

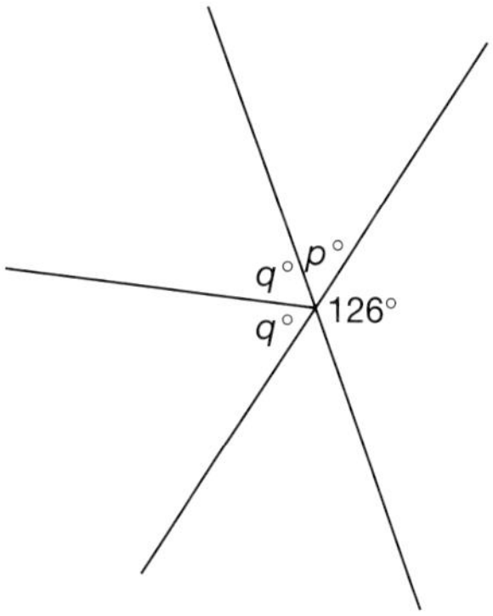

2.

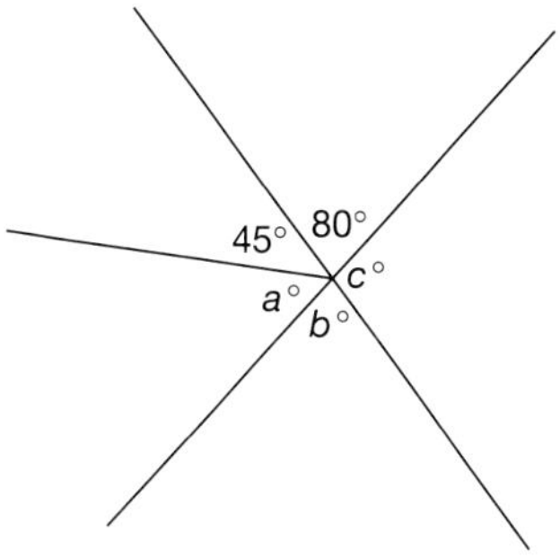

3.

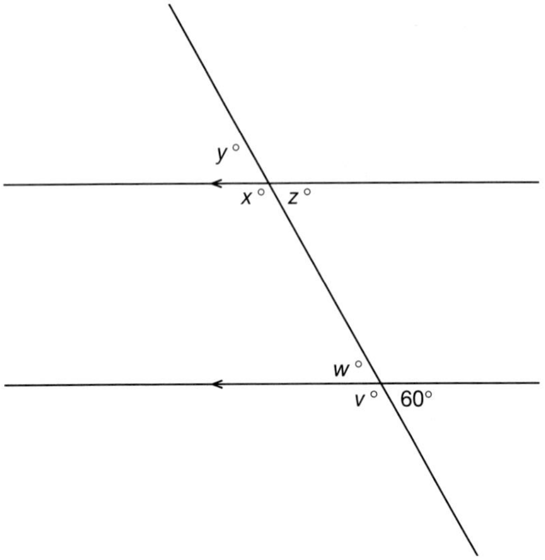

4.

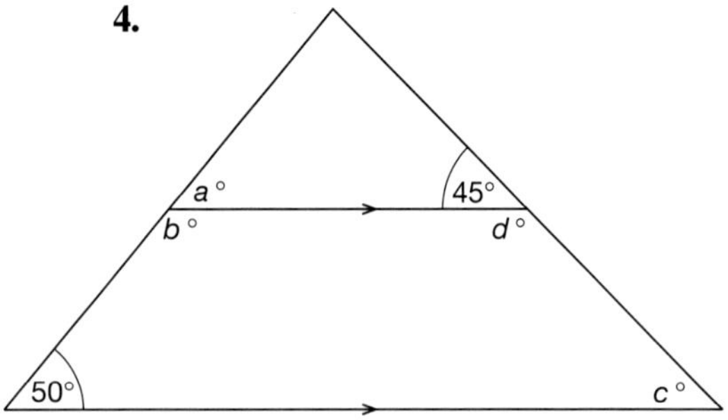

5.

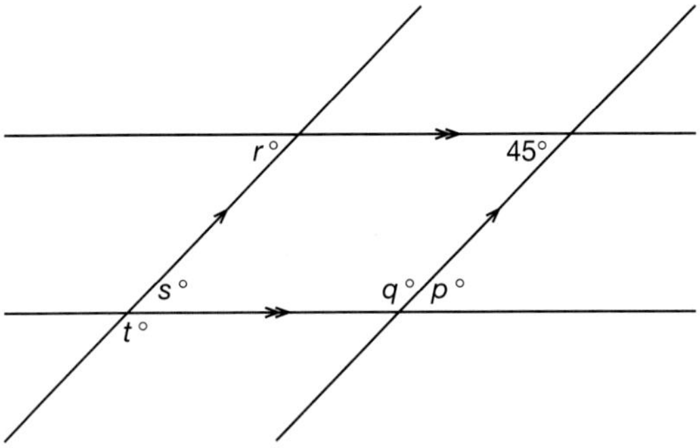

6.

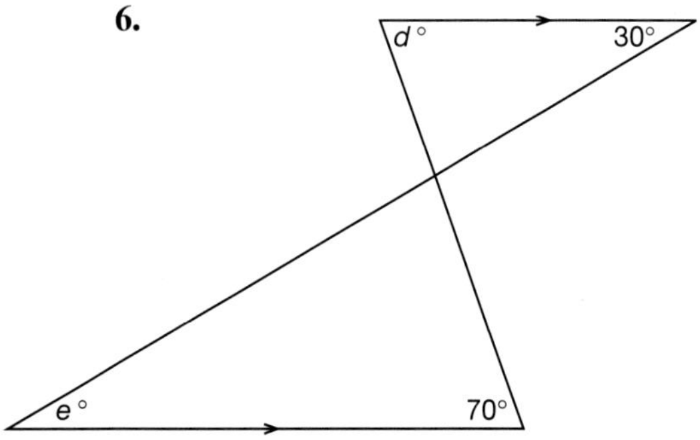

7.

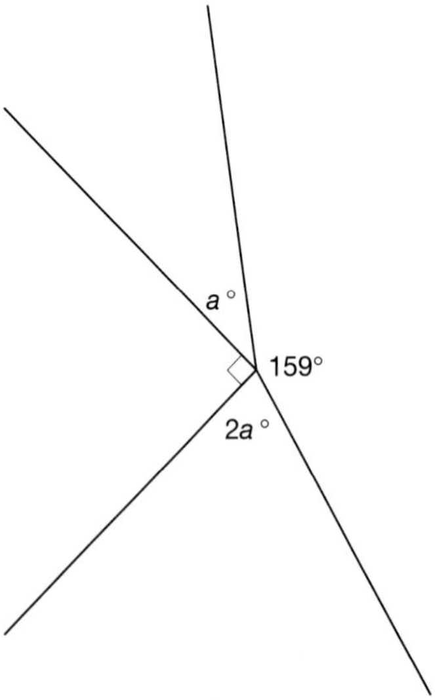

8.

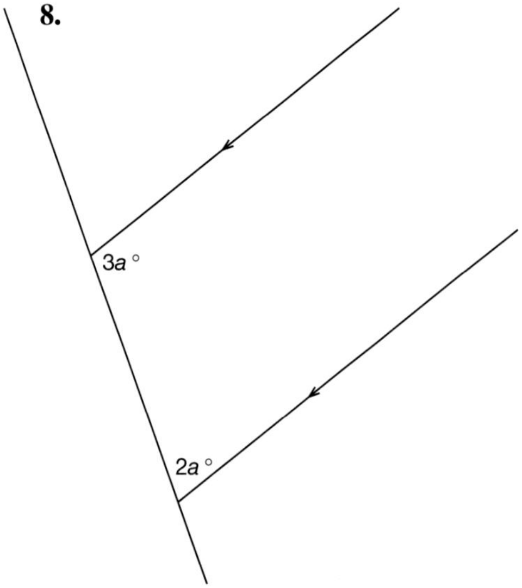

Angle properties of triangles

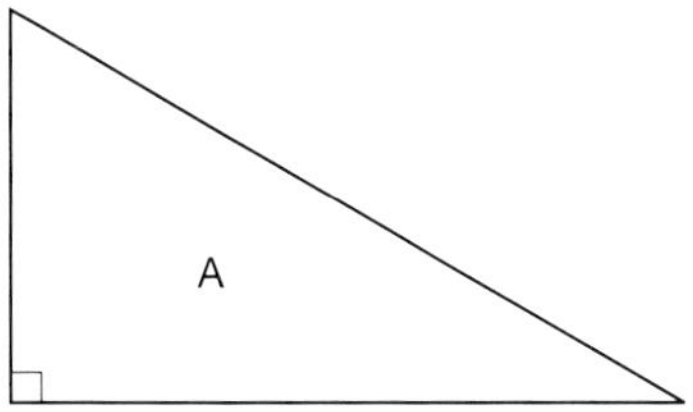

A **triangle** is a plane (two-dimensional) shape consisting of three angles and three sides. There are six main types of triangle. Their names refer to the size of their angles and/or the length of their sides, and are as follows:

Triangle A is known as a **right-angled** triangle, since one of its angles is a right angle.

Triangle B is known as an **acute-angled** triangle, since all of its angles are less than 90°.

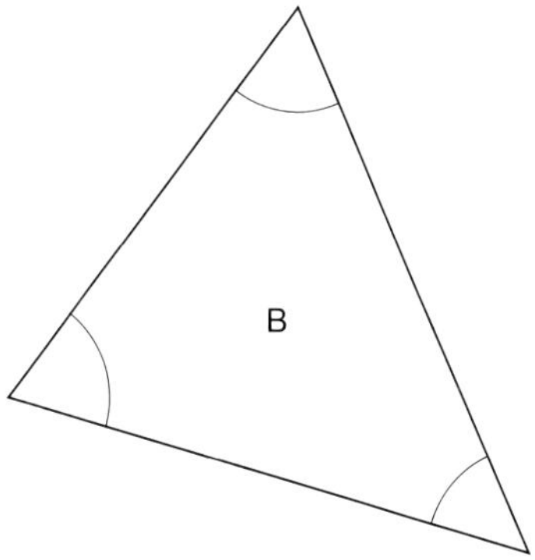

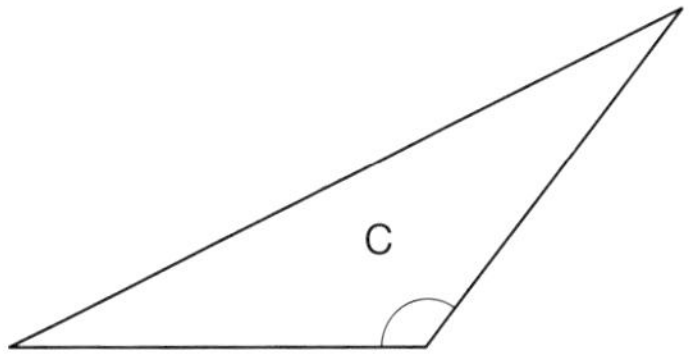

Triangle C is known as an **obtuse-angled** triangle, since one of its angles is greater than 90°.

Triangle D is known as an **equilateral** triangle. All its sides are the same length, and all its angles are the same size. Note that the dashes on the sides indicate sides of equal length.

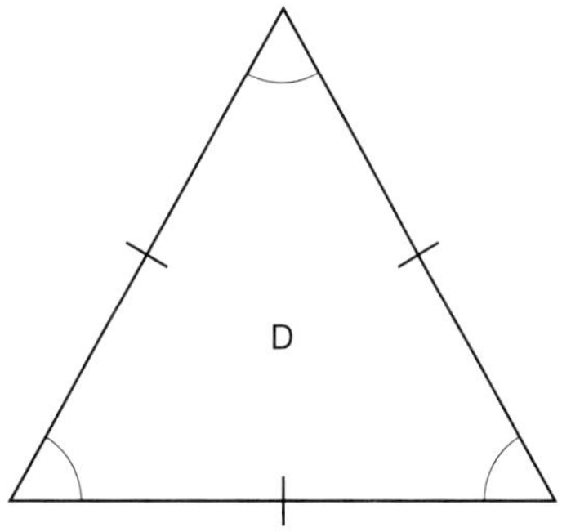

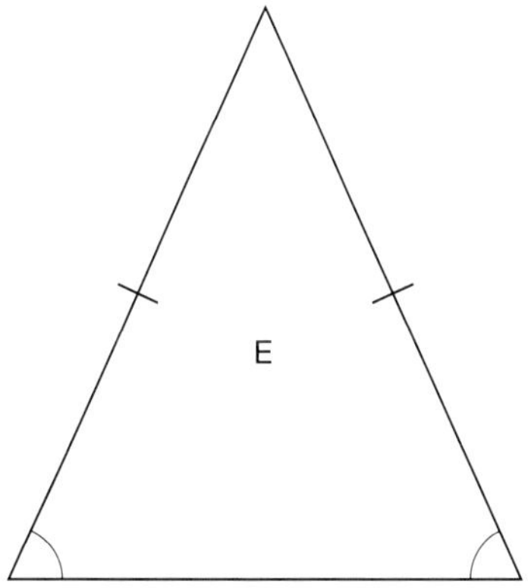

Triangle E is known as an **isosceles** triangle. Two of its sides are the same length, and it also has two angles of the same size.

Triangle F is known as a **scalene** triangle. All of its sides are of different lengths, and all of its angles are different sizes.

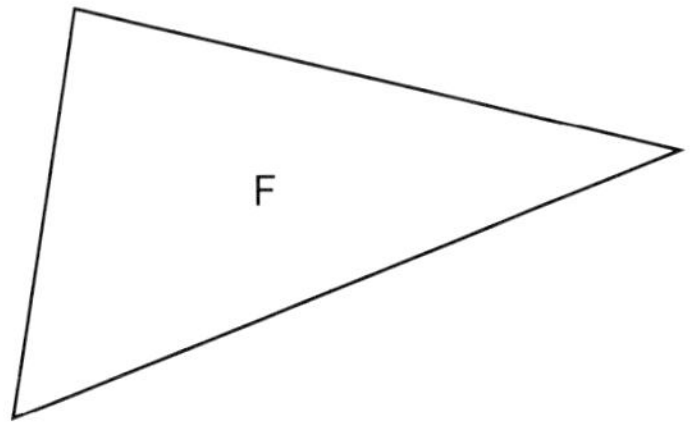

Exercise 24.5

1. a) Draw five different triangles. Label their angles x, y and z. As accurately as you can, measure the three angles of each triangle and add them together.

b) What do you notice about the sum of the three angles of each of your triangles?

2. a) Draw a triangle on a piece of paper and label the angles a, b and c. Tear off the corners of the triangle and arrange them as shown below:

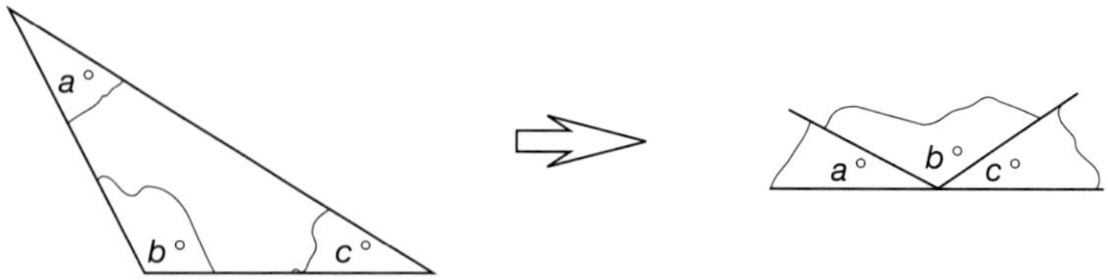

b) What do you notice about the total angle that a, b and c make?

It can be seen from the questions above that triangles of any shape have one thing in common. That is, that the sum of their three angles is constant: 180°.

Exercise 24.6

1. For each of the triangles drawn below, decide whether they are equilateral, isosceles or scalene.

a)

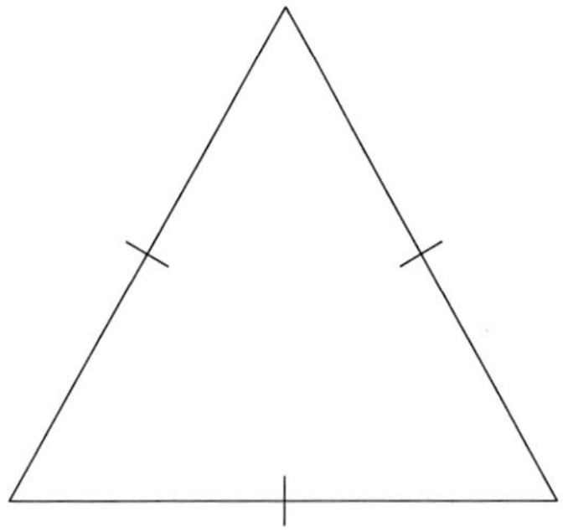

b)

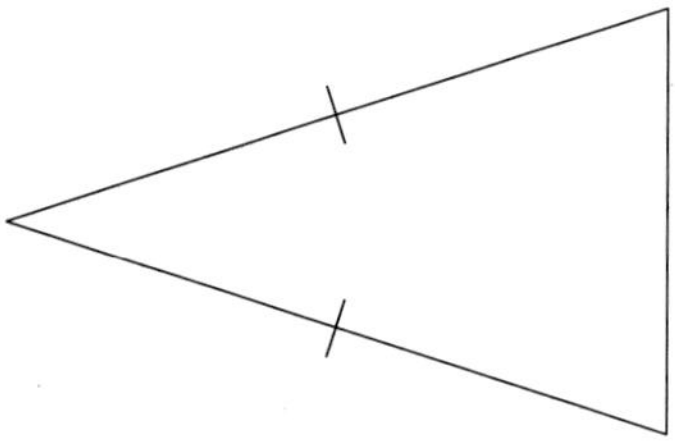

c)

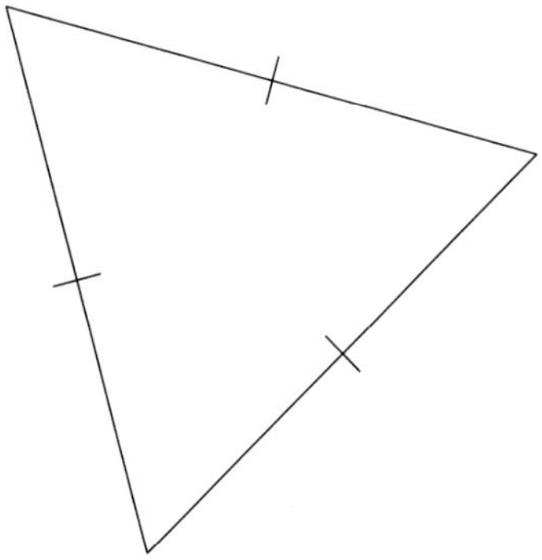

d)

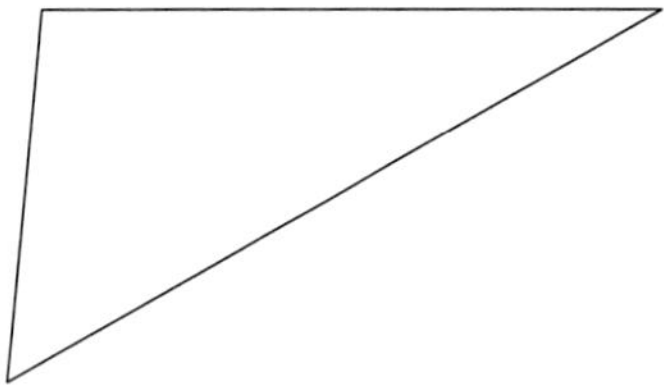

e)

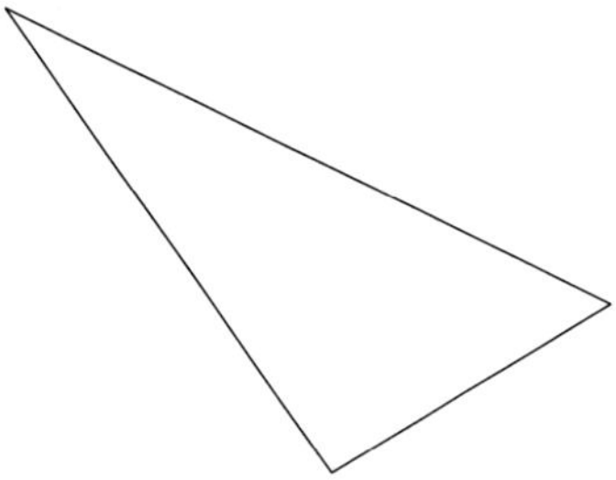

f)

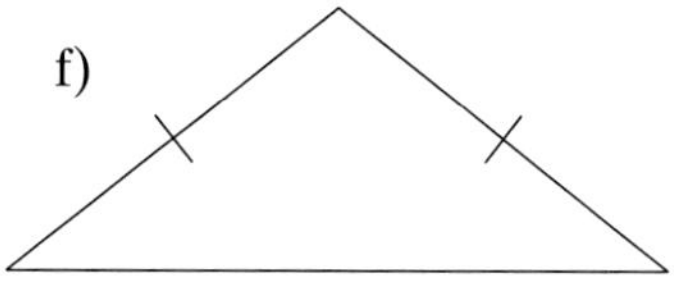

g)

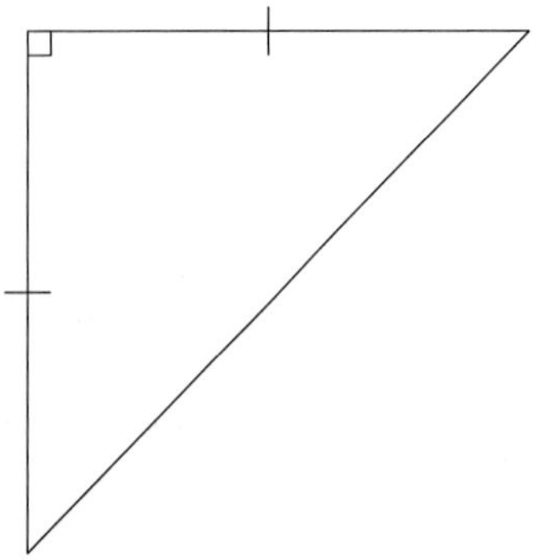

h)

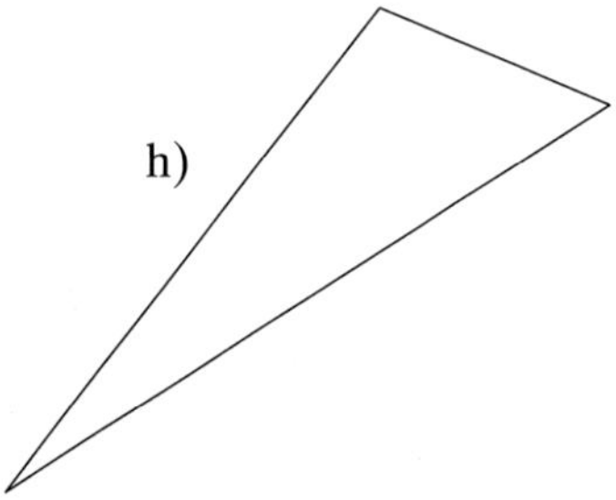

2. For each of the triangles below, use the information given to calculate the size of angle x.

a)

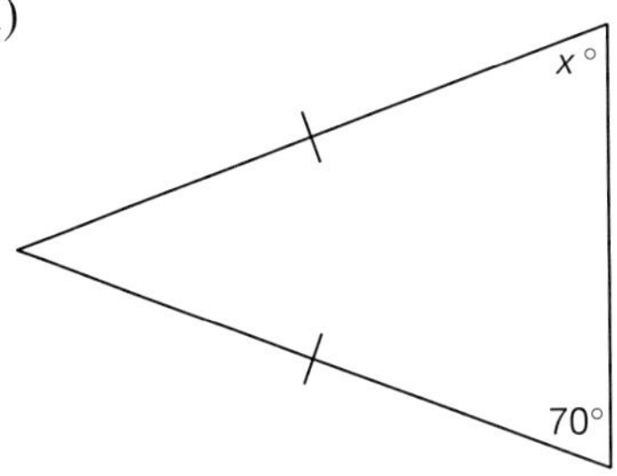

b)

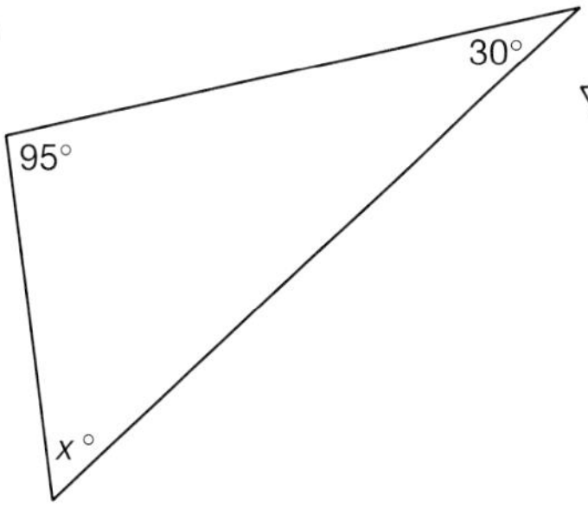

c)

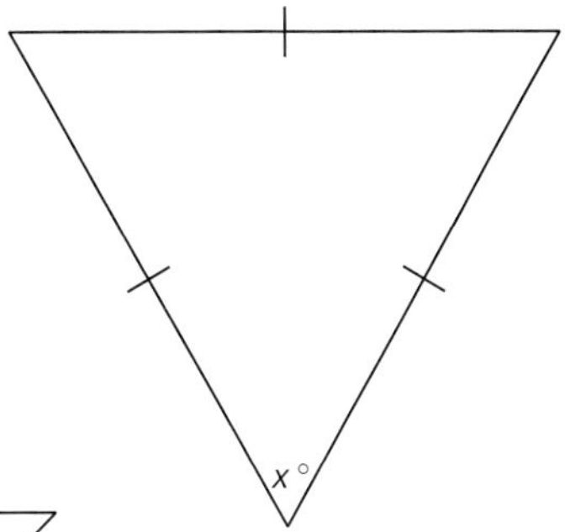

d)

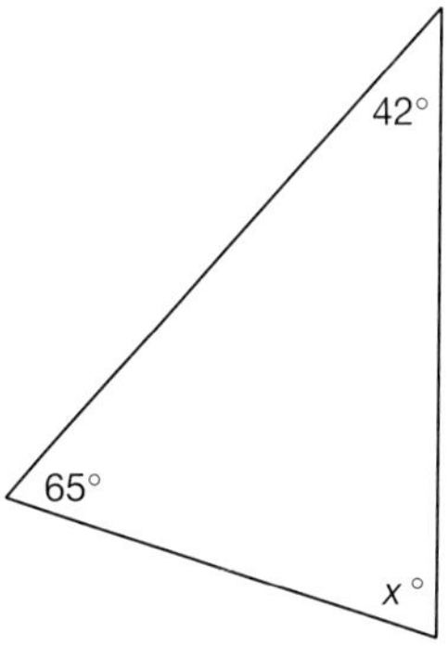

e)

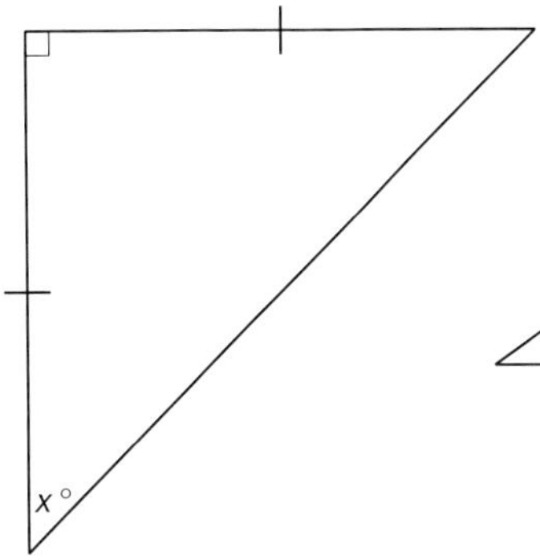

f)

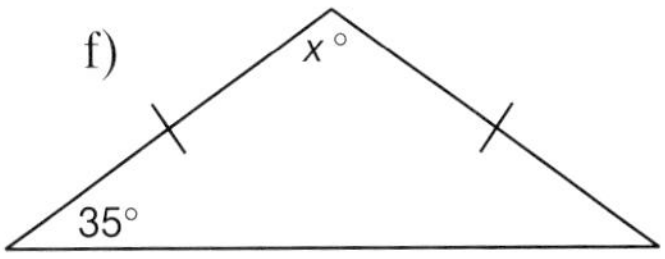

3. In each of the diagrams below, calculate the size of the labelled angles.

a)

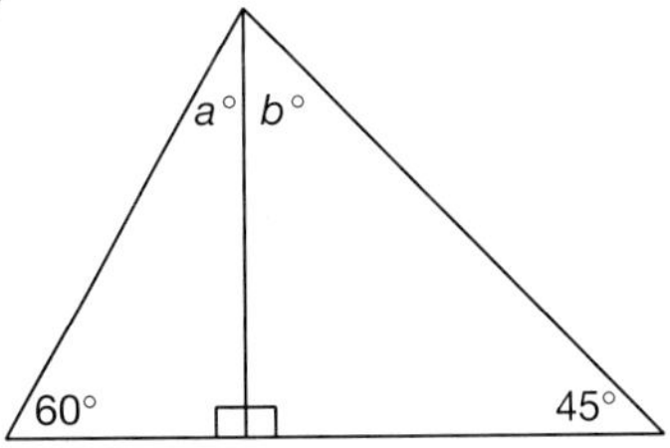

b)

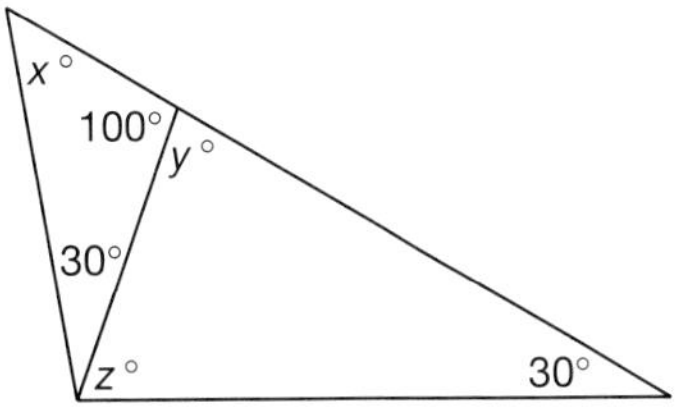

c)

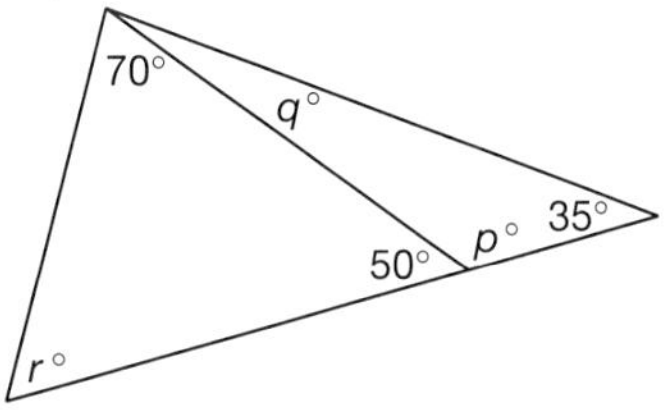

d)

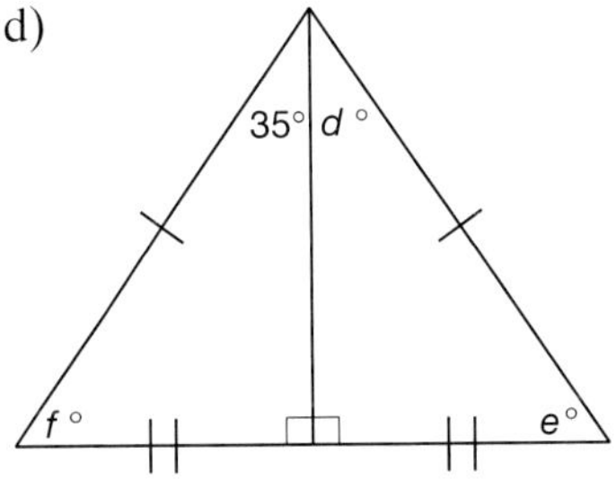

e)

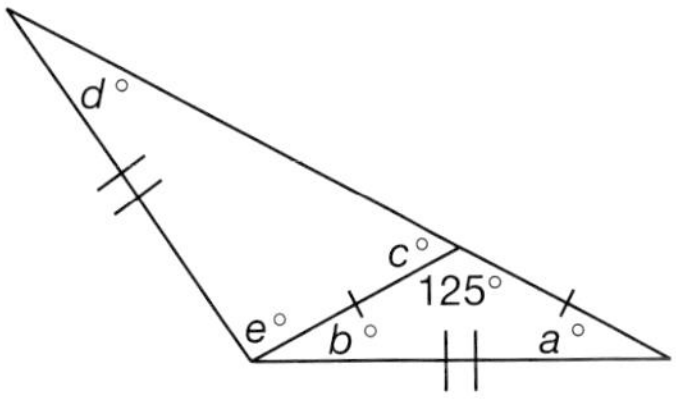

f)

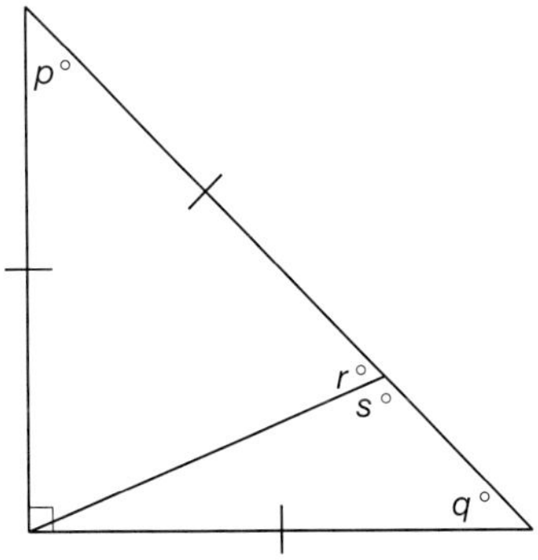

Angle properties of quadrilaterals

A **quadrilateral** is a plane shape consisting of four angles and four sides. There are several types of quadrilateral. The main ones, and their properties, are described below.

Two pairs of parallel sides.
All sides are equal.
All angles are equal.
Diagonals intersect at right angles.

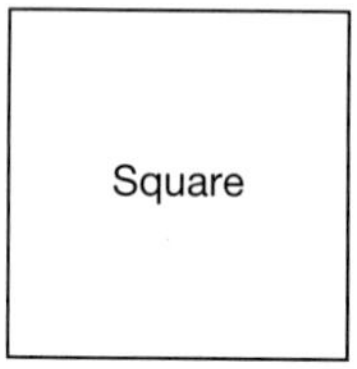

Two pairs of parallel sides.
Opposite sides are equal.
All angles are equal.

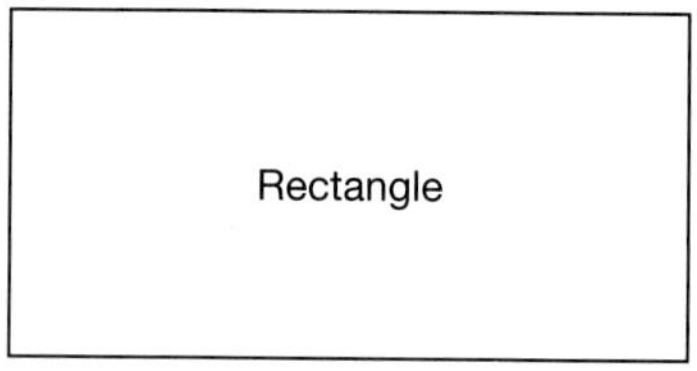

Two pairs of parallel sides.
All sides are equal.
Opposite angles are equal.
Diagonals intersect at right angles.

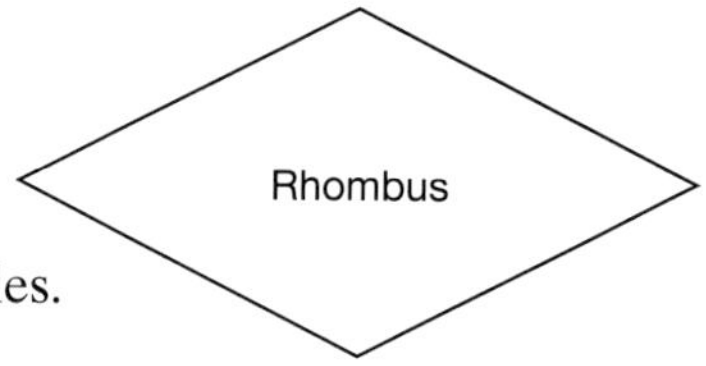

Two pairs of parallel sides.
Opposite sides are equal.
Opposite angles are equal.

One pair of parallel sides.

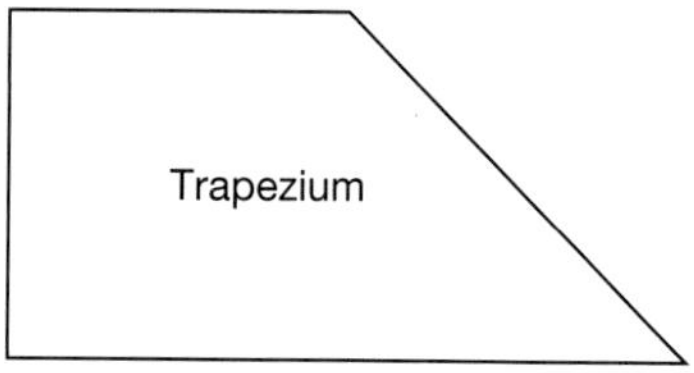

Two pairs of equal sides.
One pair of equal angles.
Diagonals intersect at right angles.

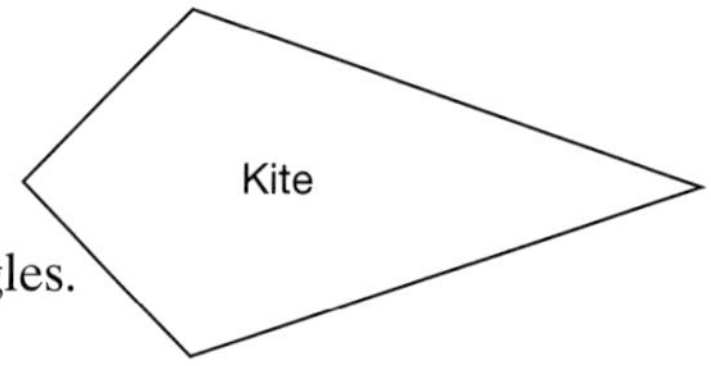

Exercise 24.7 Copy onto squared paper each of the quadrilaterals drawn below. By looking at their properties, state the type of quadrilateral that each one is.

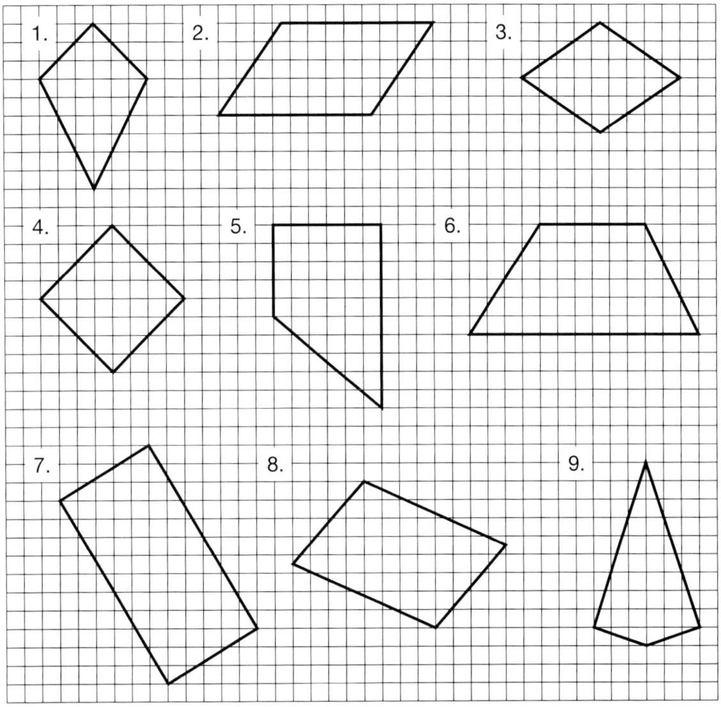

In the quadrilaterals below, a straight line is drawn from one of the corners (vertices) to the opposite corner. The result is to split the quadrilaterals into two triangles.

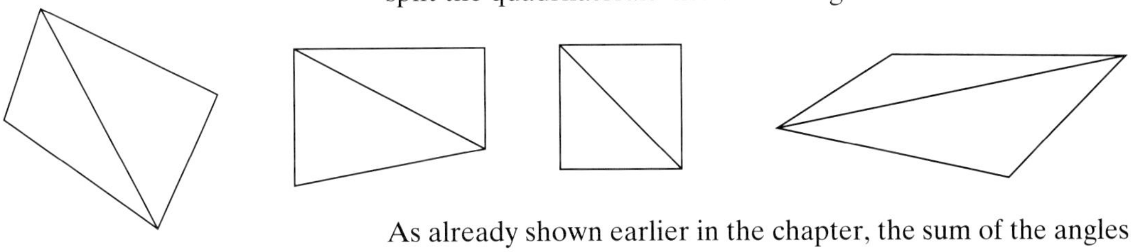

As already shown earlier in the chapter, the sum of the angles of a triangle is 180°. Therefore, as a quadrilateral can be drawn as two triangles, the sum of the four angles of any quadrilateral must be 360°.

Worked example Calculate the size of angle p in the quadrilateral below:

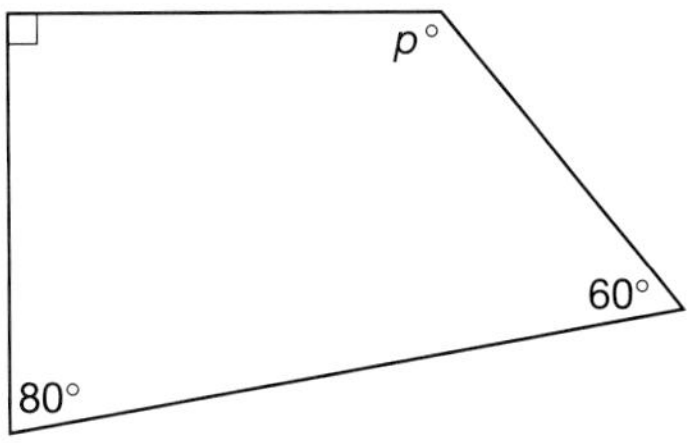

$$90 + 80 + 60 + p = 360$$
$$p = 360 - 90 - 80 - 60$$

Therefore angle p is 130°.

Exercise 24.8

In each of the diagrams below, calculate the size of the labelled angles.

1.

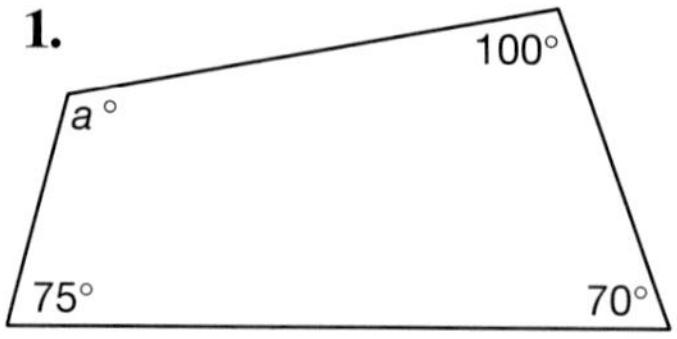

2.

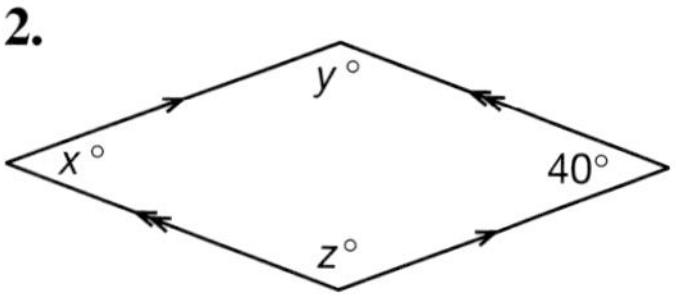

3.

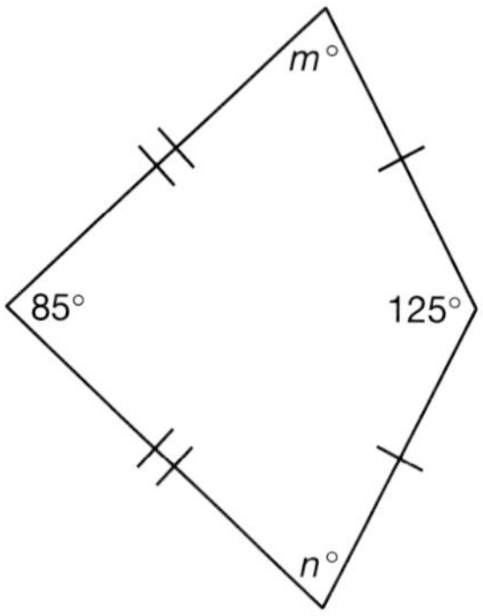

4.

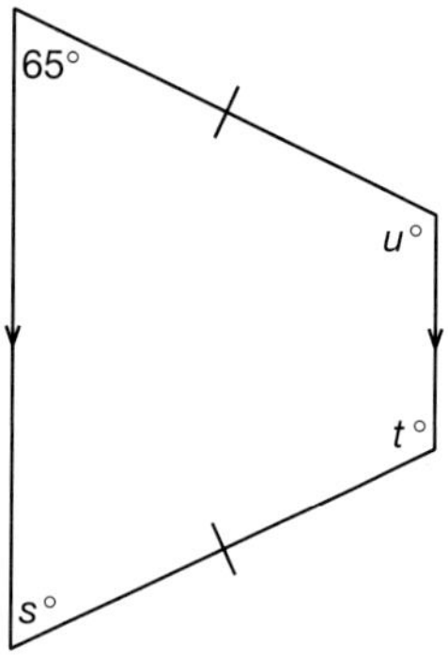

5.

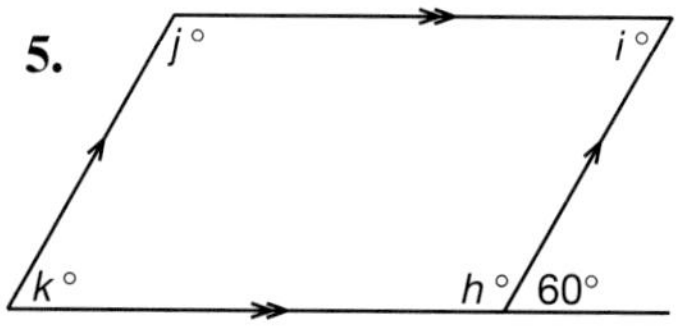

6.

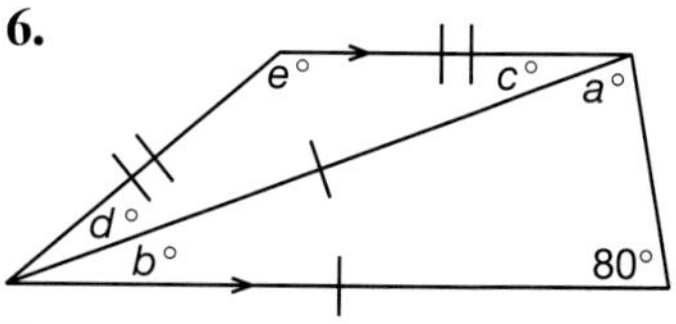

7.

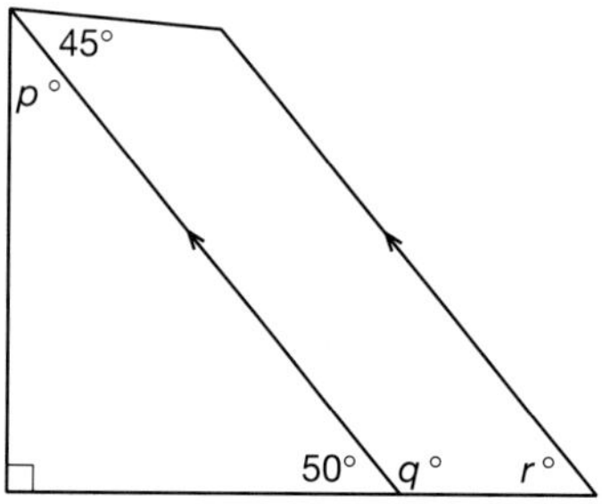

8.

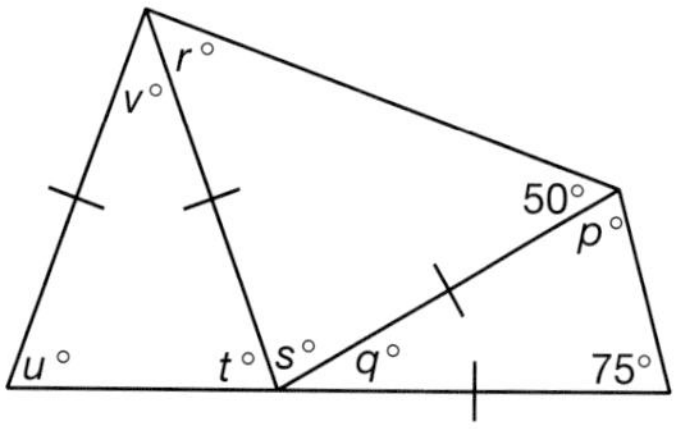

A **polygon** is a closed two-dimensional shape bounded by straight lines. Examples of polygons include triangles, quadrilaterals, pentagons and hexagons. Hence the shapes below all belong to the polygon family:

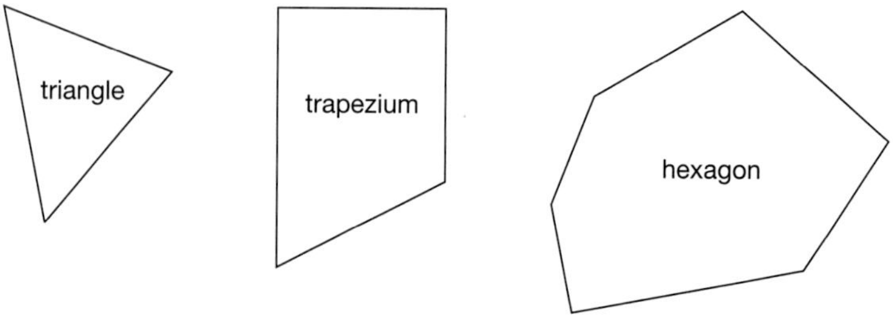

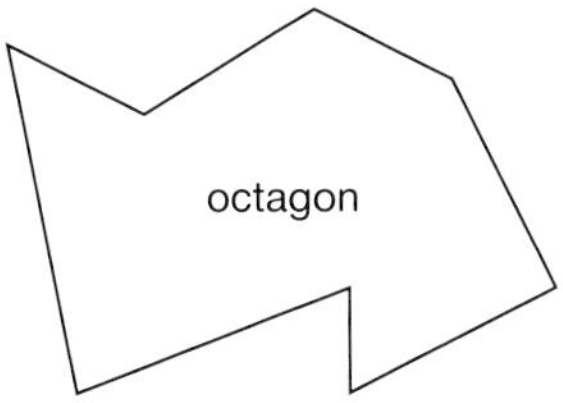

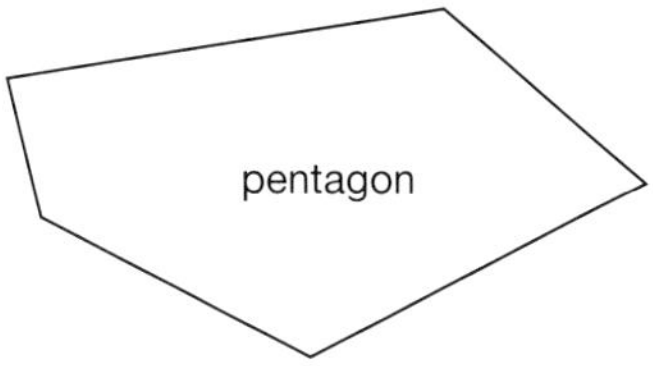

A **regular polygon** is distinctive in that all its sides are of equal length and all its angles are of equal size. Below are some examples of regular polygons.

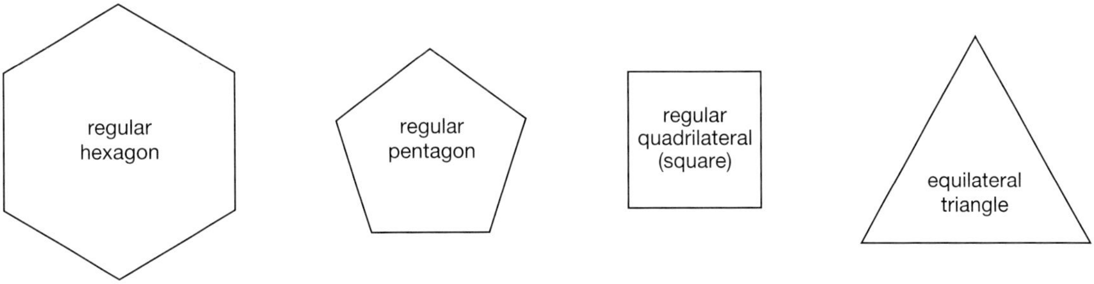

The name of each polygon is derived from the number of angles it contains. The following list identifies some of these polygons.

3 angles = **tri**angle
4 angles = **quad**rilateral (tetragon)
5 angles = **penta**gon
6 angles = **hexa**gon
7 angles = **hepta**gon
8 angles = **octa**gon
9 angles = **nona**gon
10 angles = **deca**gon
12 angles = **dodeca**gon

The sum of the interior angles of a polygon

In the polygons below a straight line is drawn from each vertex to vertex A.

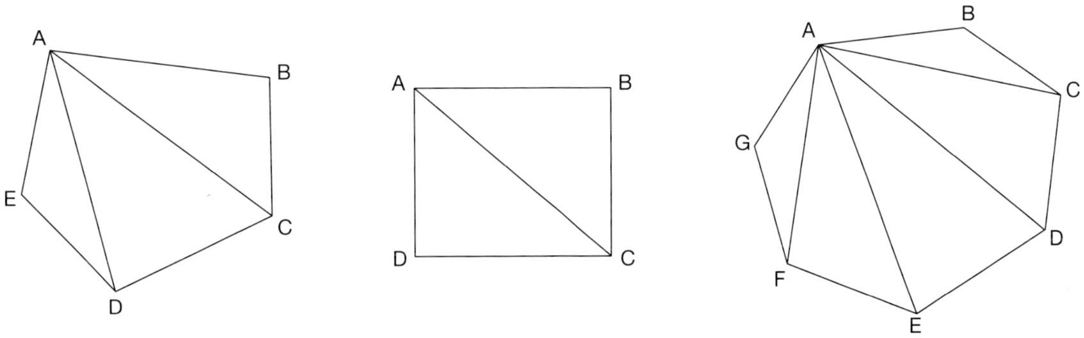

As can be seen, the number of triangles is always two less than the number of sides the polygon has, i.e. if there are n sides, there will be $(n - 2)$ triangles.

Since the angles of a triangle add up to 180°, the sum of the interior angles of a polygon is therefore $180(n - 2)$ degrees.

Worked example Find the sum of the interior angles of a regular pentagon and hence the size of each interior angle.

For a pentagon, $n = 5$.

$$\text{Therefore the sum of the interior angles} = 180(5 - 2)° = 180 \times 3° = 540°$$

For a regular pentagon the interior angles are of equal size.

$$\text{Therefore each angle} = \frac{540°}{5} = 108°.$$

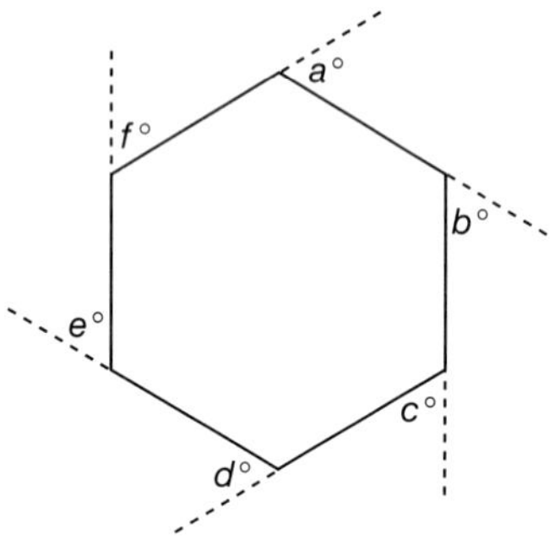

The sum of the exterior angles of a polygon

The angles marked a, b, c, d, e and f (left) represent the **exterior angles** of the regular hexagon drawn.

For any convex polygon the sum of the exterior angles is 360°.

If the polygon is regular and has n sides, then each exterior angle $= \dfrac{360°}{n}$.

Worked examples

a) Find the size of an exterior angle of a regular nonagon.

$$\frac{360°}{9} = 40°$$

b) Calculate the number of sides a regular polygon has if each exterior angle is 15°.

$$n = \frac{360}{15} = 24$$

The polygon has 24 sides.

Exercise 24.9

1. Find the sum of the interior angles of the following polygons:
   a) a hexagon b) a nonagon c) a heptagon
2. Find the value of each interior angle of the following regular polygons:
   a) an octagon b) a square
   c) a decagon d) a dodecagon
3. Find the size of each exterior angle of the following regular polygons:
   a) a pentagon b) a dodecagon c) a heptagon

4. The exterior angles of regular polygons are given below. In each case calculate the number of sides the polygon has.
a) 20° b) 36° c) 10°
d) 45° e) 18° f) 3°

5. The interior angles of regular polygons are given below. In each case calculate the number of sides the polygon has.
a) 108° b) 150° c) 162°
d) 156° e) 171° f) 179°

6. Calculate the number of sides a regular polygon has if an interior angle is five times the size of an exterior angle.

7. Copy and complete the table below for regular polygons:

Number of sides	Name	Sum of exterior angles	Size of an exterior angle	Sum of interior angles	Size of an interior angle
3					
4					
5					
6					
7					
8					
9					
10					
12					

The angle in a semi-circle

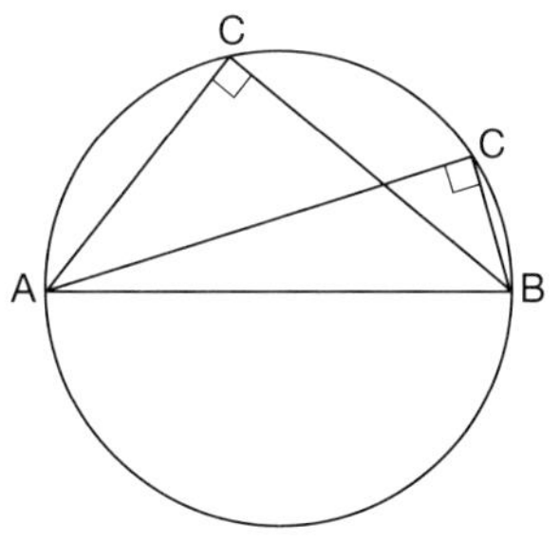

If AB represents the diameter of the circle, then the angle at C is 90°.

Exercise 24.10

In each of the following diagrams, O marks the centre of the circle. Calculate the value of x in each case.

1.

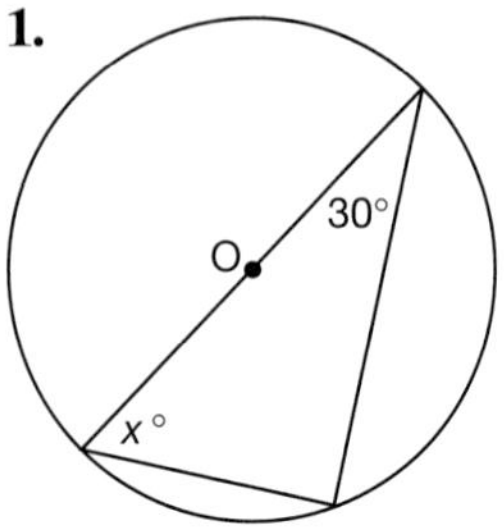

2.

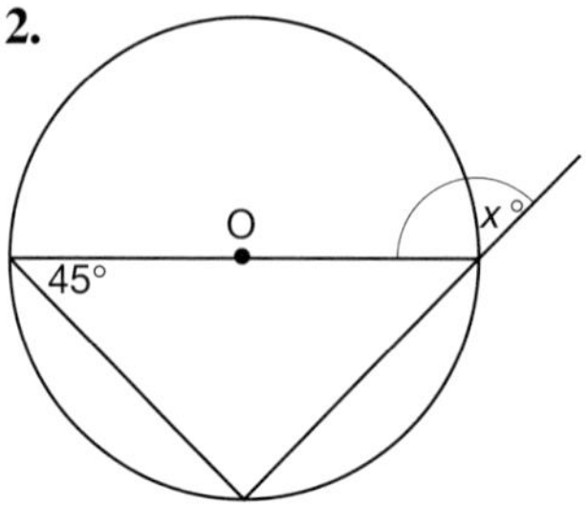

3.

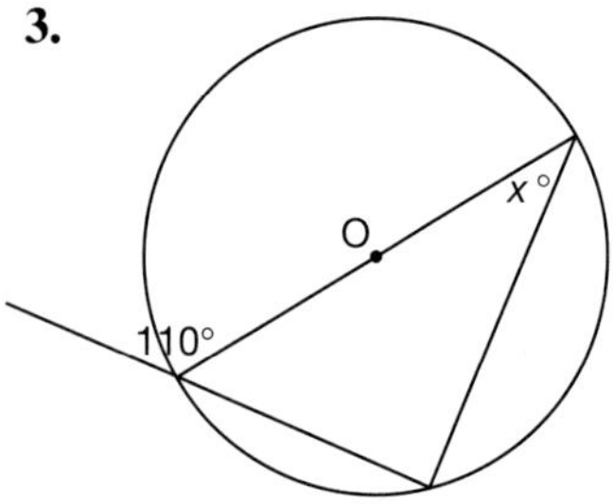

4.

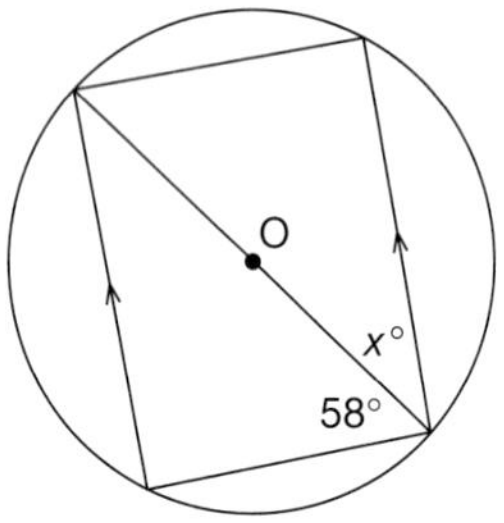

5.

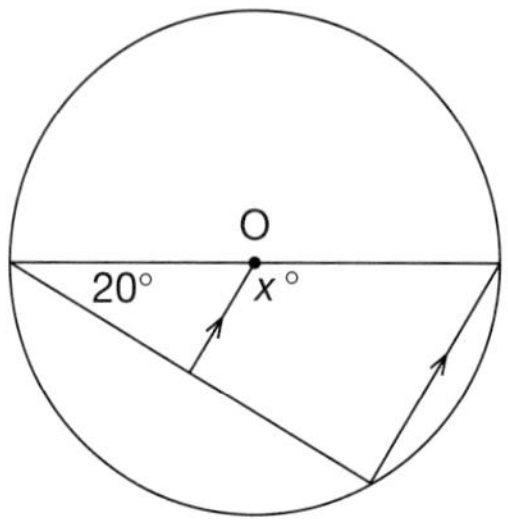

6.

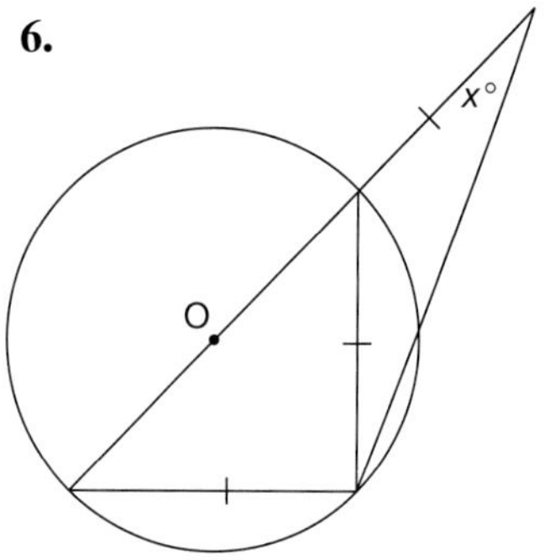

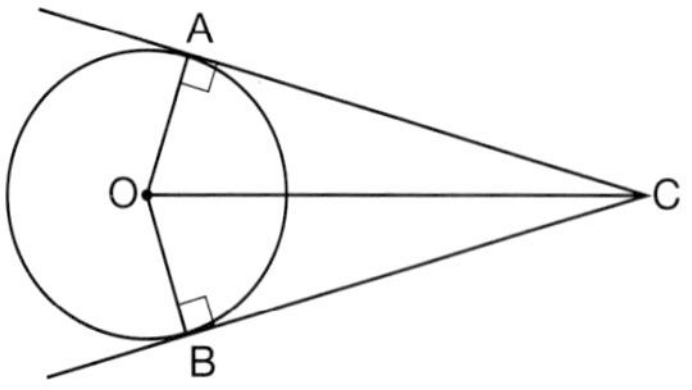

The angle between a tangent and a radius of a circle

The angle between a tangent at a point and the radius to the same point on the circle is a right angle.

Triangles OAC and OBC (left) are congruent as ∠OAC and ∠OBC are right angles, OA = OB because they are both radii and OC is common to both triangles.

Exercise 24.11

In each of the following diagrams, O marks the centre of the circle. Calculate the value of x in each case.

1.

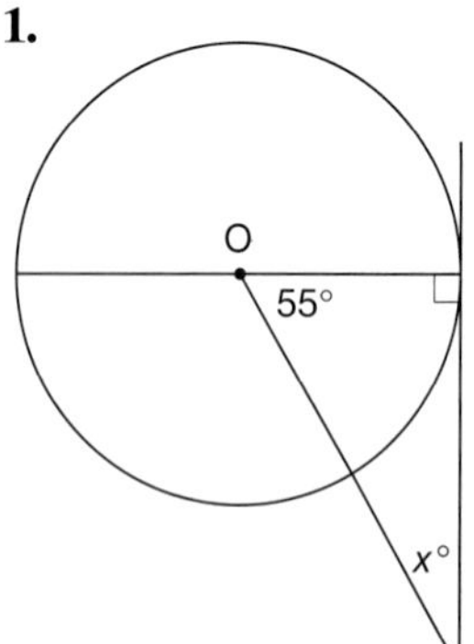

2.

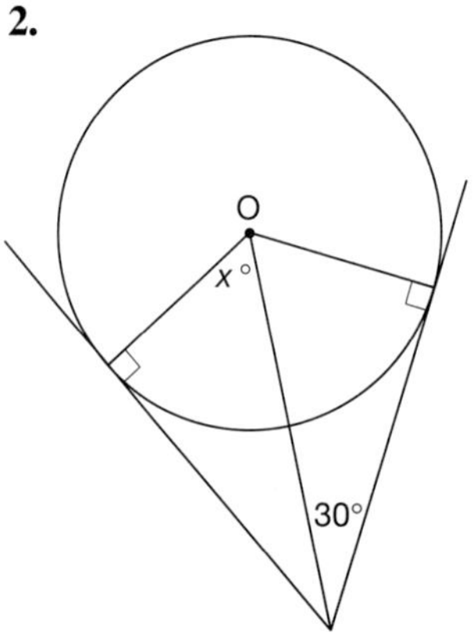

3.

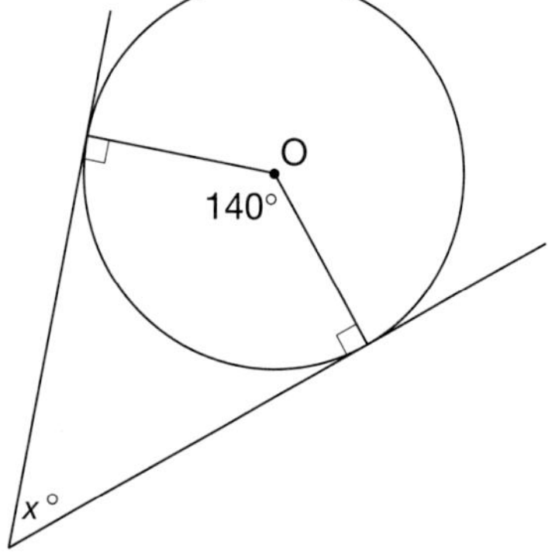

4.

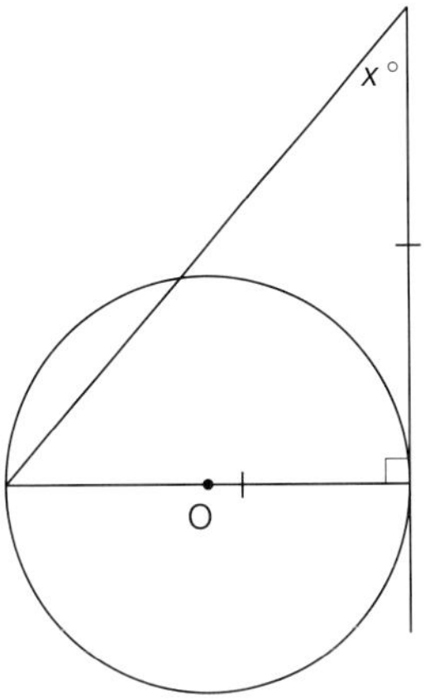

5.

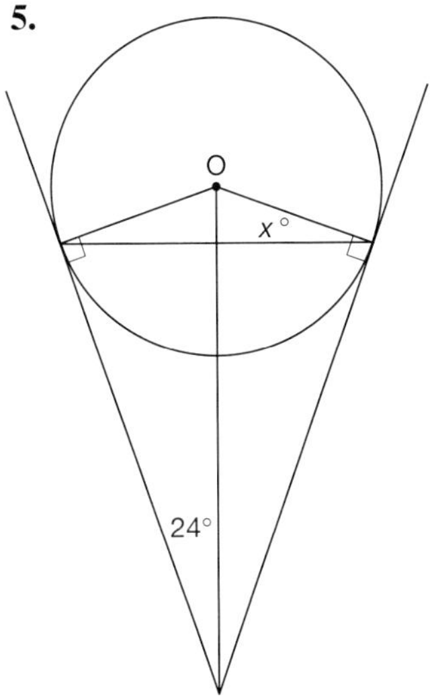

6.

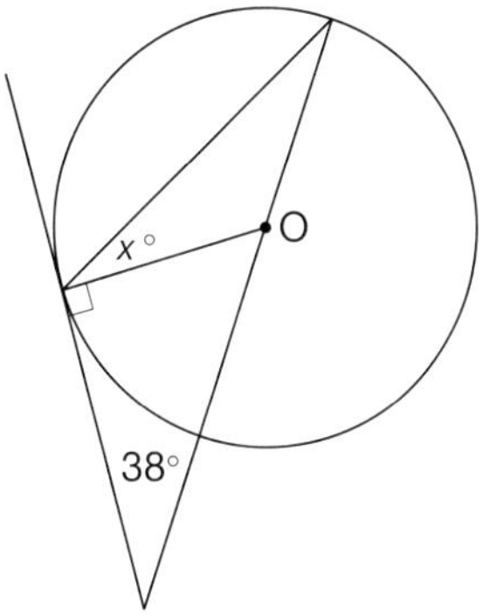

In the following diagrams, calculate the value of x.

7.

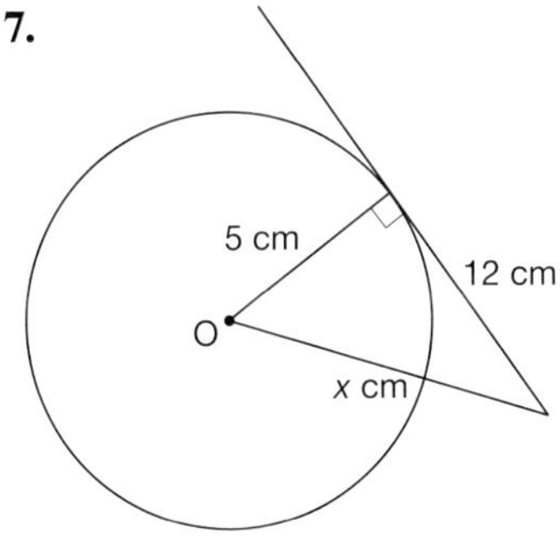

8.

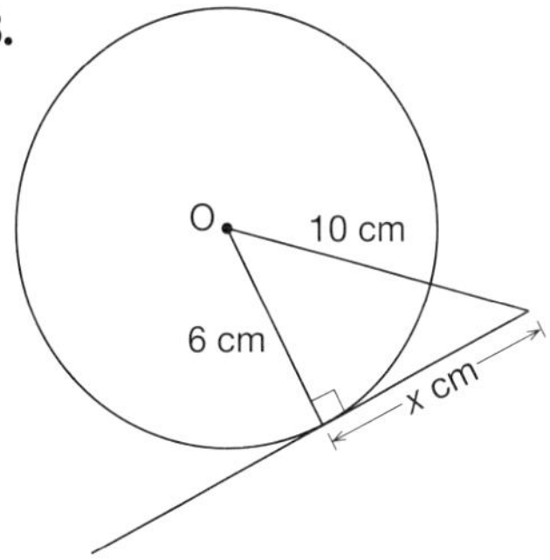

9.

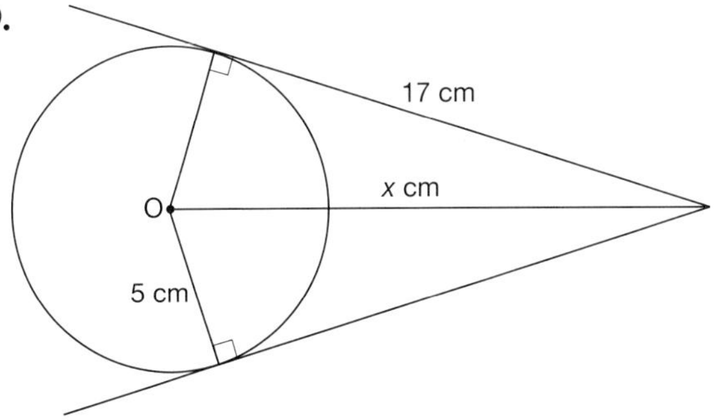

Student Assessment 1

1. For the diagrams below, calculate the size of the labelled angles.

a)

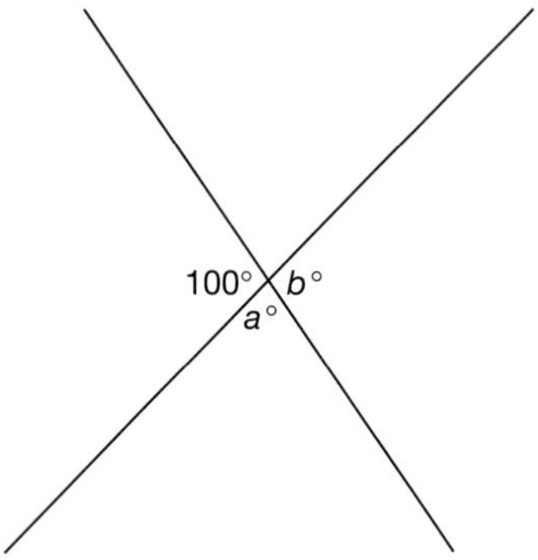

b)

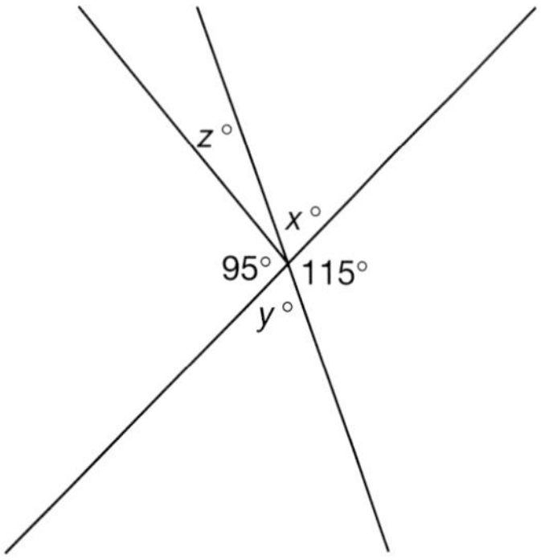

c)

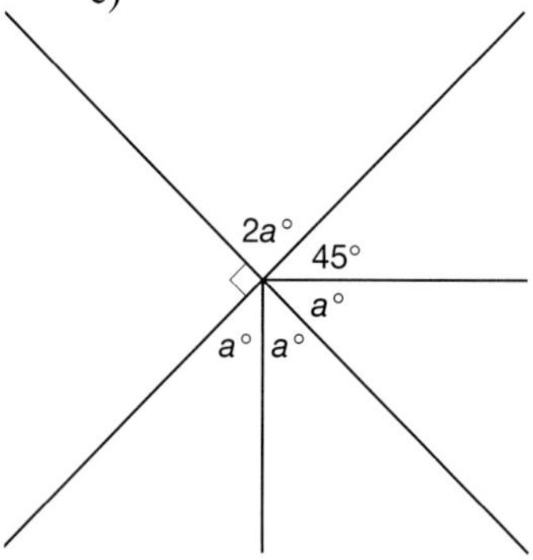

d)

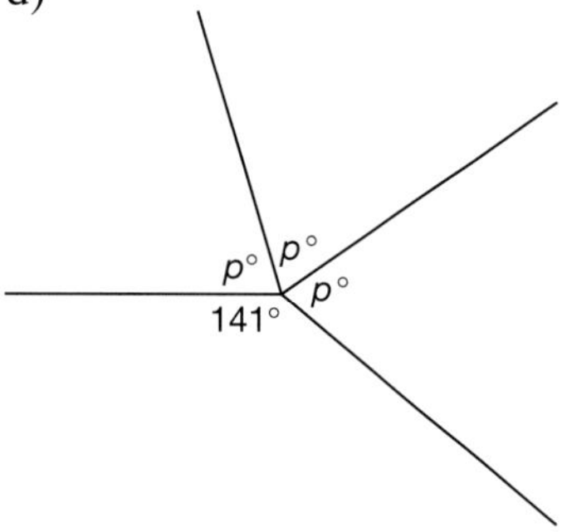

2. For the diagrams below, calculate the size of the labelled angles.

a)

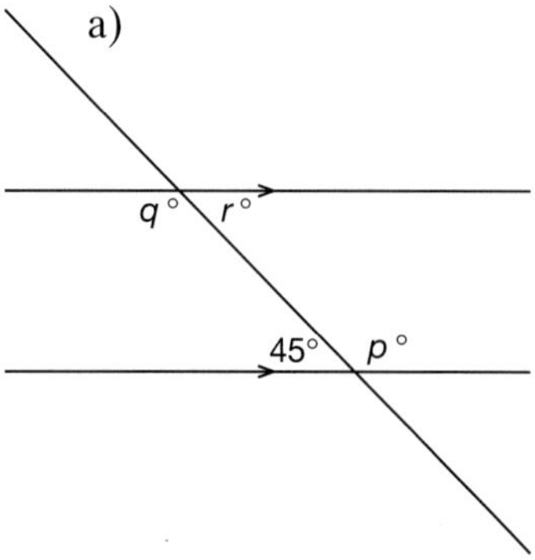

b)

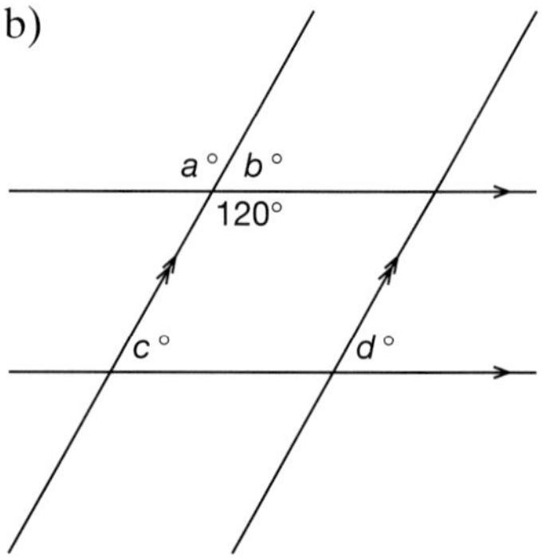

3. For the diagrams below, calculate the size of the labelled angles.

a)

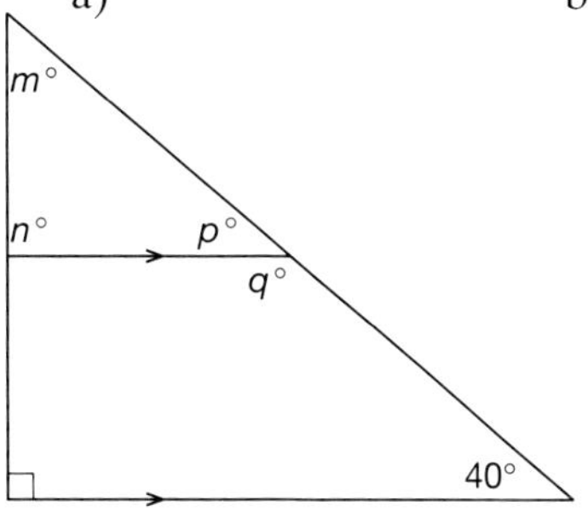

b)

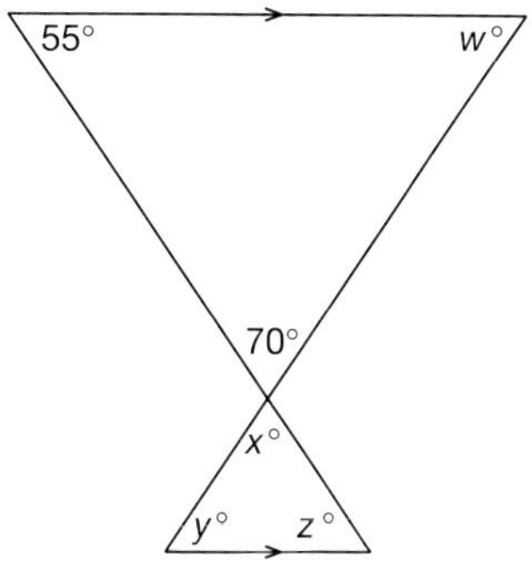

c)

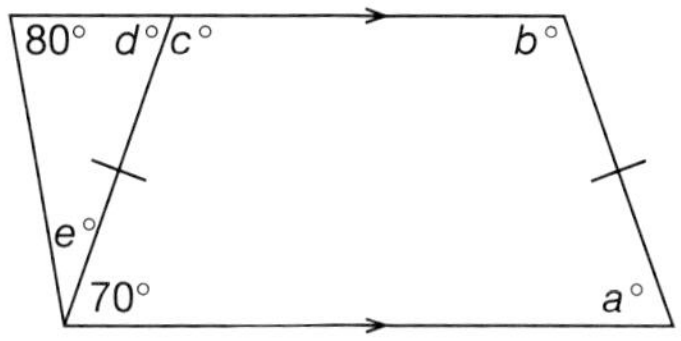

d)

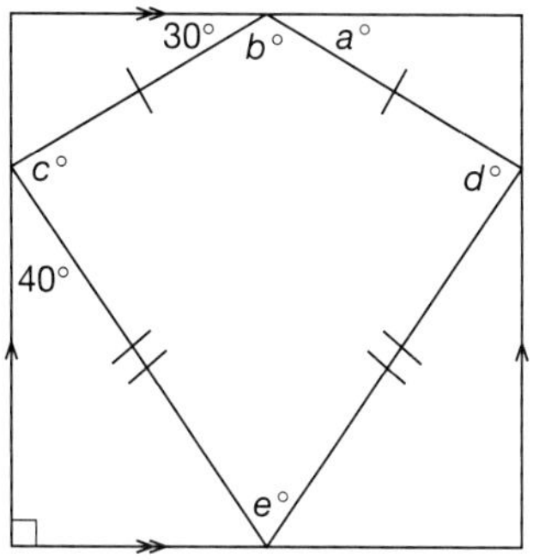

Student Assessment 2

1. For the diagrams below, calculate the size of the labelled angles.

a)

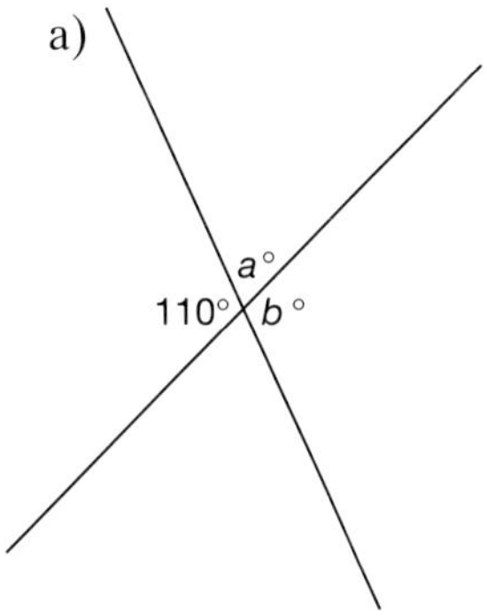

b)

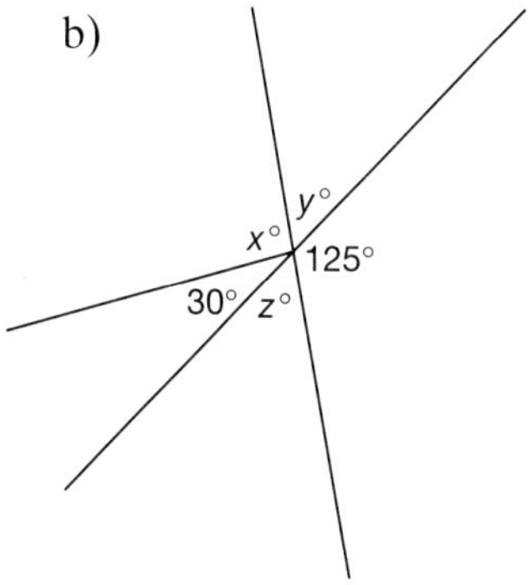

c)

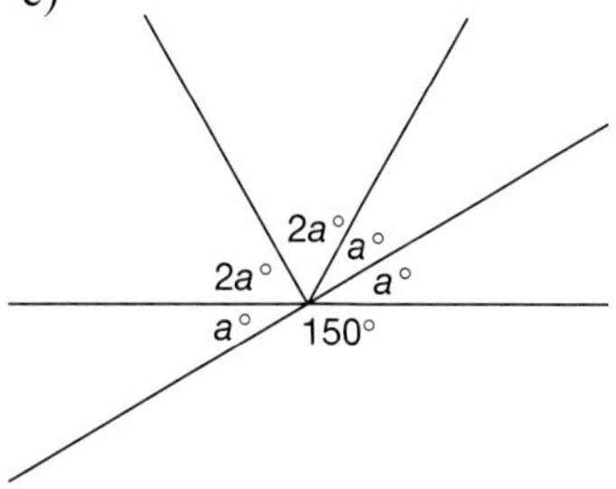

d)

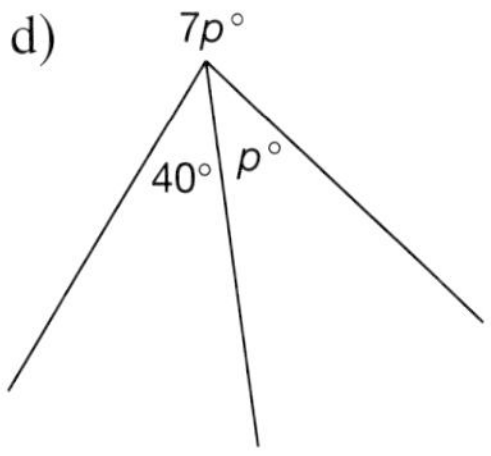

2. For the diagrams below, calculate the size of the labelled angles.

a)

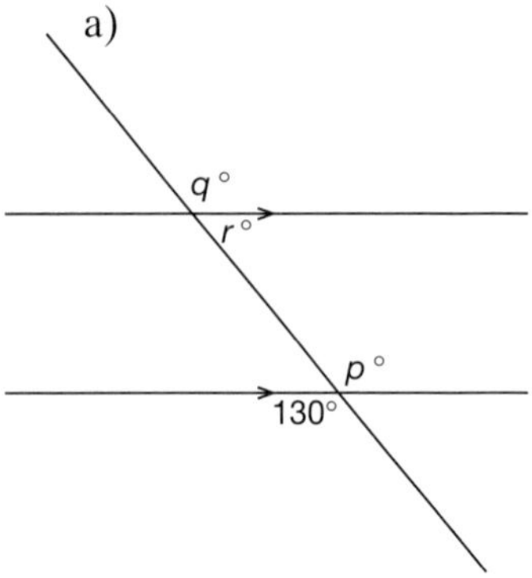

b)

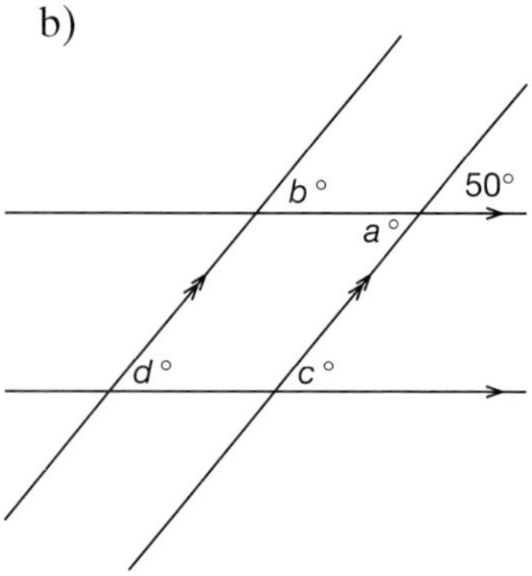

3. For the diagrams below, calculate the size of the labelled angles.

a)

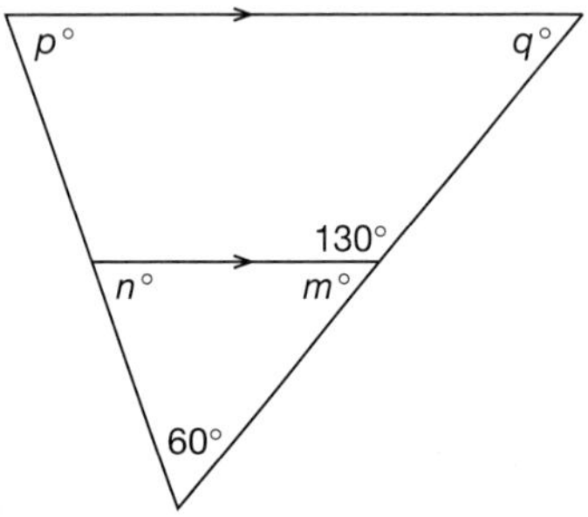

b)

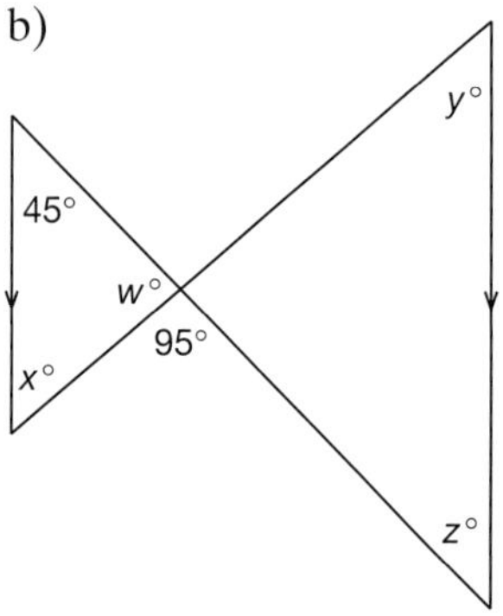

c)

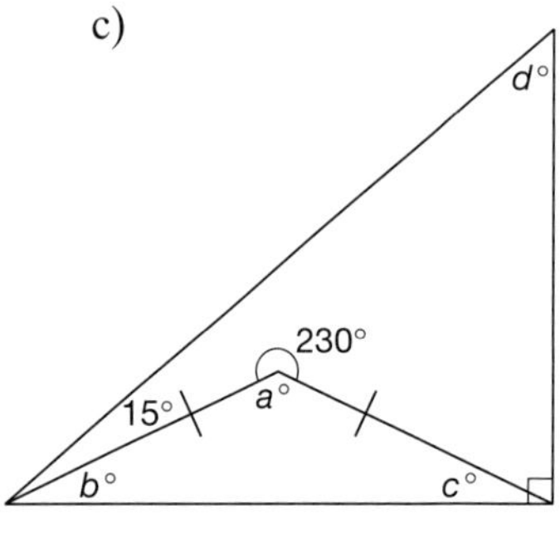

d)

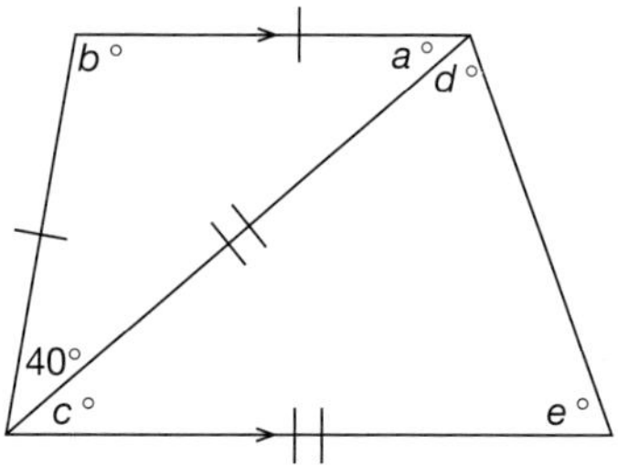

Student Assessment 3

1. Draw a diagram of an octagon to help illustrate the fact that the sum of the internal angles of an octagon is given by $180 \times (8 - 2)°$.

2. Find the size of each interior angle of a twenty-sided regular polygon.

3. What is the sum of the interior angles of a nonagon?

4. What is the sum of the exterior angles of a polygon?

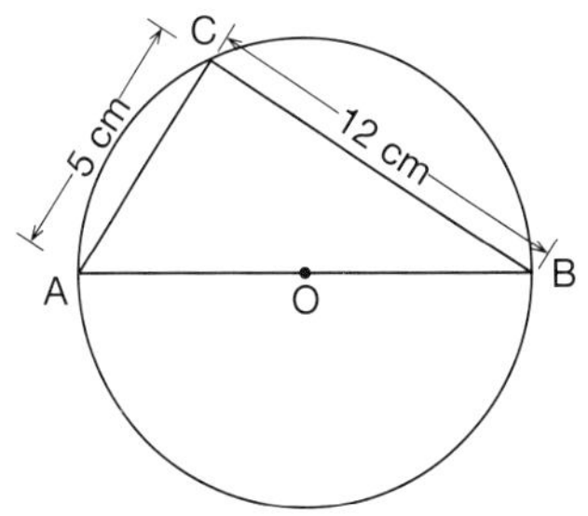

5. What is the size of the exterior angle of a regular pentagon?

6. If AB is the diameter of the circle (left) and AC = 5 cm and BC = 12 cm, calculate:
   a) the size of angle ACB,
   b) the length of the radius of the circle.

In Q.7–10, O marks the centre of the circle. Calculate the size of the angle marked x in each case.

7.

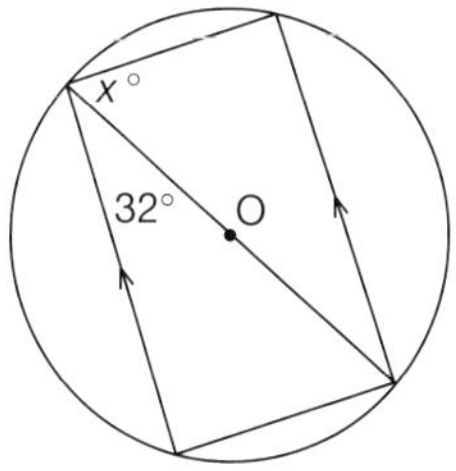

8.

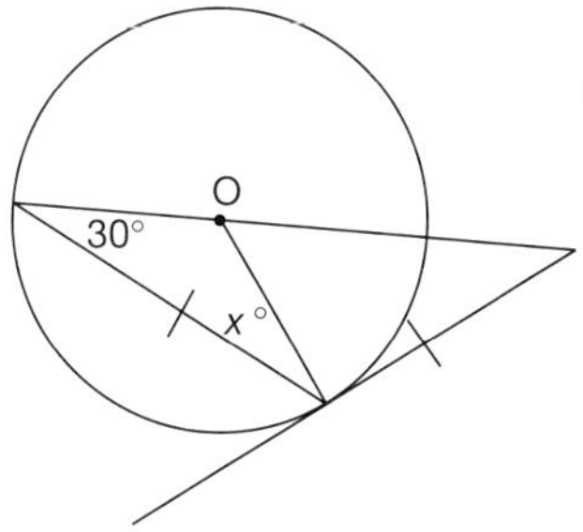

9.

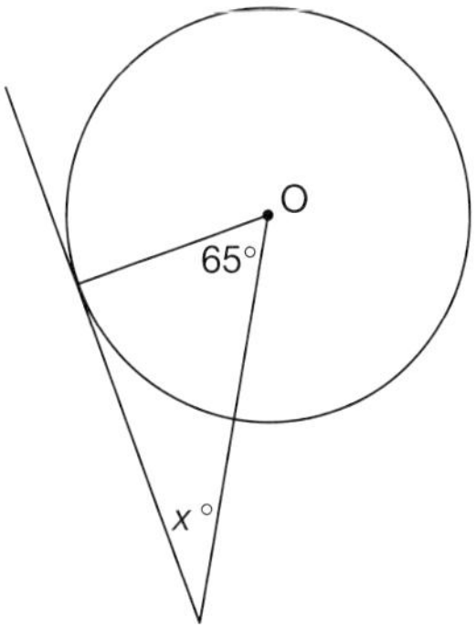

10.

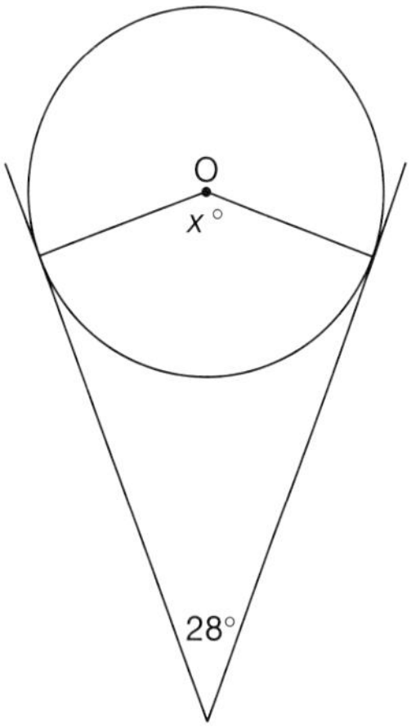

Student Assessment 4

1. Draw a diagram of a hexagon to help illustrate the fact that the sum of the internal angles of a hexagon is given by $180 \times (6 - 2)°$.

2. Find the value of each interior angle of a regular polygon with 24 sides.

3. What is the sum of the interior angles of a regular dodecagon?

4. What is the size of an exterior angle of a regular dodecagon?

5. AB and BC are both tangents to the circle centre O (below). If OA = 5 cm and AB = 12 cm, calculate:
   a) the size of angle OAB,
   b) the length OB.

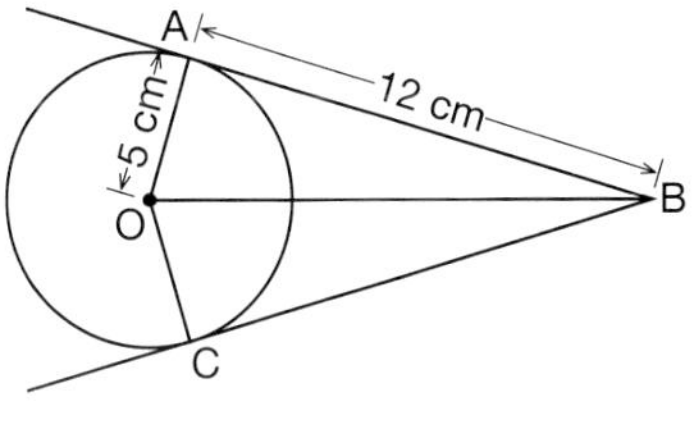

6. If OA is a radius of the circle and PB the tangent to the circle at A, calculate angle ABO.

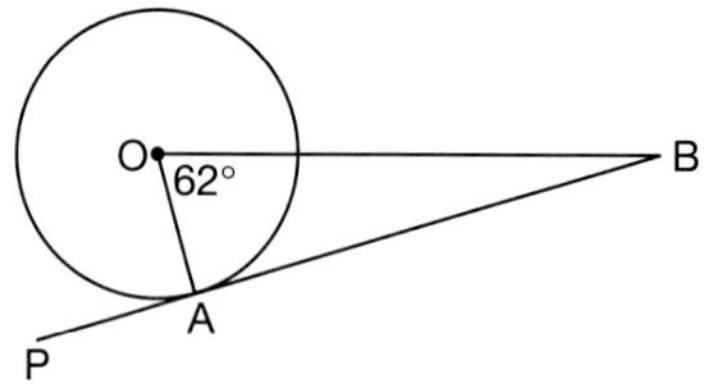

In Q.7–10, O marks the centre of the circle. Calculate the size of the angle marked x in each case.

7.

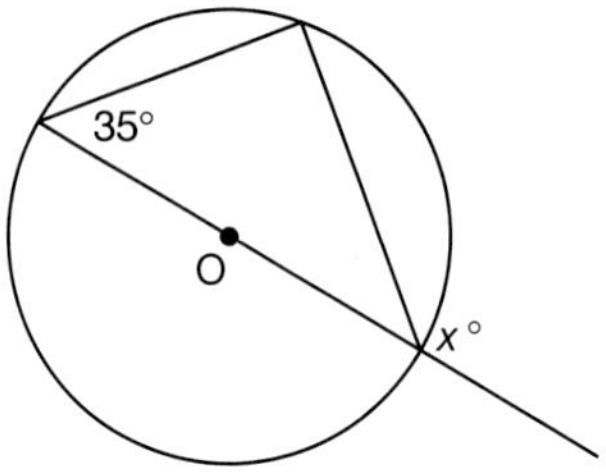

8.

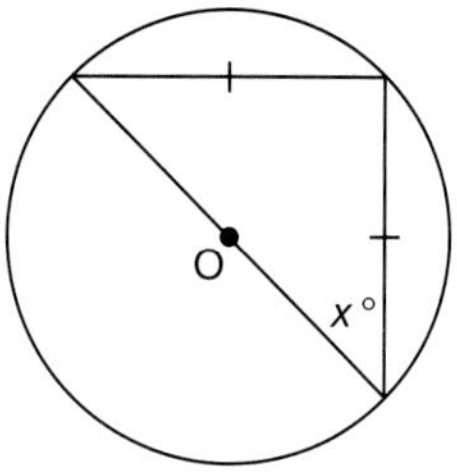

9.

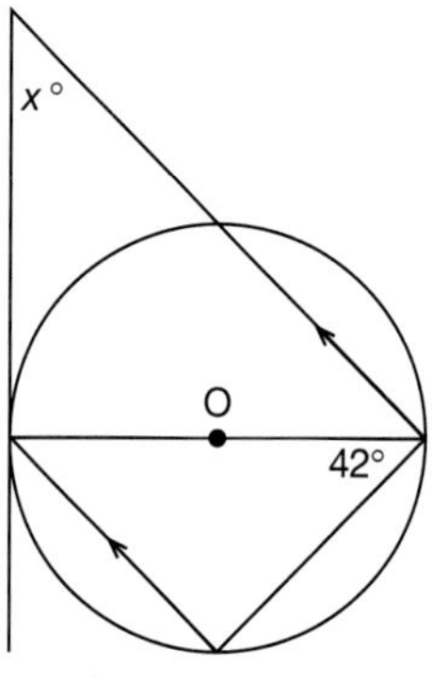

10.

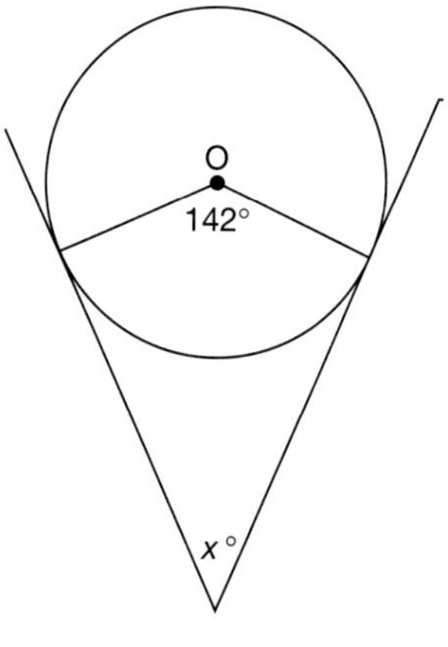

25 LOCI

Use the following loci and the method of intersecting loci for sets of points in two dimensions:

- which are at a given distance from a given point,
- which are at a given distance from a given straight line,
- which are equidistant from two given points,
- which are equidistant from two given intersecting straight lines.

NB: All diagrams are not drawn to scale.

A **locus** (plural **loci)** refers to all the points which fit a particular description. These points can either belong to a region, a line or both. The principal types of loci are explained below.

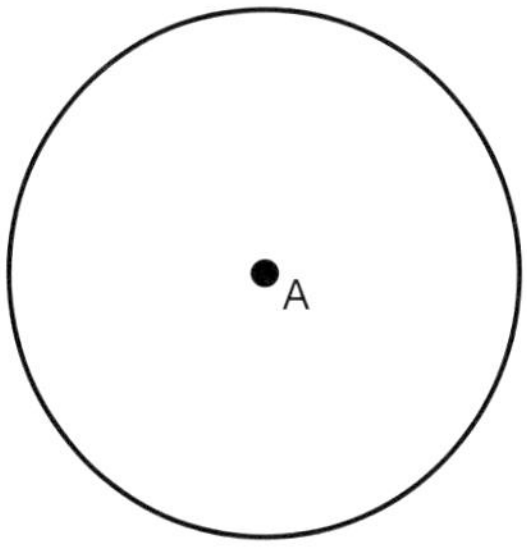

The locus of the points which are at a given distance from a given point

In the diagram (left) it can be seen that the locus of all the points equidistant from a point A is the circumference of a circle centre A. This is due to the fact that all points on the circumference of a circle are equidistant from the centre.

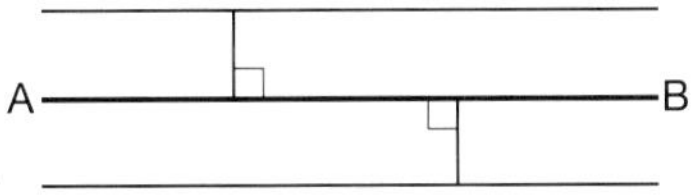

The locus of the points which are at a given distance from a given straight line

In the diagram (left) it can be seen that the locus of the points equidistant from a straight line AB runs parallel to that straight line. It is important to note that the distance of the locus from the straight line is measured at right angles to the line. This diagram, however, excludes the ends of the line. If these two points are taken into consideration then the locus takes the form shown in the diagram below.

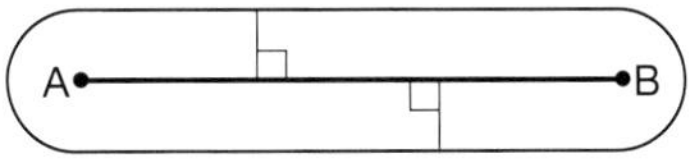

The locus of the points which are equidistant from two given points

The locus of the points equidistant from points X and Y lies on the perpendicular bisector of the line XY.

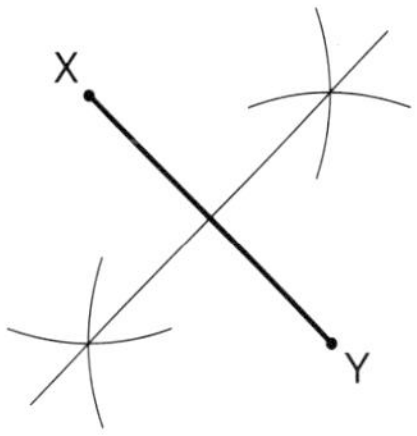

The locus of the points which are equidistant from two given intersecting straight lines

The locus in this case lies on the bisectors of both pairs of opposite angles as shown below.

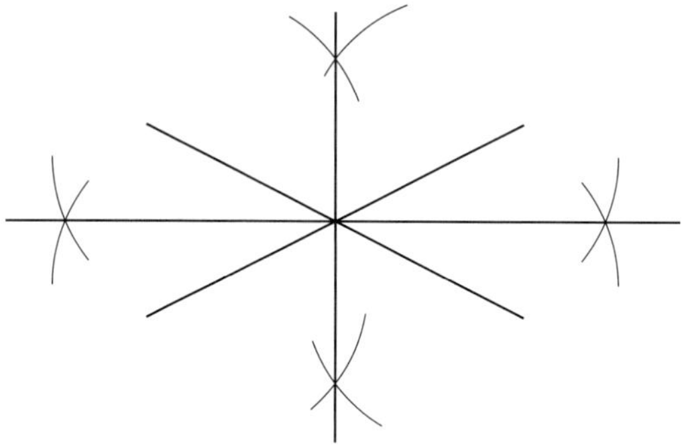

The application of the above cases will enable you to tackle problems involving loci at this level.

Worked example The diagram (left) shows a trapezoidal garden. Three of its sides are enclosed by a fence, and the fourth is adjacent to a house.

i) Grass is to be planted in the garden. However, it must be at least 2 m away from the house and at least 1 m away from the fence. Shade the region in which the grass can be planted.

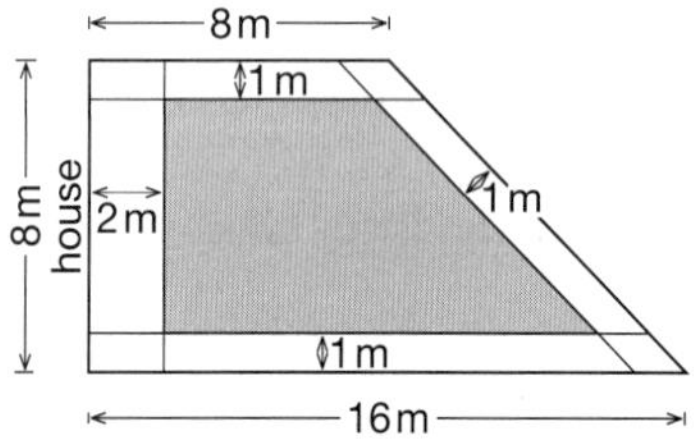

The shaded region is therefore the locus of all the points which are both at least 2 m away from the house and at least 1 m away from the surrounding fence. Note that the boundary of the region also forms part of the locus of the points.

ii) Using the same garden as before, grass must now be planted according to the following conditions: it must be **more than** 2 m away from the house and **more than** 1 m away from the fence. Shade the region in which the grass can be planted.

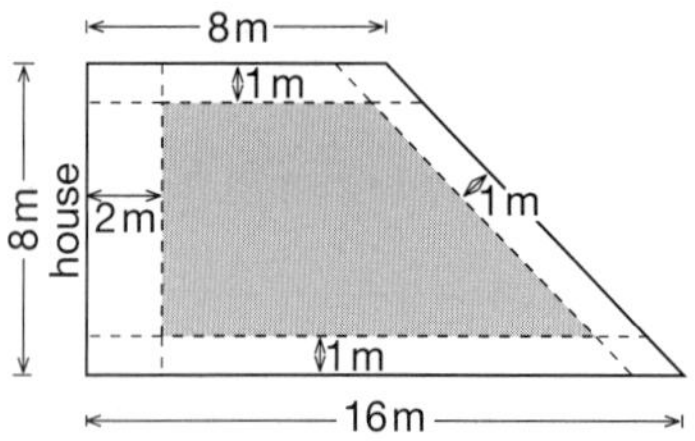

The shape of the region is the same as in the first case; however, in this instance the boundary is not included in the locus of the points as the grass cannot be exactly 2 m away from the house or exactly 1 m away from the fence.

Note: If the boundary is included in the locus points, it is represented by a **solid** line. If it is not included then it is represented by a **dashed** (broken) line.

Exercise 25.1

Q.1–4 are about a rectangular garden measuring 8 m by 6 m. For each question draw a scale diagram of the garden and identify the locus of the points which fit the criteria.

1. Draw the locus of all the points at least 1 m from the edge of the garden.

2. Draw the locus of all the points at least 2 m from each corner of the garden.

3. Draw the locus of all the points more than 3 m from the centre of the garden.

4. Draw the locus of all the points equidistant from the longer sides of the garden.

5. An airport has two radar stations at P and Q which are 20 km apart. The radar at P is set to a range of 20 km, whilst the radar at Q is set to a range of 15 km.
   a) Draw a scale diagram to show the above information.
   b) Shade the region in which an aeroplane must be flying if it is only picked up by radar P. Label this region 'a'.
   c) Shade the region in which an aeroplane must be flying if it is only picked up by radar Q. Label this region 'b'.
   d) Identify the region in which an aeroplane must be flying if it is picked up by both radars. Label this region 'c'.

6. X and Y are two ship-to-shore radio receivers. They are 25 km apart.

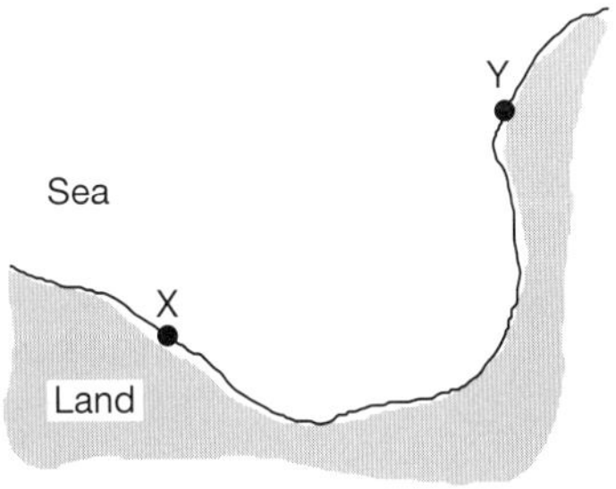

A ship sends out a distress signal. The signal is picked up by both X and Y. The radio receiver at X indicates that the ship is within a 30 km radius of X, whilst the radio receiver at Y indicates that the ship is within 20 km of Y. Draw a scale diagram and identify the region in which the ship must lie.

7. a) Mark three points L, M and N not in a straight line. By construction find the point which is equidistant from L, M and N.
   b) What would happen if L, M and N were on the same straight line?

8. Draw a line AB 8 cm long. What is the locus of a point C such that the angle ACB is always a right angle?

9. Draw a circle by drawing round a circular object (do not use a pair of compasses). By construction determine the position of the centre of the circle.

10. Three lionesses L_1, L_2 and L_3 have surrounded a gazelle. The three lionesses are equidistant from the gazelle.

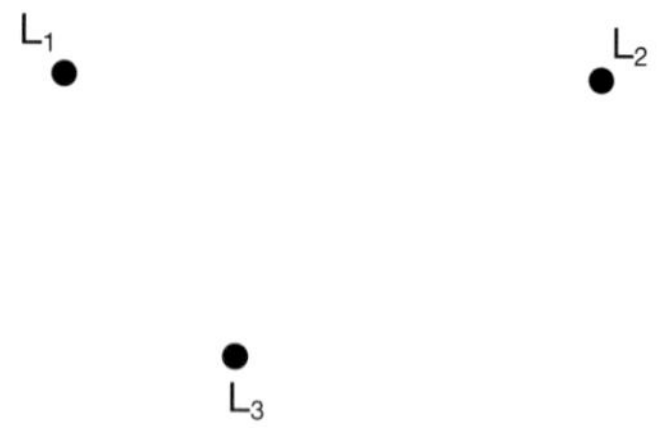

Draw a diagram with the lionesses in similar positions to those shown (above) and by construction determine the position (g) of the gazelle.

Exercise 25.2

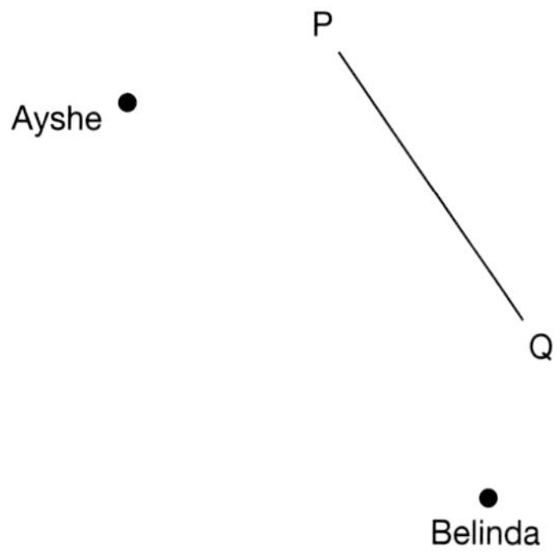

1. Three girls are playing hide and seek. Ayshe and Belinda are at the positions shown (left) and are trying to find Cristina. Cristina is on the opposite side of a wall PQ to her two friends.

   Assuming Ayshe and Belinda cannot see over the wall identify, by copying the diagram, the locus of points where Cristina could be if:
   a) Cristina can only be seen by Ayshe,
   b) Cristina can only be seen by Belinda,
   c) Cristina can not be seen by either of her two friends,
   d) Cristina can be seen by both of her friends.

2. A security guard S is inside a building in the position shown below. The building is inside a rectangular compound. If the building has three windows as shown, identify the locus of points in the compound which can be seen by the security guard.

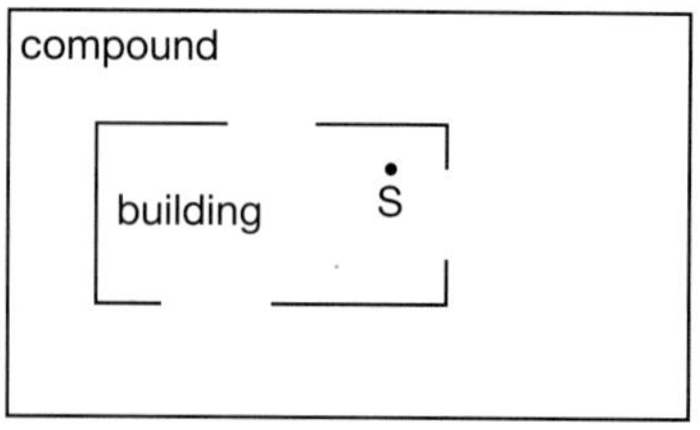

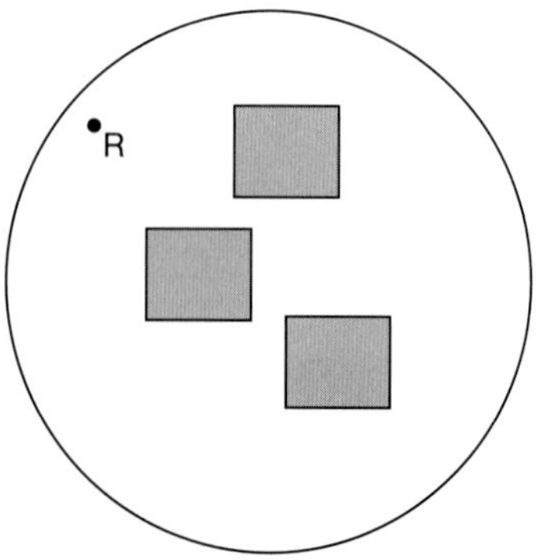

3. The circular cage shown (left) houses a snake. Inside the cage are three obstacles. A rodent is placed inside the cage at R. From where it is lying, the snake can see the rodent.

   Trace the diagram and identify the regions in which the snake could be lying.

Exercise 25.3

1. A coin is rolled in a straight line on a flat surface as shown below.

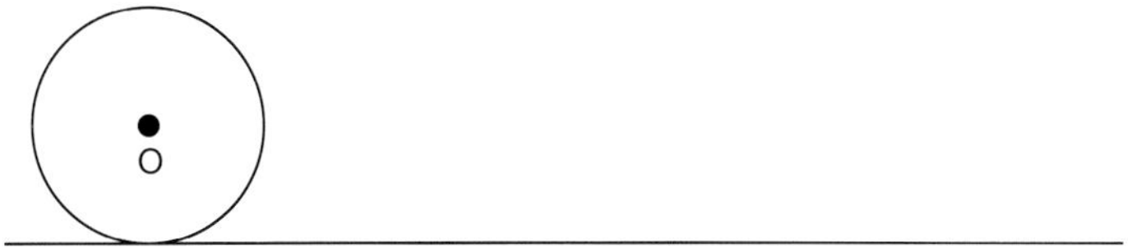

Draw the locus of the centre of the coin O as the coin rolls along the surface.

2. The diameter of the disc is the same as the width and height of each of the steps shown below. Copy the diagram and draw the locus of the centre of the disc as it rolls down the steps.

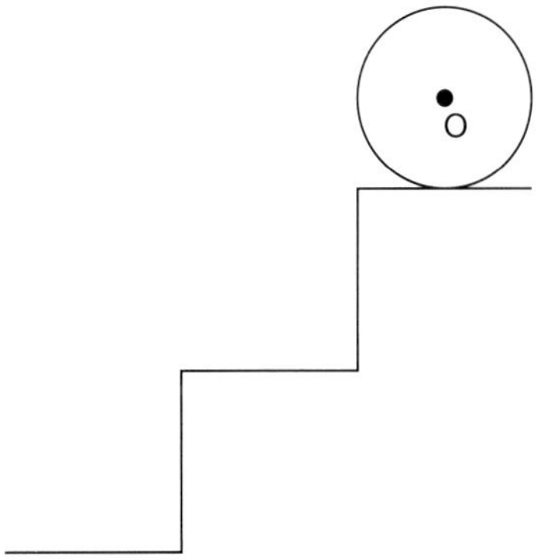

3. A stone is thrown vertically upwards. Draw the locus of its trajectory from the moment it leaves the person's hand to the moment it is caught again.

4. A stone is thrown at an angle of elevation of 45°. Sketch the locus of its trajectory.

5. X and Y are two fixed posts in the ground. The ends of a rope are tied to X and Y. A goat is attached to the rope by a ring on its collar which enables it to move freely along the rope's length.

Copy the diagram and sketch the locus of points in which the goat is able to graze.

Student Assessment I

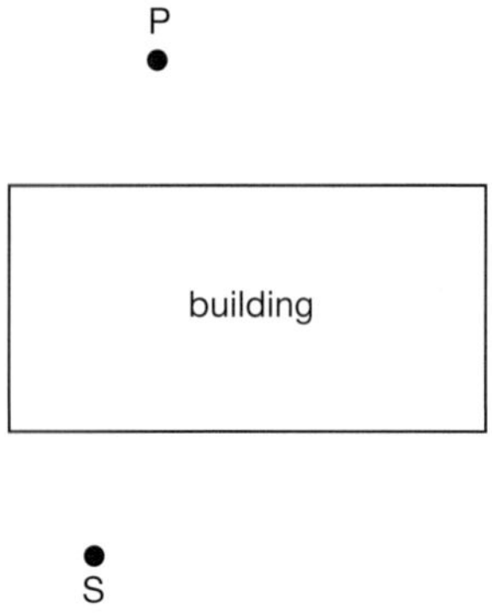

1. Pedro and Sara are on opposite sides of a building as shown (left).

Their friend Raul is standing in a place such that he cannot be seen by either Pedro or Sara. Copy the diagram and identify the locus of points at which Raul could be standing.

2. A rectangular garden measures 10 m by 6 m. A tree stands in the centre of the garden. Grass is to be planted according to the following conditions:

- it must be at least 1 m from the edge of the garden,
- it must be more than 2 m away from the centre of the tree.

a) Make a scale drawing of the garden.
b) Draw the locus of points in which the grass can be planted.

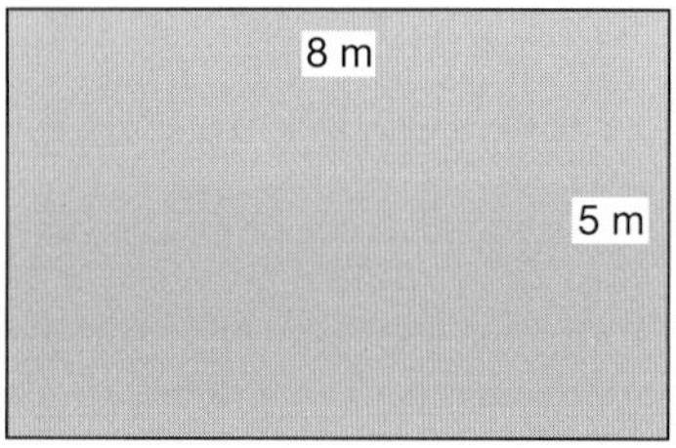

3. A rectangular rose bed in a park measures 8 m by 5 m as shown (left).

The park keeper puts a low fence around the rose bed. The fence is at a constant distance of 2 m from the rose bed.
a) Make a scale drawing of the rose bed.
b) Draw the position of the fence.

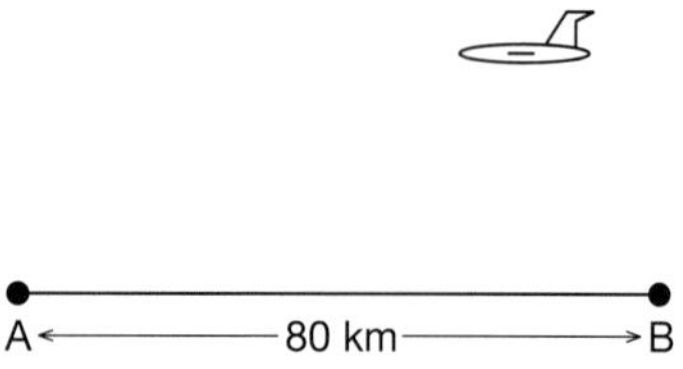

4. A and B are two radio beacons 80 km apart. A plane flies in such a way that it is always three times further from A than from B.

Showing your method of construction clearly, draw the flight path of the aeroplane.

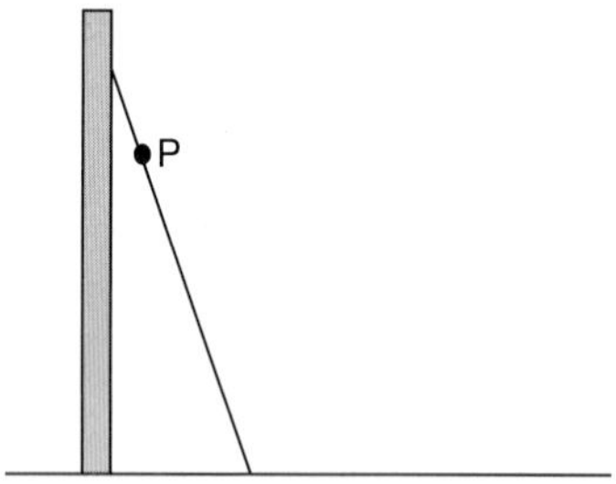

5. A ladder 10 m long is propped up against a wall as shown. A point P on the ladder is 2 m from the top.

Make a scale drawing to show the locus of point P if the ladder were to slide down the wall. Note: several positions of the ladder will need to be shown.

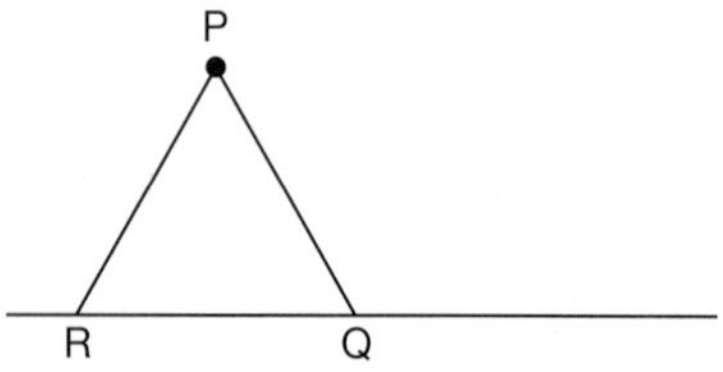

6. The equilateral triangle PQR is rolled along the line shown. At first, corner Q acts as the pivot point until P reaches the line, then P acts as the pivot point until R reaches the line, and so on.

Showing your method clearly, draw the locus of point P as the triangle makes one full rotation, assuming there is no slipping.

Student Assessment 2

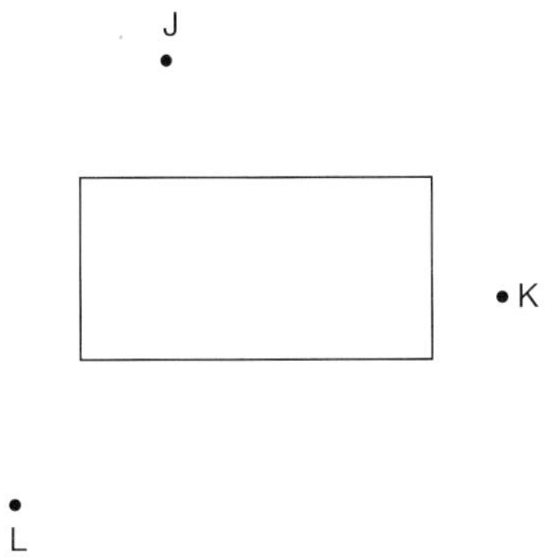

1. Jose, Katrina and Luis are standing at different points around a building as shown (left).
   Trace the diagram and show whether any of the three friends can see each other or not.

2. A rectangular courtyard measures 20 m by 12 m. A horse is tethered in the centre with a rope 7 m long. Another horse is tethered, by a rope 5 m long, to a rail which runs along the whole of the longer left-hand side of the courtyard. This rope is able to run freely along the length of the rail.
   Draw a scale diagram of the courtyard and draw the locus of points which can be reached by both horses.

3. The view in the diagram (left) is of two walls which form part of an obstacle course. A girl decides to ride her bicycle in between the two walls in such a way that she is always equidistant from them.
   Copy the diagram and, showing your construction clearly, draw the locus of her path.

4. A ball is rolling along the line shown in the diagram (below). Copy the diagram and draw the locus of the centre, O, of the ball as it rolls.

5. A square ABCD is 'rolled' along the flat surface shown below. Initially corner C acts as a pivot point until B touches the surface, then B acts as a pivot point until A touches the surface, and so on.

Assuming there is no slipping, draw the locus of point A as the square makes one complete rotation. Show your method clearly.

26 SYMMETRY

Recognise rotational and line symmetry (including order of rotational symmetry) in two dimensions, properties of circles, triangles and quadrilaterals directly related to their symmetries; recognise symmetry properties of the prism.

A **line of symmetry** divides a two-dimensional (flat) shape into two congruent (identical) shapes.
e.g.

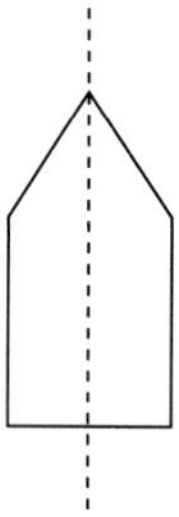

1 line of symmetry

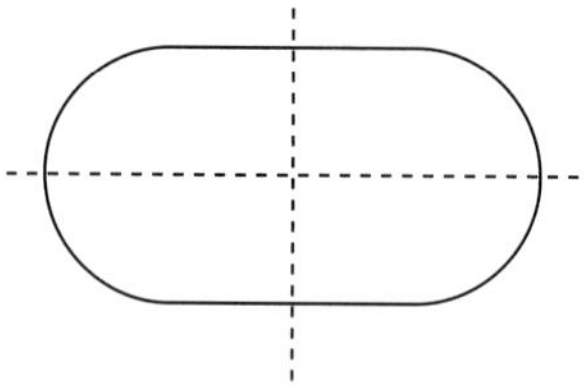

2 lines of symmetry

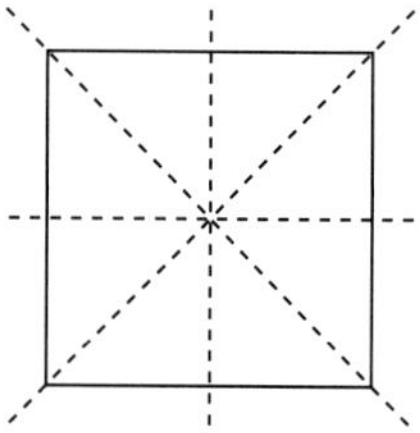

4 lines of symmetry

A **plane of symmetry** divides a three-dimensional (solid) shape into two congruent solid shapes.
e.g.

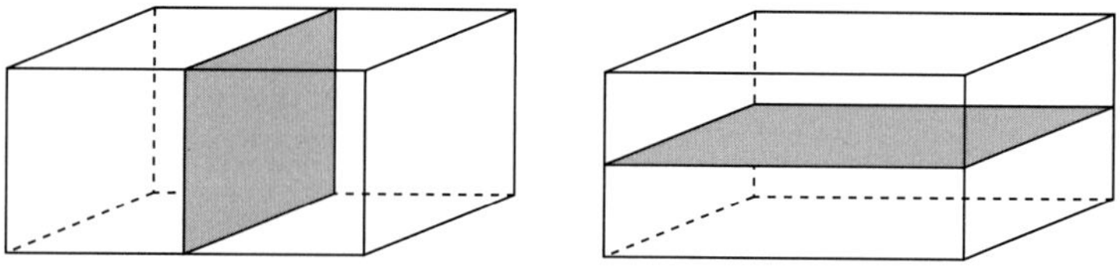

A cuboid has at least three planes of symmetry, two of which are shown above.

A shape has **reflective symmetry** if it has one or more lines or planes of symmetry.

A two-dimensional shape has **rotational symmetry** if, when rotated about a central point, it fits its outline. The number of times it fits its outline during a complete revolution is called the **order of rotational symmetry**.
e.g.

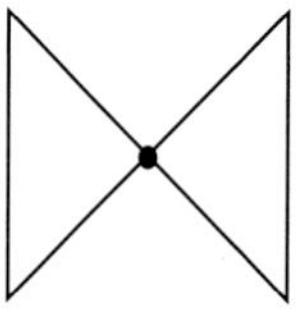

rotational symmetry of order 2

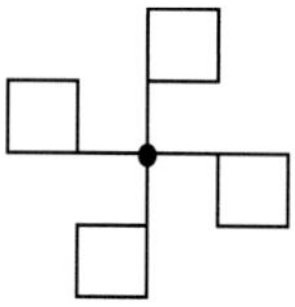

rotational symmetry of order 4

A three-dimensional shape has **rotational symmetry** if, when rotated about a central axis, it looks the same at certain intervals.
e.g.

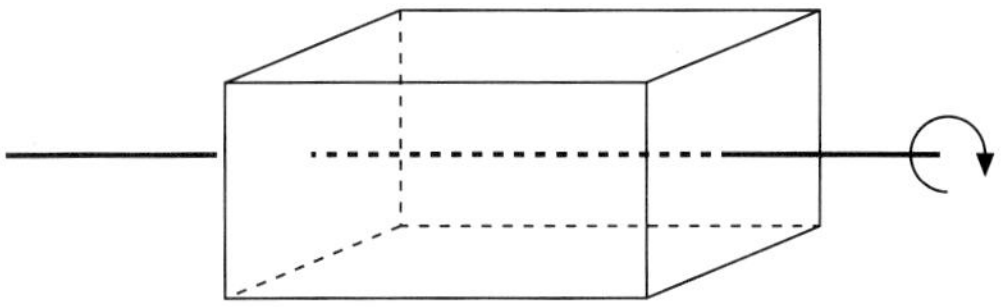

This cuboid has rotational symmetry of order 2 about the axis shown.

Exercise 26.1

1. Draw the following shapes and show all their lines of symmetry:

a) square	b) rectangle
c) equilateral triangle	d) isosceles triangle
e) kite	f) regular hexagon
g) regular octagon	h) regular pentagon
i) isosceles trapezium	

2. Copy the shapes below and where possible show all their lines of symmetry:

a)
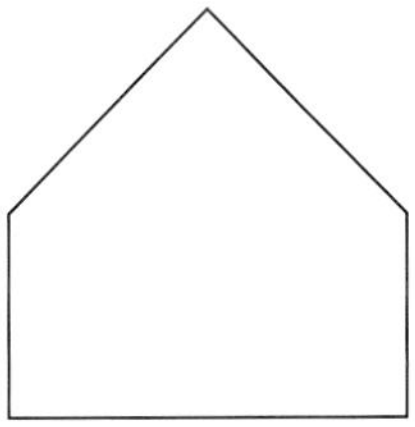
b)
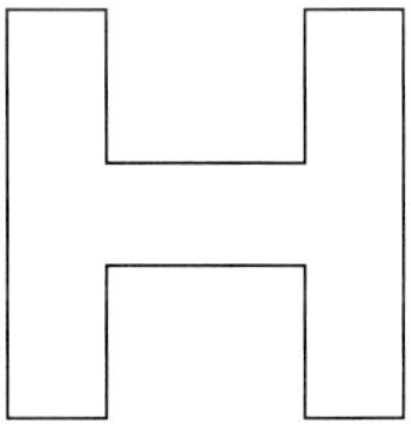
c)
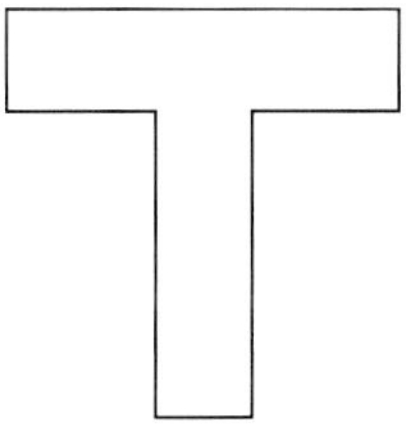
d)
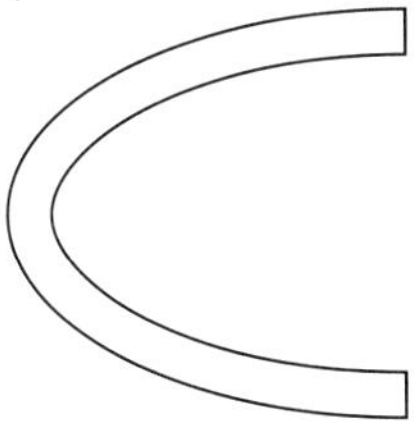
e)
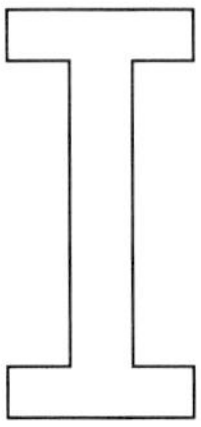
f)
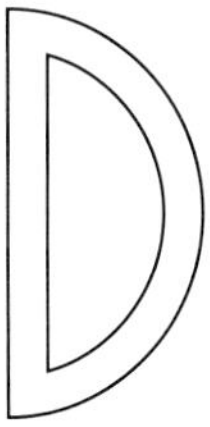
g)
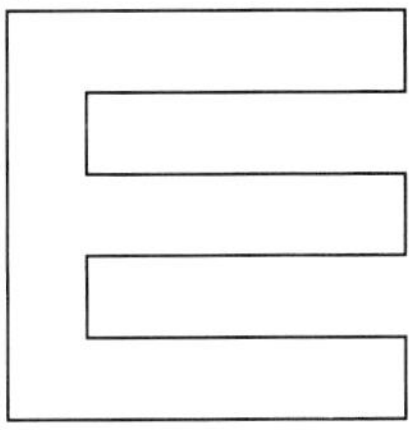
h)
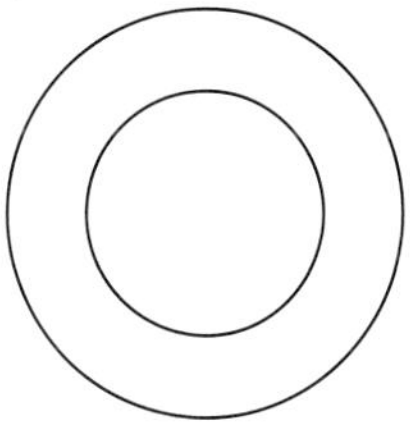
i)
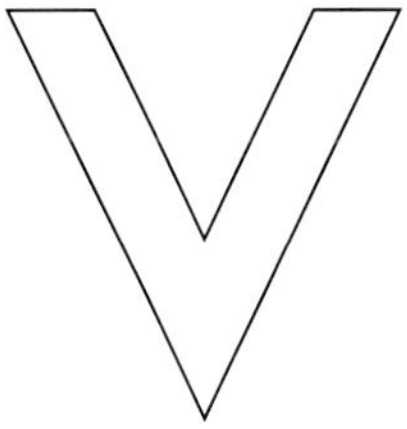

3. Copy the shapes below and complete them so that the **bold** line becomes a line of symmetry:

a) b) c) d) e) f)

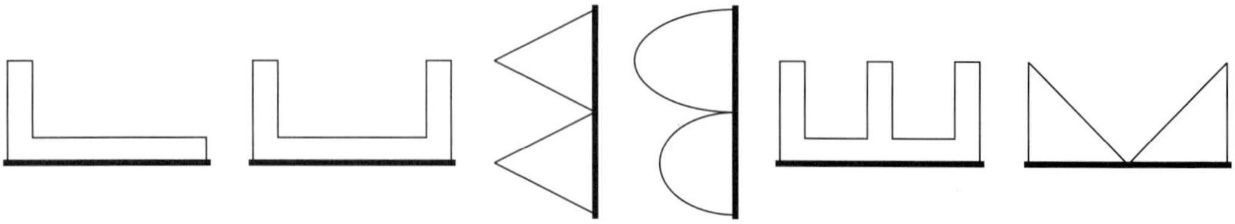

4. Copy the shapes below and complete them so that the **bold** lines become lines of symmetry:

a) b) c) d)

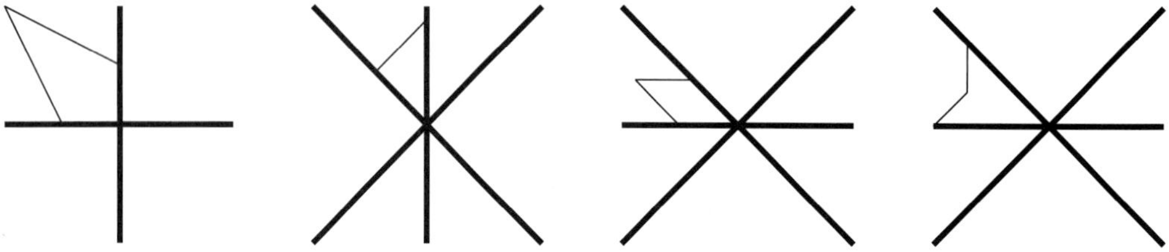

Exercise 26.2

1. Draw the following shapes. Identify the centre of rotation, and find the order of rotational symmetry:
   - a) square
   - b) equilateral triangle
   - c) regular pentagon
   - d) parallelogram
   - e) rectangle
   - f) rhombus
   - g) regular hexagon
   - h) regular octagon
2. Copy the shapes below. Indicate the centre of rotation, and find the order of rotational symmetry:

a) b) c)

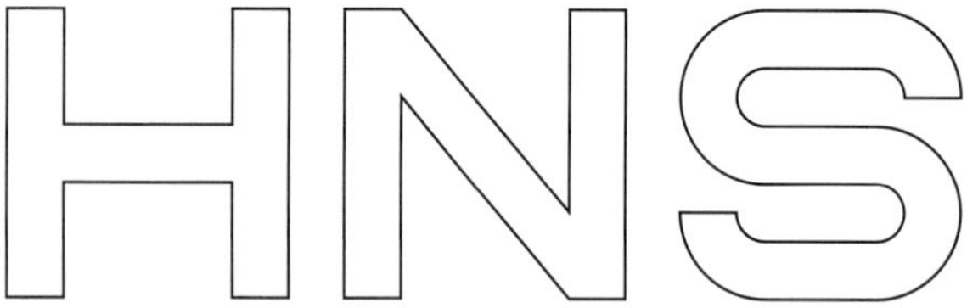

d) e)

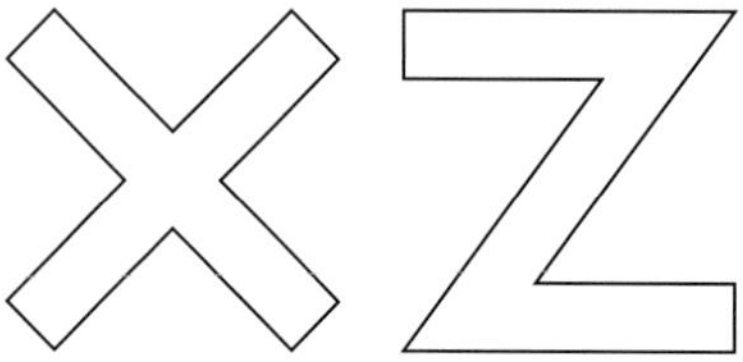

3. Draw each of the solid shapes below, then:
   i) draw a plane of symmetry,
   ii) state how many planes of symmetry the shape has in total.

a)

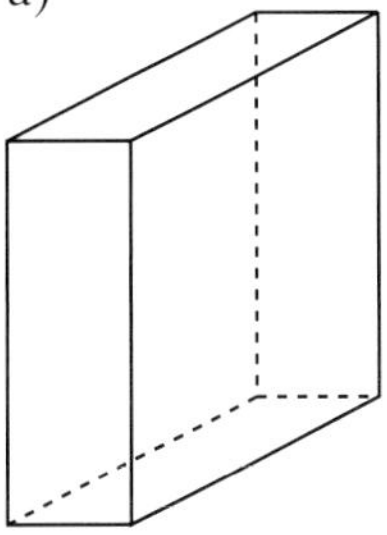

cuboid

b)

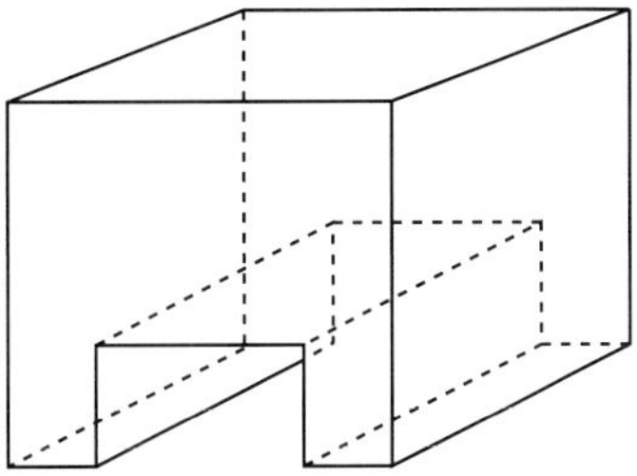

prism

c)

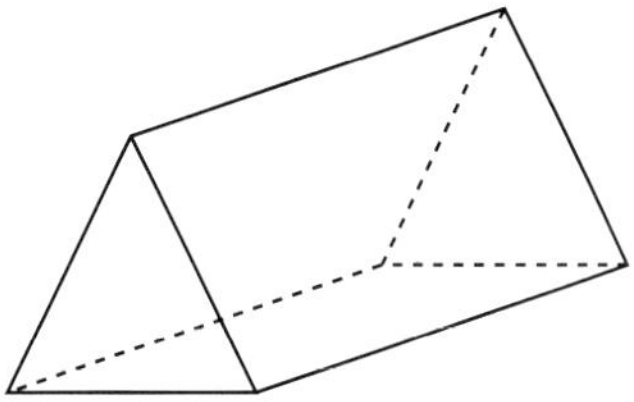

equilateral triangular prism

d)

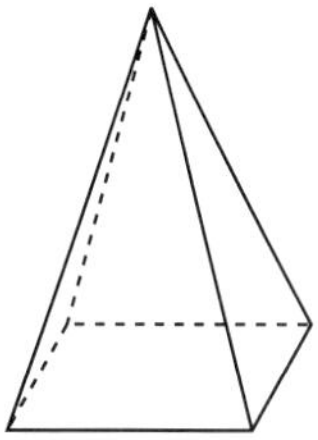

square-based pyramid

e)

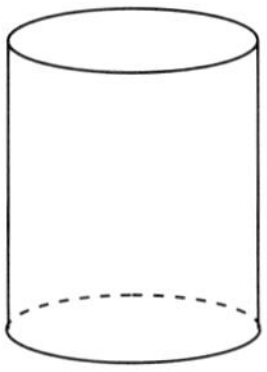

cylinder

f)

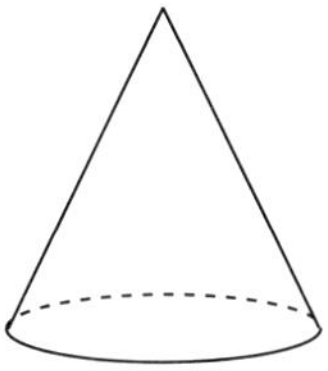

cone

g)

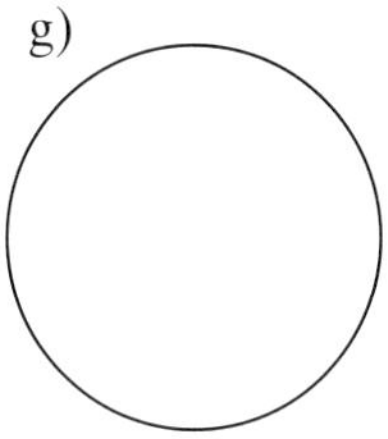

sphere

h)

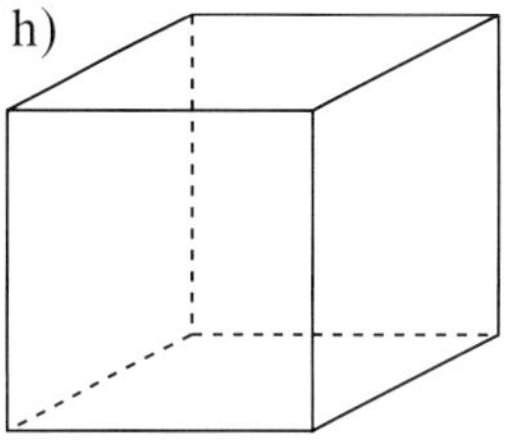

cube

4. For each of the solid shapes shown below, determine the order of rotational symmetry about the axis shown.

a)

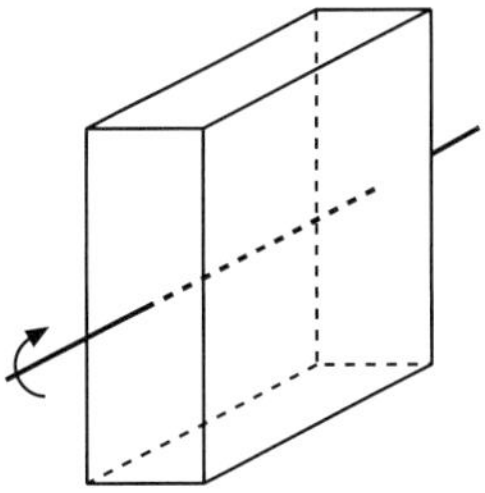

cuboid

b)

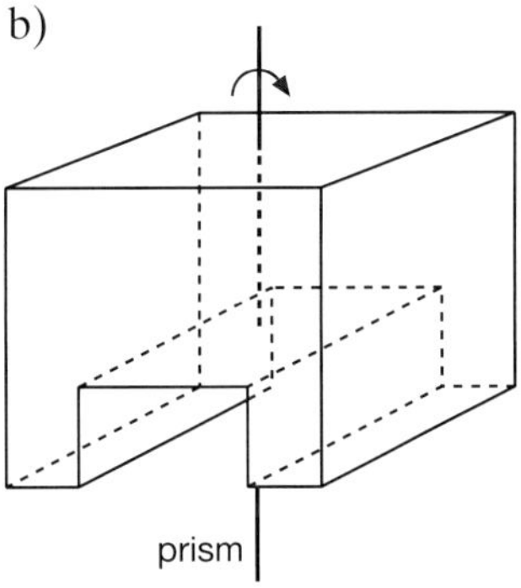

prism

c)

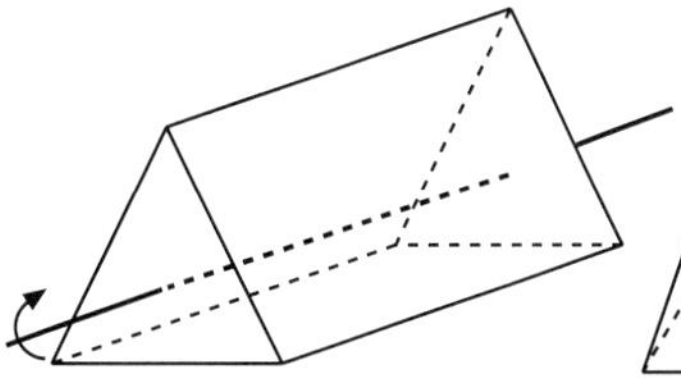

equilateral triangular prism

d)

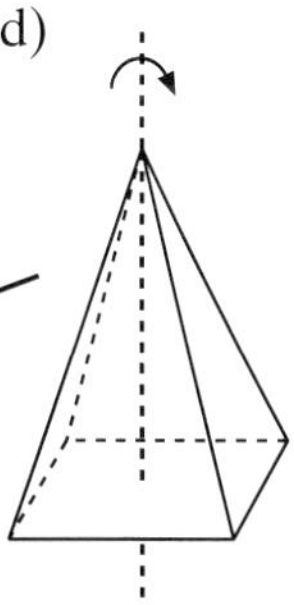

square-based pyramid

e)

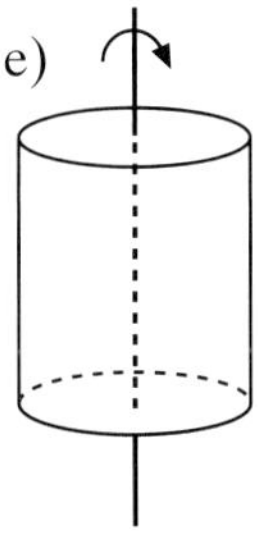

cylinder

f)

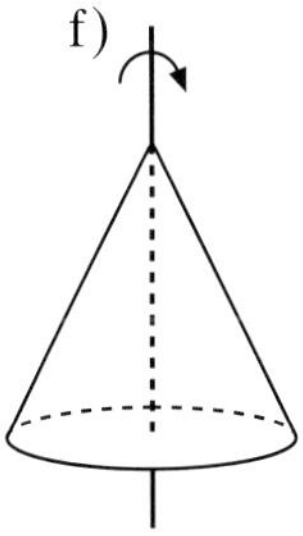

cone

g)

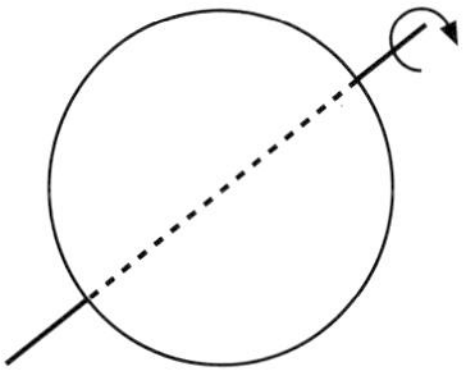

sphere

h)

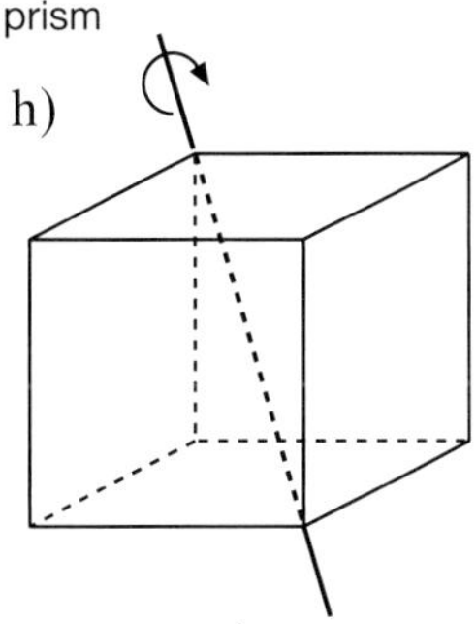

cube

Student Assessment 1

1. Draw a shape with exactly:
   a) one line of reflective symmetry,
   b) two lines of reflective symmetry,
   c) three lines of reflective symmetry.
   Mark the lines of symmetry on each diagram.

2. Draw a shape with:
   a) rotational symmetry of order exactly 2,
   b) rotational symmetry of order exactly 3.
   Mark the position of the centre of rotation on each diagram.

3. Copy and complete the following shapes so that the **bold** lines become the lines of symmetry:

a)

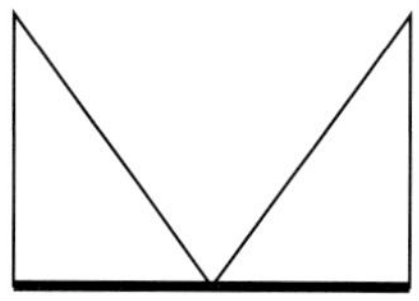

b)

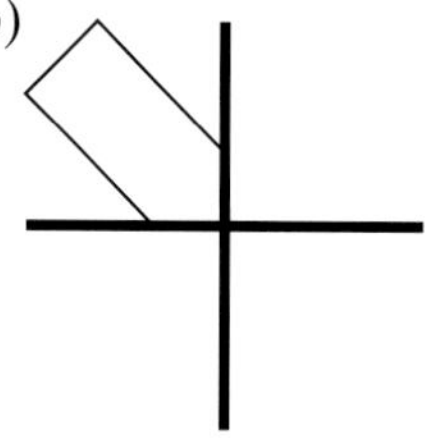

c)

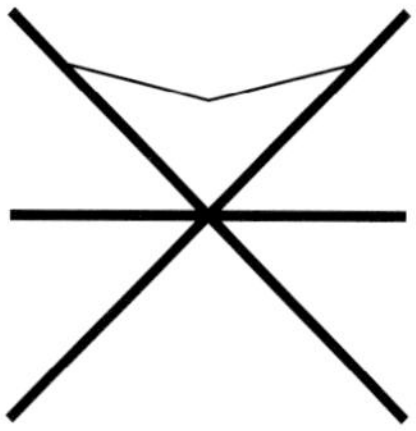

d)

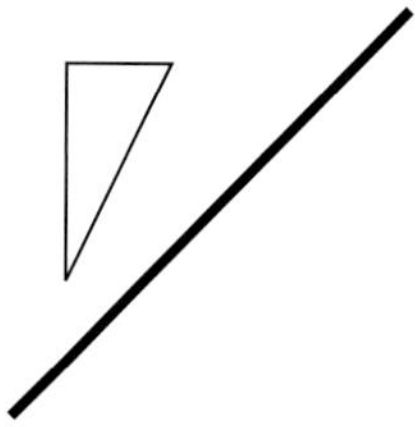

4. For the completed drawings in question 3:
   i) identify the centre of rotation,
   ii) state the order of rotational symmetry.

Student Assessment 2

1. Draw a two-dimensional shape with:
   a) rotational symmetry of order exactly 4,
   b) rotational symmetry of order exactly 8.
   Mark the position of the centre of rotation on each diagram.

2. Draw and name a three-dimensional shape with the following orders of rotational symmetry. Mark the position of the centre of rotation on each diagram.
   a) order 2 b) order 3

3. Copy and complete the following shapes so that the **bold** lines become lines of symmetry:

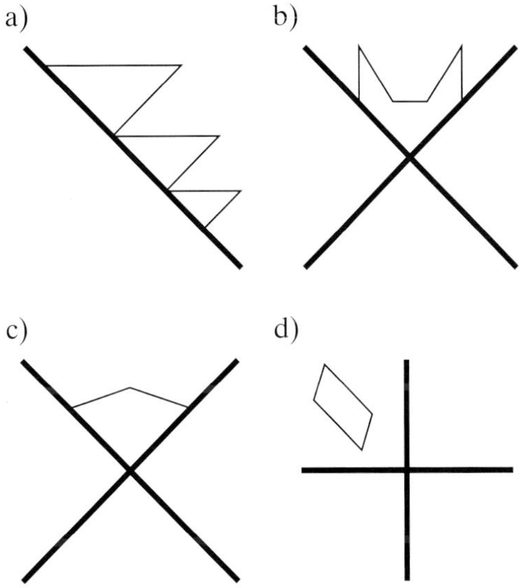

4. For the completed drawings in question 3:
   i) identify the centre of rotation,
   ii) state the order of rotational symmetry.

4.

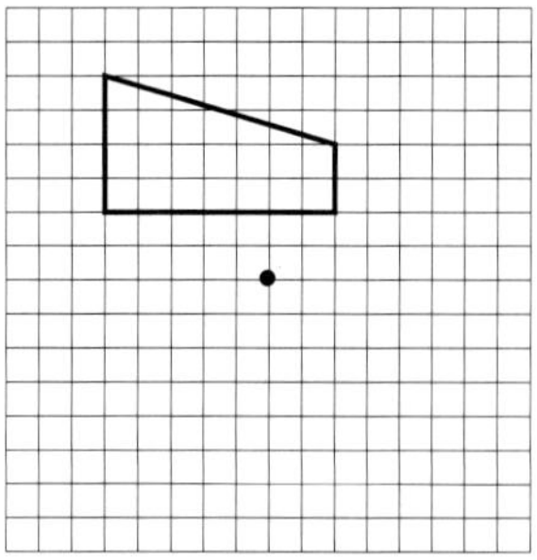

rotation 90° clockwise

5.

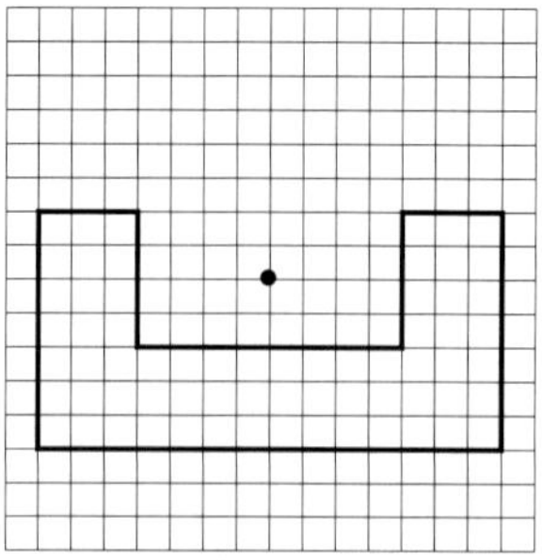

rotation 90° anti-clockwise

6.

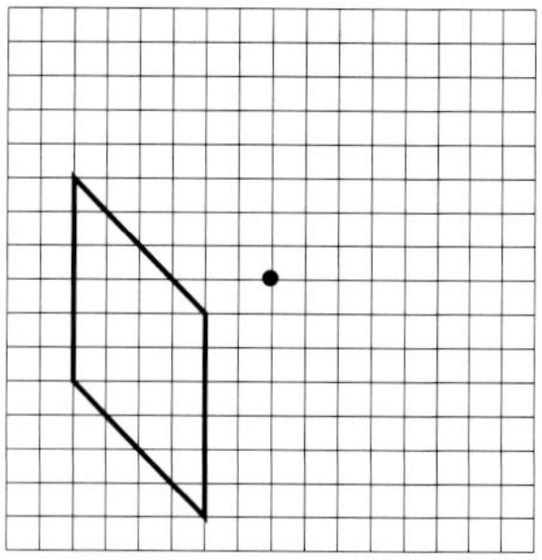

rotation 90° clockwise

Exercise 27.4 In the following, the object (unshaded) and image (shaded) have been drawn. Copy each diagram.

i) Mark the centre of rotation.

ii) Calculate the angle and direction of rotation.

1.

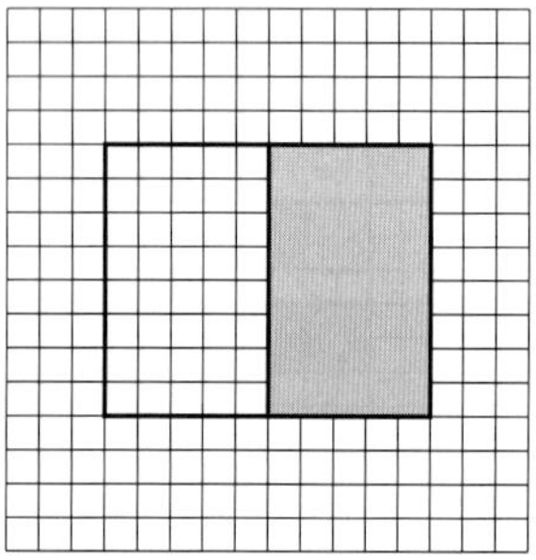

2.

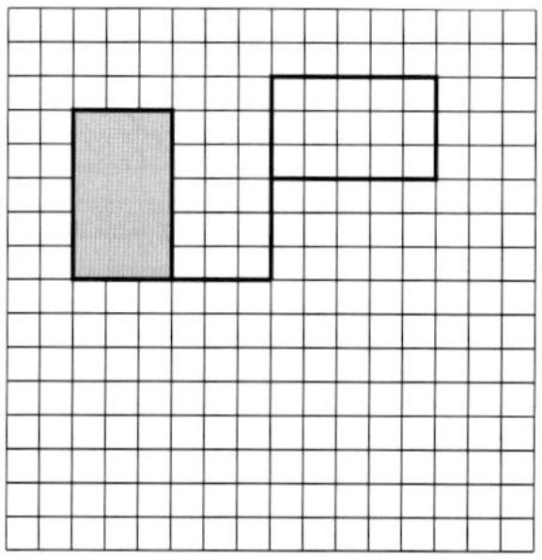

3.

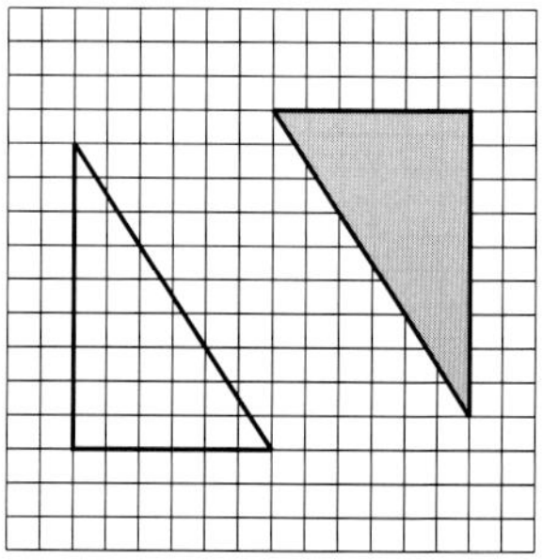

4.

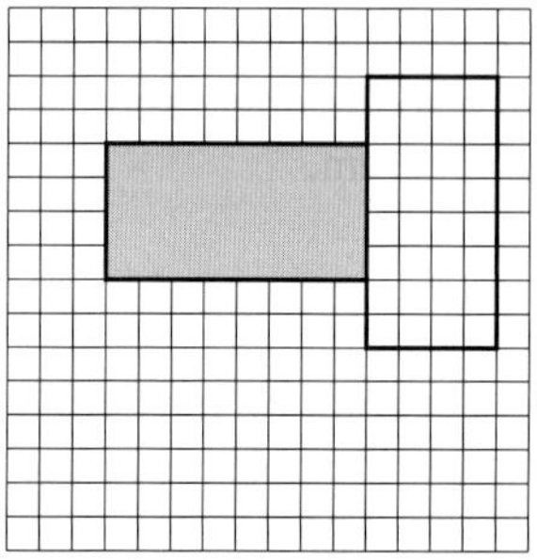

5.

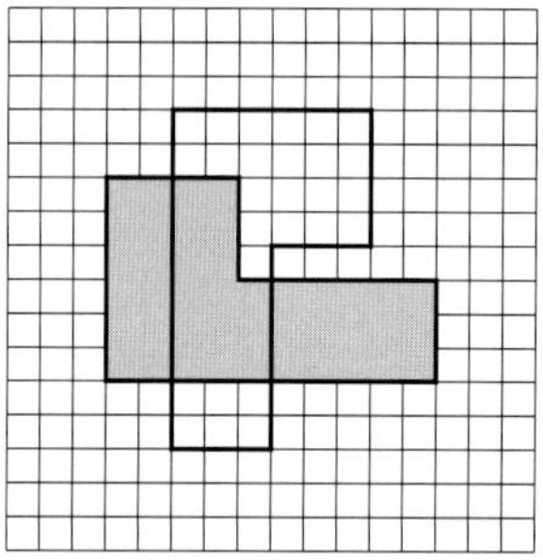

6.

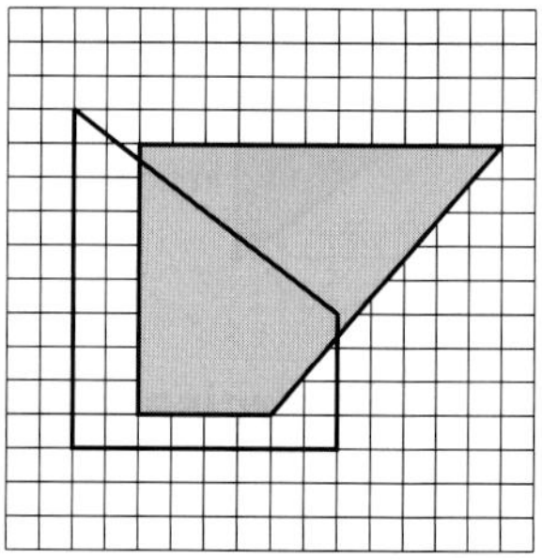

Translation

If an object is translated, it undergoes a 'straight sliding' movement. When describing a translation it is necessary to give the **translation vector**. As no rotation is involved, each point on the object moves in the same way to its corresponding point on the image, e.g.

1.

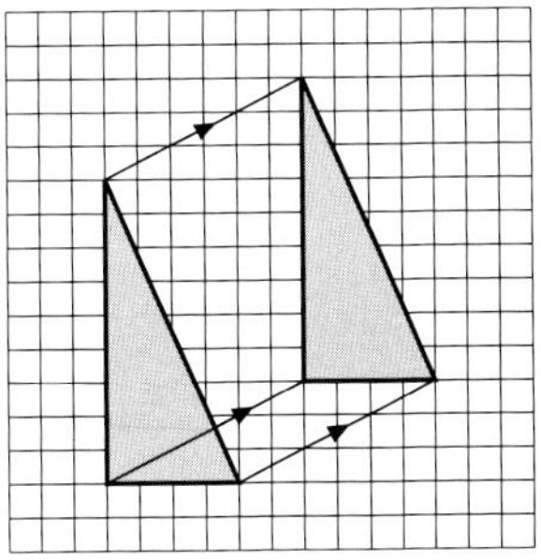

Vector $= \begin{pmatrix} 6 \\ 3 \end{pmatrix}$

2.

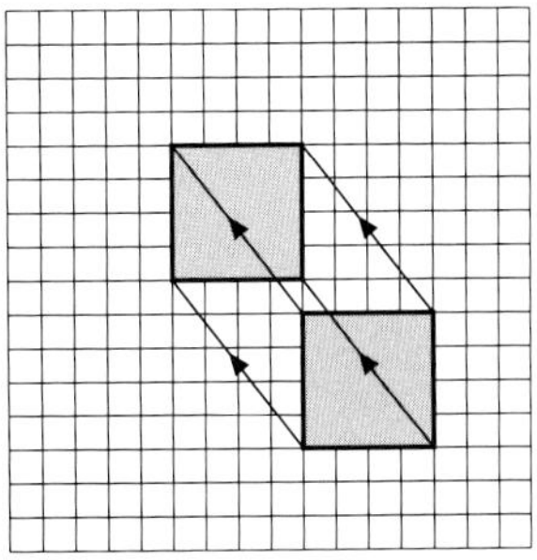

Vector $= \begin{pmatrix} -4 \\ 5 \end{pmatrix}$

Exercise 27.5

In the following diagrams, object A has been translated to each of images B and C. Give the translation vector in each case.

1.

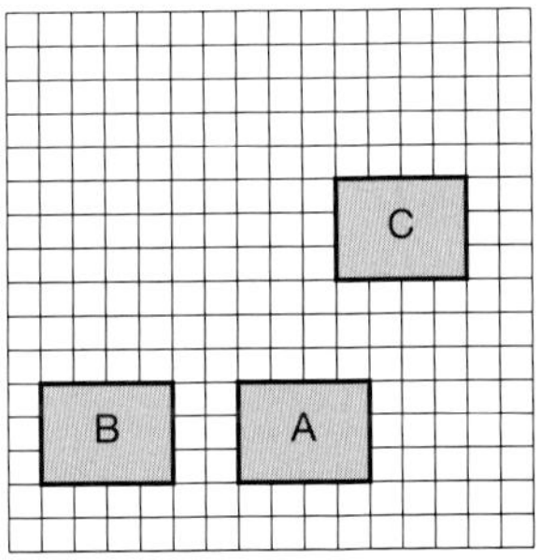

2.

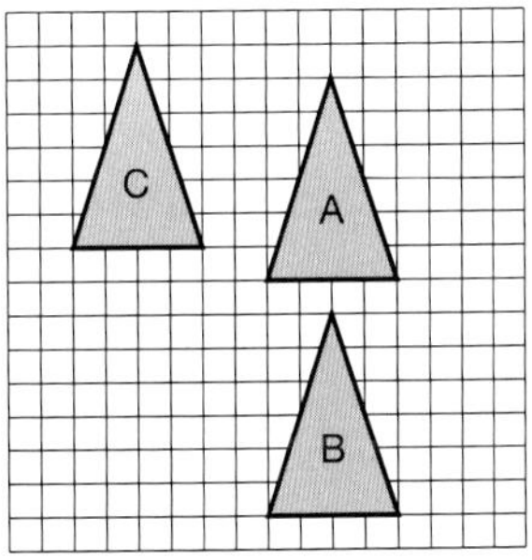

3.

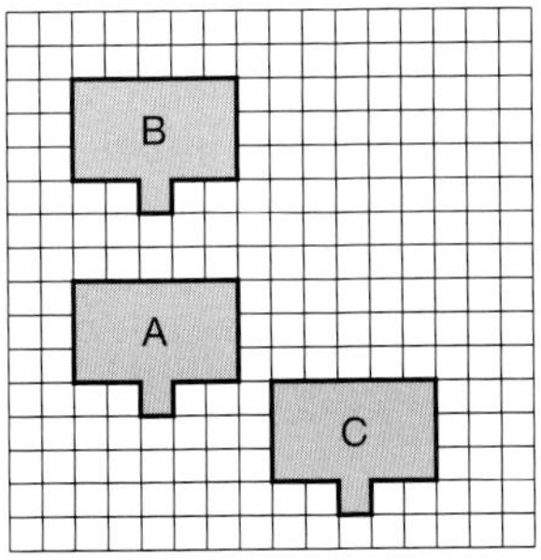

4.

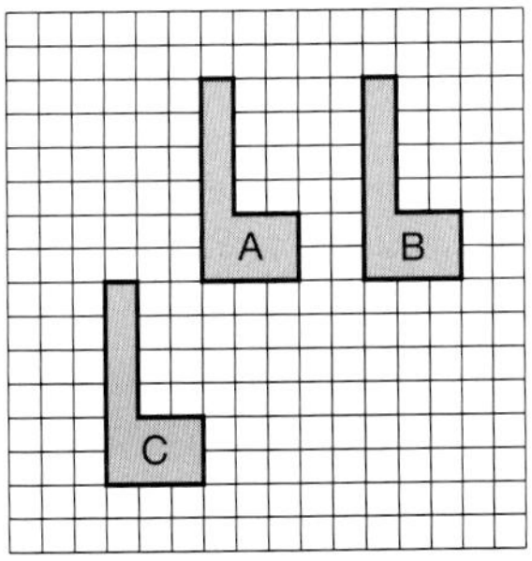

Exercise 27.6

Copy each of the following diagrams and draw the object. Translate the object by the vector given in each case and draw the image in its position. (Note that a bigger grid than the one shown may be needed.)

1.

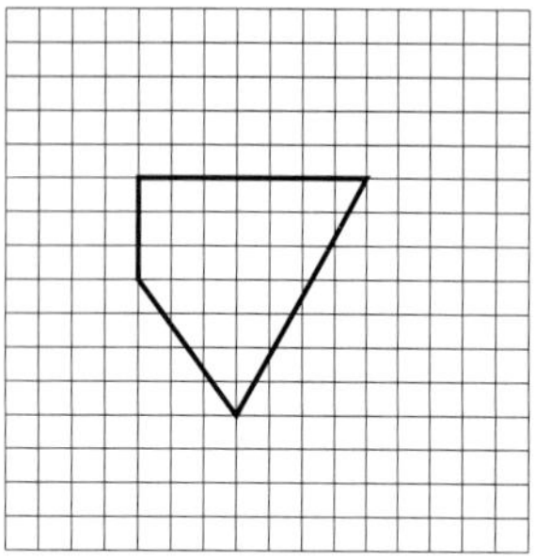

Vector = $\begin{pmatrix} 3 \\ 5 \end{pmatrix}$

2.

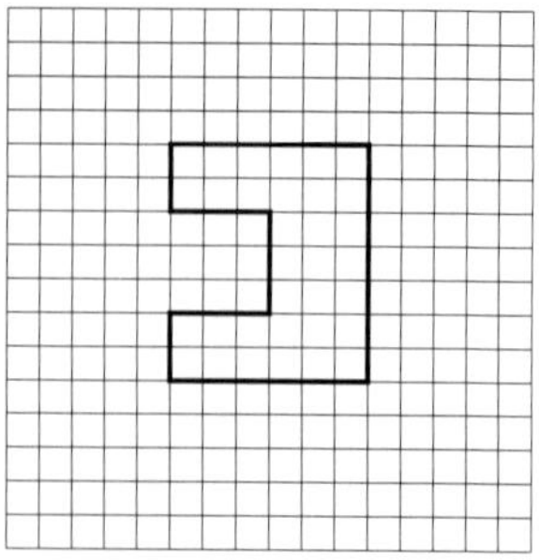

Vector = $\begin{pmatrix} 5 \\ -4 \end{pmatrix}$

3.

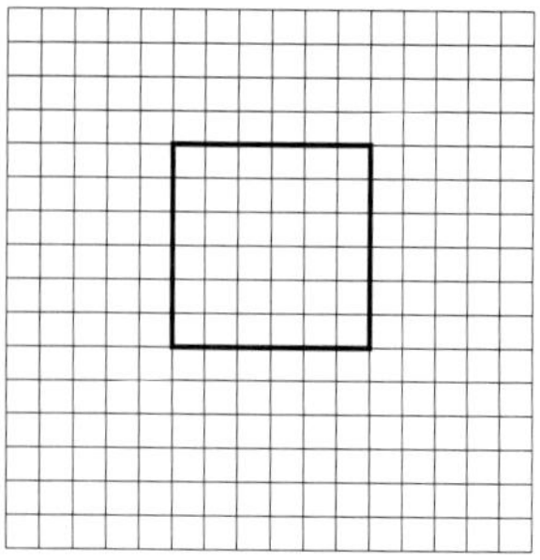

Vector = $\begin{pmatrix} -4 \\ 6 \end{pmatrix}$

4.

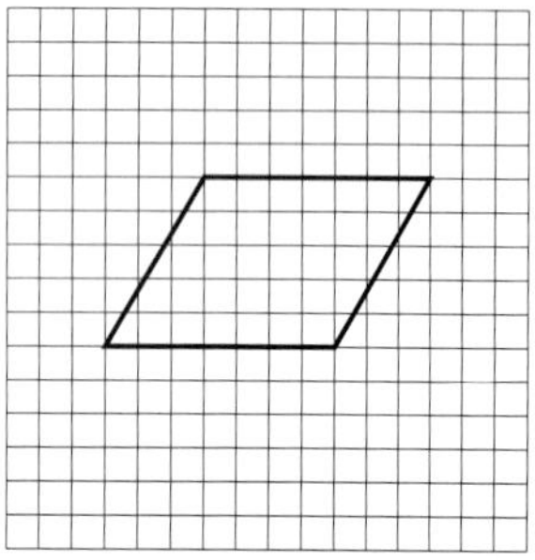

Vector = $\begin{pmatrix} -2 \\ -5 \end{pmatrix}$

5.

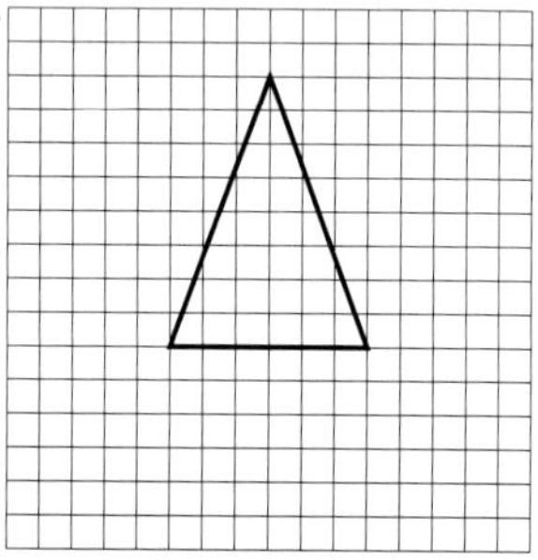

Vector = $\begin{pmatrix} -6 \\ 0 \end{pmatrix}$

6.

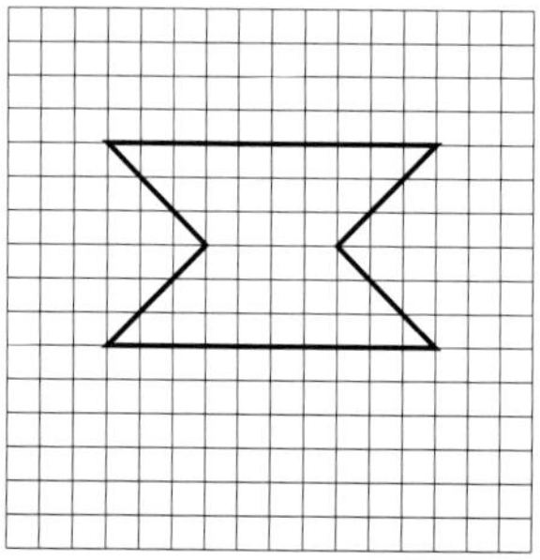

Vector = $\begin{pmatrix} 0 \\ -1 \end{pmatrix}$

Enlargement

If an object is enlarged, the result is an image which is mathematically similar to the object but of a different size. The image can be either larger or smaller than the original object. When describing an enlargement two pieces of information need to be given, the position of the **centre of enlargement** and the **scale factor of enlargement**.

Worked examples

a) In the diagram below, triangle ABC is enlarged to form triangle A'B'C'.

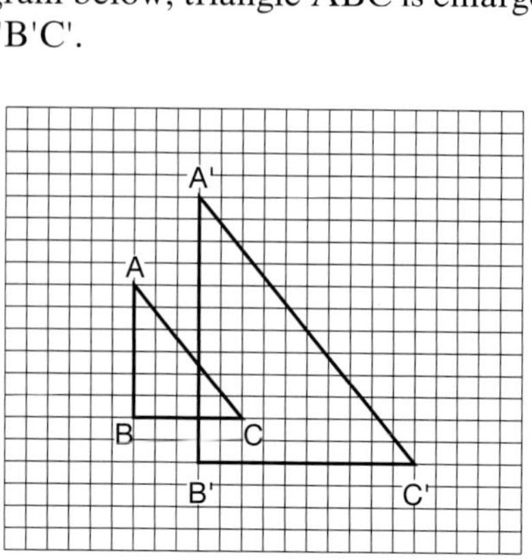

i) Find the centre of enlargement.

The centre of enlargement is found by joining corresponding points on the object and image with a straight line. These lines are then extended until they meet. The point at which they meet is the centre of enlargement O.

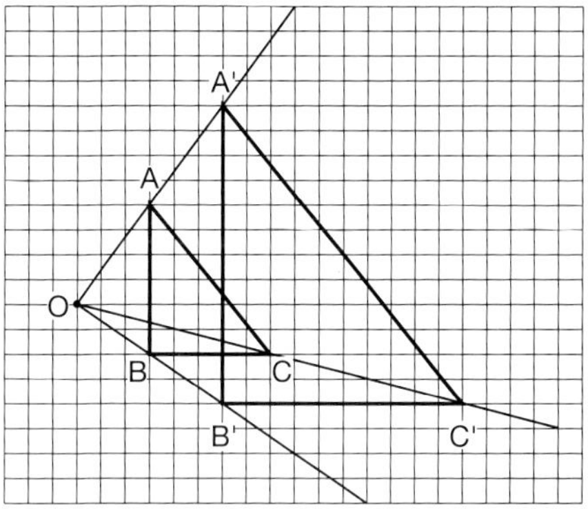

ii) Calculate the scale factor of enlargement.

The scale factor of enlargement can be calculated in one of two ways. From the diagram above it can be seen that the distance OA' is twice the distance OA. Similarly OC' and OB' are both twice OC and OB, respectively, hence the scale factor of enlargement is 2.

Alternatively the scale factor can be found by considering the ratio of the length of a side on the image to the length of the corresponding side on the object. i.e.

$$\frac{A'B'}{AB} = \frac{12}{6} = 2$$

Hence the scale factor of enlargement is 2.

b) In the diagram below, the rectangle ABCD undergoes a transformation to form rectangle A'B'C'D'.

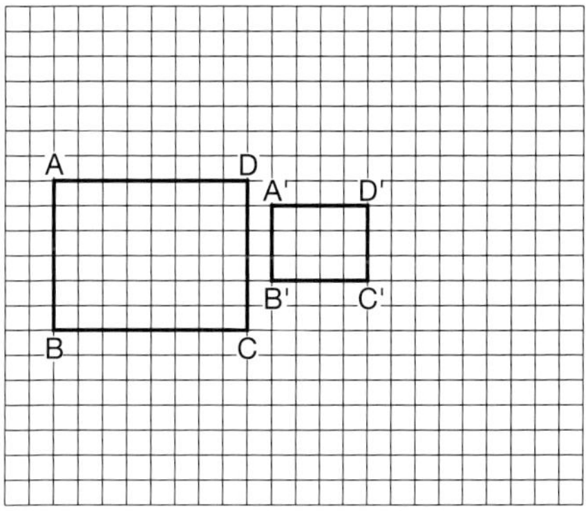

i) Find the centre of enlargement.

By joining corresponding points on both the object and the image the centre of enlargement is found at O.

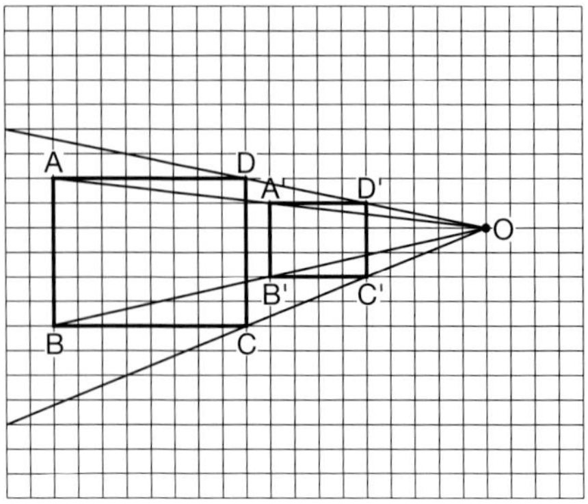

ii) Calculate the scale factor of enlargement.

$$\text{The scale factor of enlargement} = \frac{A'B'}{AB} = \frac{3}{6} = \frac{1}{2}$$

Note: If the scale factor of enlargement is greater than 1, then the image is larger than the object. If the scale factor lies between 0 and 1, then the resulting image is smaller than the object. In these cases, although the image is smaller than the object, the transformation is still known as an enlargement.

Exercise 27.7 Copy the following diagrams and find:

a) the centre of enlargement,
b) the scale factor of enlargement.

1.

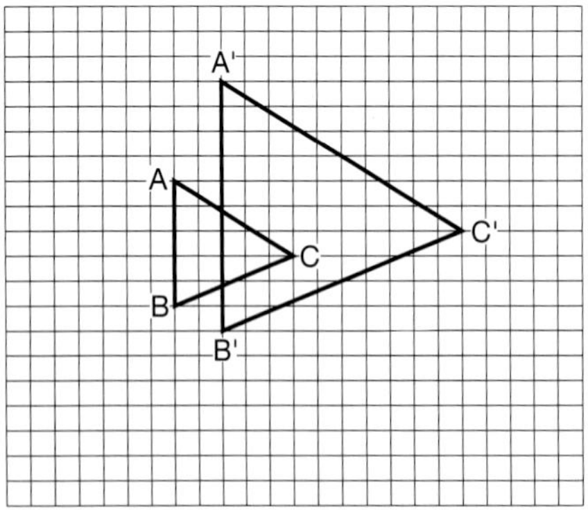

2.

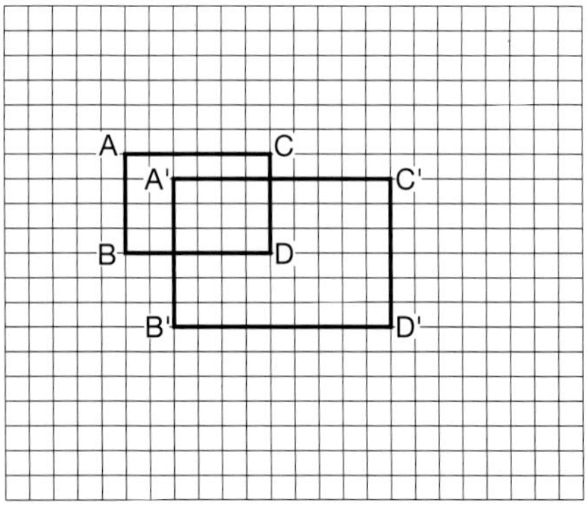

3.

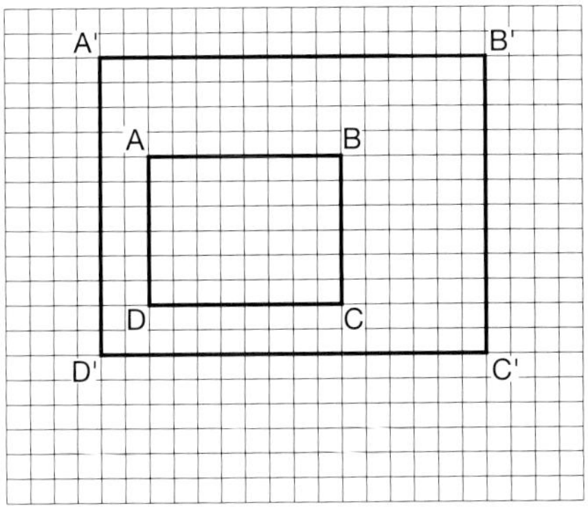

4.

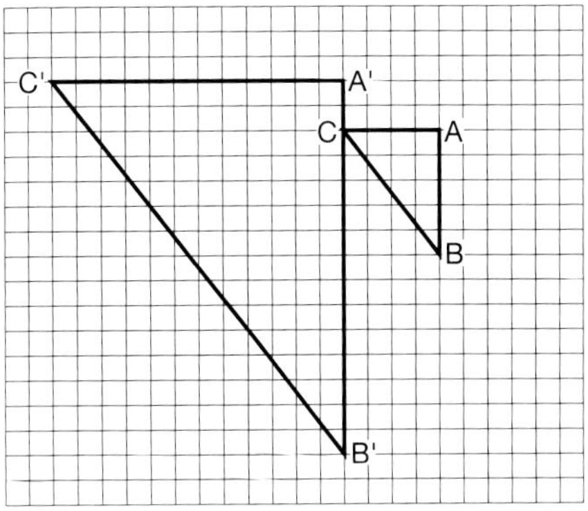

5.

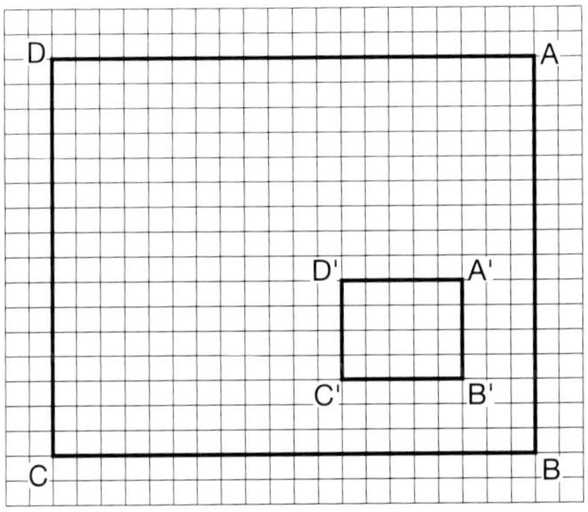

Exercise 27.8

Copy the following diagrams and enlarge the objects by the scale factor given and from the centre of enlargement shown. Grids larger than those shown may be needed.

1.

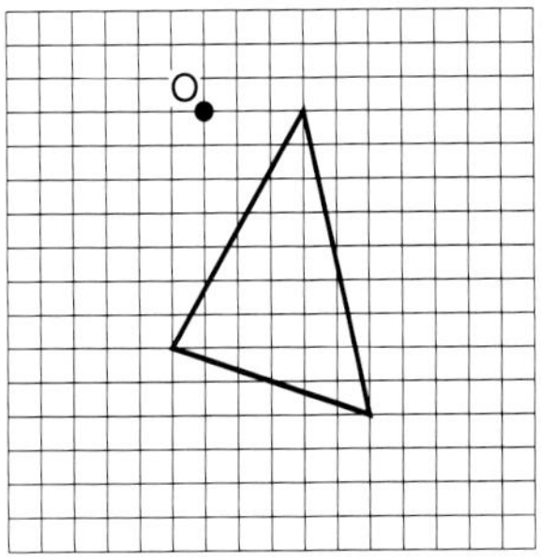

scale factor 2

2.

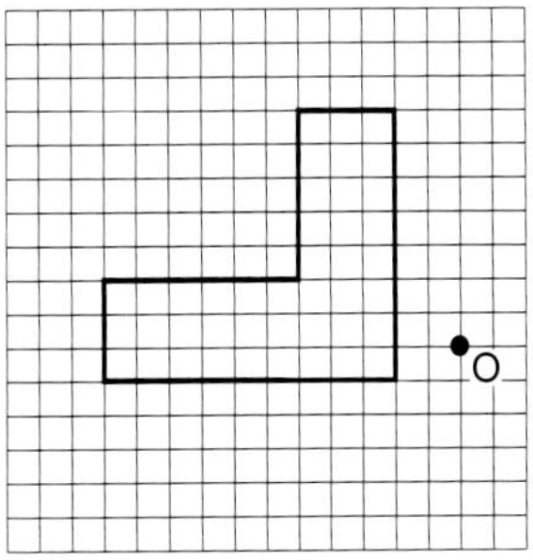

scale factor 2

3.

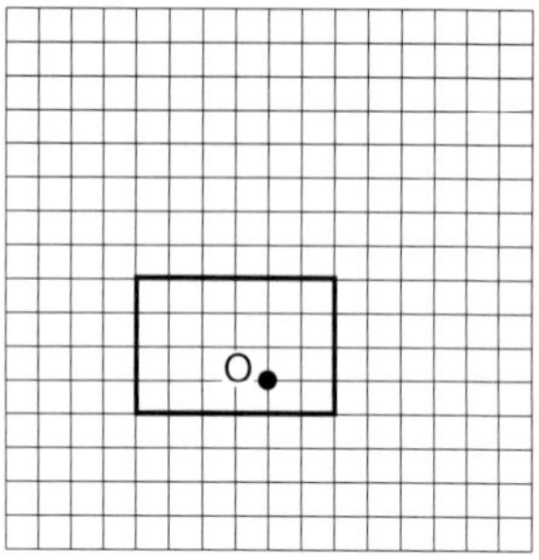

scale factor 3

4.

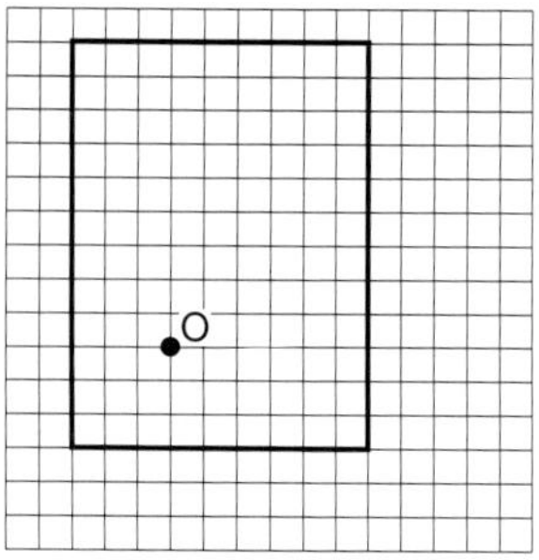

scale factor $\frac{1}{3}$

The diagram below shows an example of **negative enlargement**.

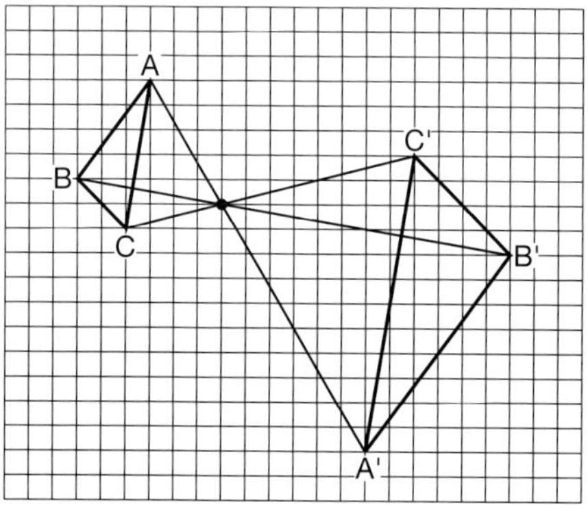

scale factor of enlargement is −2

With negative enlargement each point and its image are on opposite sides of the centre of enlargement. The scale factor of enlargement is calculated in the same way, remembering, however, to write a '−' sign before the number.

Exercise 27.9

1. Copy the following diagram and then calculate the scale factor of enlargement and show the position of the centre of enlargement.

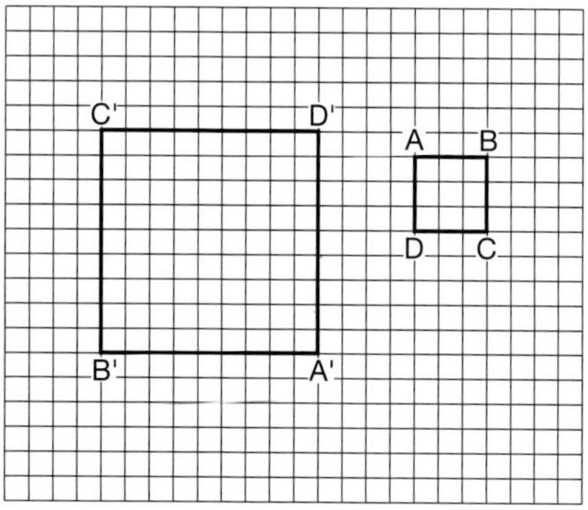

2. The scale factor of enlargement and centre of enlargement are both given. Copy and complete the diagram.

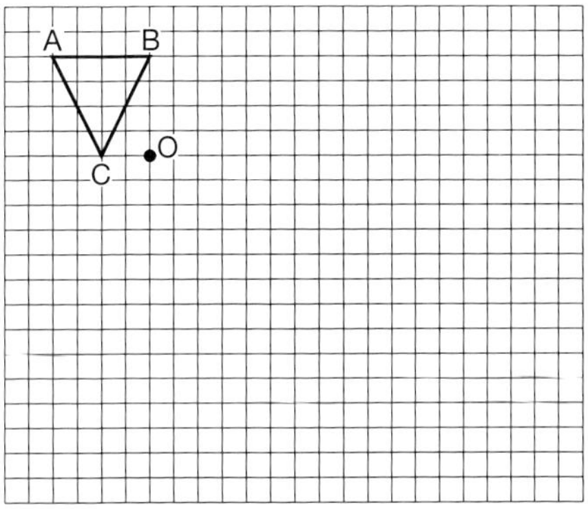

scale factor of enlargement is −2.5

3. The scale factor of enlargement and centre of enlargement are both given. Copy and complete the diagram.

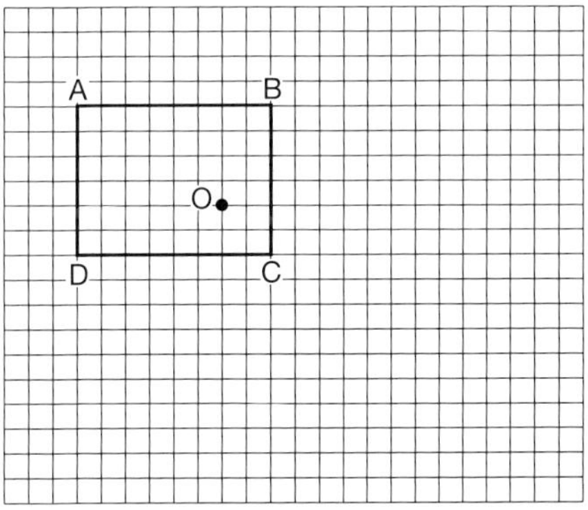

scale factor of enlargement is −2

4. Copy the following diagram and then calculate the scale factor of enlargement and show the position of the centre of enlargement.

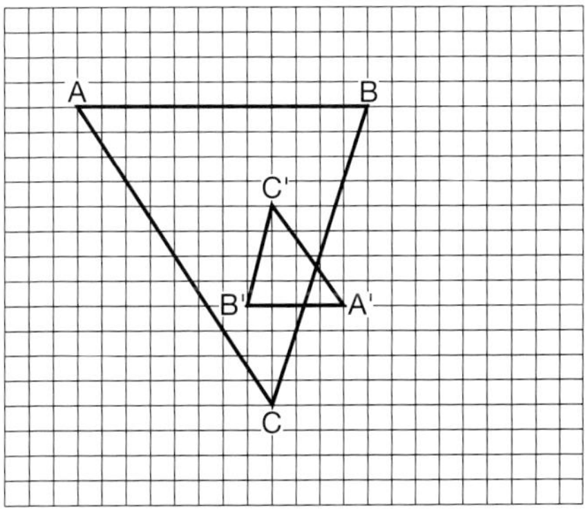

5. An object and part of its image under enlargement are given in the diagram below. Copy the diagram and complete the image. Also find the centre of enlargement and calculate the scale factor of enlargement.

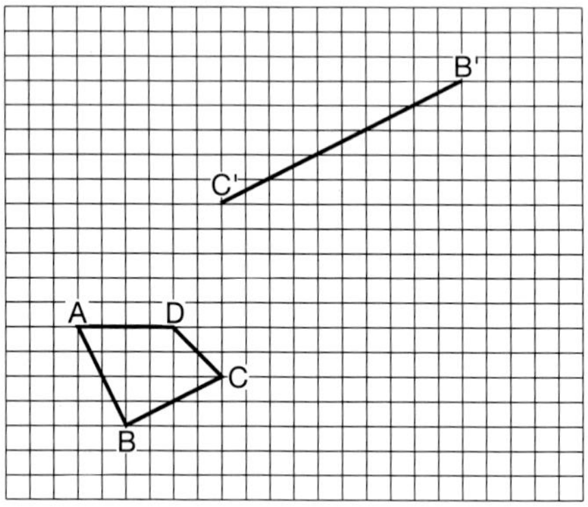

6. In the diagram below, part of an object in the shape of a quadrilateral and its image under enlargement are drawn. Copy and complete the diagram. Also find the centre of enlargement and calculate the scale factor of enlargement.

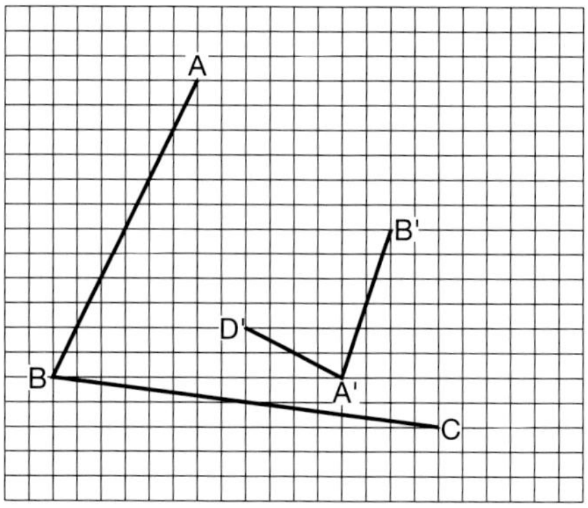

Student Assessment 1

1. Reflect the object below in the mirror line shown.

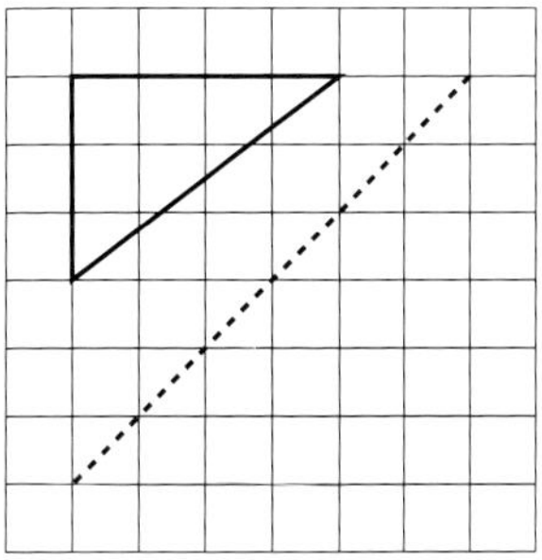

2. Rotate the object below 90° anti-clockwise about the centre of rotation O.

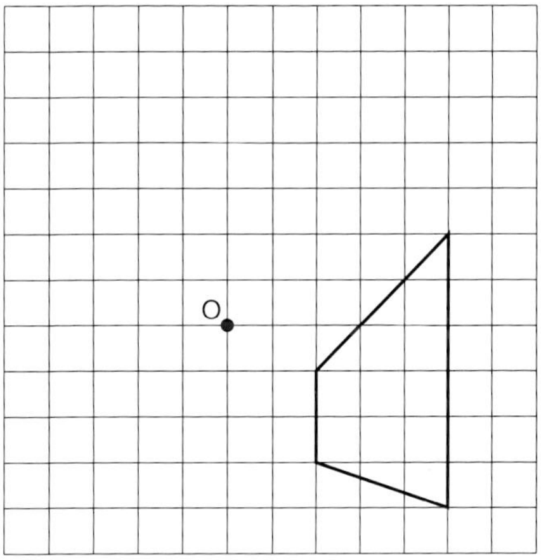

3. Write down the column vector of the translation which maps:
   a) rectangle A to rectangle B,
   b) rectangle B to rectangle C.

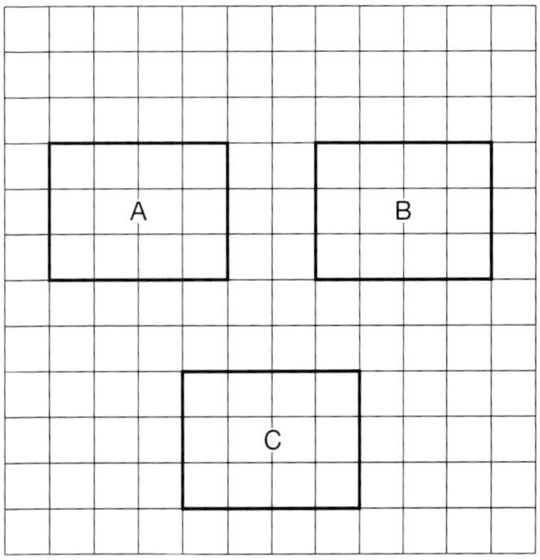

4. Enlarge the triangle below by a scale factor 2 and from the centre of enlargement O.

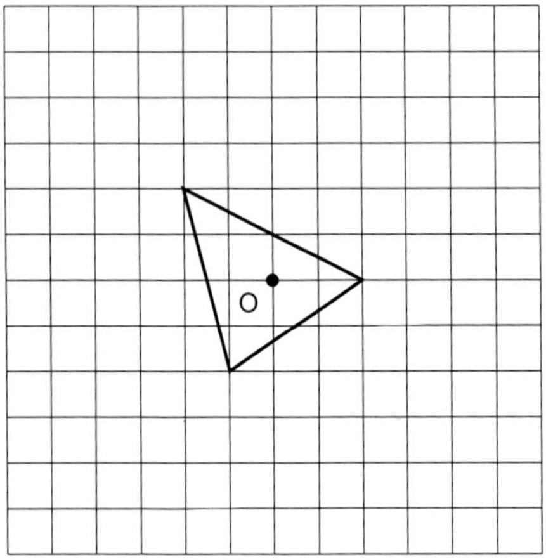

5. An object ABCD and its image A'B'C'D' are shown below.
   a) Find the position of the centre of enlargement.
   b) Calculate the scale factor of enlargement.

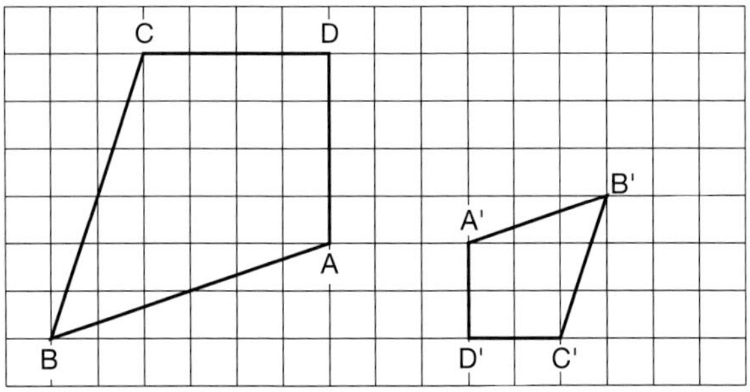

Student Assessment 2

1. Reflect the object below in the mirror line shown.

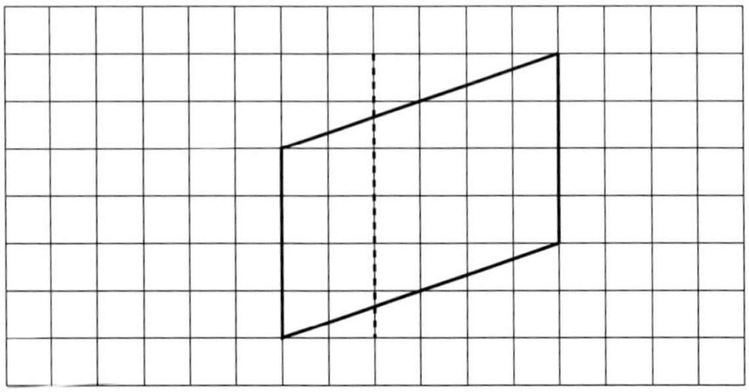

2. Rotate the object below 180° about the centre of rotation O.

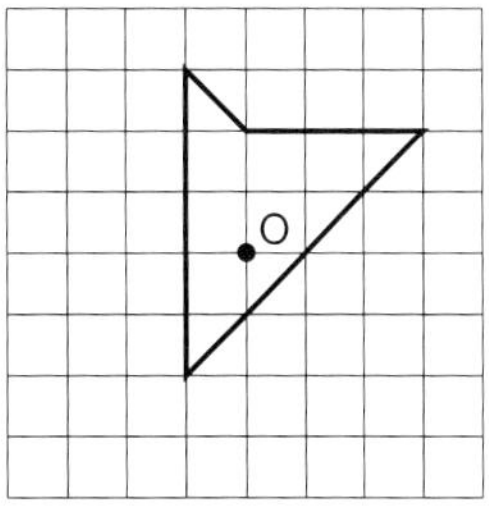

3. Write down the column vector of the translation which maps:

a) triangle A to triangle B,

b) triangle B to triangle C.

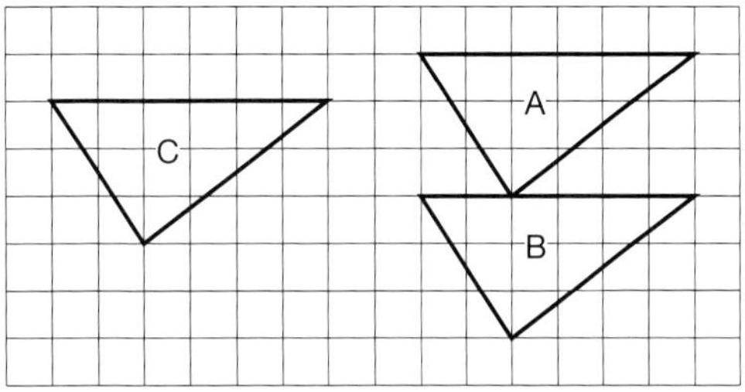

4. Enlarge the rectangle below by a scale factor 1.5 and from the centre of enlargement O.

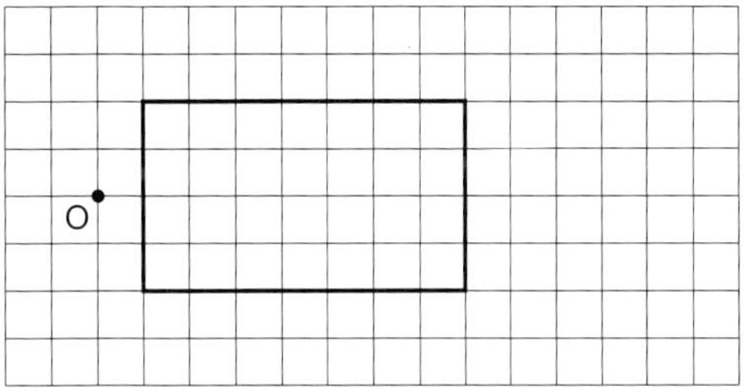

5. An object WXYZ and its image W'X'Y'Z' are shown below.

a) Find the position of the centre of enlargement.

b) Calculate the scale factor of enlargement.

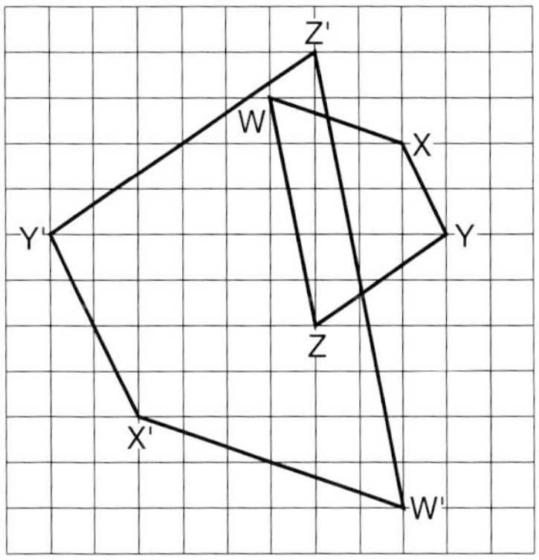

28 *MENSURATION*

Carry out calculations involving the perimeter and area of a rectangle and triangle, the circumference and area of a circle, the area of a parallelogram and a trapezium, the volume of a cuboid, prism and cylinder, and the surface area of a cuboid and a cylinder.

NB: All diagrams are not drawn to scale.

The perimeter and area of a rectangle

The **perimeter** of a shape is the distance around the outside of the shape. Perimeter can be measured in mm, cm, m, km, etc.

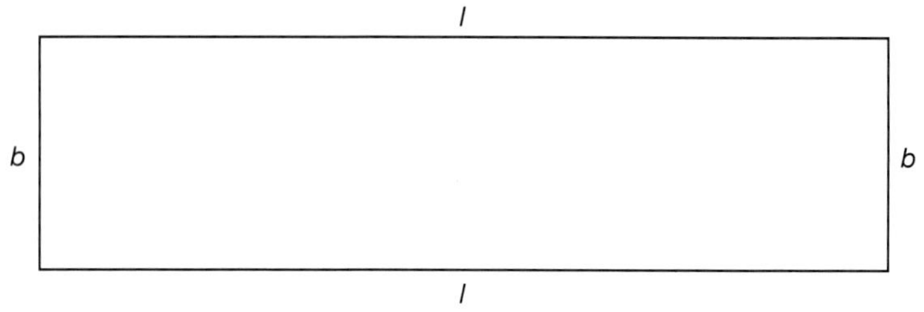

The perimeter of the rectangle above of length l and breadth b is therefore:

$$\text{Perimeter} = l + b + l + b$$

This can be rearranged to give:

$$\text{Perimeter} = 2l + 2b$$

This in turn can be factorised to give:

$$\text{Perimeter} = 2(l + b)$$

The **area** of a shape is the amount of surface that it covers. Area is measured in mm^2, cm^2, m^2, km^2, etc.

The area A of the rectangle above is given by the formula:

$$A = lb$$

Worked example Find the area of the rectangle below.

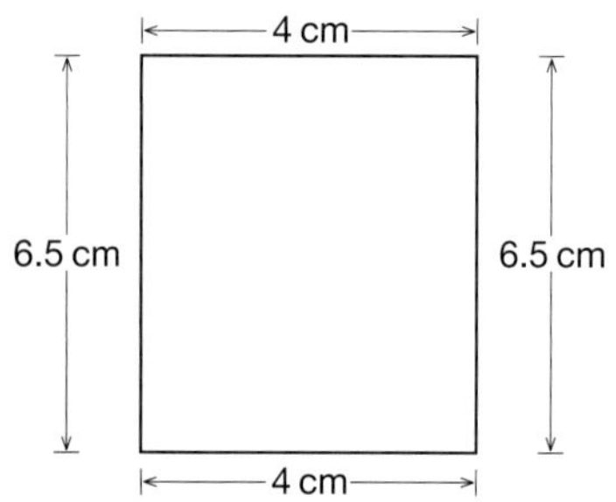

$A = lb$
$A = 6.5 \times 4$
$A = 26$
Area is 26 cm^2.

Exercise 28.1

1. Calculate the area and perimeter of the rectangles described below.

	Length	Breadth	Area	Perimeter
a)	6 cm	4 cm		
b)	5 cm	9 cm		
c)	4.5 cm	6 cm		
d)	3.8 m	10 m		
e)	5 m	4.2 m		
f)	3.75 cm	6 cm		
g)	3.2 cm	4.7 cm		
h)	18.7 m	5.5 cm		
i)	85 cm	1.2 m		
j)	3.3 m	75 cm		

Worked example Calculate the breadth of a rectangle of area 200 cm^2 and length 25 cm.

$A = lb$

$200 = 25b$

$b = 8$ so the breadth is 8 cm.

2. Use the formula for the area of a rectangle to find the value of A, l or b as indicated in the table below.

	Length	Breadth	Area
a)	8.5 cm	7.2 cm	A
b)	25 cm	b cm	250 cm^2
c)	l cm	25 cm	400 cm^2
d)	7.8 m	b m	78 cm^2
e)	l cm	8.5 cm	102 cm^2
f)	22 cm	b cm	330 cm^2
g)	l cm	7.5 cm	187.5 cm^2

3. Find the area and perimeter of each of the following squares or rectangles:
   a) the floor of a room which is 8 m long by 7.5 m broad
   b) a stamp which is 35 mm long by 25 mm broad
   c) a wall which is 8.2 m long by 2.5 m high
   d) a field which is 130 m long by 85 m wide
   e) a chess board of side 45 cm
   f) a book which is 25 cm broad by 35 cm long
   g) an airport runway which is 3.5 km long by 800 m wide
   h) a street which is 1.2 km long by 25 m wide
   i) a sports hall 65 m long by 45 m wide
   j) a compact disc box which is a square of side 125 mm

Area of a triangle

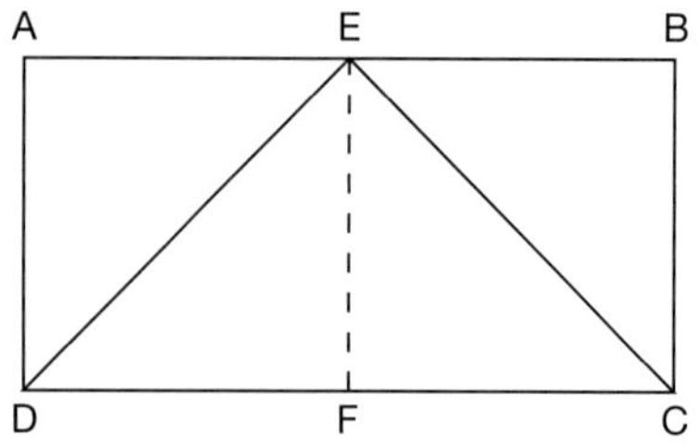

Rectangle ABCD has a triangle CDE drawn inside it.

Point E is said to be a **vertex** of the triangle.
EF is the **height** or **altitude** of the triangle.
CD is the **length** of the rectangle, but is called the **base** of the triangle.

It can be seen from the diagram that triangle DEF is half the area of the rectangle AEFD.

Also triangle CFE is half the area of rectangle EBCF.

It follows that **triangle CDE is half the area of rectangle ABCD.**

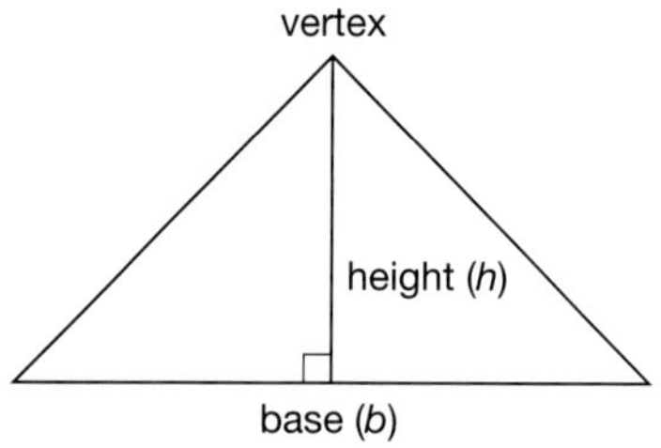

Area of a triangle $A = \frac{1}{2}bh$, where b is the base and h is the height.

Note: it does not matter which side is called the base, but the height must be measured at right angles from the base to the opposite vertex.

Exercise 28.2

1. Calculate the areas of the triangles below:

a)

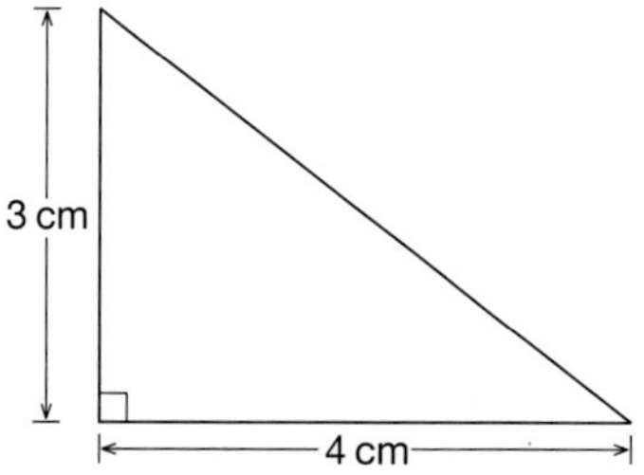

b)

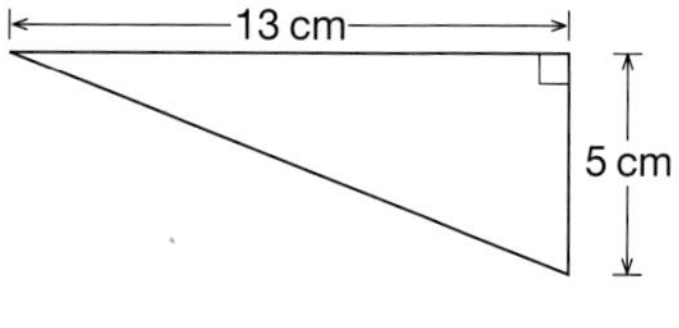

c)

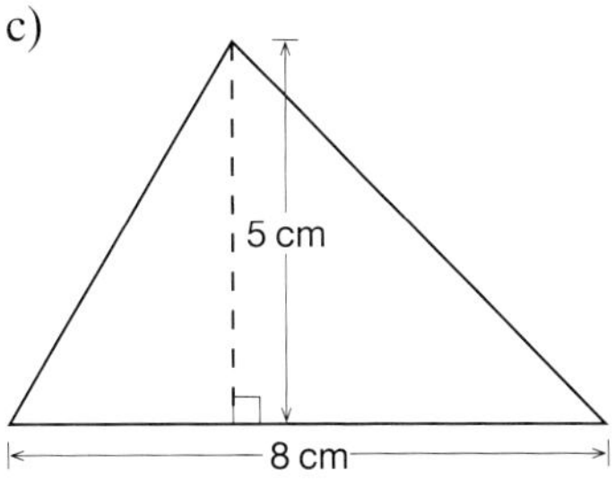

d)

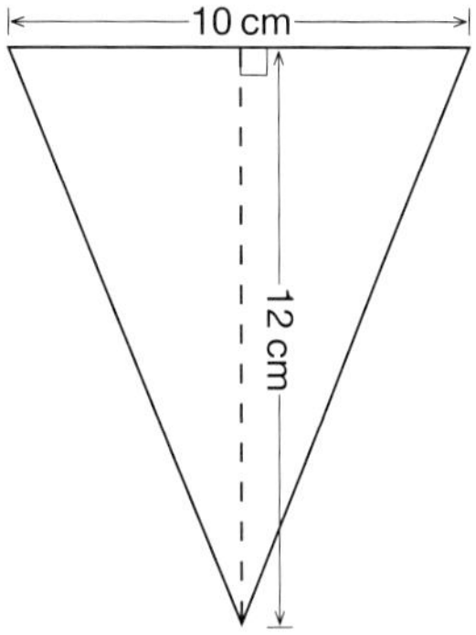

e)

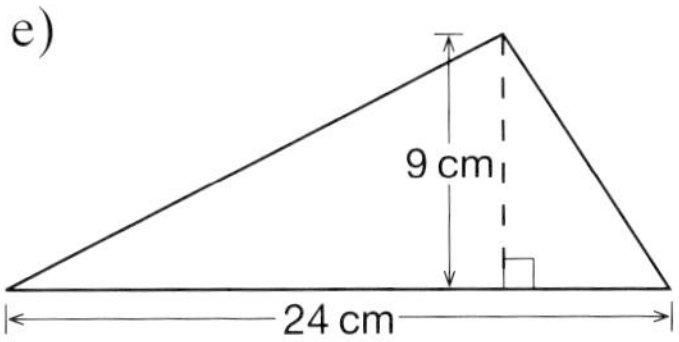

f)

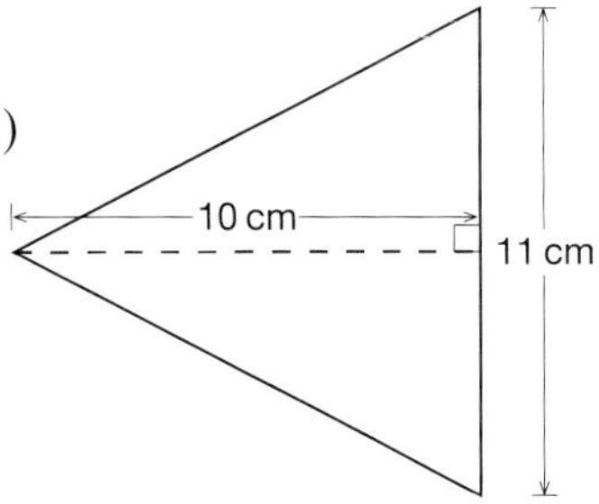

2. Calculate the areas of the shapes below:

a)

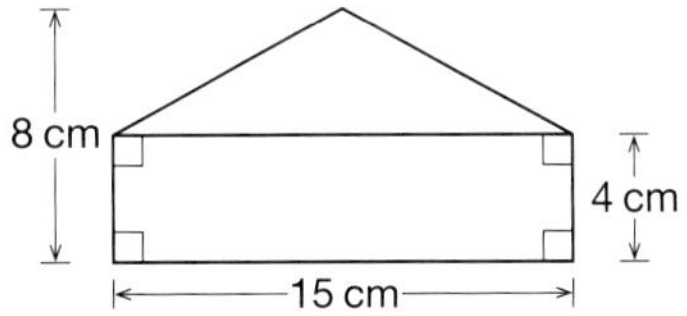

b)

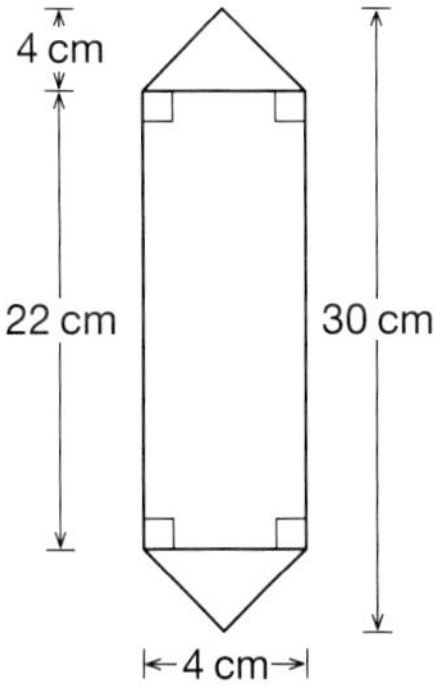

c)

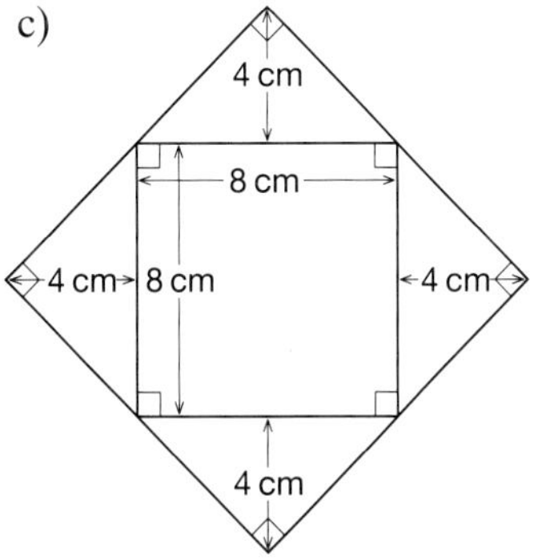

d)

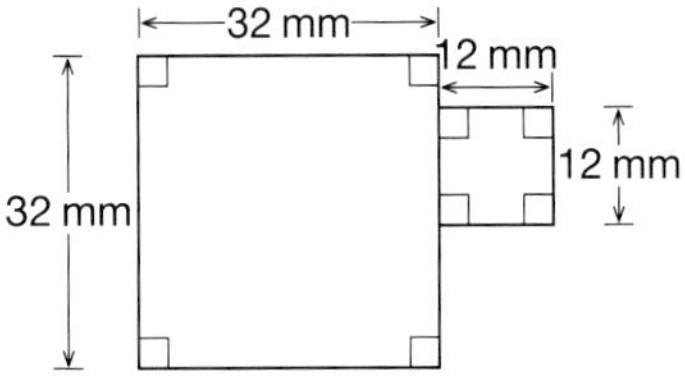

e)

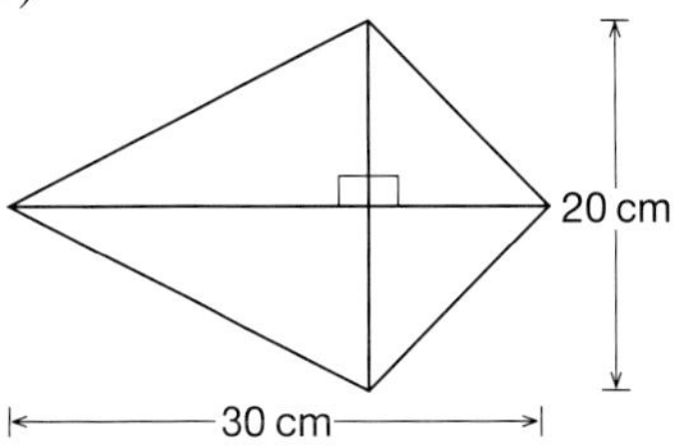

f)

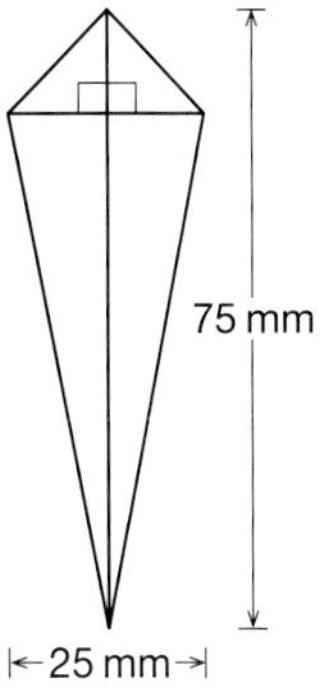

The circumference and area of a circle

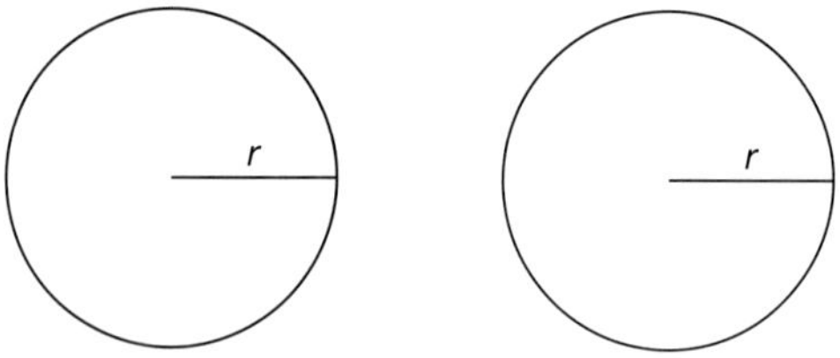

The circumference is $2\pi r$. The area is πr^2.

$C = 2\pi r$ $\quad$ $A = \pi r^2$

Worked examples

a) Calculate the circumference of this circle, giving your answer to 2 d.p.

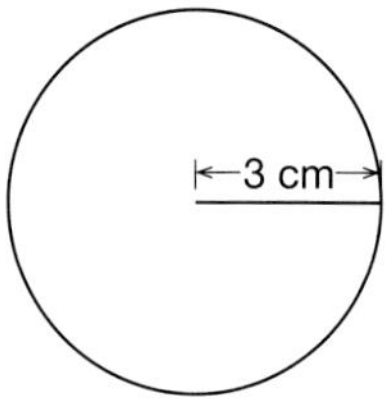

$$\begin{aligned} C &= 2\pi r \\ &= 2\pi \times 3 \\ &= 18.85 \end{aligned}$$

The circumference is 18.85 cm.

b) If the circumference of this circle is 12 cm, calculate the radius, giving your answer to 2 d.p.

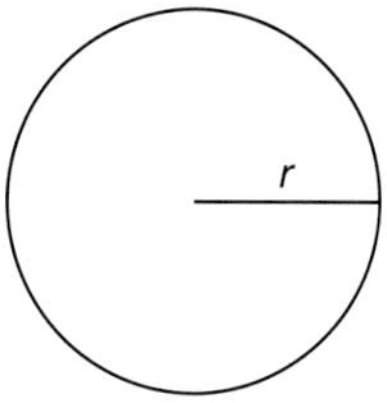

$$C = 2\pi r$$

$$r = \frac{C}{2\pi}$$

$$r = \frac{12}{2\pi}$$

$$= 1.91$$

The radius is 1.91 cm.

c) Calculate the area of this circle, giving your answer to 2 d.p.

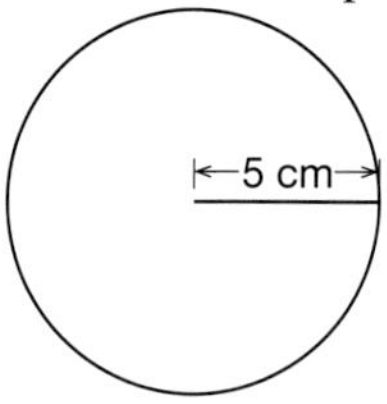

$$\begin{aligned} A &= \pi r^2 \\ &= \pi \times 5^2 \\ &= 78.54 \end{aligned}$$

The area is 78.54 cm^2.

d) If the area of this circle is 34 cm^2, calculate the radius, giving your answer to 2 d.p.

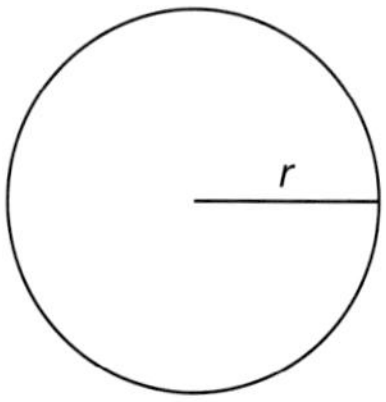

$$A = \pi r^2$$

$$r = \sqrt{\frac{A}{\pi}}$$

$$r = \sqrt{\frac{34}{\pi}}$$

$$= 3.29$$

The radius is 3.29 cm.

Exercise 28.3

1. Calculate the circumference of each circle below, giving your answers to 2 d.p.

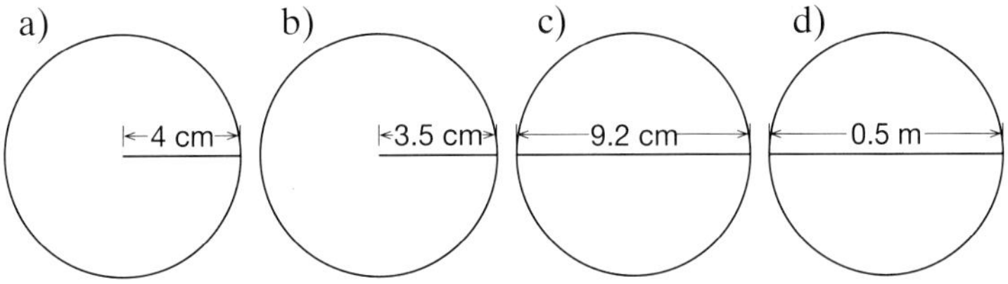

2. Calculate the area of each of the circles in question 1. Give your answers to 2 d.p.

3. Calculate the radius of a circle when the circumference is:
   a) 15 cm b) π cm
   c) 4 m d) 8 mm

4. Calculate the diameter of a circle when the area is:
   a) 16 cm^2 b) 9π cm^2
   c) 8.2 m^2 d) 14.6 mm^2

Exercise 28.4

1. The wheel of a car has an outer radius of 25 cm. Calculate:
   a) how far the car has travelled after one complete turn of the wheel,
   b) how many times the wheel turns for a journey of 1 km.

2. If the wheel of a bicycle has a diameter of 60 cm, calculate how far a cyclist will have travelled after the wheel has rotated 100 times.

3. A circular ring has a cross-section as shown below. If the outer radius is 22 mm and the inner radius 20 mm, calculate the cross-sectional area of the ring.

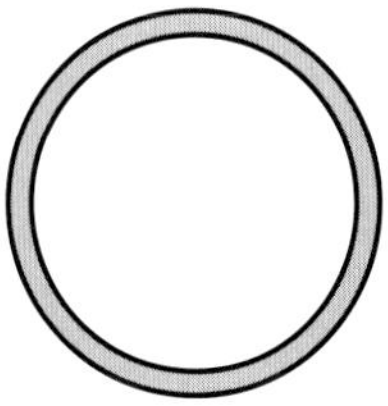

4. 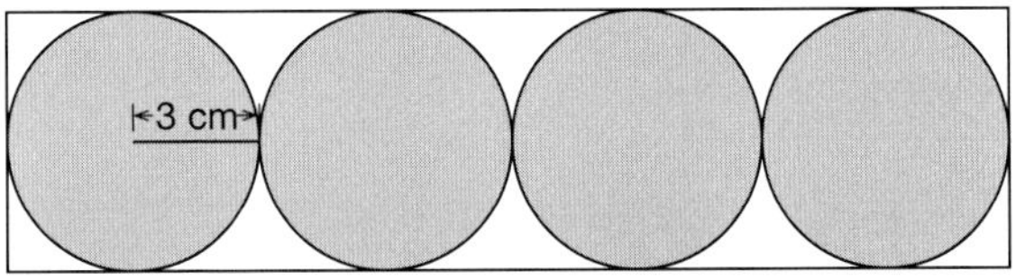

   Four circles are drawn in a line and enclosed by a rectangle as shown. If the radius of each circle is 3 cm, calculate:
   a) the area of the rectangle,
   b) the area of each circle,
   c) the unshaded area within the rectangle.

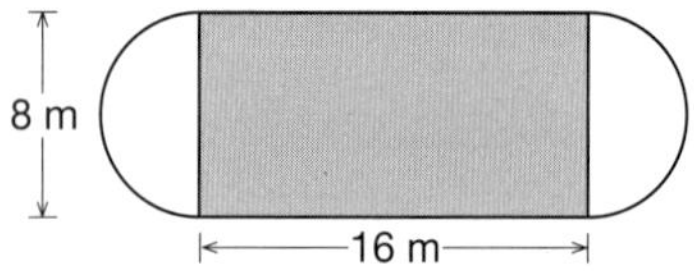

5. A garden is made up of a rectangular patch of grass and two semi-circular vegetable patches. If the length and width of the rectangular patch are 16 m and 8 m respectively, calculate:
   a) the perimeter of the garden,
   b) the total area of the garden.

The area of a parallelogram and a trapezium

A **parallelogram** can be rearranged to form a rectangle in the following way:

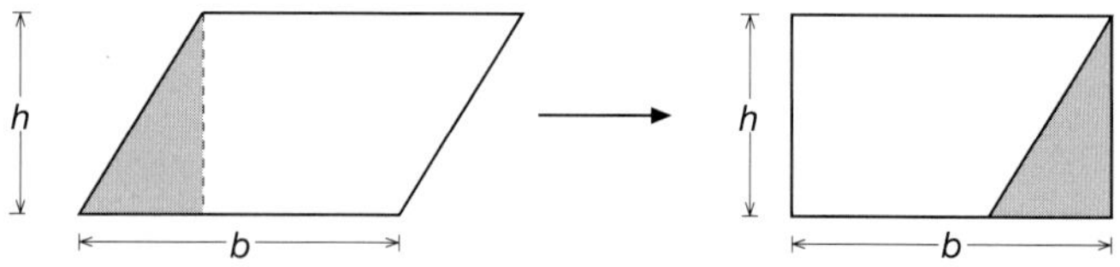

Therefore: area of parallelogram
= base length × perpendicular height.

A **trapezium** can be visualised as being split into two triangles as shown on the left:

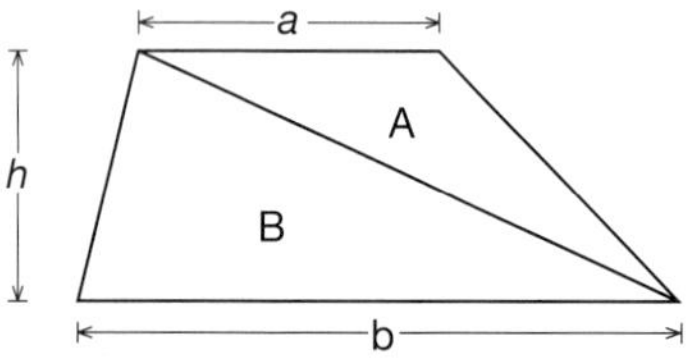

Area of triangle A $= \frac{1}{2} \times a \times h$
Area of triangle B $= \frac{1}{2} \times b \times h$
Area of the trapezium
= area of triangle A + area of triangle B
$= \frac{1}{2}ah + \frac{1}{2}bh$
$= \frac{1}{2}h(a + b)$

Worked examples

a) Calculate the area of the parallelogram (left):

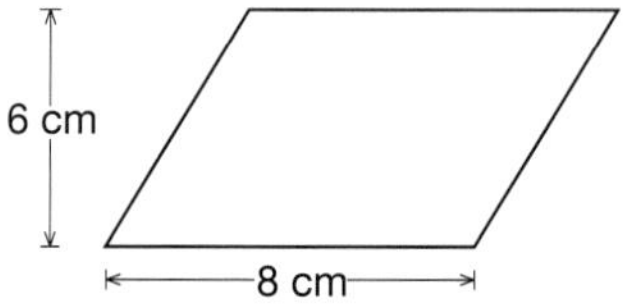

Area = base length × perpendicular height
$= 8 \times 6$
$= 48\ \text{cm}^2$

b) Calculate the shaded area in the shape (left):

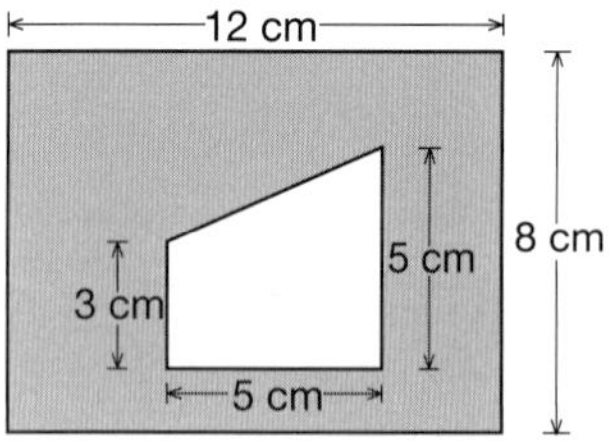

Area of rectangle $= 12 \times 8$
$= 96\ \text{cm}^2$

Area of trapezium $= \frac{1}{2} \times 5(3 + 5)$
$= 2.5 \times 8$
$= 20\ \text{cm}^2$

Shaded area $= 96 - 20$
$= 76\ \text{cm}^2$

Exercise 28.5

Find the area of each of the following shapes:

1.

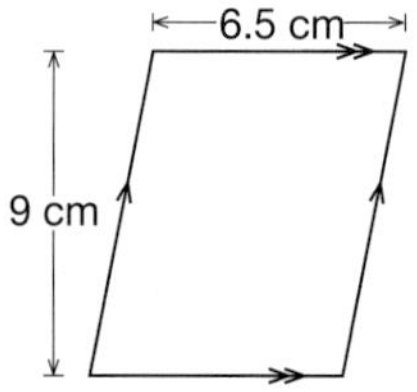

2.

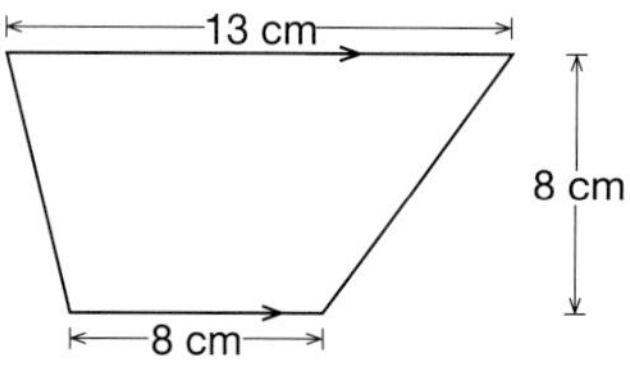

3. 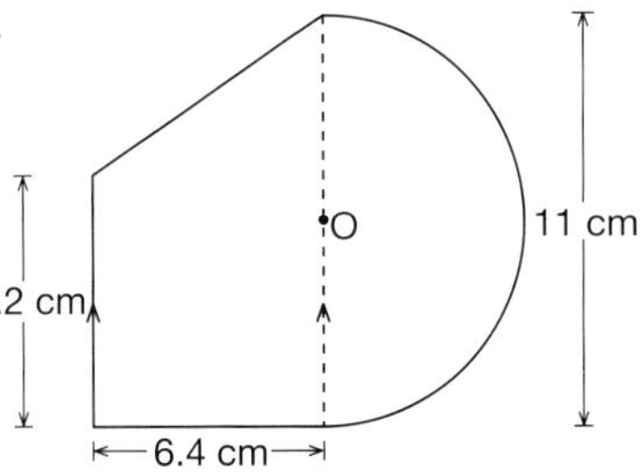

4. 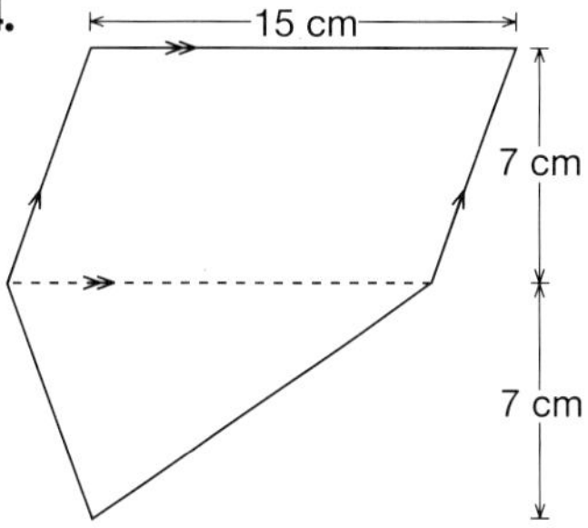

Exercise 28.6

1. Calculate a.

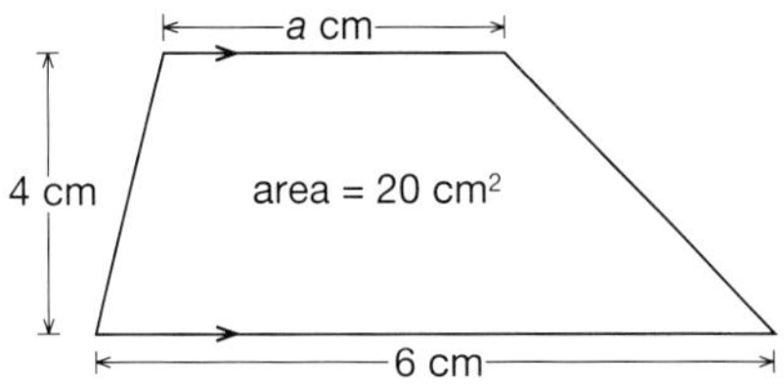

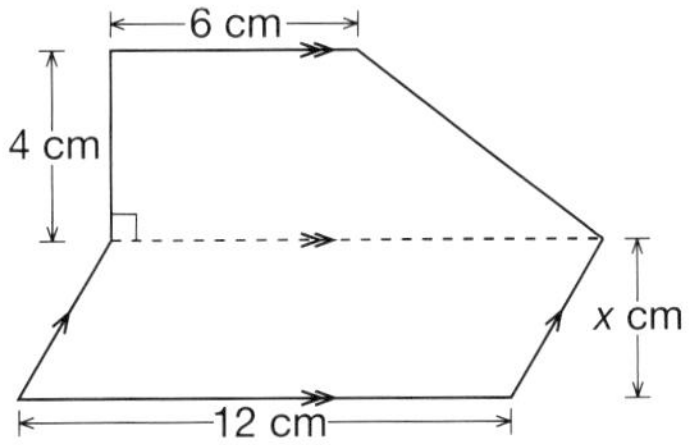

2. If the areas of the trapezium and parallelogram in the diagram (left) are equal, calculate x.

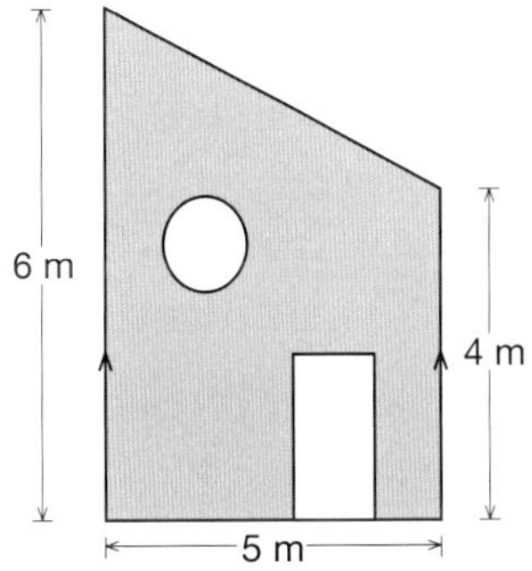

3. The end view of a house is as shown in the diagram (left). If the door has a width and height of 0.75 m and 2 m, respectively and the circular window has a diameter of 0.8 m, calculate the area of brickwork.

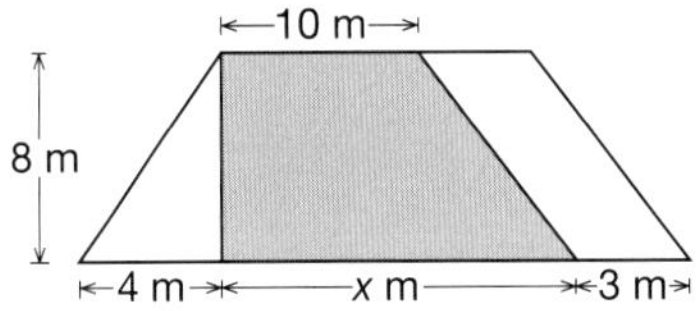

4. A garden in the shape of a trapezium (left) is split into three parts: flower beds in the shape of a triangle and a parallelogram; and a section of grass in the shape of a trapezium. The area of the grass is two and a half times the total area of flower beds. Calculate:
   a) the area of each flower bed,
   b) the area of grass,
   c) the value of x.

The surface area of a cuboid and cylinder

To calculate the surface area of a **cuboid** start by looking at its individual faces. These are either squares or rectangles. The surface area of a cuboid is the sum of the areas of its faces.

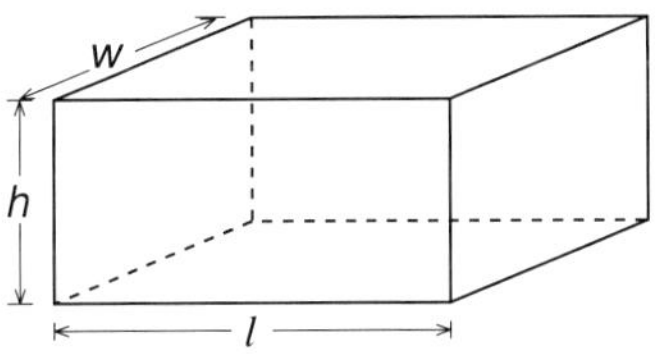

Area of top $= wl$ Area of bottom $= wl$
Area of front $= lh$ Area of back $= lh$
Area of one side $= wh$ Area of other side $= wh$

Total surface area
$= 2wl + 2lh + 2wh$
$= 2(wl + lh + wh)$

For the surface area of a **cylinder** it is best to visualise the net of the solid: it is made up of one rectangular piece and two circular pieces.

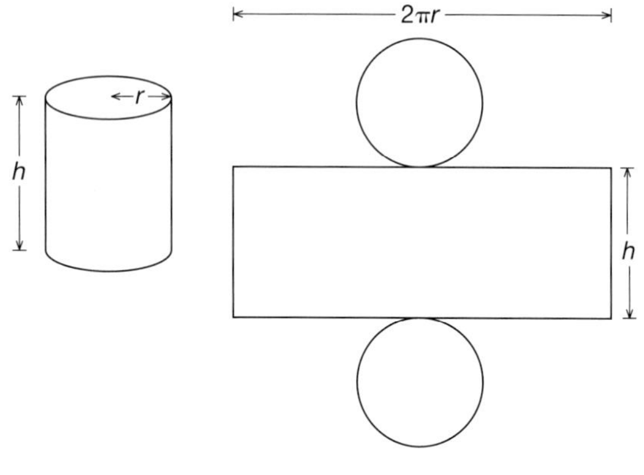

Area of circular pieces $= 2 \times \pi r^2$
Area of rectangular piece $= 2\pi r \times h$
Total surface area $= 2\pi r^2 + 2\pi rh$
$= 2\pi r(r + h)$

Worked examples

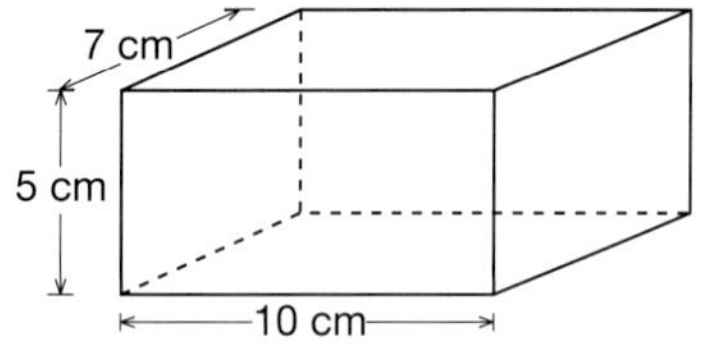

a) Calculate the surface area of the cuboid shown (left).

Total area of top and bottom $= 2 \times 7 \times 10 = 140\ \text{cm}^2$
Total area of front and back $= 2 \times 5 \times 10 = 100\ \text{cm}^2$
Total area of both sides $= 2 \times 5 \times 7 = 70\ \text{cm}^2$

Total surface area $= 140 + 100 + 70 = 310\ \text{cm}^2$

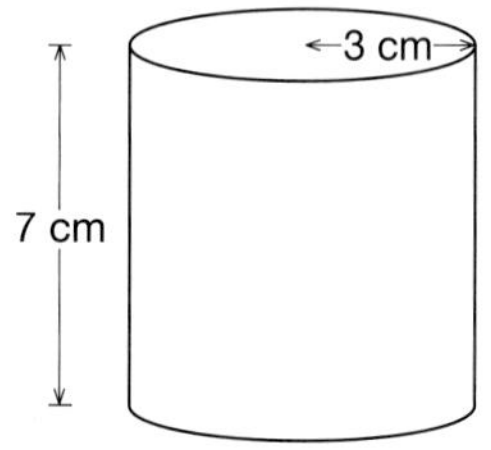

b) If the height of a cylinder is 7 cm and the radius of its circular top is 3 cm, calculate its surface area.

Total surface area $= 2\pi r(r + h)$
$= 2\pi \times 3 \times (3 + 7)$
$= 6\pi \times 10$
$= 60\pi$
$= 188.50\ \text{cm}^2$ (2 d.p.)

The total surface area is $188.50\ \text{cm}^2$.

Exercise 28.7

1. Calculate the surface area of each of the following cuboids if:
a) $l = 12$ cm, $w = 10$ cm, $h = 5$ cm
b) $l = 4$ cm, $w = 6$ cm, $h = 8$ cm
c) $l = 4.2$ cm, $w = 7.1$ cm, $h = 3.9$ cm
d) $l = 5.2$ cm, $w = 2.1$ cm, $h = 0.8$ cm

2. Calculate the height of each of the following cuboids if:
a) $l = 5$ cm, $w = 6$ cm, surface area $= 104\ \text{cm}^2$
b) $l = 2$ cm, $w = 8$ cm, surface area $= 112\ \text{cm}^2$
c) $l = 3.5$ cm, $w = 4$ cm, surface area $= 118\ \text{cm}^2$
d) $l = 4.2$ cm, $w = 10$ cm, surface area $= 226\ \text{cm}^2$

3. Calculate the surface area of each of the following cylinders if:
   a) $r = 2$ cm, $h = 6$ cm b) $r = 4$ cm, $h = 7$ cm
   c) $r = 3.5$ cm, $h = 9.2$ cm d) $r = 0.8$ cm, $h = 4.3$ cm

4. Calculate the height of each of the following cylinders. Give your answers to 1 d.p.
   a) $r = 2.0$ cm, surface area $= 40$ cm^2
   b) $r = 3.5$ cm, surface area $= 88$ cm^2
   c) $r = 5.5$ cm, surface area $= 250$ cm^2
   d) $r = 3.0$ cm, surface area $= 189$ cm^2

Exercise 28.8

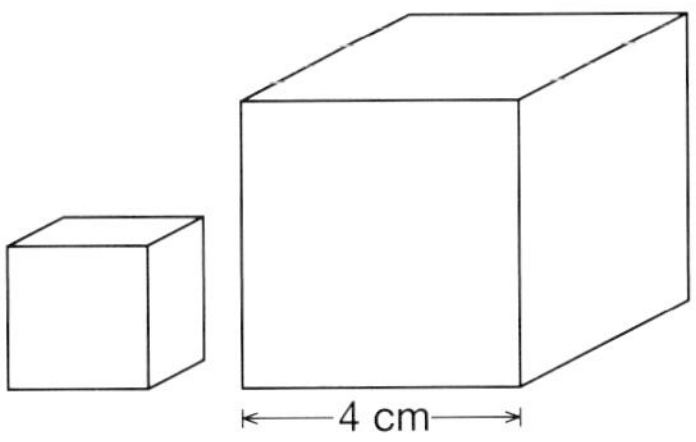

1. Two cubes (left) are placed next to each other. The length of each of the edges of the larger cube is 4 cm. If the ratio of their surface areas is 1 : 4, calculate:
   a) the surface area of the small cube,
   b) the length of an edge of the small cube.

2. A cube and a cylinder have the same surface area. If the cube has an edge length of 6 cm and the cylinder a radius of 2 cm, calculate:
   a) the surface area of the cube,
   b) the height of the cylinder.

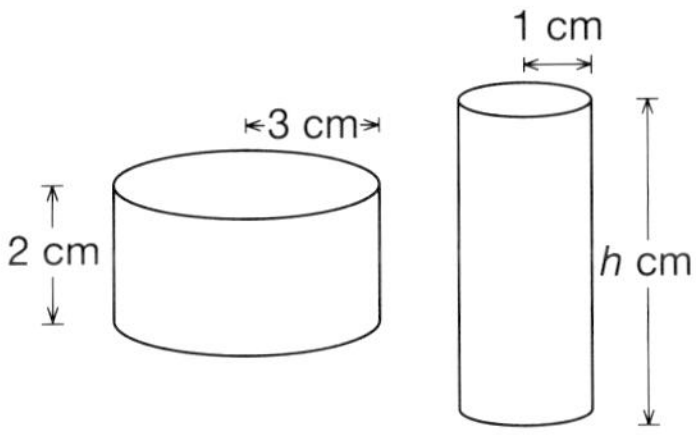

3. Two cylinders (left) have the same surface area. The shorter of the two has a radius of 3 cm and a height of 2 cm, and the taller cylinder has a radius of 1 cm. Calculate:
   a) the surface area of one of the cylinders,
   b) the height of the taller cylinder.

4. Two cuboids have the same surface area. The dimensions of one of them are: length = 3 cm, width = 4 cm and height = 2 cm.

   Calculate the height of the other cuboid if its length is 1 cm and width is 4 cm.

The volume of a prism

A **prism** is any three-dimensional object which has a constant cross-sectional area.

Below are a few examples of some of the more common types of prism:

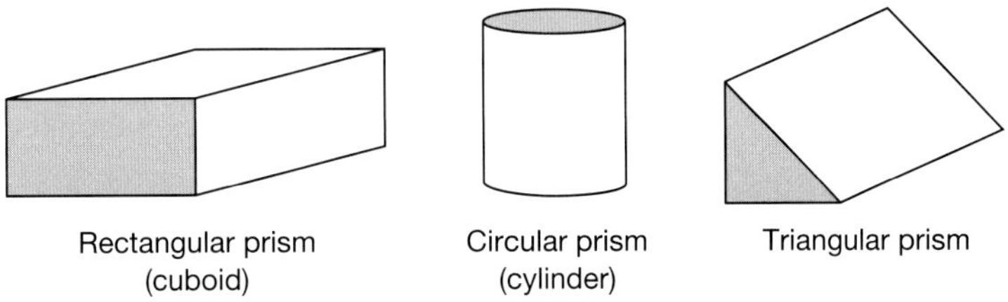

When each of the shapes is cut parallel to the shaded face, the cross-section is constant and the shape is therefore classified as a prism.

Volume of a prism = area of cross-section × length

Worked examples

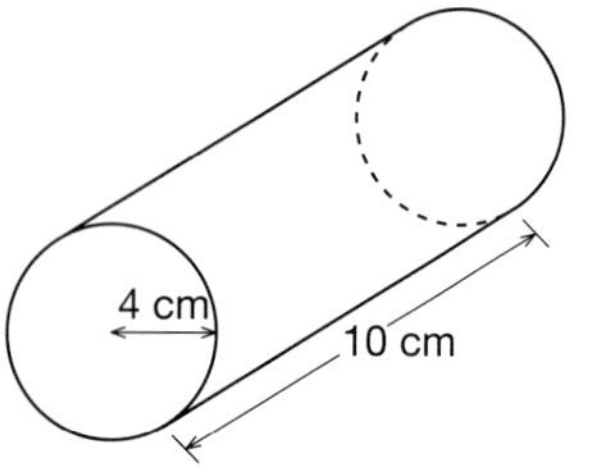

a) Calculate the volume of the cylinder shown in the diagram (left):

Volume = cross-sectional area × length
$= \pi \times 4^2 \times 10$
Volume = 502.7 cm^3 (1 d.p.)

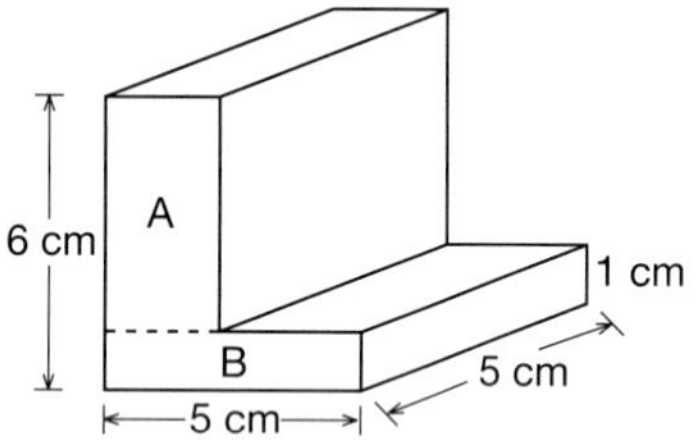

b) Calculate the volume of the 'L' shaped prism shown in the diagram (left):
The cross-sectional area can be split into two rectangles:

Area of rectangle A $= 5 \times 2$
$= 10\ cm^2$
Area of rectangle B $= 5 \times 1$
$= 5\ cm^2$
Total cross-sectional area $= 10\ cm^2 + 5\ cm^2 = 15\ cm^2$
Volume of prism $= 15 \times 5$
$= 75\ cm^3$

Exercise 28.9

1. Calculate the volume of each of the following cuboids, where w, l and h represent the width, length and height, respectively.
a) $w = 2$ cm, $l = 3$ cm, $h = 4$ cm
b) $w = 6$ cm, $l = 1$ cm, $h = 3$ cm
c) $w = 6$ cm, $l = 23$ mm, $h = 2$ cm
d) $w = 42$ mm, $l = 3$ cm, $h = 0.007$ m

2. Calculate the volume of each of the following cylinders where r represents the radius of the circular face and h the height of the cylinder.
a) $r = 4$ cm, $h = 9$ cm
b) $r = 3.5$ cm, $h = 7.2$ cm
c) $r = 25$ mm, $h = 10$ cm
d) $r = 0.3$ cm, $h = 17$ mm

3. Calculate the volume of each of the following triangular prisms where b represents the base length of the triangular face, h its perpendicular height and l the length of the prism.
a) $b = 6$ cm, $h = 3$ cm, $l = 12$ cm
b) $b = 4$ cm, $h = 7$ cm, $l = 10$ cm
c) $b = 5$ cm, $h = 24$ mm, $l = 7$ cm
d) $b = 62$ mm, $h = 2$ cm, $l = 0.01$ m

4. Calculate the volume of each of the following prisms. All dimensions are given in centimetres.

a)

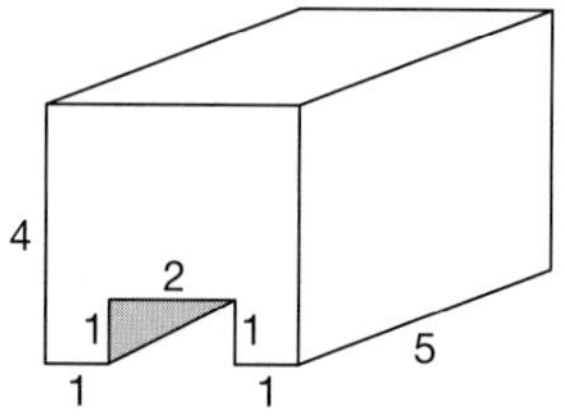

b)

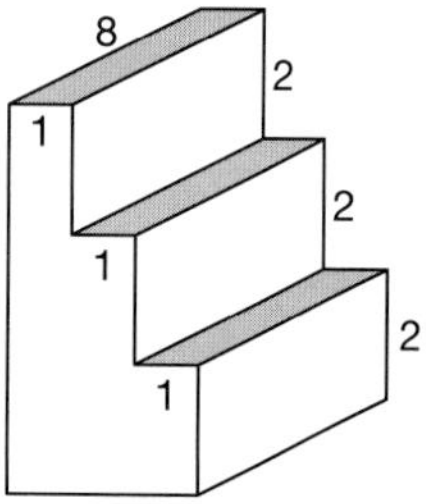

c)

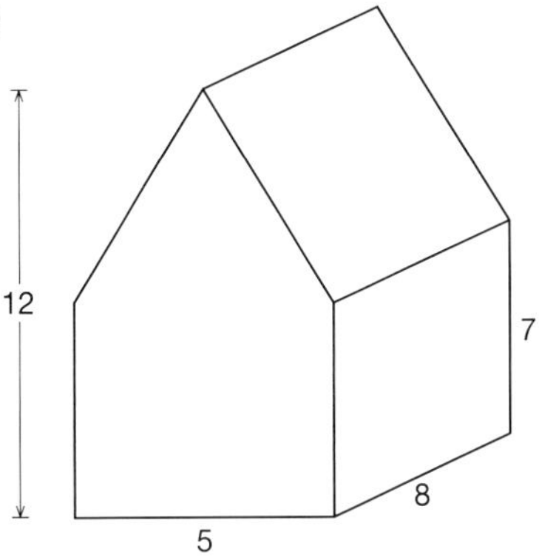

d)

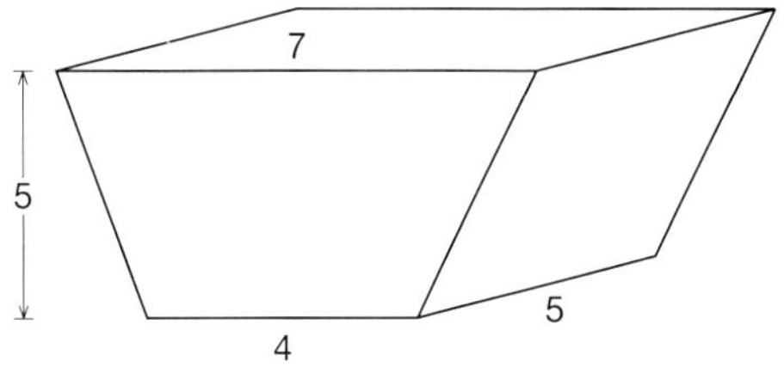

Exercise 28.10

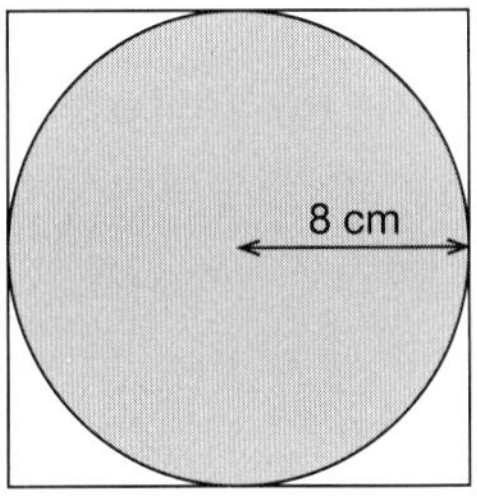

1. The diagram (left) shows a plan view of a cylinder inside a box the shape of a cube. If the radius of the cylinder is 8 cm, calculate:
   a) the height of the cube,
   b) the volume of the cube,
   c) the volume of the cylinder,
   d) the percentage volume of the cube not occupied by the cylinder.

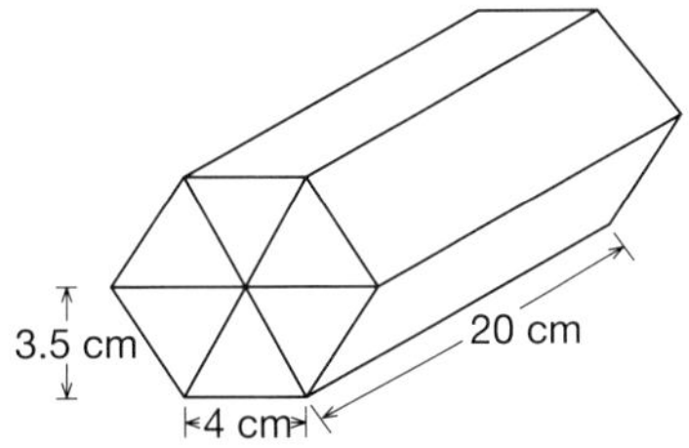

2. A chocolate bar is made in the shape of a triangular prism. The triangular face of the prism is equilateral and has an edge length of 4 cm and a perpendicular height of 3.5 cm. The manufacturer also sells these in special packs of six bars arranged as a hexagonal prism.
   If the prisms are 20 cm long, calculate:
   a) the cross-sectional area of the pack,
   b) the volume of the pack.

3. A cuboid and a cylinder have the same volume. The radius and height of the cylinder are 2.5 cm and 8 cm, respectively. If the length and width of the cuboid are each 5 cm, calculate its height to 1 d.p.

4. A section of steel pipe is shown in the diagram below. The inner radius is 35 cm and the outer radius 36 cm. Calculate the volume of steel used in making the pipe if it has a length of 130 m.

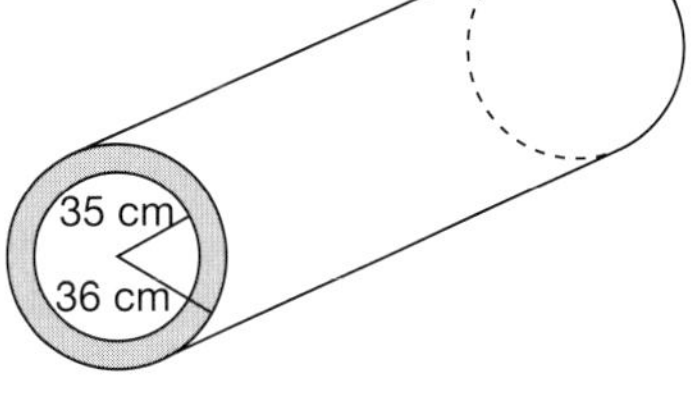

Student Assessment 1

1. A swimming pool is 50 m long by 20 m wide. Find:
   a) the surface area,
   b) the perimeter of the pool.
   Square tiles of side 25 cm are to be placed around the edge of the pool.
   c) How many tiles will it take to fit around it?

2. A floor measures 8 m by 6 m and is to be covered in square tiles of side 50 cm. How many tiles are needed?

3. A carpet is 2 m 40 cm by 3 m 80 cm. Calculate its area and perimeter.

4. A set square has a base of 40 cm and a height of 25 cm. Calculate its area.

5. The plan of a rocket shows a rectangle 70 cm by 50 cm with a triangular nose cone of height 8 cm.
   a) Calculate the area of the plan of the rocket.
   b) If the scale of the plan is 1 : 20, calculate the area of a full-size diagram of the rocket.

Student Assessment 2

1. Calculate the circumference and area of each of the following circles. Give your answers to 1 d.p.

a) 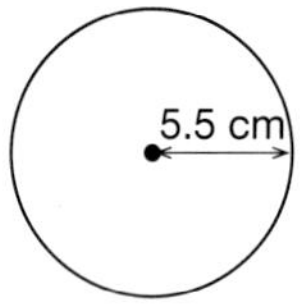

b) 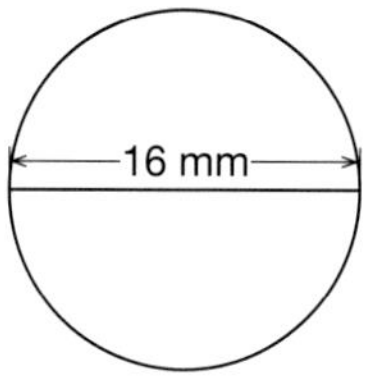

2. A semi-circular shape is cut out of the side of a rectangle as shown. Calculate the shaded area to 1 d.p.

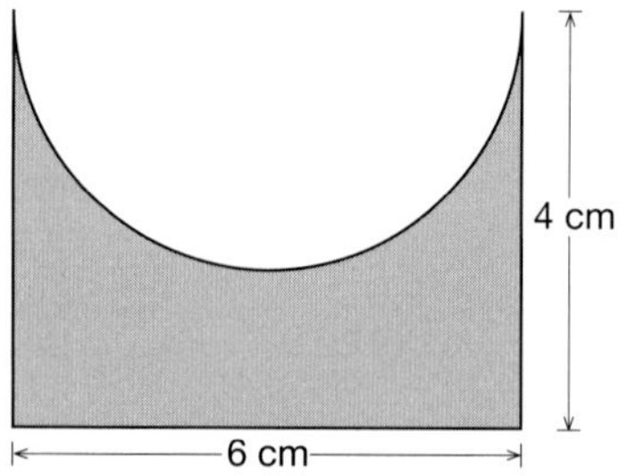

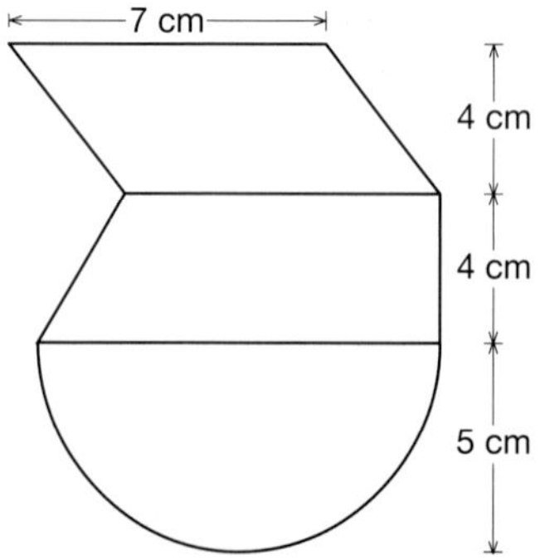

3. For the shape shown in the diagram (left), calculate the area of:
   a) the semi-circle,
   b) the trapezium,
   c) the whole shape.

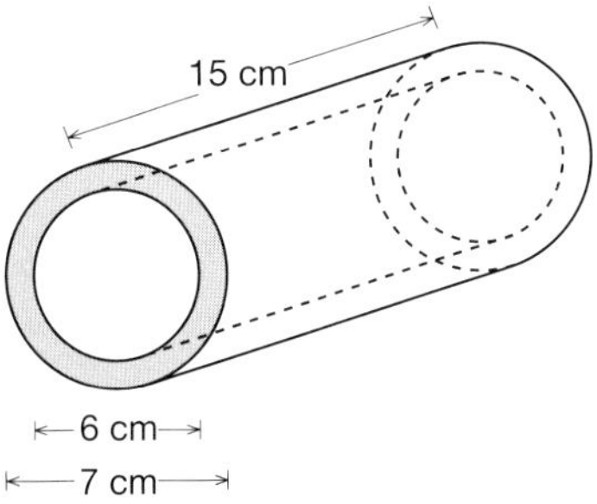

4. A cylindrical tube has an inner diameter of 6 cm, an outer diameter of 7 cm and a length of 15 cm. Calculate the following to 1 d.p.:
   a) the surface area of the shaded end,
   b) the inside surface area of the tube,
   c) the total surface area of the tube.

5. Calculate the volume of each of the following cylinders:

a) 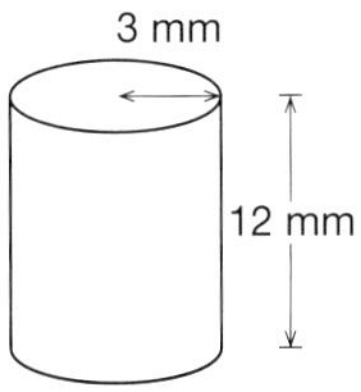

b) 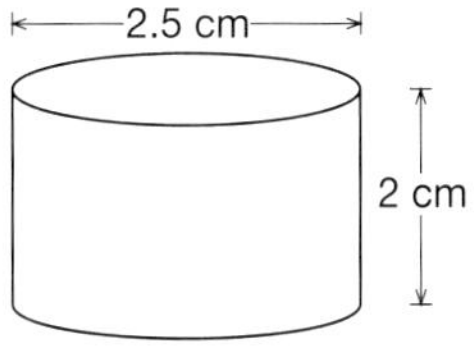

Student Assessment 3

1. A rowing lake, rectangular in shape, is 2.5 km long by 500 m wide. Calculate the surface area of the water in km^2.

2. A rectangular floor 12 m long by 8 m wide is to be covered in ceramic tiles 40 cm long by 20 cm wide.
   a) Calculate the number of tiles required to cover the floor.
   b) The tiles are bought in boxes of 24 at a cost of £70 per box. What is the cost of the tiles needed to cover the floor?

3. A flower bed is in the shape of a right-angled triangle of sides 3 m, 4 m and 5 m. Sketch the flower bed, and calculate its area and perimeter.

4. The plan of a church shows a rectangle 50 cm high and 10 cm wide with a triangular tower 20 cm high and 10 cm wide at the base on top of it. Find the area of the plan of the church.

5. The squares of a chess board are each of side 7.5 cm. What is the area of the chess board?

Student Assessment 4

1. Calculate the circumference and area of each of the following circles. Give your answers to 1 d.p.

a) 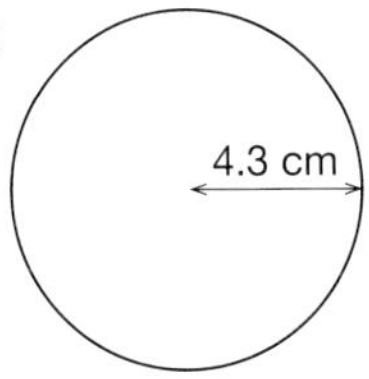

b) 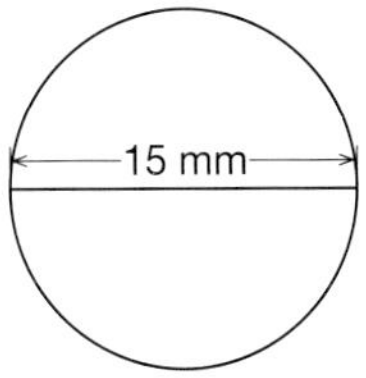

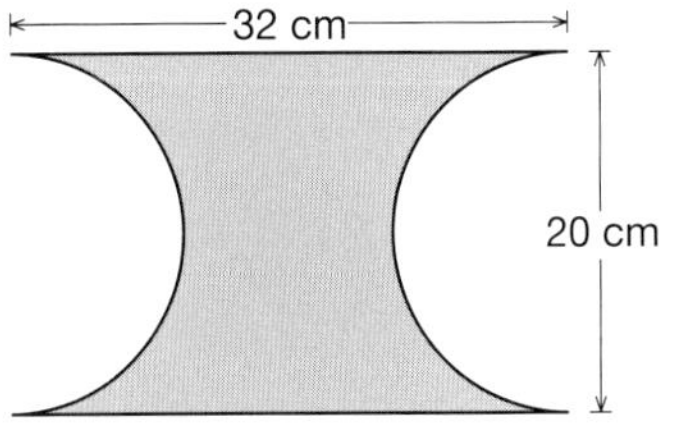

2. A rectangle of length 32 cm and width 20 cm has a semi-circle cut out of two of its sides as shown (left). Calculate the shaded area to 1 d.p.

3. In the diagram below, calculate the area of:
   a) the semi-circle,
   b) the parallelogram,
   c) the whole shape.

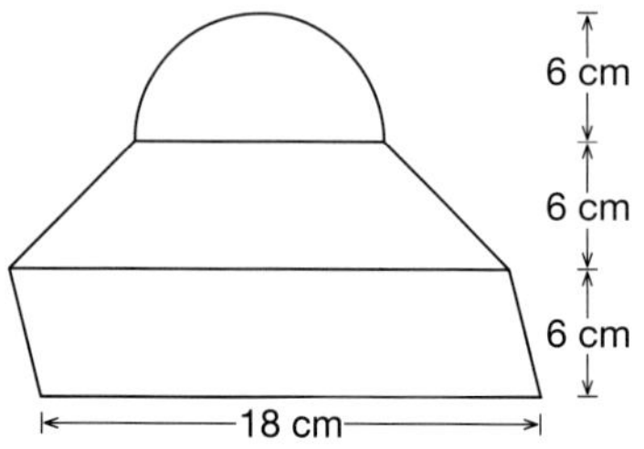

4. A prism in the shape of a hollowed-out cuboid has dimensions as shown below. If the end is square, calculate the volume of the prism.

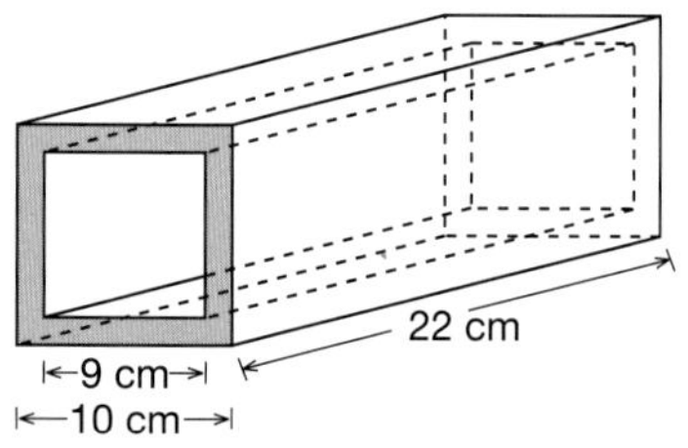

5. Calculate the surface area of each of the following cylinders:

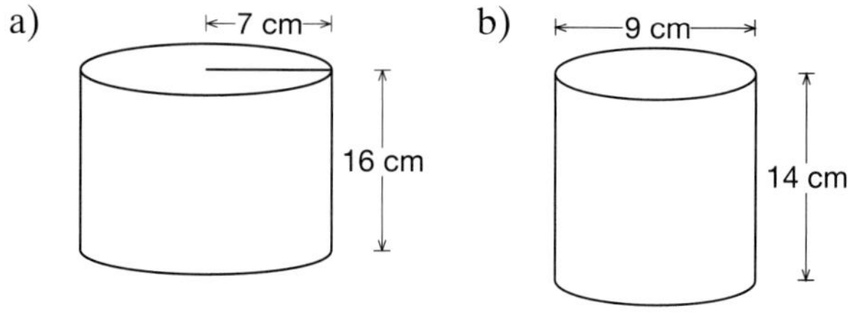

29 TRIGONOMETRY

Interpret and use three figure bearings; apply Pythagoras' theorem and the sine, cosine and tangent ratios for acute angles to the calculation of a side or an angle of a right-angled triangle.

NB: All diagrams are not drawn to scale.

Bearings

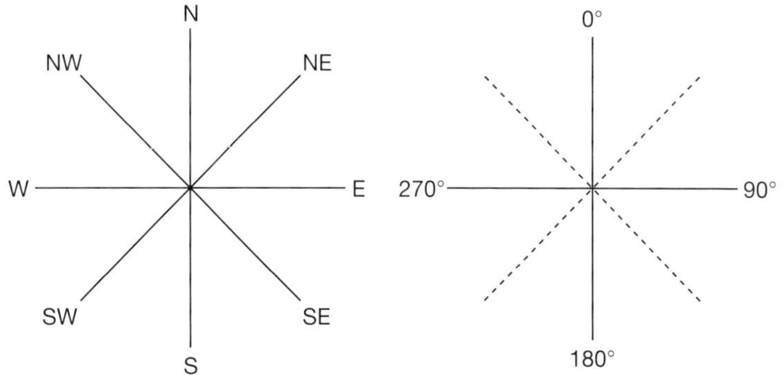

In the days when sailing ships travelled the oceans of the world, compass bearings like the ones in the diagram above were used.

As the need for more accurate direction arose, extra points were added to N, S, E, W, NE, SE, SW and NW. Midway between North and North East was North North East, and midway between North East and East was East North East, and so on. This gave thirty two points of the compass. This was later extended even further, to sixty four points.

As the speed of travel increased, a new system was required. The new system was the **three figure bearing** system. North was given the bearing zero. 360° in a clockwise direction was one full rotation.

Exercise 29.1

1. Copy the three figure bearing diagram at the top of the page, on the right-hand side. Mark on your diagram the compass points North East, South East, South West and North West.

2. Draw diagrams to show the following compass bearings and journeys. Use a scale of 1 cm : 1 km. North can be taken to be a line vertically up the page.
   a) Start at point A. Travel a distance of 7 km on a bearing of 135° to point B. From B, travel 12 km on a bearing of 250° to point C. Measure the distance and bearing of A from C.

b) Start at point P. Travel a distance of 6.5 km on a bearing of 225° to point Q. From Q, travel 7.8 km on a bearing of 105° to point R. From R, travel 8.5 km on a bearing of 090° to point S. What is the distance and bearing of P from S?

c) Start from point M. Travel a distance of 11.2 km on a bearing of 270° to point N. From point N, travel 5.8 km on a bearing of 170° to point O. What is the bearing and distance of P from O?

Back Bearings

Worked examples

a) The bearing of B from A is 135° and the distance from A to B is 8 cm, as shown below. The bearing of A from B is called the **back bearing**.

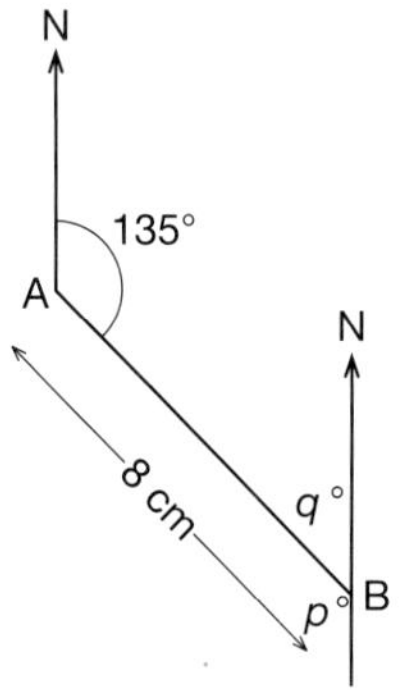

Since the two North lines are parallel:
$p = 135°$ (alternate angles), so the back bearing is $(180 + 135)°$.
That is, 315°.
(There are a number of methods of solving this type of problem.)

b) The bearing of B from A is 245°. What is the bearing of A from B?

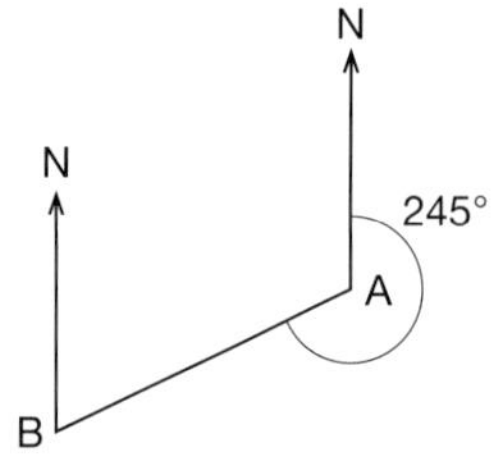

Since the two North lines are parallel:
$b = 65°$ (alternate angles), so the bearing is $(245 – 180)°$.
That is, 065°.

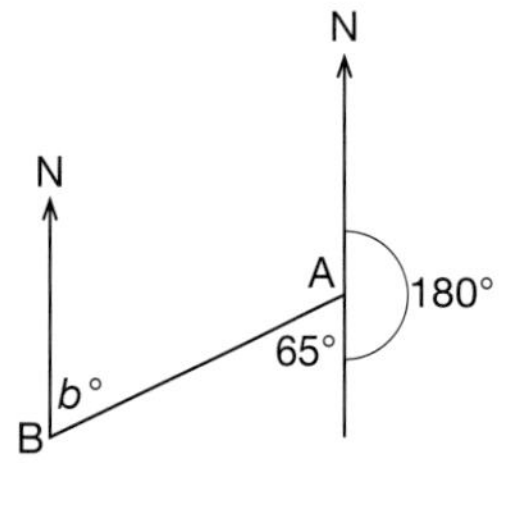

3. Given the following bearings of point B from point A, draw diagrams and use them to calculate the bearings of A from B.

a) bearing 130° b) bearing 145°
c) bearing 220° d) bearing 200°
e) bearing 152° f) bearing 234°
f) bearing 163° h) bearing 214°

4. Given the following bearings of point D from point C, draw diagrams and use them to calculate the bearings of C from D.

a) bearing 300° b) bearing 320°
c) bearing 290° d) bearing 282°

Trigonometry

There are three basic trigonometric ratios: sine, cosine and tangent.

Each of these relates an angle of a right-angled triangle to a ratio of the lengths of two of its sides.

The sides of the triangle have names, two of which are dependent on their position in relation to a specific angle.

The longest side (always opposite the right angle) is called the **hypotenuse**. The side opposite the angle is called the **opposite** side and the side next to the angle is called the **adjacent** side.

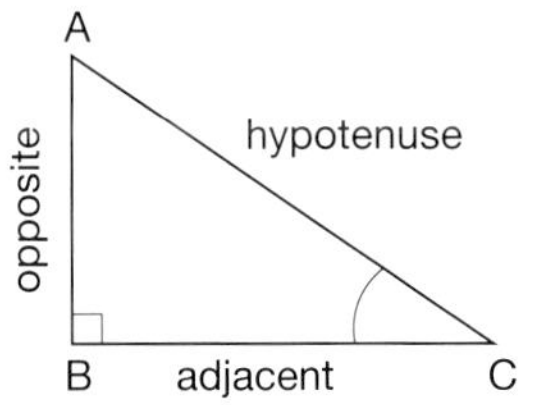

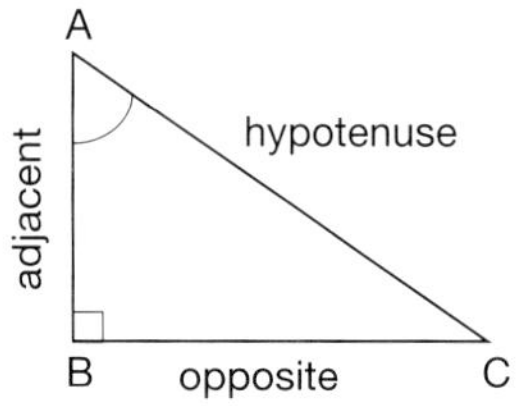

Note that, when the chosen angle is at A, the sides labelled opposite and adjacent change (above right).

Tangent

$$\tan C = \frac{\text{length of opposite side}}{\text{length of adjacent side}}$$

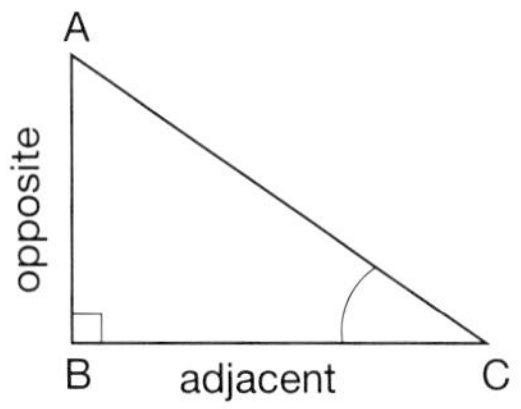

Worked examples **a)** Calculate the size of angle BAC in each of the following triangles.

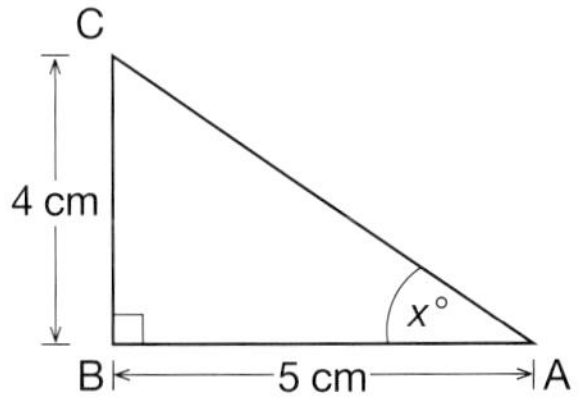

i) $\tan x° = \frac{\text{opposite}}{\text{adjacent}} = \frac{4}{5}$

$x = \tan^{-1}\left(\frac{4}{5}\right)$

$x = 38.7$ (3 s.f.)

$\angle BAC = 38.7°$ (3 s.f.)

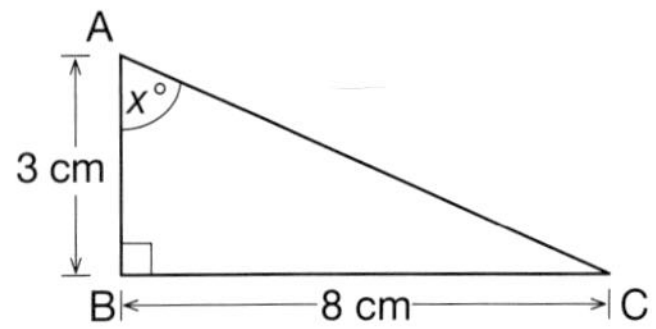

ii) $\tan x° = \frac{8}{3}$

$x = \tan^{-1}\left(\frac{8}{3}\right)$

$x = 69.4$ (3 s.f.)
$\angle BAC = 69.4°$ (3 s.f.)

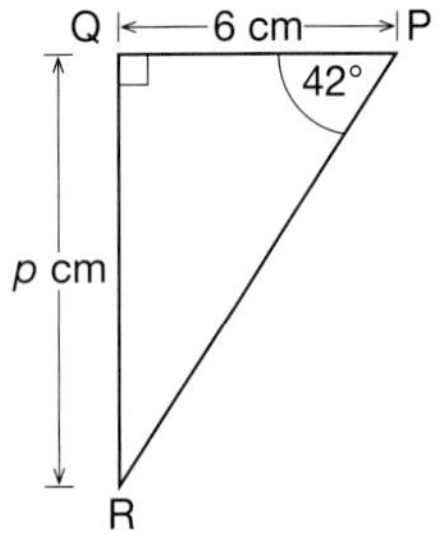

b) Calculate the length of the opposite side QR.

$\tan 42° = \frac{p}{6}$

$6 \times \tan 42° = p$
$p = 5.40$ (3 s.f.)
QR = 5.40 cm (3 s.f.)

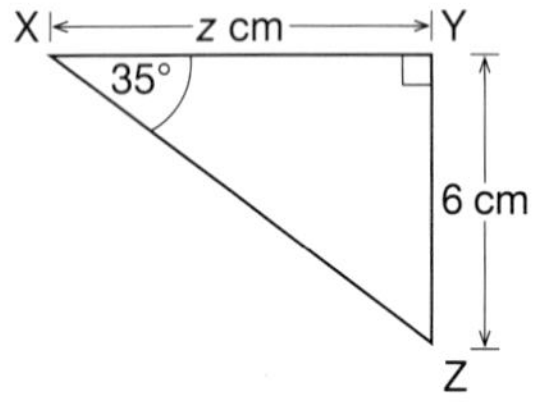

c) Calculate the length of the adjacent side XY.

$\tan 35° = \frac{6}{z}$

$z \times \tan 35° = 6$

$z = \frac{6}{\tan 35°}$

$z = 8.57$ (3 s.f.)
XY = 8.57 cm (3 s.f.)

Exercise 29.2

Calculate the length of the side marked x cm in each of the diagrams in Q.1 and 2. Give your answers to 1 d.p.

1.

a)

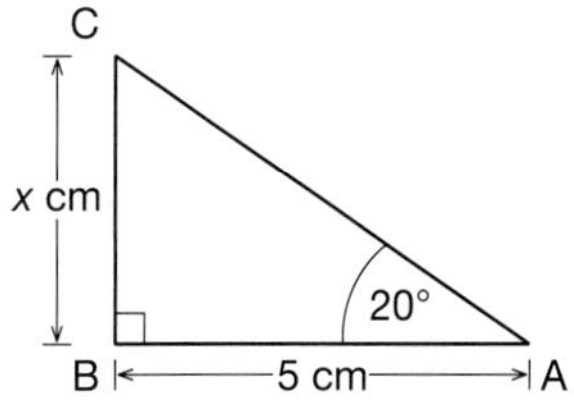

b)

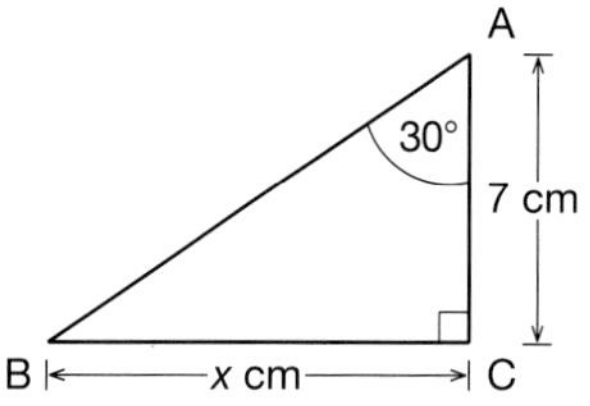

c)

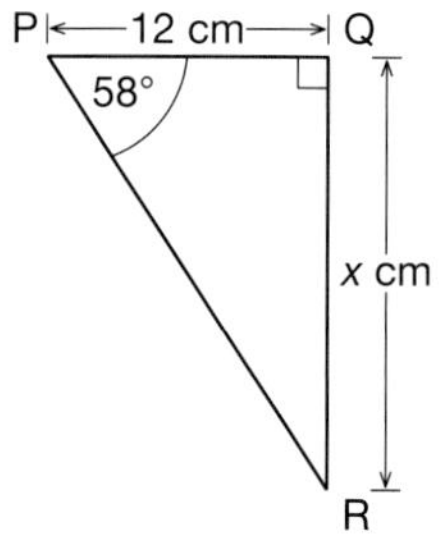

d)

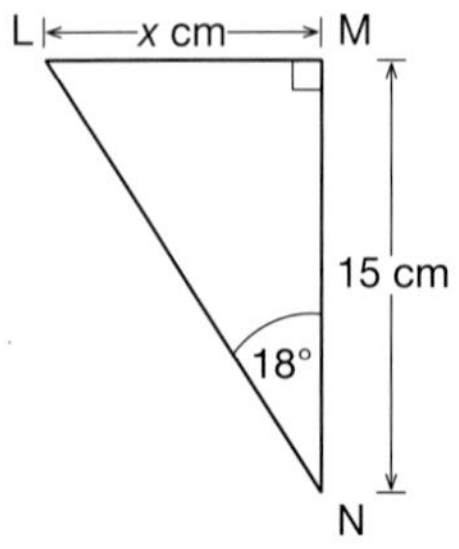

e)

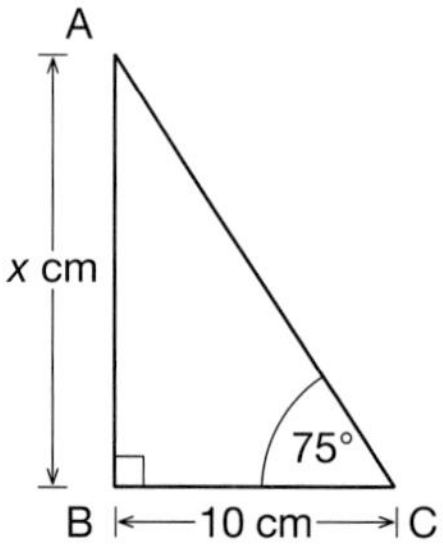

f)

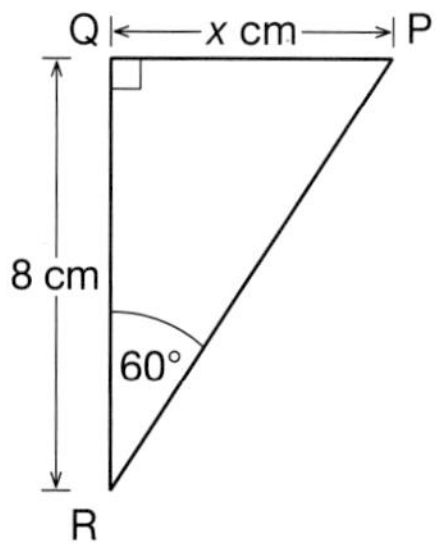

2.

a)

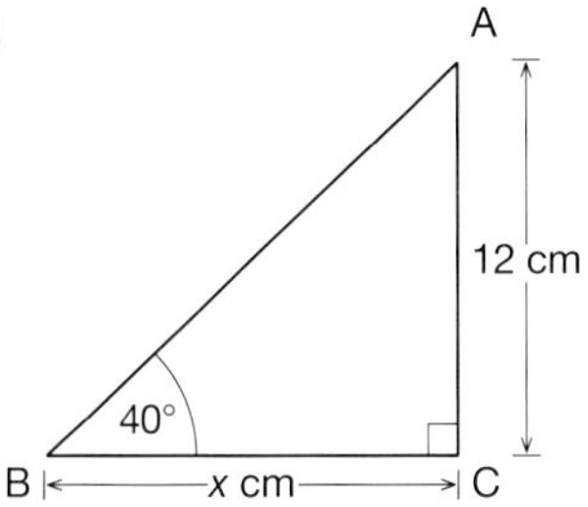

b)

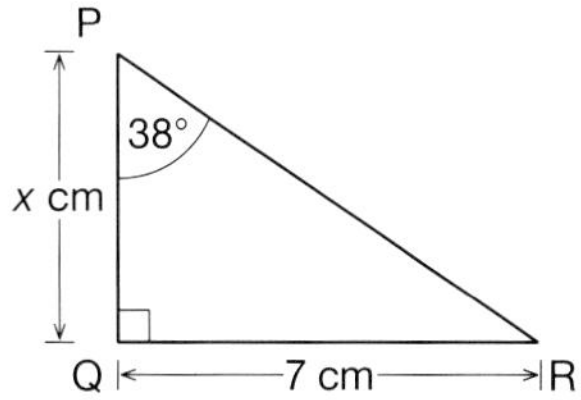

c)

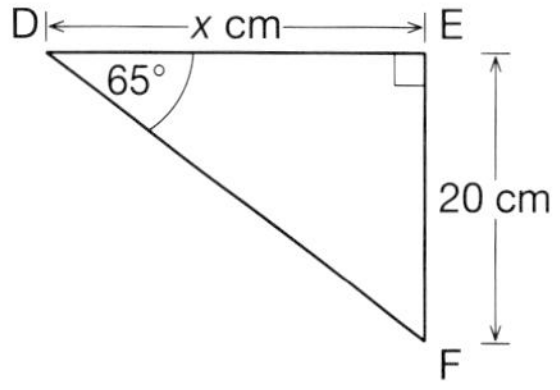

d)

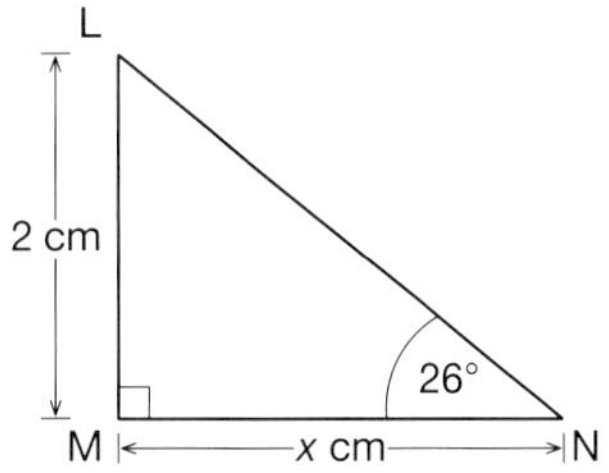

e)

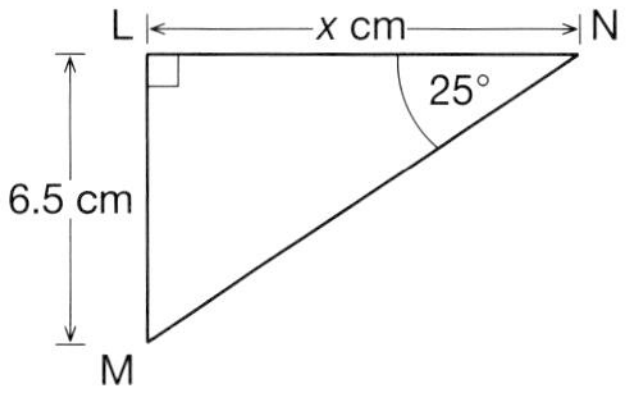

f)

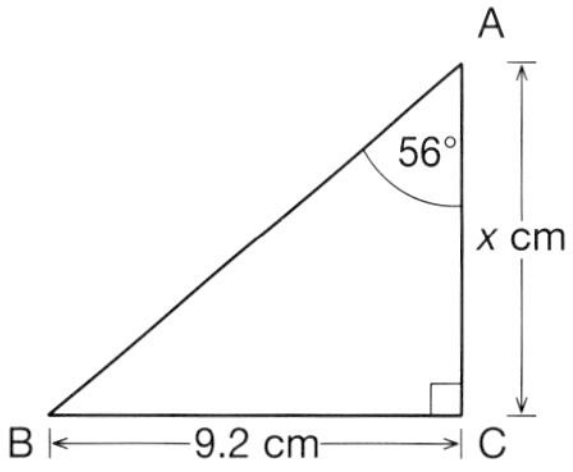

3. Calculate the size of the marked angle $x°$ in each of the following diagrams. Give your answer to 1 d.p.

a)

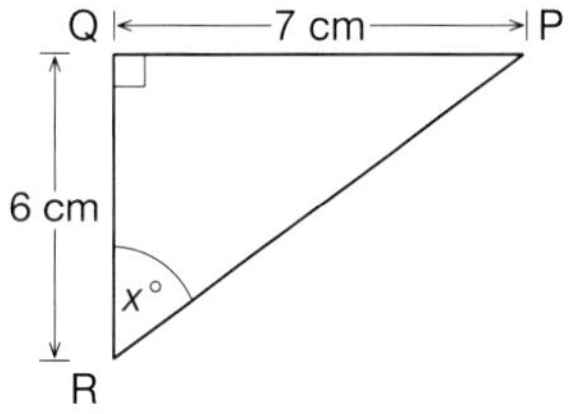

b)

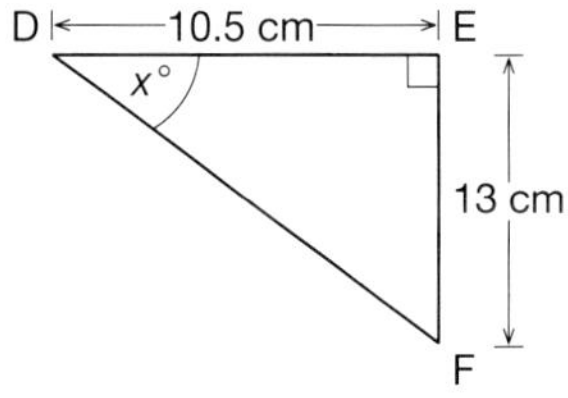

c)

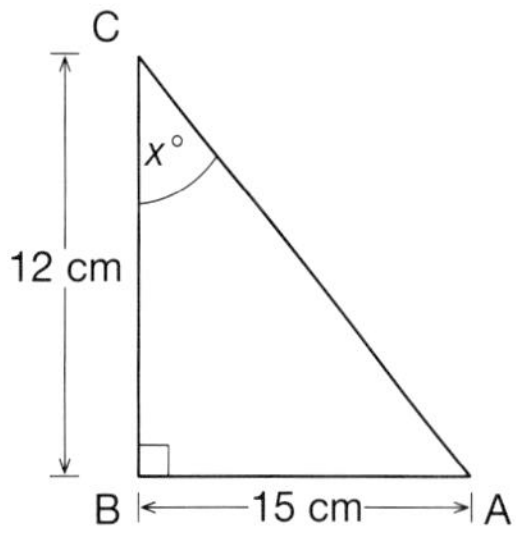

d)

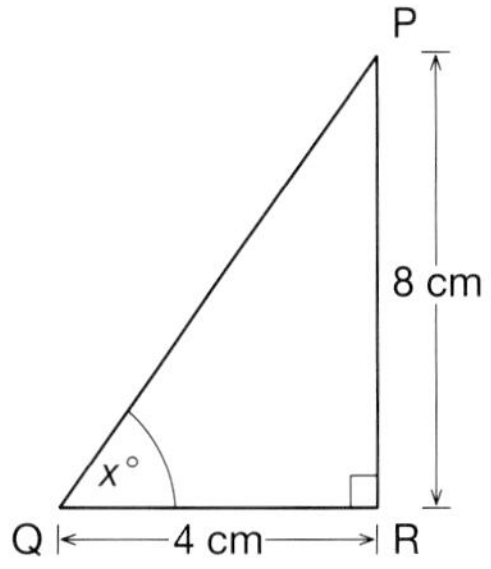

e)

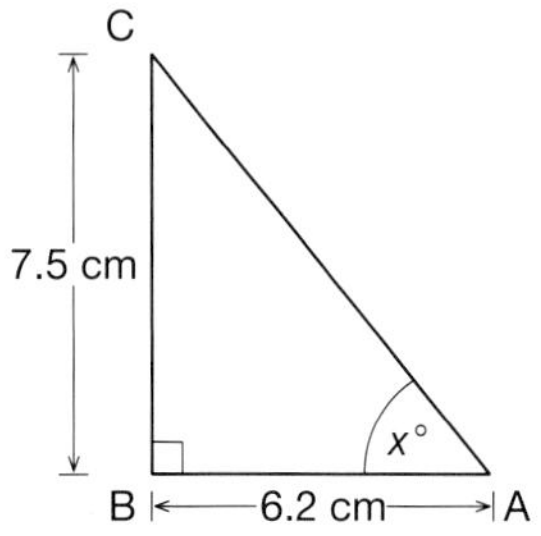

f)

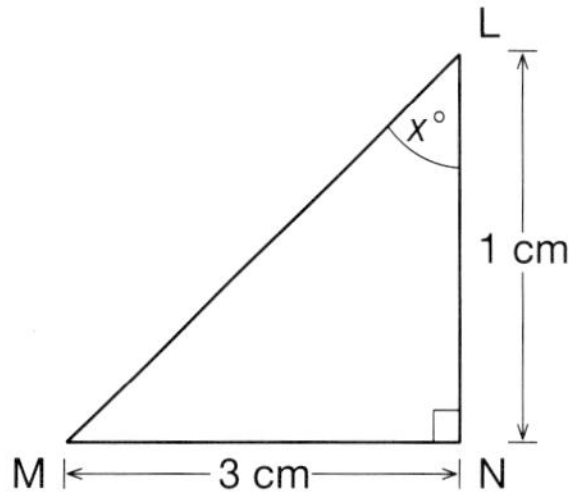

Sine

$$\sin N = \frac{\text{length of opposite side}}{\text{length of hypotenuse}}$$

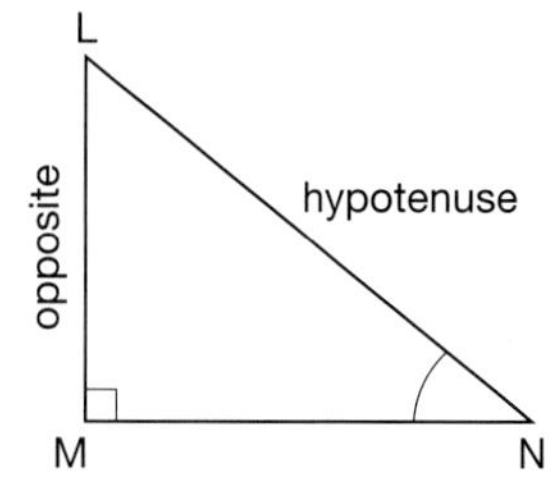

Worked examples

a) Calculate the size of angle BAC.

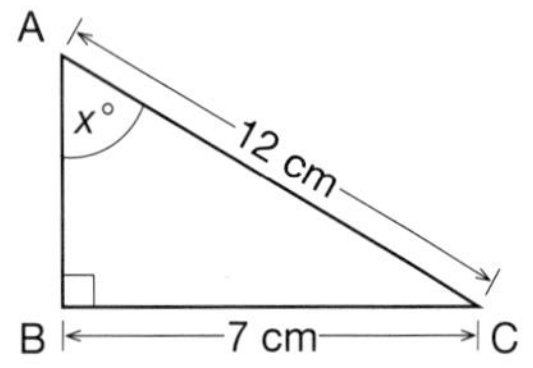

$$\sin x = \frac{\text{opposite}}{\text{hypotenuse}} = \frac{7}{12}$$

$$x = \sin^{-1}\left(\frac{7}{12}\right)$$

$x = 35.7$ (1 d.p.)
$\angle BAC = 35.7°$ (1 d.p.)

b) Calculate the length of the hypotenuse PR.

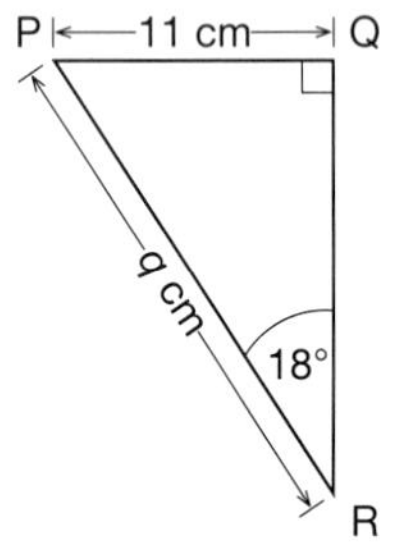

$$\sin 18° = \frac{11}{q}$$

$$q \times \sin 18° = 11$$

$$q = \frac{11}{\sin 18°}$$

$q = 35.6$ (1 d.p.)

PR = 35.6 cm (1 d.p.)

Exercise 29.3

1. Calculate the length of the marked side in each of the following diagrams. Give your answers to 1 d.p.

a) 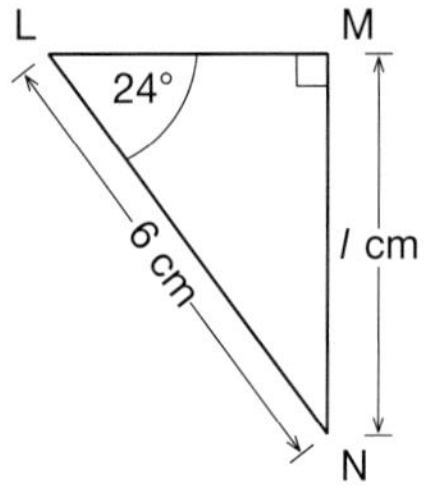

b) 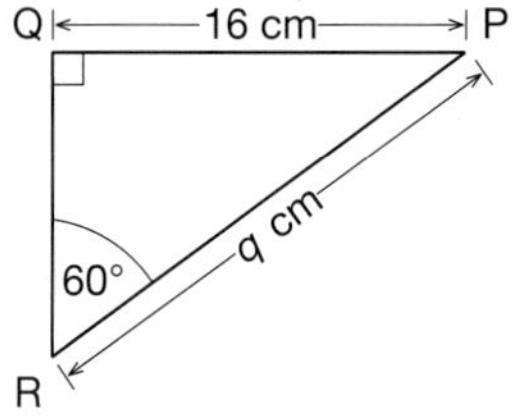

c) 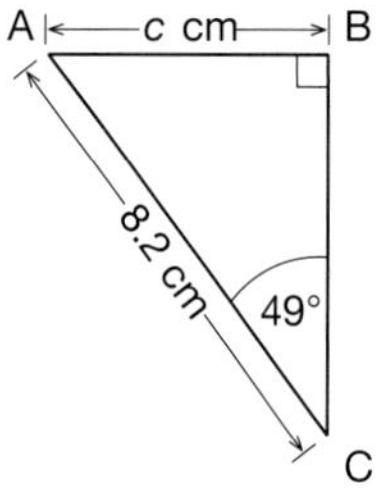

d) 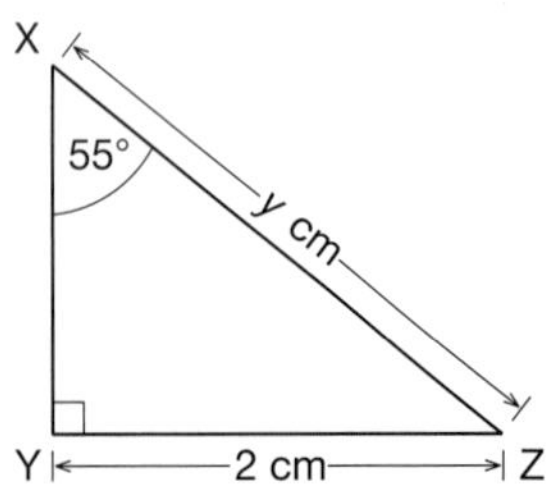

e) 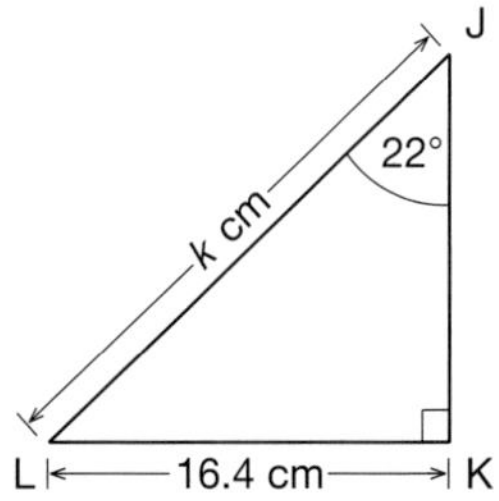

f) 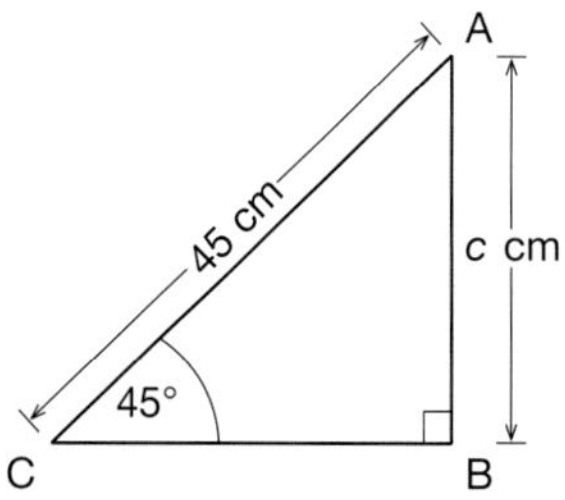

2. Calculate the size of the angle marked x in each of the following diagrams. Give your answers to 1 d.p.

a)

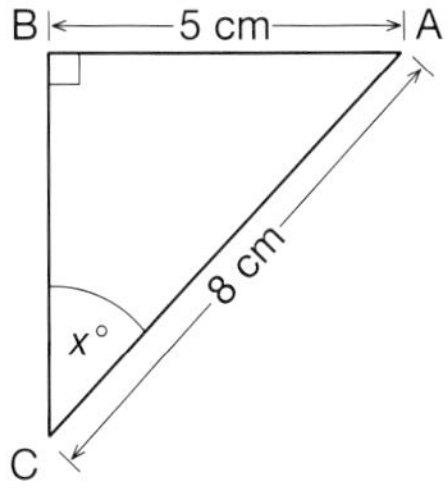

b)

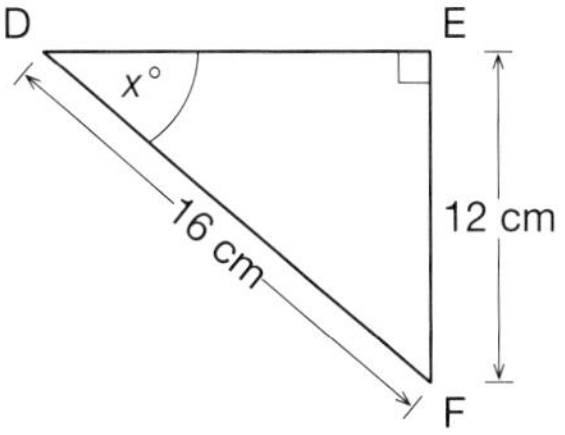

c)

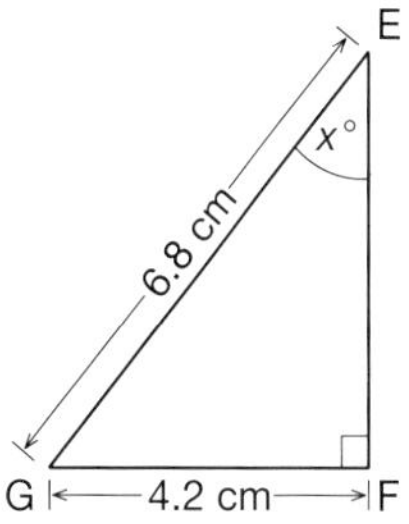

d)

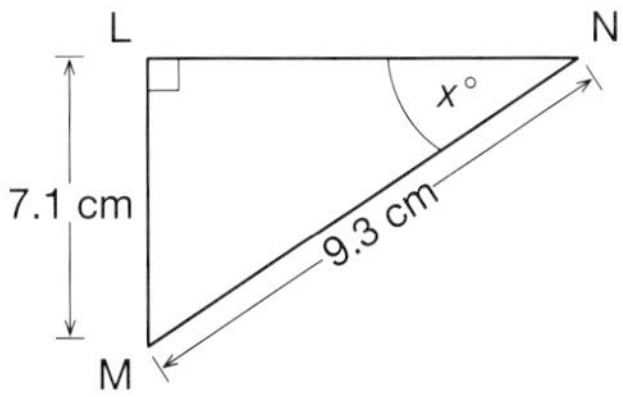

e)

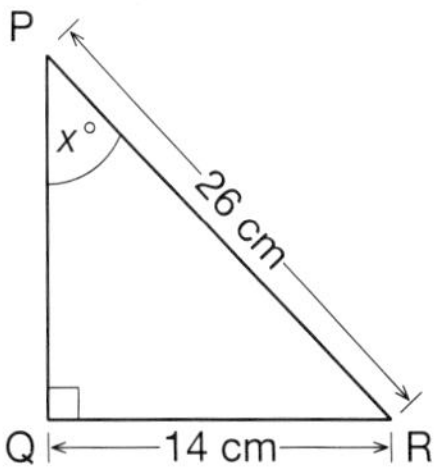

f)

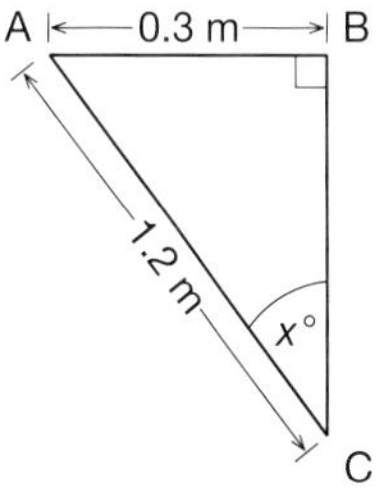

Cosine

$$\cos Z = \frac{\text{length of adjacent side}}{\text{length of hypotenuse}}$$

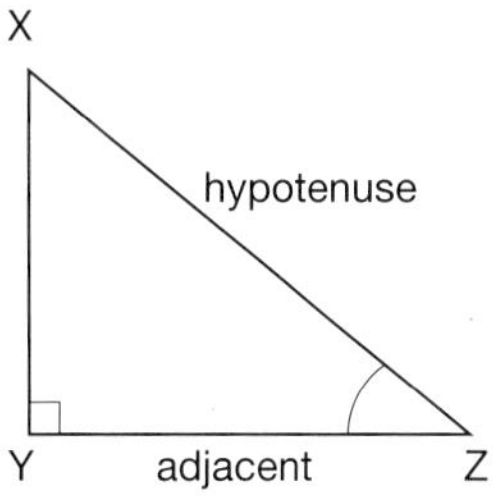

Worked examples **a)** Calculate the length XY.

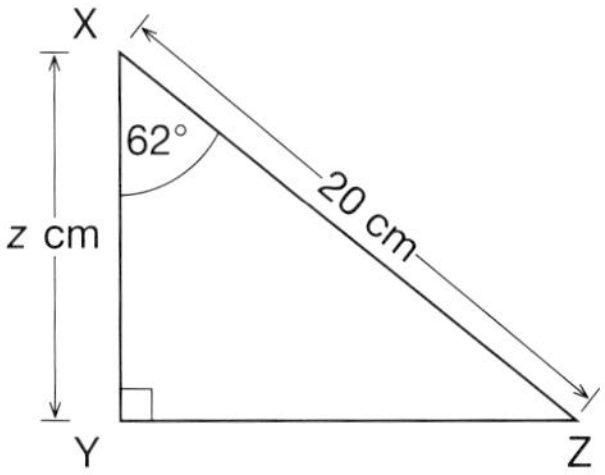

$$\cos 62° = \frac{\text{adjacent}}{\text{hypotenuse}} = \frac{z}{20}$$

$z = 20 \times \cos 62°$
$z = 9.4$ (1 d.p.)
XY $= 9.4$ cm (1 d.p.)

b) Calculate the size of angle ABC.

$$\cos x = \frac{5.3}{12}$$

$$x = \cos^{-1}\left(\frac{5.3}{12}\right)$$

$x = 63.8$ (1 d.p.)
$\angle$ABC $= 63.8°$ (1 d.p.)

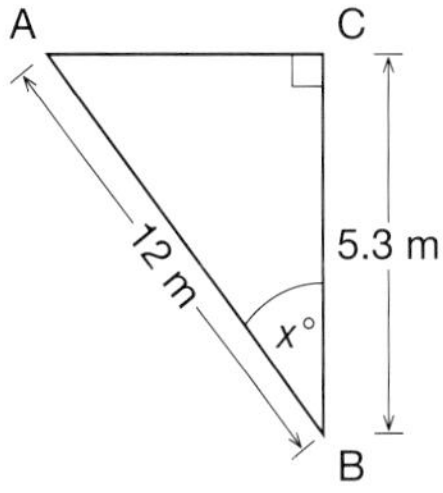

Exercise 29.4 Calculate the marked side or angle in each of the following diagrams. Give your answers to 1 d.p.

a)

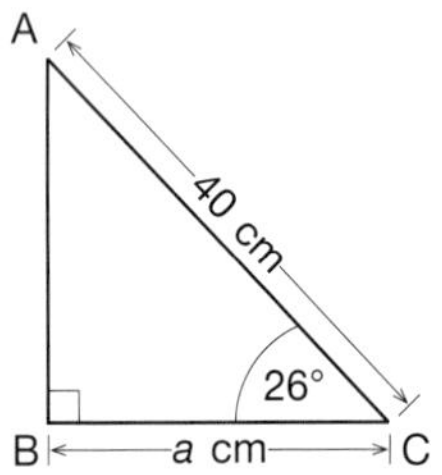

b)

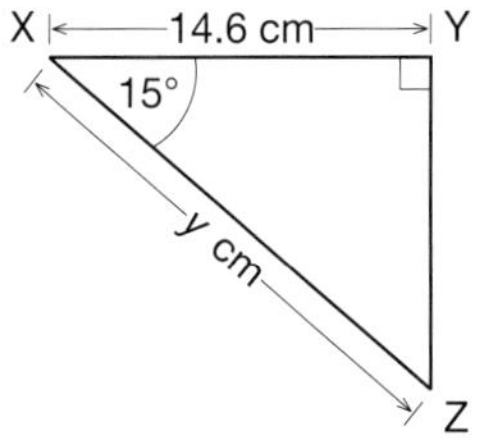

c)

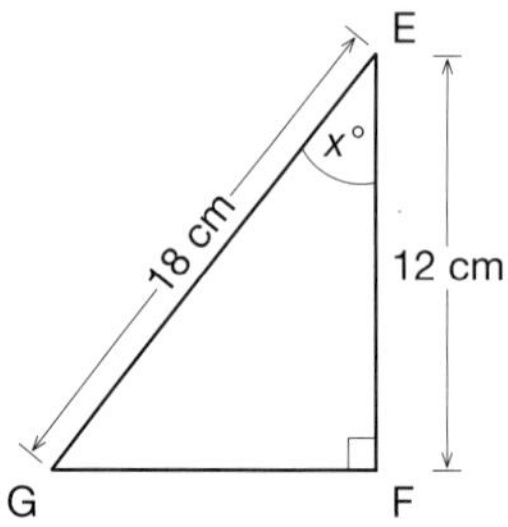

d)

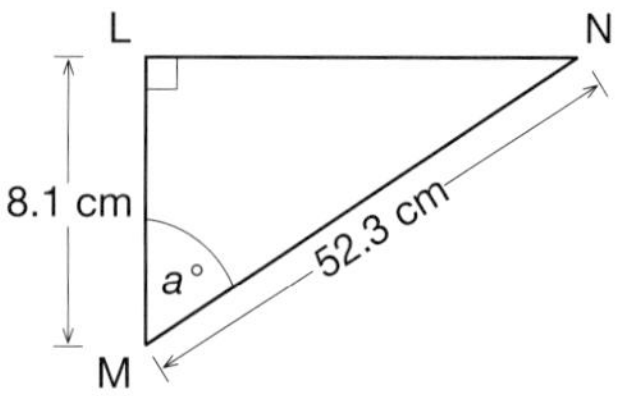

e)

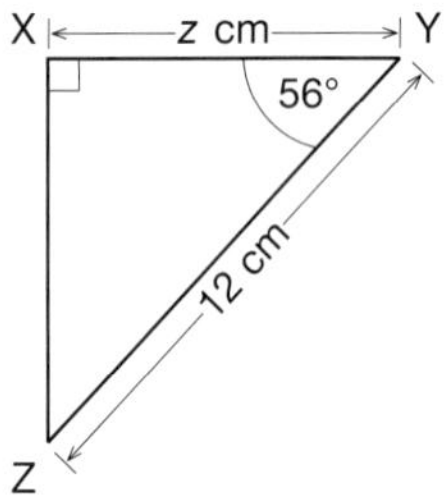

f)

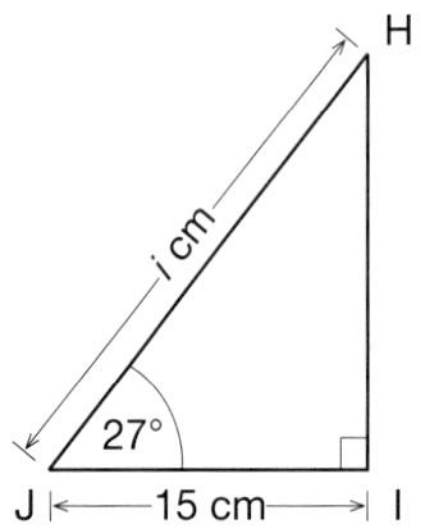

g)

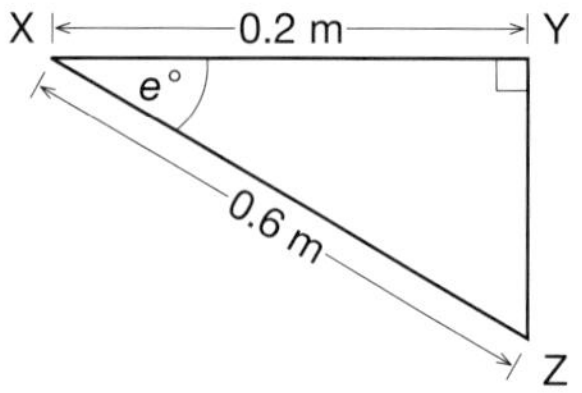

h)

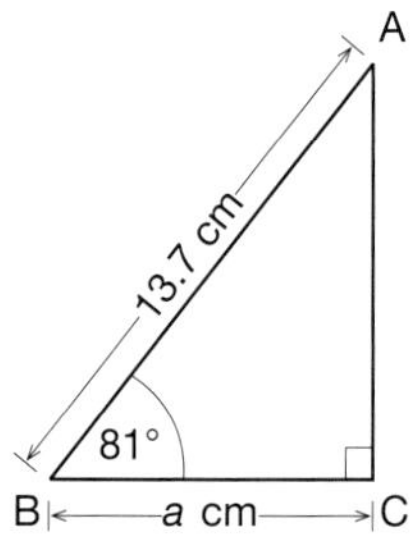

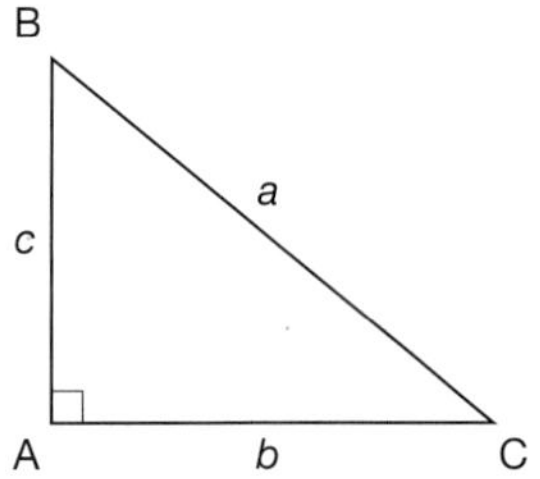

Pythagoras' theorem

Pythagoras' theorem states the relationship between the lengths of the three sides of a right-angled triangle.

Pythagoras' theorem states that:

$$a^2 = b^2 + c^2$$

Worked examples

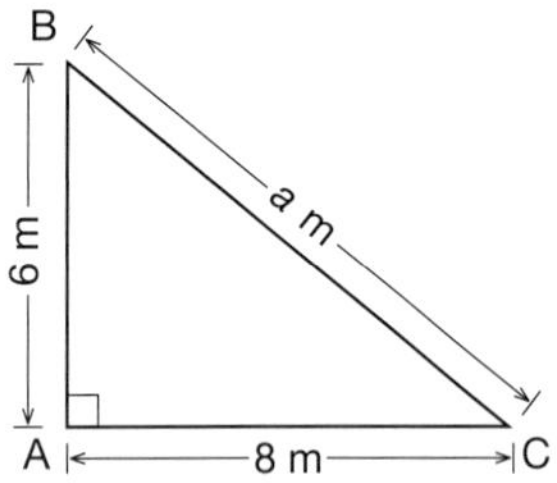

a) Calculate the length of the side BC.

Using Pythagoras:

$$a^2 = b^2 + c^2$$
$$a^2 = 8^2 + 6^2$$
$$a^2 = 64 + 36 = 100$$
$$a = \sqrt{100}$$
$$a = 10$$
$$BC = 10 \text{ m}$$

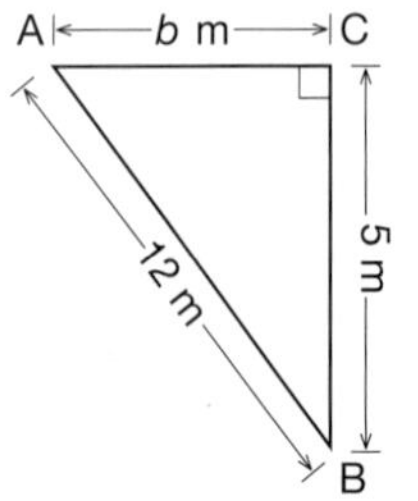

b) Calculate the length of the side AC.

Using Pythagoras:

$$a^2 = b^2 + c^2$$
$$a^2 - c^2 = b^2$$
$$b^2 = 144 - 25 = 119$$
$$b = \sqrt{119}$$
$$b = 10.9 \text{ (1 d.p.)}$$
$$AC = 10.9 \text{ m (1 d.p.)}$$

Exercise 29.5

In each of the diagrams in Q.1 and 2, use Pythagoras' theorem to calculate the length of the marked side.

1. a)

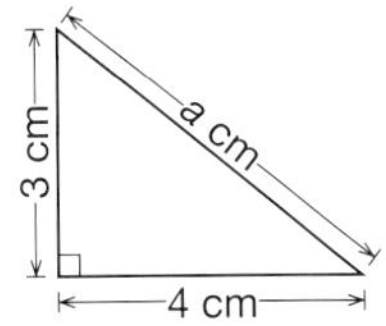

b)

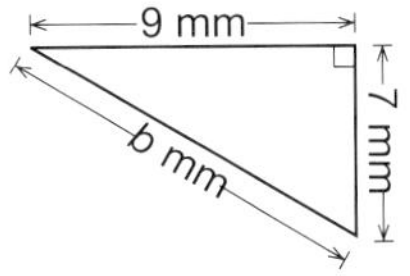

c)

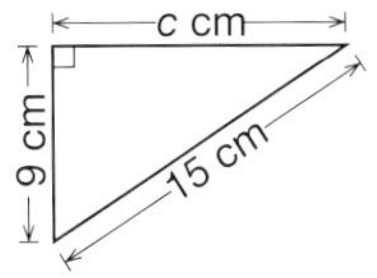

d)

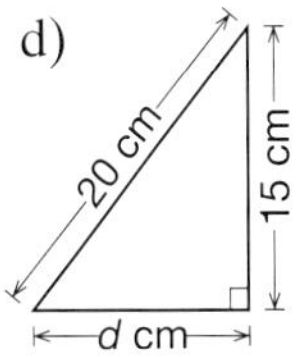

2. a)

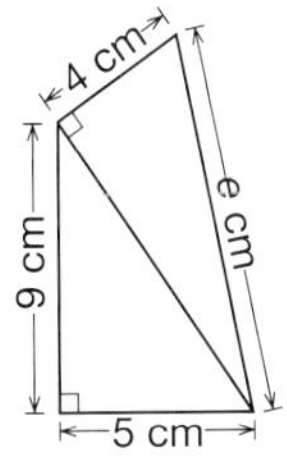

b)

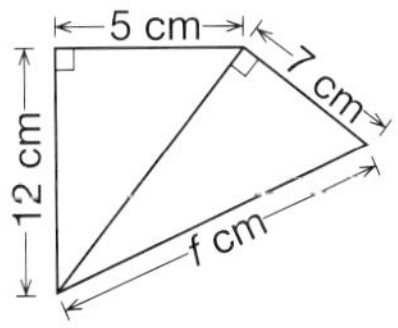

c)

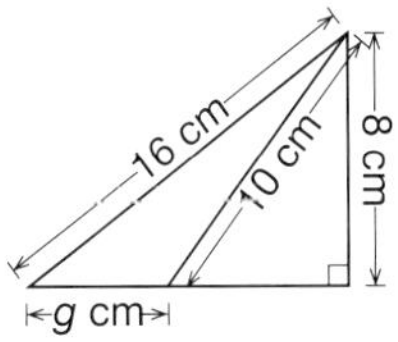

d)

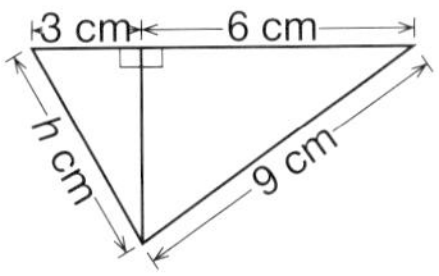

e)

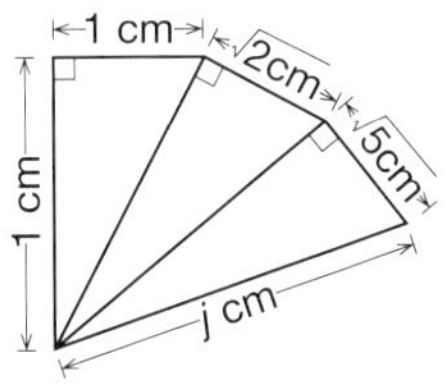

f)

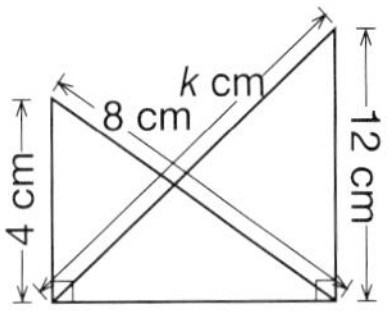

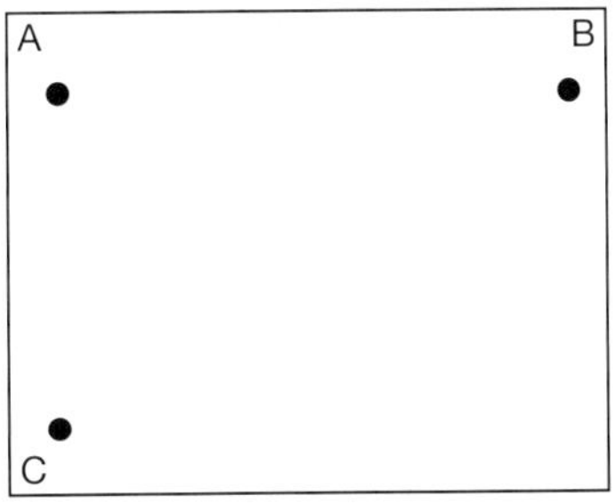

3. Villages A, B and C (left) lie on the edge of the Namib desert. Village A is 30 km due North of village C. Village B is 65 km due East of A.

Calculate the shortest distance between villages C and B, giving your answer to the nearest 0.1 km.

4. Town X is 54 km due West of town Y. The shortest distance between town Y and town Z is 86 km. If town Z is due South of X, calculate the distance between X and Z, giving your answer to the nearest kilometre.

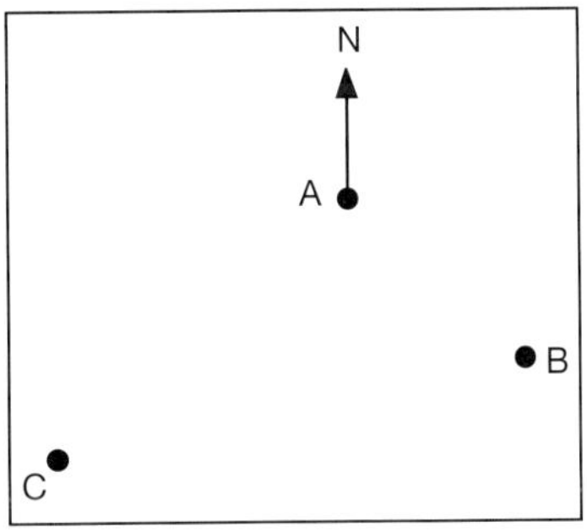

5. Village B (left) is on a bearing of 135° and at a distance of 40 km from village A. Village C is on a bearing of 225° and a distance of 62 km from village A.

a) Show that triangle ABC is right-angled.

b) Calculate the distance from B to C, giving your answer to the nearest 0.1 km.

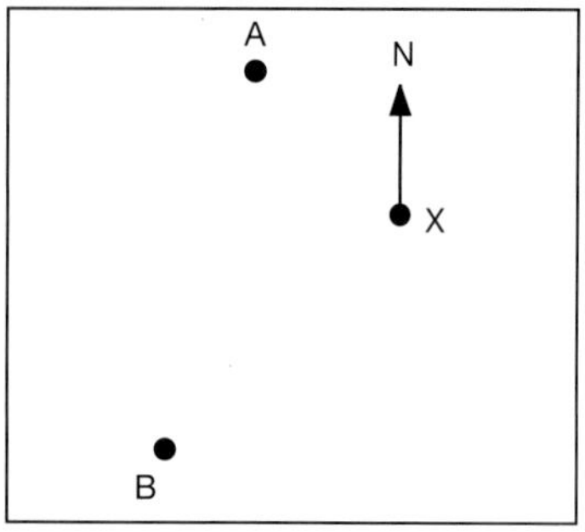

6. Two boats set off from X at the same time. Boat A sets off on a bearing of 325° with a velocity of 14 km/h. Boat B sets off on a bearing of 235° with a velocity of 18 km/h.

   Calculate the distance between the boats after they have been travelling for 2.5 hours. Give your answer to the nearest metre.

7. A boat sets off on a trip from S. It heads towards B, a point 6 km away and due North. At B it changes direction and heads towards point C, also 6 km away and due East of B. At C it changes direction once again and heads on a bearing of 135° towards D which is 13 km from C.
   a) Calculate the distance between S and C to the nearest 0.1 km.
   b) Calculate the distance the boat will have to travel if it is to return to S from D.

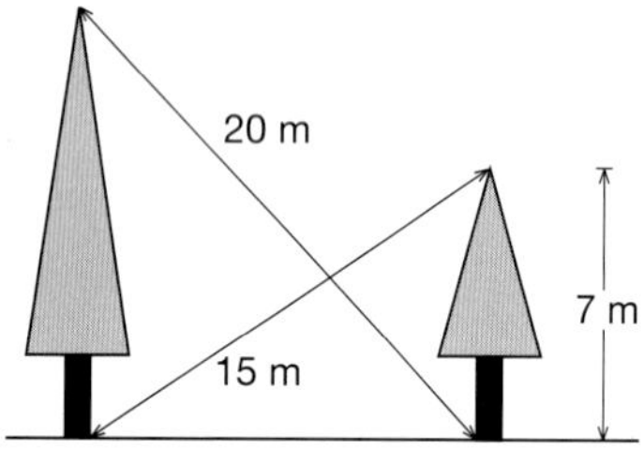

8. Two trees are standing on flat ground. The height of the smaller tree is 7 m. The distance between the top of the smaller tree and the base of the taller tree is 15 m. The distance between the top of the taller tree and the base of the smaller tree is 20 m.
   a) Calculate the horizontal distance between the two trees.
   b) Calculate the height of the taller tree.

Exercise 29.6

1. By using Pythagoras' theorem, trigonometry or both, calculate the marked value in each of the following diagrams. In each case give your answer to 1 d.p.

a)

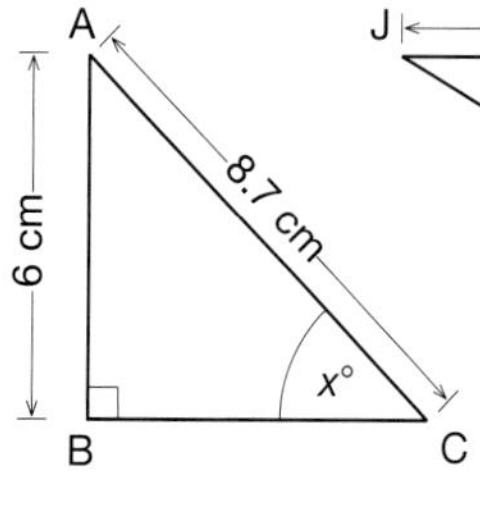

b)

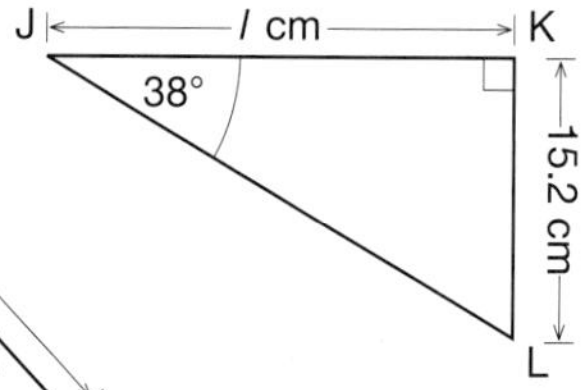

c)

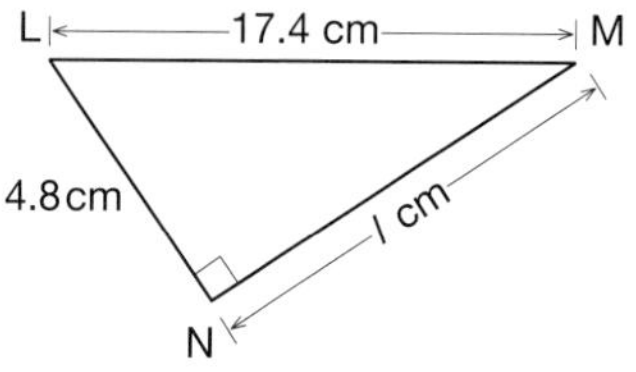

d)

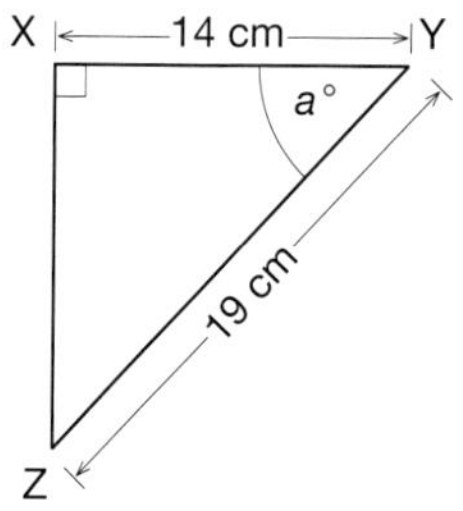

2. a) A sailing boat sets off from a point X and heads towards Y, a point 17 km North. At point Y it changes direction and heads towards point Z, a point 12 km away on a bearing of 090°. Once at Z the crew want to sail back to X. Calculate:
   i) the distance ZX,
   ii) the bearing of X from Z.

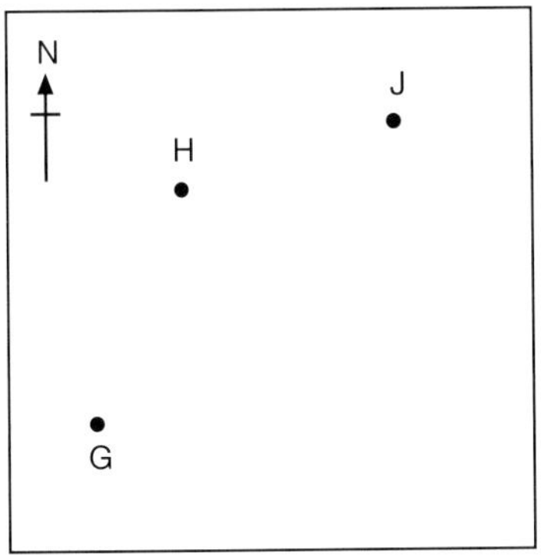

b) An aeroplane sets off from G on a bearing of 024° towards H, a point 250 km away. At H it changes course and heads towards J on a bearing of 055° and a distance of 180 km away.
   i) How far is H to the North of G?
   ii) How far is H to the East of G?
   iii) How far is J to the North of H?
   iv) How far is J to the East of H?
   v) What is the shortest distance between G and J?
   vi) What is the bearing of G from J?

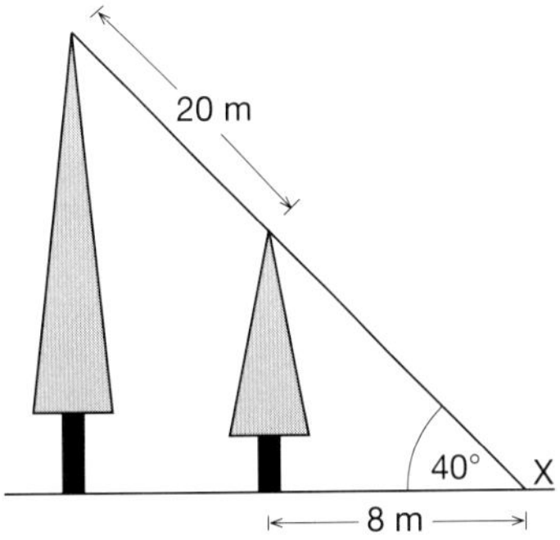

c) Two trees are standing on flat ground. The angle of elevation of their tops from a point X on the ground is 40°. If the horizontal distance between X and the small tree is 8 m and the distance between the tops of the two trees is 20 m, calculate:
   i) the height of the small tree,
   ii) the height of the tall tree,
   iii) the horizontal distance between the trees.

d) PQRS is a quadrilateral. The sides RS and QR are the same length. The sides QP and RS are parallel.

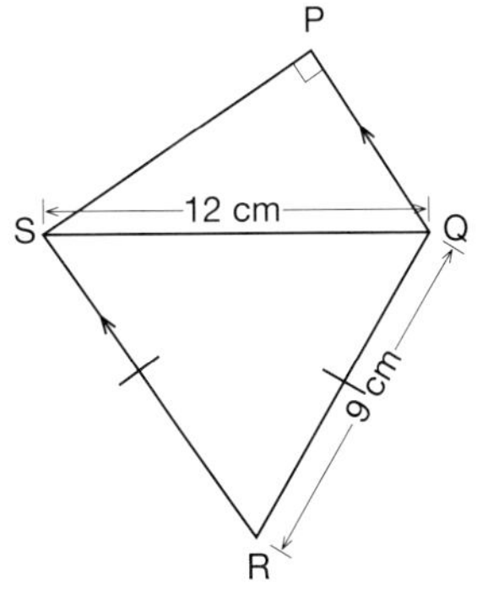

Calculate:
i) angle SQR,
ii) angle PSQ,
iii) length PQ,
iv) length PS,
v) the area of PQRS.

Student Assessment I

1. From a village church it is possible to see the spires of five neighbouring churches. The bearing and distance of each one is given below.

Bourn 8 km	bearing 070°
Comberton 12 km	bearing 135°
Duxford 9 km	bearing 185°
Eversden 7.5 km	bearing 250°
Foxton 11 km	bearing 310°

Choose an appropriate scale and draw a diagram to show the position of each church. What is the distance and bearing of the following?

a) Bourn from Duxford

b) Eversden from Comberton

2. A coastal radar station picks up a distress call from a ship. It is 50 km away on a bearing of 345°. The radar station contacts a lifeboat at sea which is 20 km away on a bearing of 220°.

Make a scale drawing and use it to find the distance and bearing of the ship from the lifeboat.

3. Calculate the side marked with a letter in each of the following diagrams. Give your answers correct to 1 d.p.

a)

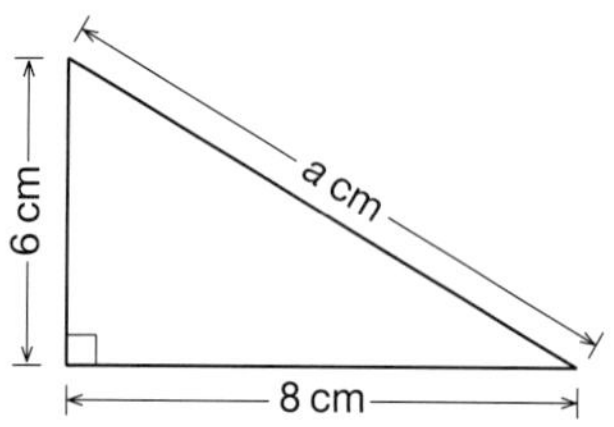

b)

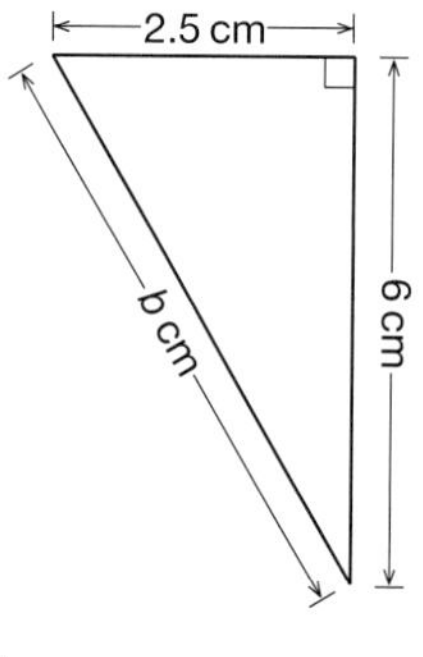

c)

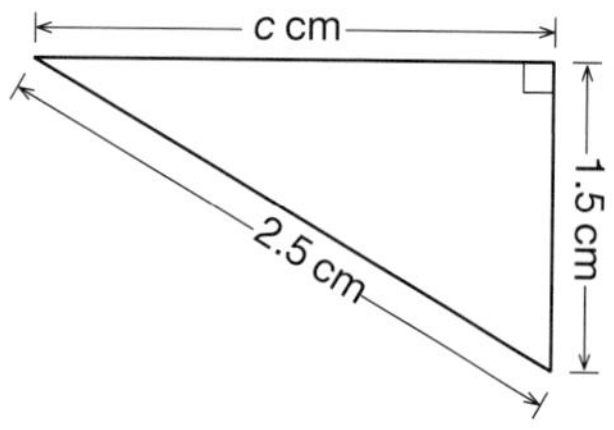

d)

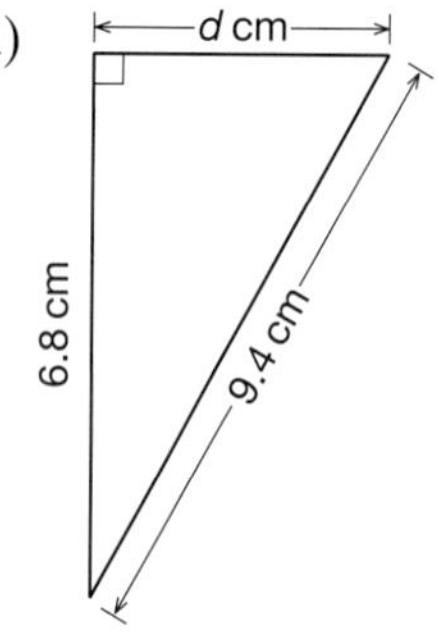

e)

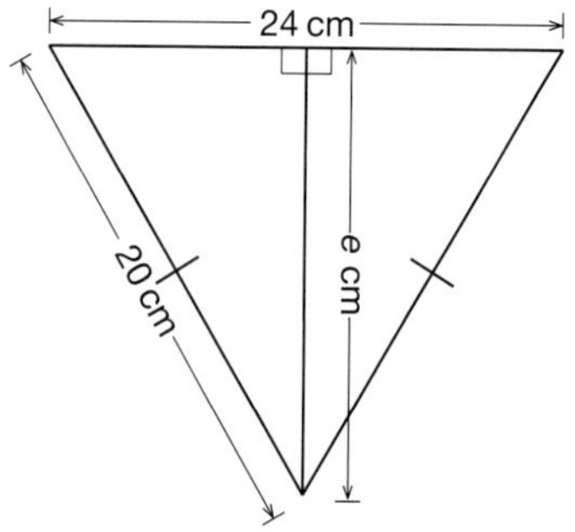

4. A rectangular swimming pool measures 50 m long by 15 m wide. Calculate the length of the diagonal of the pool. Give your answer correct to 1 d.p.

Student Assessment 2

1. A climber gets to the top of Mont Blanc. He can see in the distance a number of ski resorts. He uses his map to find the bearing and distance of the resorts, and records them as shown below:

Val d'Isere 30 km	bearing 082°
Les Arcs 40 km	bearing 135°
La Plagne 45 km	bearing 205°
Meribel 35 km	bearing 320°

Choose an appropriate scale and draw a diagram to show the position of each resort. What is the distance and bearing of the following?

a) Val d'Isere from La Plagne
b) Meribel from Les Arcs

2. An aircraft is seen on radar at Milan airport. The aircraft is 210 km away from the airport on a bearing of 065°. The aircraft is diverted to Rome airport, which is 130 km away from Milan on a bearing of 215°. Use an appropriate scale and make a scale drawing to find the distance and bearing of Rome airport from the aircraft.

3. Calculate the side marked with a letter in each of the following diagrams. Give your answers correct to 1 d.p.

a)
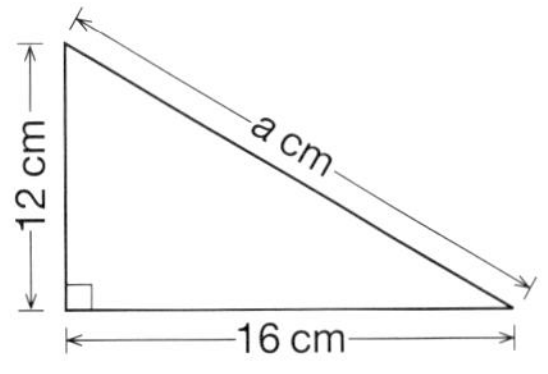

b)
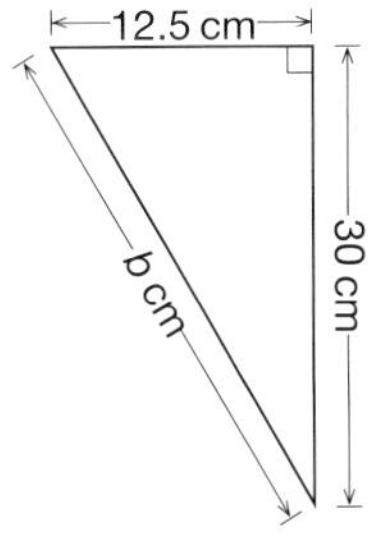

c)
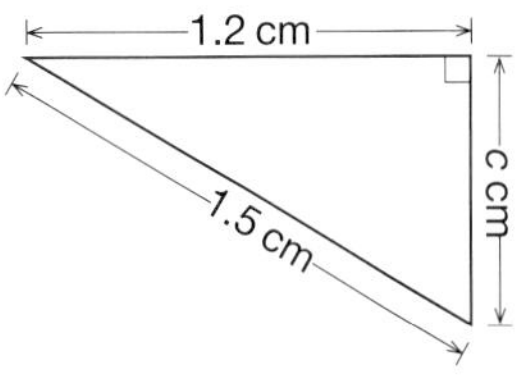

d)
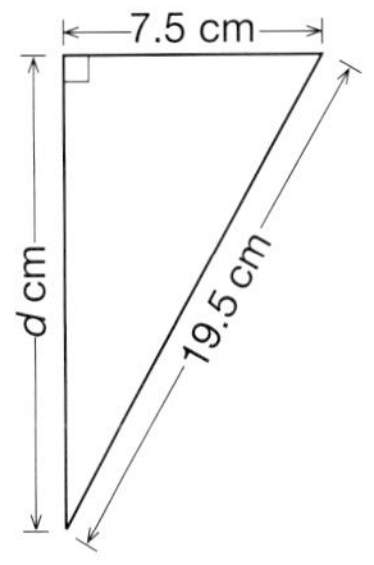

e)
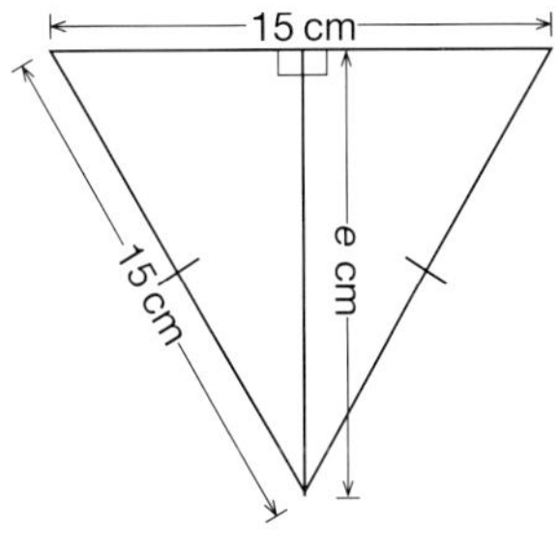

4. A snooker table measures 3.9 m by 2.4 m. Calculate the distance between the opposite corner pockets. Give your answer correct to 1 d.p.

Student Assessment 3

1. Calculate the length of the side marked x cm in each of the following. Give your answers correct to 1 d.p.

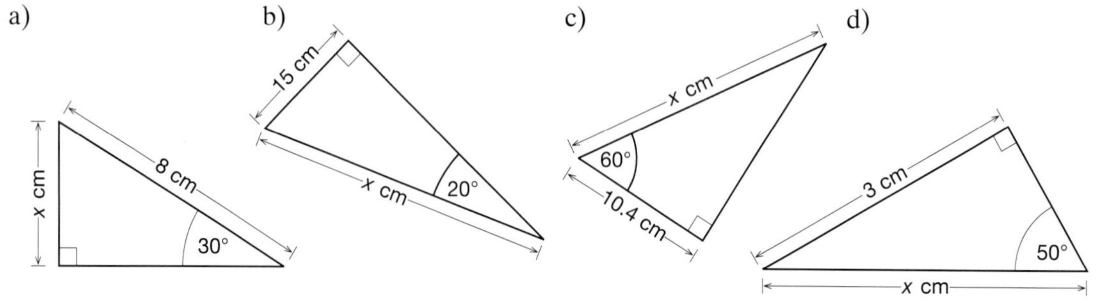

2. Calculate the angle marked $\theta°$ in each of the following. Give your answers correct to the nearest degree.

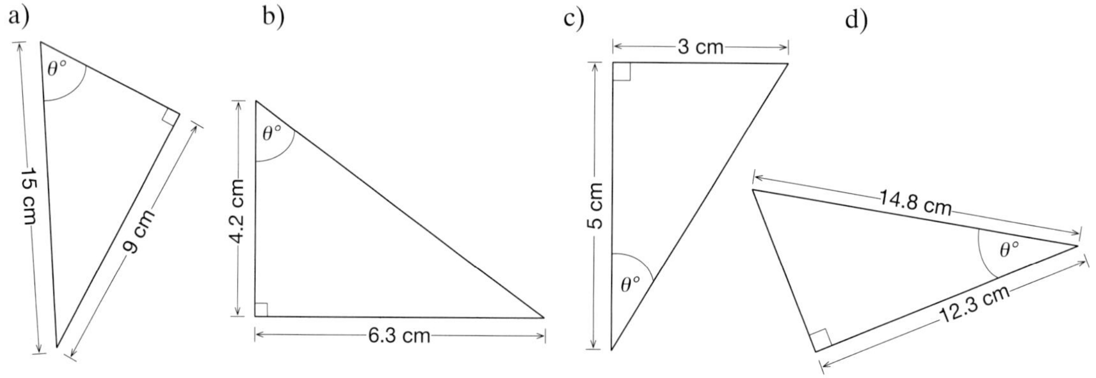

3. Calculate the length of the side marked q cm in each of the following. Give your answers correct to 1 d.p.

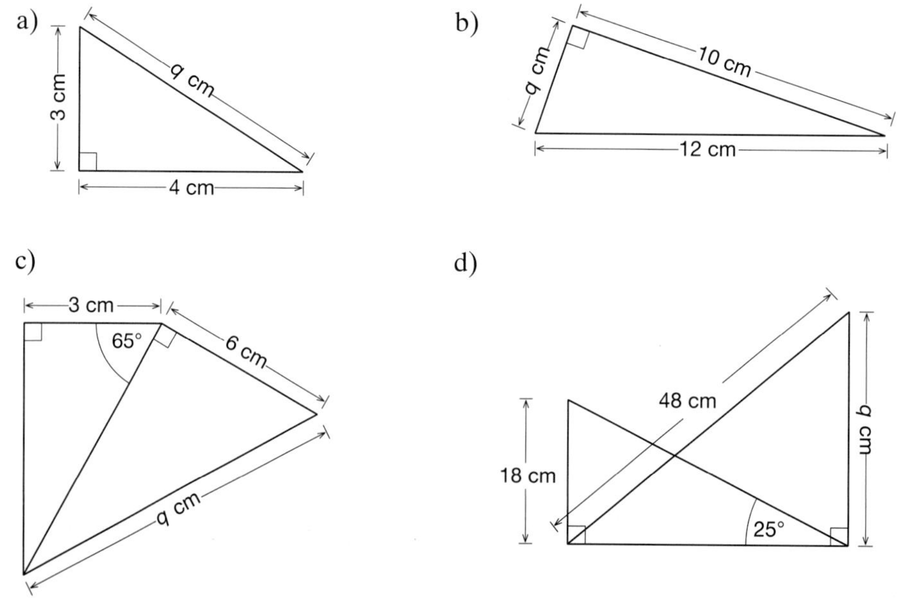

Student Assessment 4

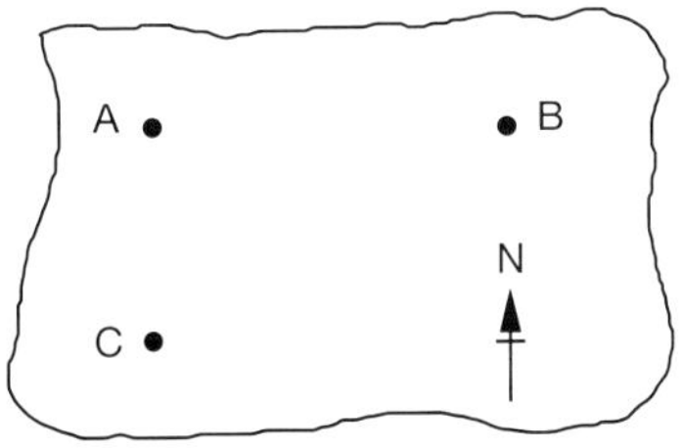

1. A map shows three towns A, B and C. Town A is due North of C. Town B is due East of A. The distance AC is 75 km and the bearing of C from B is 245°. Calculate, giving your answers to the nearest 100 m:
   a) the distance AB,
   b) the distance BC.
2. Two trees stand 16 m apart. Their tops make an angle of $\theta°$ at point A on the ground.
   a) Express $\theta°$ in terms of the height of the shorter tree and its distance x metres from point A.
   b) Express $\theta°$ in terms of the height of the taller tree and its distance from A.
   c) Form an equation in terms of x.
   d) Calculate the value of x.
   e) Calculate the value θ.

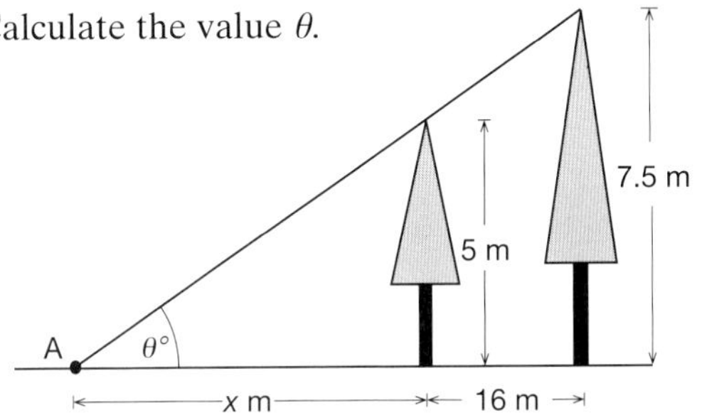

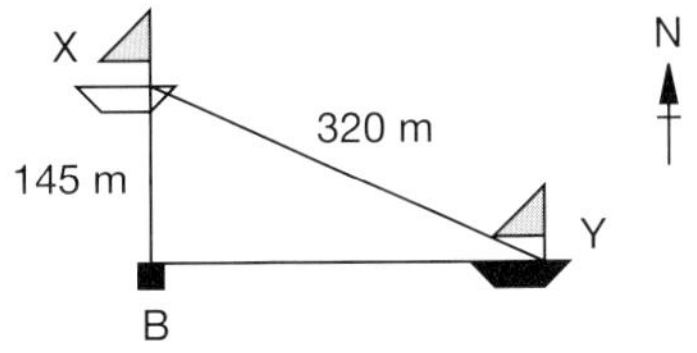

3. Two boats X and Y, sailing in a race, are shown in the diagram (left). Boat X is 145 m due North of a buoy B. Boat Y is due East of buoy B. Boats X and Y are 320 m apart. Calculate:
   a) the distance BY,
   b) the bearing of Y from X,
   c) the bearing of X from Y.
4. Hawk Q is flying 250 m vertically above hawk P. They both spot a snake at R.

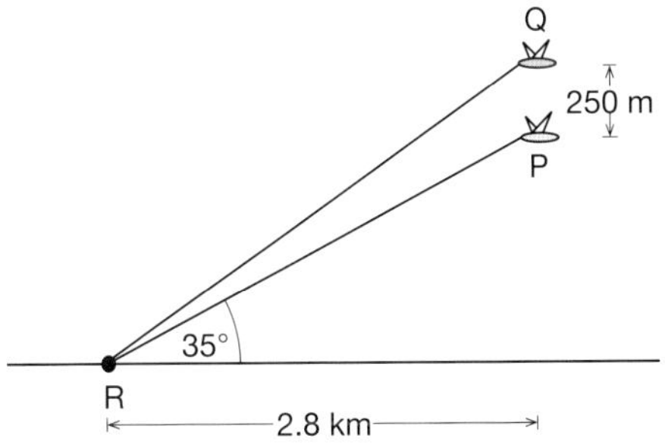

Using the information given, calculate:
a) the height of P above the ground,
b) the distance between P and R,
c) the distance between Q and R.

30 VECTORS

Describe a translation by using a column vector, $\overrightarrow{AB}$ or **a**; add vectors and multiply a vector by a scalar.

A **translation** (a sliding movement) can be described using **column vectors**. A column vector describes the movement of the object in both the x direction and the y direction.

Worked example

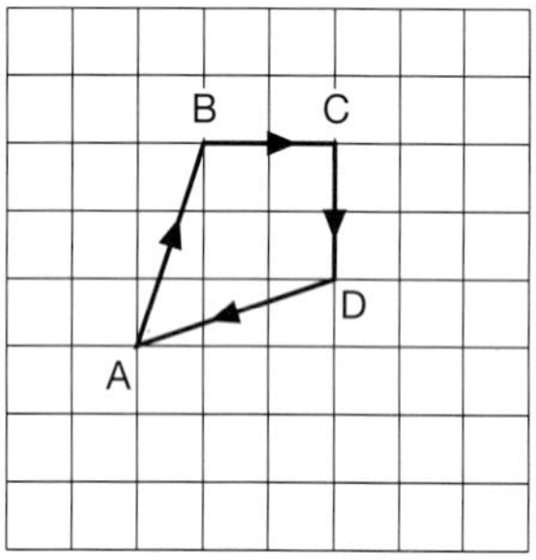

i) Describe the translation from A to B in the diagram (left) in terms of a column vector.

$$\overrightarrow{AB} = \begin{pmatrix} 1 \\ 3 \end{pmatrix}$$

i.e. 1 unit in the x direction, 3 units in the y direction

ii) Describe $\overrightarrow{BC}$ in terms of a column vector.

$$\overrightarrow{BC} = \begin{pmatrix} 2 \\ 0 \end{pmatrix}$$

iii) Describe $\overrightarrow{CD}$ in terms of a column vector.

$$\overrightarrow{CD} = \begin{pmatrix} 0 \\ -2 \end{pmatrix}$$

iv) Describe $\overrightarrow{DA}$ in terms of a column vector.

$$\overrightarrow{DA} = \begin{pmatrix} -3 \\ -1 \end{pmatrix}$$

Translations can also be named by a single letter. The direction of the arrow indicates the direction of the translation.

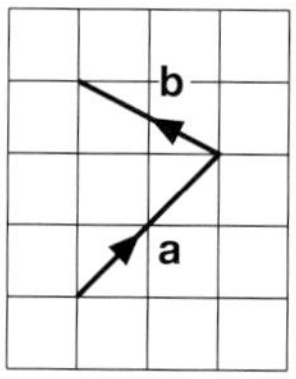

Worked example

Define **a** and **b** in the diagram above using column vectors.

$$\mathbf{a} = \begin{pmatrix} 2 \\ 2 \end{pmatrix} \qquad \mathbf{b} = \begin{pmatrix} -2 \\ 1 \end{pmatrix}$$

Note: When you represent vectors by single letters, i.e. **a**, in handwritten work, you should write them as $\underline{a}$.

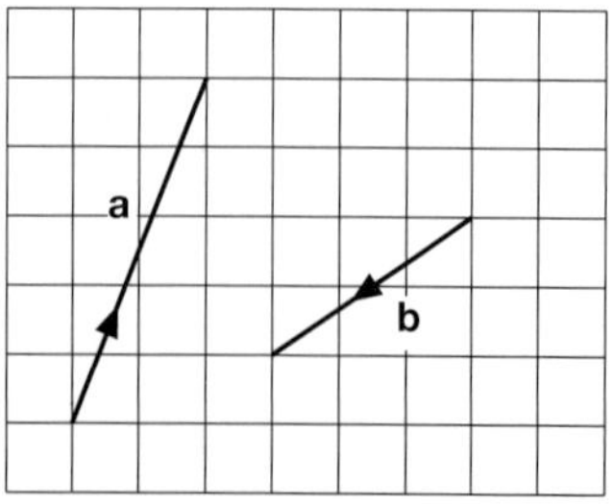

If $\mathbf{a} = \begin{pmatrix} 2 \\ 5 \end{pmatrix}$ and $\mathbf{b} = \begin{pmatrix} -3 \\ -2 \end{pmatrix}$, they can be represented diagrammatically as shown (left).

The diagrammatic representation of $-\mathbf{a}$ and $-\mathbf{b}$ is shown below.

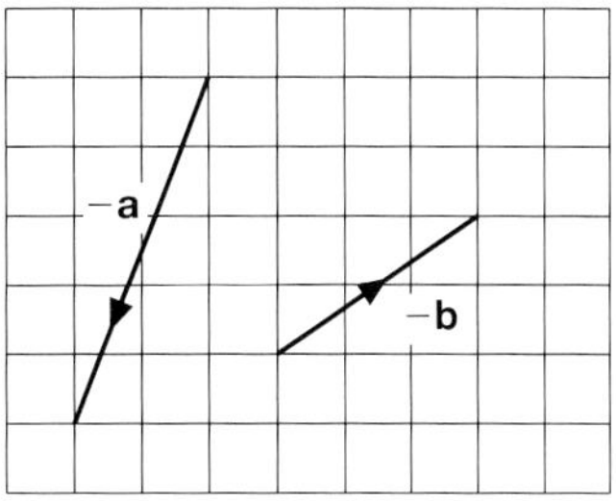

It can be seen from the diagram above that

$$-\mathbf{a} = \begin{pmatrix} -2 \\ -5 \end{pmatrix} \text{ and } -\mathbf{b} = \begin{pmatrix} 3 \\ 2 \end{pmatrix}.$$

Exercise 30.1

In Q.1 and 2, describe each translation using a column vector.

1. a) $\overrightarrow{AB}$
b) $\overrightarrow{BC}$
c) $\overrightarrow{CD}$
d) $\overrightarrow{DE}$
e) $\overrightarrow{EA}$
f) $\overrightarrow{AE}$
g) $\overrightarrow{DA}$
h) $\overrightarrow{CA}$
i) $\overrightarrow{DB}$

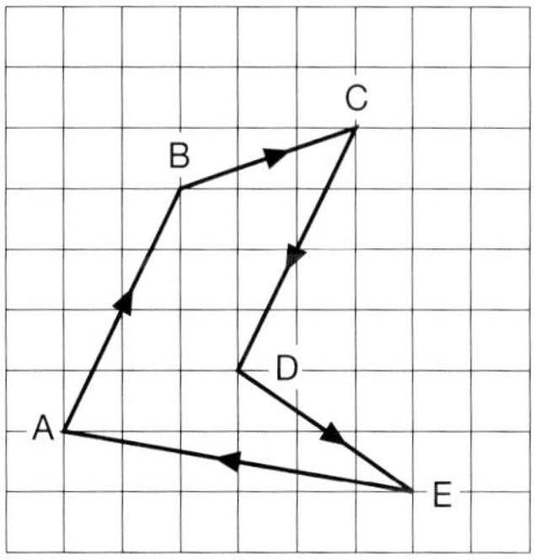

2.

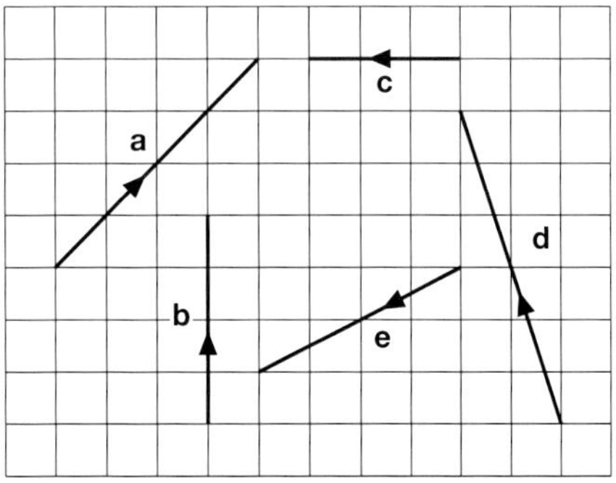

a) **a**	b) **b**	c) **c**
d) **d**	e) **e**	f) $-\mathbf{b}$
g) $-\mathbf{c}$	h) $-\mathbf{d}$	i) $-\mathbf{a}$

3. Draw and label the following vectors on a square grid:

a) $\mathbf{a} = \begin{pmatrix} 2 \\ 4 \end{pmatrix}$ b) $\mathbf{b} = \begin{pmatrix} -3 \\ 6 \end{pmatrix}$ c) $\mathbf{c} = \begin{pmatrix} 3 \\ -5 \end{pmatrix}$

d) $\mathbf{d} = \begin{pmatrix} -4 \\ -3 \end{pmatrix}$ e) $\mathbf{e} = \begin{pmatrix} 0 \\ -6 \end{pmatrix}$ f) $\mathbf{f} = \begin{pmatrix} -5 \\ 0 \end{pmatrix}$

g) $\mathbf{g} = -\mathbf{c}$ h) $\mathbf{h} = -\mathbf{b}$ i) $\mathbf{i} = -\mathbf{f}$

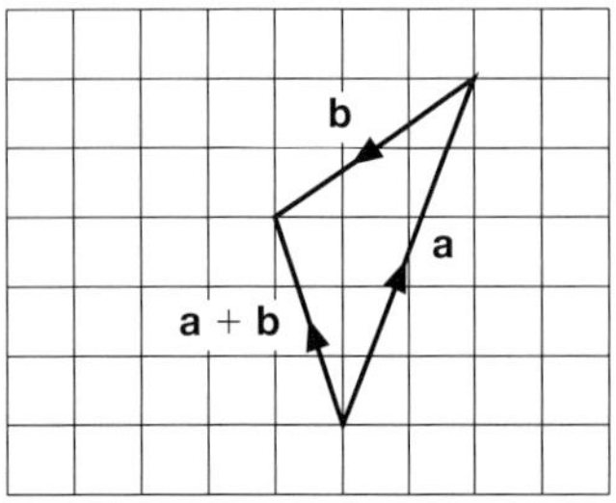

Addition of vectors

Vectors can be added together and represented diagrammatically as shown (left).

The translation represented by **a** followed by **b** can be written as a single transformation **a** + **b**:

i.e. $\begin{pmatrix} 2 \\ 5 \end{pmatrix} + \begin{pmatrix} -3 \\ -2 \end{pmatrix} = \begin{pmatrix} -1 \\ 3 \end{pmatrix}$

Exercise 30.2

In the following questions,

$$\mathbf{a} = \begin{pmatrix} 3 \\ 4 \end{pmatrix} \quad \mathbf{b} = \begin{pmatrix} -2 \\ 1 \end{pmatrix} \quad \mathbf{c} = \begin{pmatrix} -4 \\ -3 \end{pmatrix} \quad \mathbf{d} = \begin{pmatrix} 3 \\ -2 \end{pmatrix}$$

1. Draw vector diagrams to represent the following:

a) **a** + **b** b) **b** + **a** c) **a** + **d**
d) **d** + **a** e) **b** + **c** f) **c** + **b**

2. What conclusions can you draw from your answers to Q.1 above?

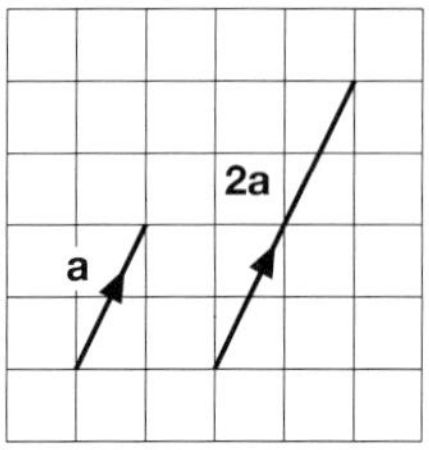

Multiplying a vector by a scalar

Look at the two vectors in the diagram (left).

$\mathbf{a} = \begin{pmatrix} 1 \\ 2 \end{pmatrix} \quad 2\mathbf{a} = 2\begin{pmatrix} 1 \\ 2 \end{pmatrix} = \begin{pmatrix} 2 \\ 4 \end{pmatrix}$

Worked example If $\mathbf{a} = \begin{pmatrix} 2 \\ -4 \end{pmatrix}$, express the vectors **b**, **c**, **d** and **e** in terms of **a**.

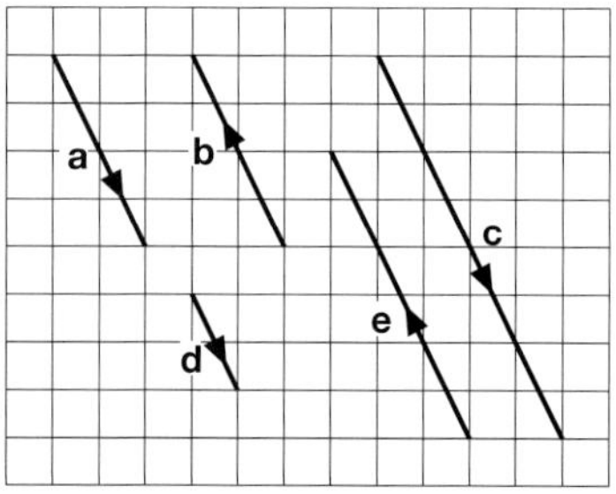

$\mathbf{b} = -\mathbf{a} \quad \mathbf{c} = 2\mathbf{a} \quad \mathbf{d} = \tfrac{1}{2}\mathbf{a} \quad \mathbf{e} = -\tfrac{3}{2}\mathbf{a}$

Exercise 30.3

1. $\mathbf{a} = \begin{pmatrix} 1 \\ 4 \end{pmatrix}$ $\mathbf{b} = \begin{pmatrix} -4 \\ -2 \end{pmatrix}$ $\mathbf{c} = \begin{pmatrix} -4 \\ 6 \end{pmatrix}$

   Express the following vectors in terms of either **a**, **b** or **c**.

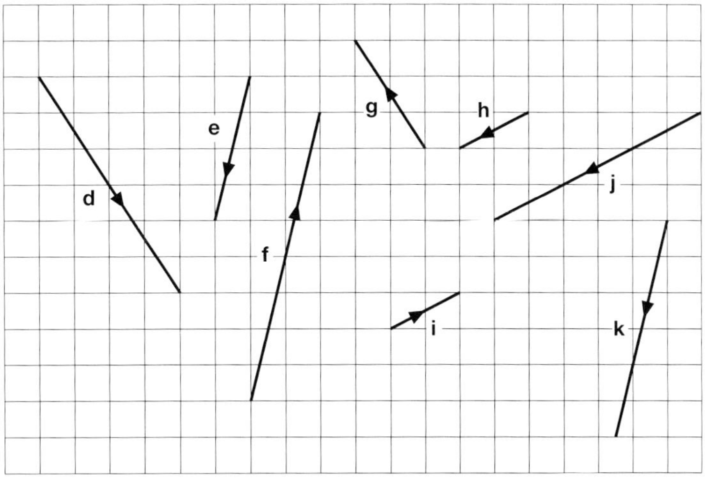

2. $\mathbf{a} = \begin{pmatrix} 2 \\ 3 \end{pmatrix}$ $\mathbf{b} = \begin{pmatrix} -4 \\ -1 \end{pmatrix}$ $\mathbf{c} = \begin{pmatrix} -2 \\ 4 \end{pmatrix}$

   Represent each of the following as single column vectors:

   a) $2\mathbf{a}$ b) $3\mathbf{b}$ c) $-\mathbf{c}$
   d) $\mathbf{a} + \mathbf{b}$ e) $\mathbf{b} + \mathbf{c}$ f) $3\mathbf{c} + \mathbf{a}$
   g) $2\mathbf{b} + \mathbf{a}$ h) $\frac{1}{2}(\mathbf{a} + \mathbf{b})$ i) $2\mathbf{a} + 3\mathbf{c}$

3. $\mathbf{a} = \begin{pmatrix} -2 \\ 3 \end{pmatrix}$ $\mathbf{b} = \begin{pmatrix} 0 \\ -3 \end{pmatrix}$ $\mathbf{c} = \begin{pmatrix} 4 \\ -1 \end{pmatrix}$

   Express each of the following vectors in terms of **a**, **b** and **c**:

   a) $\begin{pmatrix} -4 \\ 6 \end{pmatrix}$ b) $\begin{pmatrix} 0 \\ 3 \end{pmatrix}$ c) $\begin{pmatrix} 4 \\ -4 \end{pmatrix}$ d) $\begin{pmatrix} 8 \\ -2 \end{pmatrix}$

Student Assessment 1

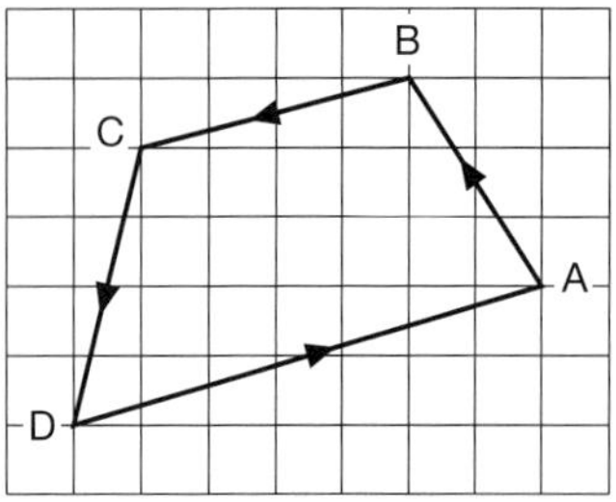

1. Using the diagram (left), describe the following translations using column vectors.

   a) $\overrightarrow{AB}$ b) $\overrightarrow{DA}$ c) $\overrightarrow{CA}$

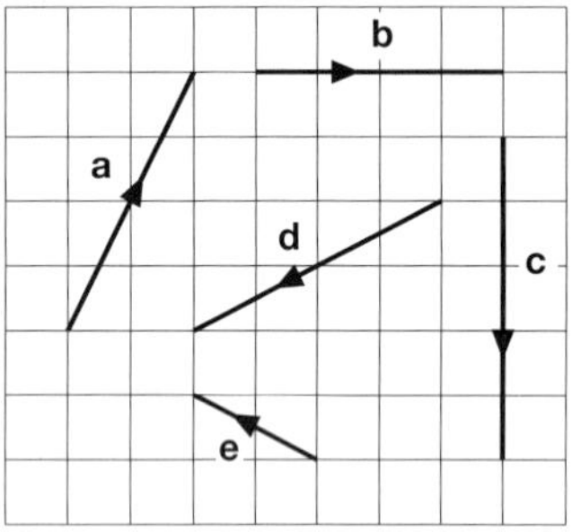

2. Describe each of the translations **a** to **e** shown in the diagram (left) using column vectors.

3. Using the vectors from Q.2 above, draw diagrams to represent:

   a) $\mathbf{a} + \mathbf{b}$ b) $\mathbf{e} + \mathbf{d}$ c) $\mathbf{c} + \mathbf{e}$ d) $2\mathbf{e} + \mathbf{b}$

4. In the following,

$$\mathbf{a} = \begin{pmatrix} 2 \\ 6 \end{pmatrix} \quad \mathbf{b} = \begin{pmatrix} -3 \\ -1 \end{pmatrix} \quad \mathbf{c} = \begin{pmatrix} -2 \\ 4 \end{pmatrix}$$

   Calculate:

   a) $\mathbf{a} + \mathbf{b}$ b) $\mathbf{c} + \mathbf{b}$ c) $2\mathbf{a} + \mathbf{b}$ d) $3\mathbf{c} + 2\mathbf{b}$

Student Assessment 2

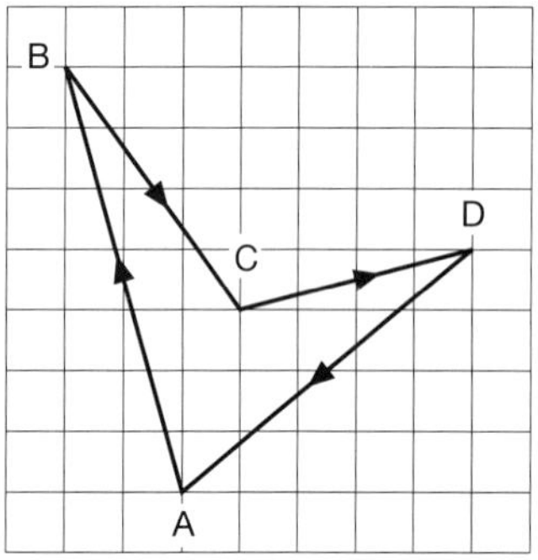

1. Using this diagram (left), describe the following translations with column vectors.

   a) $\overrightarrow{AB}$ b) $\overrightarrow{DA}$ c) $\overrightarrow{CA}$

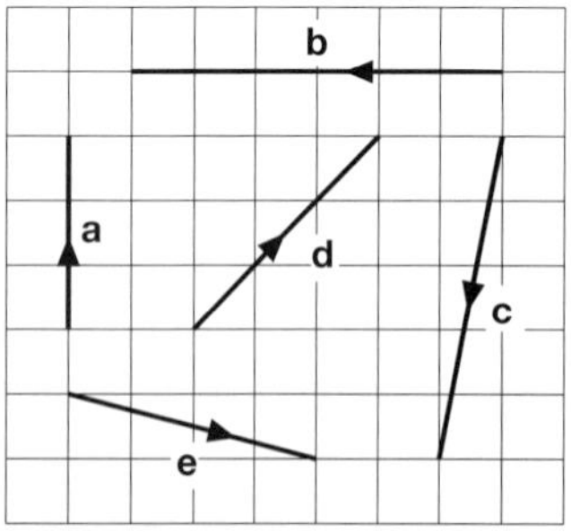

2. Describe each of the translations **a** to **e** in the diagram (left) using column vectors.

3. Using the vectors from Q.2 above, draw diagrams to represent:

   a) $\mathbf{a} + \mathbf{e}$ b) $\mathbf{c} + \mathbf{d}$ c) $-\mathbf{c} + \mathbf{e}$ d) $-\mathbf{b} + 2\mathbf{a}$

4. In the following,

$$\mathbf{a} = \begin{pmatrix} 3 \\ -5 \end{pmatrix} \quad \mathbf{b} = \begin{pmatrix} 0 \\ 4 \end{pmatrix} \quad \mathbf{c} = \begin{pmatrix} -4 \\ 6 \end{pmatrix}$$

   Calculate:

   a) $\mathbf{a} + \mathbf{c}$ b) $\mathbf{b} + \mathbf{a}$ c) $2\mathbf{a} + \mathbf{b}$ d) $3\mathbf{c} + 2\mathbf{a}$

ICT SECTION

Shape and Space Activity 1

In this activity, you will use a spreadsheet to calculate the sizes of interior and exterior angles of regular polygons.

Set up a spreadsheet as shown below:

	A	B	C	D	E	F
1	Regular Polygons					
2	Number of sides	Name	Sum of exterior angles	Size of an exterior angle	Size of an interior angle	Sum of interior angles
3	3					
4	4					
5	5					
6	6					
7	7					
8	8					
9	9					
10	10					
11	12					
12	20					
13						
14						
15			Use formulae to generate the results in these columns			
16						

1. By using formulae, use the spreadsheet to generate the results for the sizes of the interior and exterior angles.
2. Write down the general formulae you would use to calculate the sizes of the interior and exterior angles of an n-sided regular polygon.

Shape and Space Activity 2

In this activity, you will be using a geometry package to investigate enlargements.

1. Using a geometry package, draw an object and enlarge it by a scale factor of 2. An example is shown below:

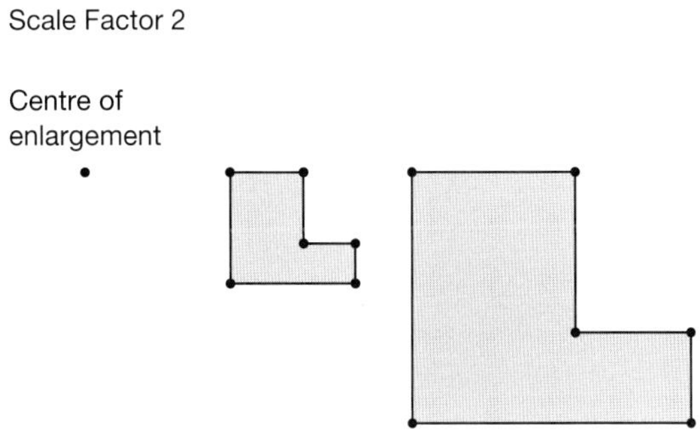

2. Describe the position of the centre of enlargement used to produce the following diagrams:

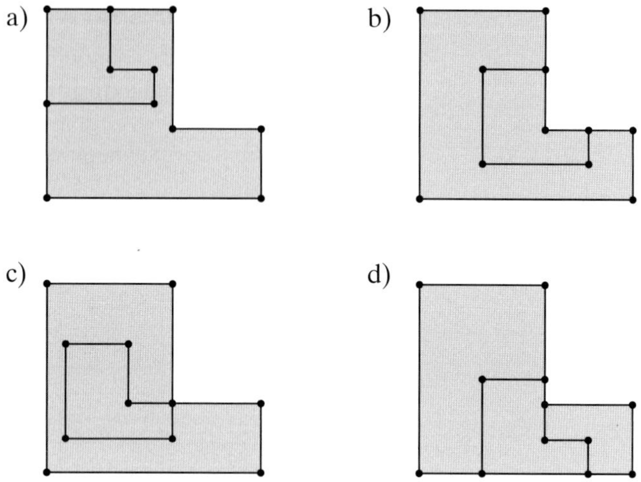

3. Move the position of the centre of enlargement to test your answers to Q.2 above.

4. By changing the scale factor of enlargement, calculate the value that will produce the diagram below.

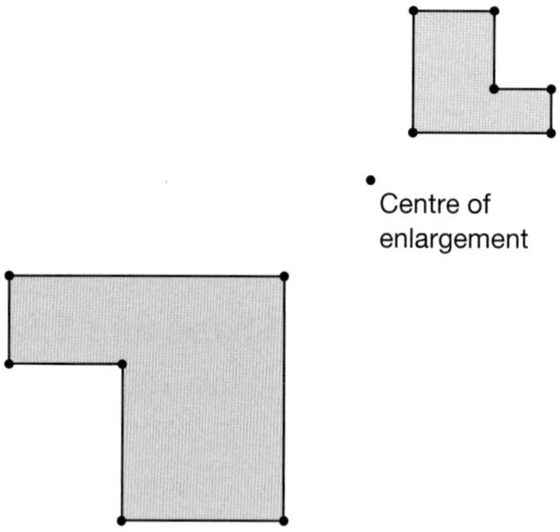

5. What conclusions can you make from your answer to Q.4 above?

REVIEWS FOR CHAPTERS 22–30

Review 19

1. Draw an obtuse angle of approximately 120°. Then, using a protractor, measure the angle exactly.

2. Construct an isosceles triangle of base length 5.2 cm and two sides of 8 cm. Measure the base angles.

3. Calculate the size of each exterior angle of a regular pentagon.

4. The diagram below shows a trapezoidal garden. Three of its sides are enclosed by a fence, whilst the fourth is adjacent to a house. Grass seed is to be sown in the garden. However, the grass must be at least 3 m away from the house and at least 1.5 m away from the fence. Copy the diagram and shade the region in which the seed can be sown.

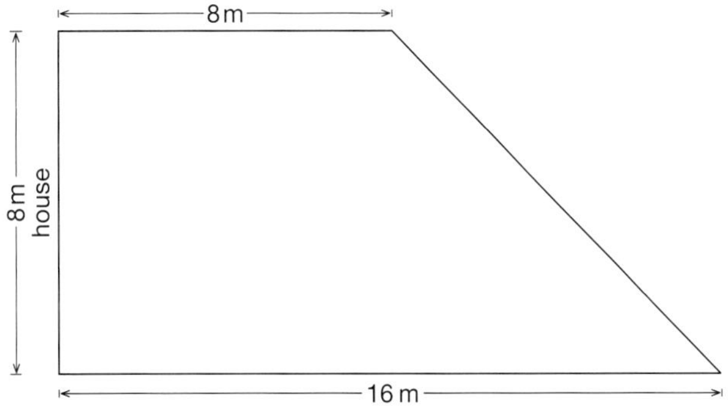

5. Draw the five vowels as capital letters and show, if possible, their lines of symmetry.

6. Copy the diagram below and draw the object's image when reflected in the dotted line.

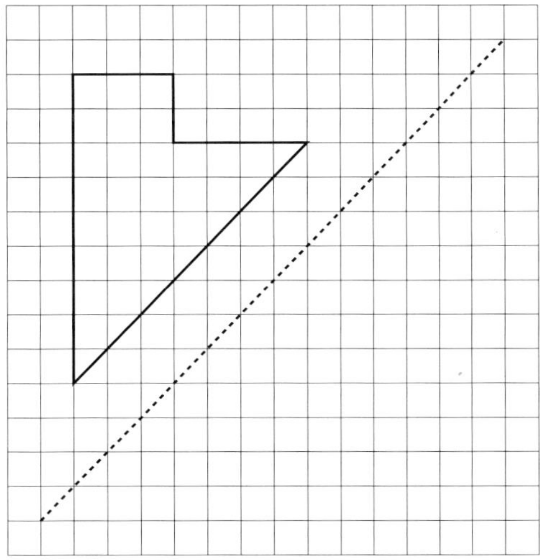

Review 20

1. Calculate the area of the kite below:

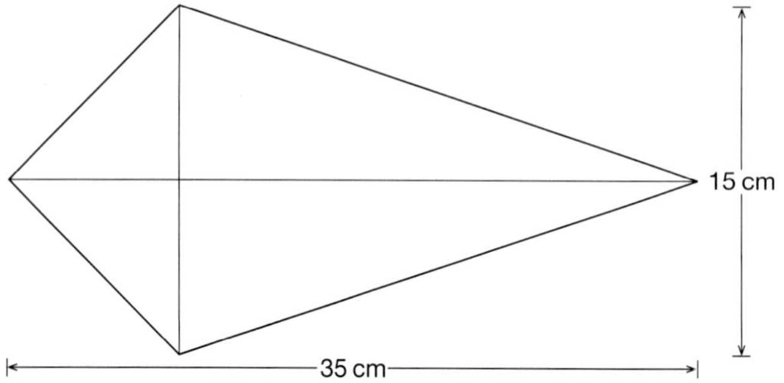

2. The wheel of a bicycle has a radius of 32 cm. Calculate the distance, in km, the bicycle has travelled after 2000 rotations. Give your answer to 2 s.f.

3. Use a scale of 1 cm to 1 km to show the following journey of a yacht. Starting at point A, it sails 8 km on a bearing of 135° to point B. From B, it sails 10 km on a bearing of 280° to point C. Find the distance and bearing of A from C.

4. Calculate the size of the angle ABC in the triangle below. Give your answer to 2 s.f.

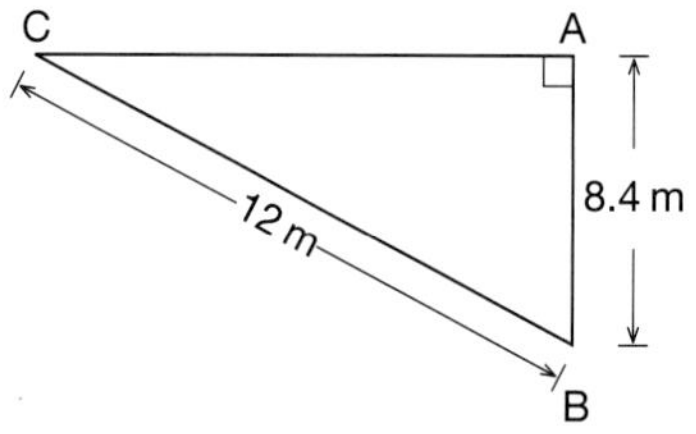

5. Using the diagram below, describe the following translations using column vectors:

a) $\overrightarrow{AB}$ b) $\overrightarrow{CD}$ c) $\overrightarrow{DE}$ d) $\overrightarrow{EA}$

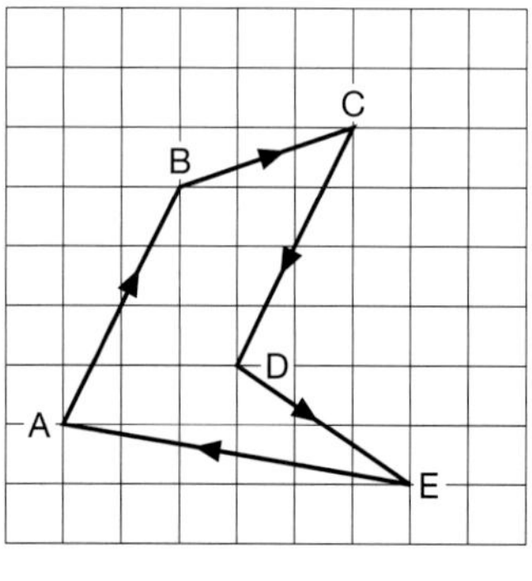

Review 21

1. a) What is the supplement of 60°?
b) What is the complement of 60°?

2. Draw a line AB of length 8.8 cm. Construct the perpendicular bisector of AB.

3. The interior angle of a regular polygon is 150°. Name the polygon and sketch it.

4. Draw a circle of radius 5 cm. Inside this circle, label points A, B and C, ensuring that they do not lie in a straight line.
a) By construction, find the point which is equidistant from A, B and C. Label this point O.
b) Mark two points D and E on the circumference of the circle, such that OD = OE = OA = OB = OC. If no such points exist in your diagram, explain why not.

5. Draw a two-dimensional shape with rotational symmetry of order 6. Mark on the shape all its lines of symmetry.

6. In the diagram below, triangle A'B'C' is an enlargement of triangle ABC.

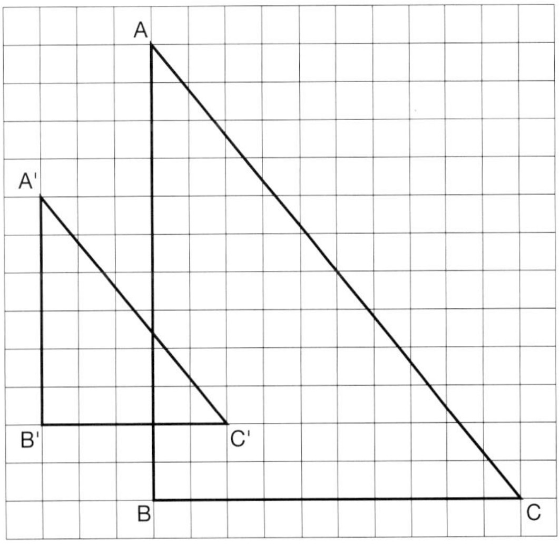

Copy the diagram and:
a) find the position of the centre of enlargement,
b) calculate the scale factor of enlargement.

Review 22

1. Calculate the height of a trapezium whose area is 20 cm^2, if its parallel sides measure 6 cm and 4 cm.

2. Calculate the volume, in cm^3, of a cylinder of radius 4 cm and length 12 cm. Give your answer to 3 s.f.

3. Use Pythagoras' theorem to calculate the length of the side marked x cm in the diagram (right). Give your answer to 1 d.p.

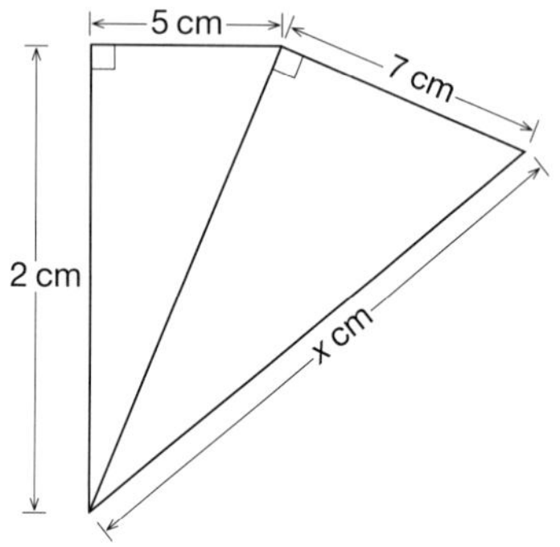

4. Use trigonometry to calculate the size of the angle marked $y°$ in the diagram (right).

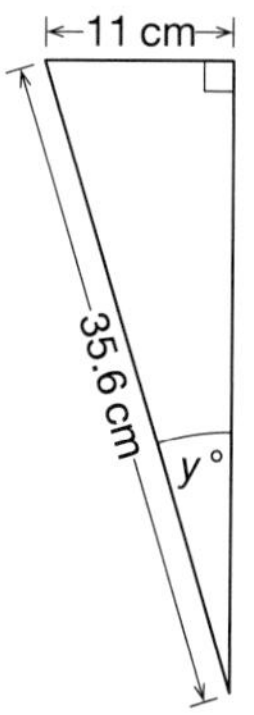

5. **p** is represented by the column vector $\begin{pmatrix} 2 \\ 5 \end{pmatrix}$

   q is represented by the column vector $\begin{pmatrix} -4 \\ -2 \end{pmatrix}$

   a) Draw a diagram to show the single transformation $\mathbf{p} + \mathbf{q}$.
   b) What column vector is represented by $\mathbf{p} + \mathbf{q}$?

Review 23

1. Use a protractor and a ruler to draw a triangle ABC with AB = 8.4 cm, ∠BAC = 40° and ∠ABC = 80°. Measure BC, AC and ∠BCA.

2. Identify the two quadrilaterals described below:
   a) Opposite sides are equal in length and diagonals bisect each other. Adjacent angles are not equal.
   b) All sides are equal in length. Diagonals, however, are not equal in length.

3. In the diagram below, O marks the centre of the circle. Calculate the value of x.

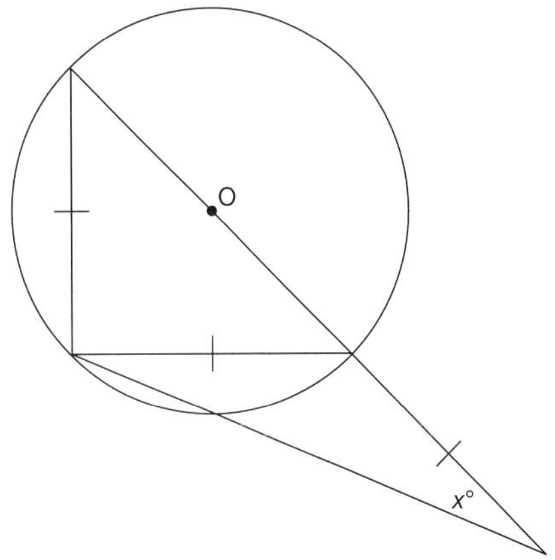

4. Draw a scale diagram of a rectangular garden 8 m by 12 m. Identify the locus of points which fulfil all the following criteria:
a) at least 1.5 m from the edge of the garden,
b) at least 3.5 m from the corners of the garden,
c) at least 2 m from the centre of the garden.

5. Draw a shape made from four congruent, equilateral triangles and one square, which has rotational symmetry of order 2 and only two lines of symmetry.

6. Copy the diagram below.

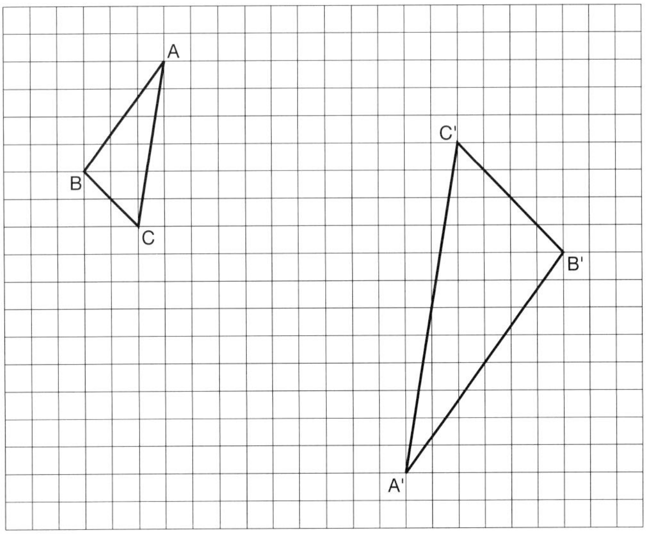

a) Find the centre of enlargement which transforms triangle ABC into triangle A'B'C'.
b) Calculate the scale factor of enlargement.

Review 24

1. A circular ring has a cross-section of two concentric circles. If the radius of the inner circle is 20 mm and the radius of the outer circle is 25 mm, calculate the cross-sectional area of the ring. Give your answer to 3 s.f.

2. A cuboid measures 15 cm by 10 cm by 5 cm. Calculate its surface area in cm^2.

3. A dinghy at point P is 50 m due North of a buoy at point O. A second dinghy at point Q is 120 m due East of the same buoy. What is the shortest distance, in metres, from P to Q?

4. In Q.3 above, what is the bearing of the dinghy at Q from the dinghy at P?

5. In this question:

   $\mathbf{p} = \begin{pmatrix} 1 \\ 3 \end{pmatrix} \quad \mathbf{q} = \begin{pmatrix} -2 \\ -4 \end{pmatrix} \quad \mathbf{r} = \begin{pmatrix} -3 \\ 1 \end{pmatrix}$

   Represent the following as single column vectors:
   a) $\mathbf{p} + \mathbf{q} + \mathbf{r}$
   b) $2\mathbf{q} + 3\mathbf{p}$
   c) $2\mathbf{r} + 3\mathbf{q} + \mathbf{p}$

Review 25

1. An amphitheatre is an open cylinder of radius 5 m and height 5 m. Inside are two square-based pillars of side length 2 m and height 3 m. Draw a plan view to a scale of 1 cm : 1 m. Choose one corner of each square and label the points A and B. Shade the area of the floor of the amphitheatre that would not be visible to either of two people standing at A and B.

2. A box, whose interior is in the shape of a hollow cube of side 5 cm, holds inside it a cylinder which just fits the box. Calculate:
   a) the volume of the cylinder,
   b) the curved surface area of the cylinder.

3. In the diagram below, calculate the length of the side marked x cm.

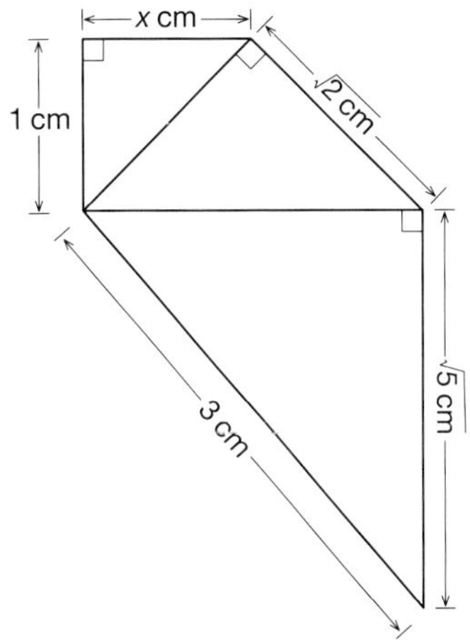

4. Identify the following two-dimensional shapes from the properties given:

a) rotational symmetry of order 3 and three lines of symmetry

b) a quadrilateral with no rotational symmetry but one line of symmetry

c) a quadrilateral with rotational symmetry of order 2 but no line of symmetry

STATISTICS AND PROBABILITY

31 HANDLING DATA

Collect, classify and tabulate statistical data; read, interpret and draw inferences from tables and statistical diagrams; construct and use bar charts, pie charts, pictograms, simple frequency distributions and histograms with equal intervals; calculate mean, median, mode and range, and understand the purposes for which they are used; draw and interpret scatter diagrams and understand positive, negative and zero correlation.

Tally charts and frequency tables

Worked example The figures in the list below give the total number of chocolate buttons in each of twenty packets of buttons.

35 36 38 37 35 36 38 36 37 35 36 36 38 36 35
38 37 38 36 38

The figures can be shown on a tally chart:

Number	Tally	Frequency
35	IIII	4
36	~~IIII~~ II	7
37	III	3
38	~~IIII~~ I	6

When the tallies are added up to get the frequency, the chart is usually called a **frequency table**. The information can then be displayed in a variety of ways.

Pictogram

@ = 1 packet of chocolate buttons

Buttons per packet	
35	@@@@
36	@@@@@@@
37	@@@
38	@@@@@@

Block graph

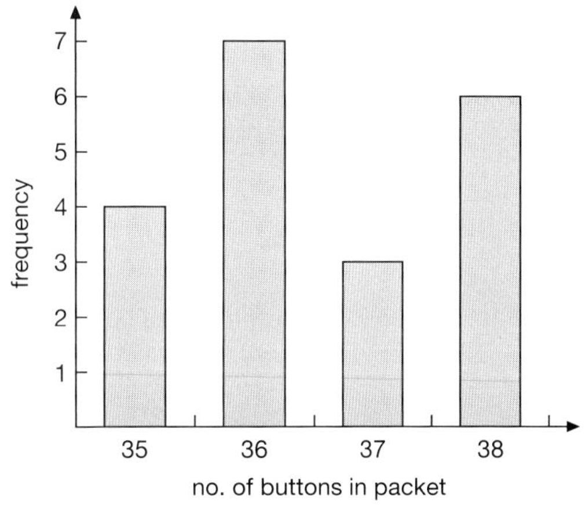

Exercise 31.1

1. A group of students was asked to say which animal they would most like to have as a pet. The results are listed below:

   dog cat cat fish cat hamster dog cat hamster
   horse cat dog dog dog cat snake fish hamster
   horse cat dog cat cat dog hamster cat fish
   dog

   Make a frequency table of the data collected, then display the information on either a block graph or pictogram.

2. The colours of cars parked in a station car park are listed below:

   red blue green white red white red blue white
   red white green white red black red white
   blue red blue blue red white white white red
   white red white white red blue green

   Complete a frequency table of the data, and then display the information on a pictogram.

3. A survey is carried out among a group of students to find their favourite subject at school. The results are listed below:

   Art Maths Science English Maths Art English
   Maths English Art Science Science Science
   Maths Art English Art Science Maths English
   Art

   Complete a frequency table for this data, and illustrate it on a block graph.

If there is a big range in the data it is easier to group the data in a **grouped frequency table**.

The groups are arranged so that no score can appear in two groups.

Worked example The scores for the first round of a golf competition are shown below.

71 75 82 96 83 75 76 82 103 85 79 77 83 85
88 104 76 77 79 83 84 86 88 102 95 96 99 102
75 72

This data can be grouped as shown:

Score	Frequency
71–75	5
76–80	6
81–85	8
86–90	3
91–95	1
96–100	3
100–105	4
Total	30

Note: it is not possible to score 70.5 or 77.3 at golf. The scores are said to be **discrete**. If the data is **continuous**, for example when measuring time, the intervals can be shown as 0–, 10–, 20–, 30– and so on.

Histograms

A histogram displays the frequency of either continuous or grouped discrete data in the form of bars. There are several important features of a histogram which distinguish it from a bar chart.

- The bars are joined together.
- The bars can be of varying width.
- The frequency of the data is represented by the area of the bar and not the height (though in the case of bars of equal width, the area is directly proportional to the height of the bar and so the height is usually used as the measure of frequency).

Worked example The table (top of opposite page) shows the marks out of 100 in a maths test for a class of 32 students. Draw a histogram representing this data.

All the class intervals are the same. As a result, the bars of the histogram will all be of equal width, and the frequency can be plotted on the vertical axis.

Test marks	Frequency
1–10	0
11–20	0
21–30	1
31–40	2
41–50	5
51–60	8
61–70	7
71–80	6
81–90	2
91–100	1

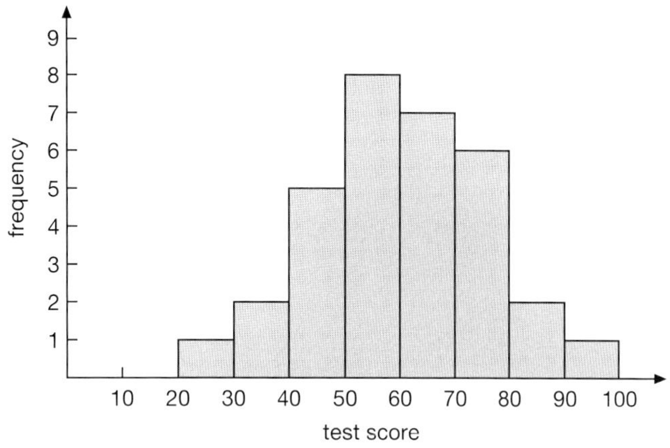

Exercise 31.2

1. The following data record the percentage scores obtained by students in an IGCSE maths examination.

   35 42 48 53 67 69 52 54 73 46 62 59 71 65
   41 55 62 78 69 56 46 53 63 67 72 83 61 53
   92 65 45 52 61 59 75 53 57 68 72 64 63 85
   69 64

   Make a grouped frequency table with values 30–39, 40–49, etc., and illustrate the results on a histogram.

2. The numbers of apples collected from fifty trees are recorded below:

   35 78 15 65 69 32 12 9 89 110 112 148 98
   67 45 25 18 23 56 71 62 46 128 7 133 96
   24 38 73 82 142 15 98 6 123 49 85 63 19
   111 52 84 63 78 12 55 138 102 53 80

   Make a grouped frequency table with the values 0–9, 10–19, 20–29, etc. Illustrate this data on a histogram.

3. The number of cars coming off an assembly line per day is recorded over a thirty-day period. The results are below:

   15 26 34 53 64 9 32 49 21 15 7 58 63 48
   36 29 41 20 27 30 51 63 39 43 60 38 19 8
   35 10

   Make a grouped frequency table with class intervals 0–9, 10–19, 20–29, etc. Illustrate this data on a histogram.

4. The heights of 50 basketball players attending a tournament are recorded in the grouped frequency table below. Note that "1.8–" means $1.8 \leqslant H < 1.9$.

Height (m)	Frequency
1.8–	2
1.9–	5
2.0–	10
2.1–	22
2.2–	7
2.3–2.4	4

Illustrate this data on a histogram.

5. The number of hours of overtime worked by employees at a factory over a period of a month is given in the table below.

Hours of overtime	Frequency
0–	12
10–	18
20–	22
30–	64
40–	32
50–60	20

Illustrate this data on a histogram.

6. The lengths of the index fingers of 30 students were measured. The results were recorded and are shown in a grouped frequency table below.

Length (cm)	Frequency
5.0–	3
5.5–	8
6.0–	10
6.5–	7
7.0–7.5	2

Illustrate this data on a histogram.

Surveys (1)

A survey requires data to be collected, organised, analysed and presented.

A survey may be carried out for interest's sake, for example to find out how many cars pass your school in an hour.
A survey could be carried out to help future planning – information about traffic flow could lead to the building of new roads, or the placing of traffic lights or a pedestrian crossing.

Below is a suggested list of subjects on which you could base a survey.

1. The number of people in cars/buses passing your school.

2. The number/type of pens or pencils in the possession of students in your class.

3. The number of CDs/videos/books owned by students in your class.

4. The number of rooms in the homes of students in your class.

5. The number of family members living in the homes of students in your class.

6. The number of words on a page of a book in the possession of each student in your class.

7. The amount of advertising in various newspapers and magazines.

Pie charts

Data can be displayed on a **pie chart** – a circle divided into sectors. The size of the sector is in direct proportion to the frequency of the data. The sector size does not show the actual frequency. The actual frequency can be calculated easily from the size of the sector.

Worked examples

a)

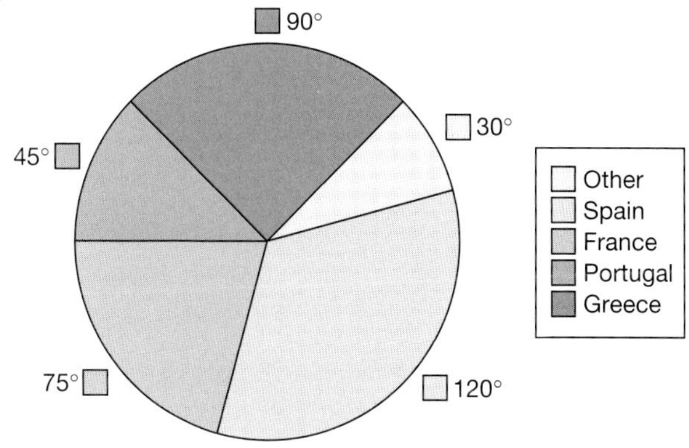

A survey of 240 English children asked them to vote for their favourite holiday destination. The results are shown on the pie chart above (previous page). Calculate the actual number of votes for each destination.

The total 240 votes are represented by 360°.
It follows that if 360° represents 240 votes:

There were $240 \times (\frac{120}{360})$ votes for Spain
so, 80 votes for Spain.
There were $240 \times (\frac{75}{360})$ votes for France
so, 50 votes for France.
There were $240 \times (\frac{45}{360})$ votes for Portugal
so, 30 votes for Portugal.
There were $240 \times (\frac{90}{360})$ votes for Greece
so, 60 votes for Greece.
Other destinations received $240 \times (\frac{30}{360})$ votes
so, 20 votes for other destinations.

Note: it is worth checking your result by adding them:

$80 + 50 + 30 + 60 + 20 = 240$ total votes

b) The table shows the percentage of votes cast for various political parties in an election. If a total of 5 million votes were cast, how many votes were cast for each party?

Party	Percentage of vote
Social Democrats	45%
Liberal Democrats	36%
Green Party	15%
Others	4%

The Social Democrats received $\frac{45}{100} \times 5$ million votes
so, 2.25 million votes.
The Liberal Democrats received $\frac{36}{100} \times 5$ million votes
so, 1.8 million votes.
The Green Party received $\frac{15}{100} \times 5$ million votes
so, 750 000 votes.
Other parties received $\frac{4}{100} \times 5$ million votes
so, 200 000 votes.

Check total:

$2.25 + 1.8 + 0.75 + 0.2 = 5$ (million votes)

c) The table (left) shows the results of a survey among 72 students to find their favourite sport. Display this data on a pie chart.

Sport	Frequency
Football	35
Tennis	14
Volleyball	10
Hockey	6
Basketball	5
Other	2

72 votes are represented by 360°, so 1 vote is represented by $\frac{360}{72}$ degrees. Therefore the size of each sector can be calculated as shown:

Football	$35 \times \frac{360}{72}$ degrees	i.e., 175°
Tennis	$14 \times \frac{360}{72}$ degrees	i.e., 70°
Volleyball	$10 \times \frac{360}{72}$ degrees	i.e., 50°
Hockey	$6 \times \frac{360}{72}$ degrees	i.e., 30°
Basketball	$5 \times \frac{360}{72}$ degrees	i.e., 25°
Other sports	$2 \times \frac{360}{72}$ degrees	i.e., 10°

Check total:

$$175 + 70 + 50 + 30 + 25 + 10 = 360$$

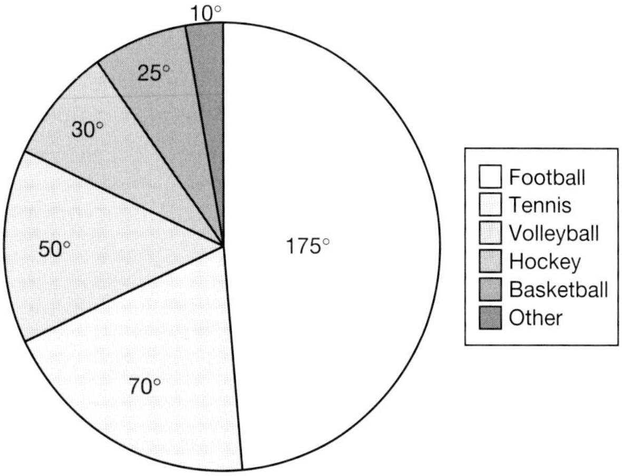

Exercise 31.3

1. The pie charts below show how a girl and her brother spent one day. Calculate how many hours they spent on each activity. The diagrams are to scale.

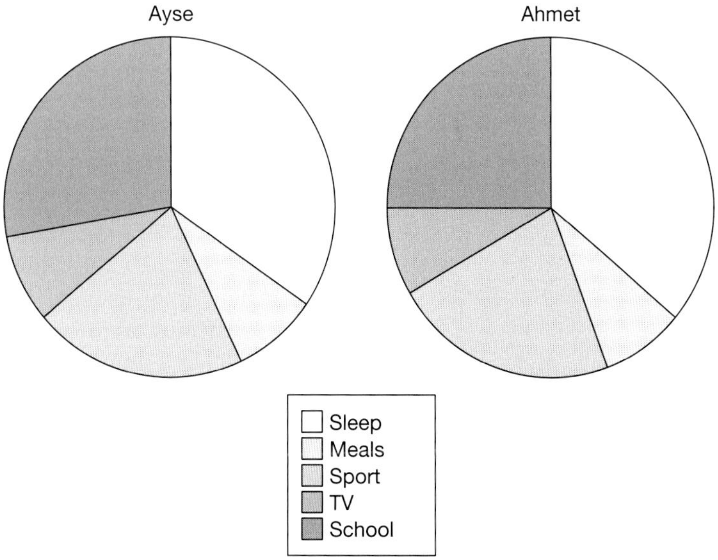

2. A survey was carried out among a class of 40 students. The question asked was, 'How would you spend a gift of £15?'. The results are shown below:

Choice	Frequency
CDs	14
Books	6
Clothes	18
Cinema	2

Illustrate these results on a pie chart.

3. A student works during the holidays. He earns a total of £2400. He estimates that the money earned has been used as follows: clothes $\frac{1}{3}$, transport $\frac{1}{5}$, entertainment $\frac{1}{4}$. He has saved the rest.

Calculate how much he has spent on each category, and illustrate this information on a pie chart.

4. The table below shows the number of passengers (in millions) arriving at four British airports in one year. Show this information on a pie chart.

Heathrow	27 million
Gatwick	6 million
Stansted	4.5 million
Luton	2.5 million

5. A girl completes a survey on the number of advertisements on each page of a 240-page magazine:

None – 15% One – 35% Two – 20%
Three – 22% More than three – 8%

a) Illustrate this information on a pie chart.
b) How many pages had no advertisements?

6. 300 students in a school of 1000 students play one musical instrument. Of these, 48% play the piano, 32% play the guitar, 12% play the violin and 5% play the drums. The rest play other instruments.

a) Illustrate this information on a pie chart.
b) Calculate how many students play the guitar.
c) Calculate the percentage of the students in the whole school who play the piano.

7. A research project looking at the careers of men and women in Spain produced the following results:

	Male (percentage)	Female (percentage)
Clerical	22	38
Professional	16	8
Skilled craft	24	16
Non-skilled craft	12	24
Social	8	10
Managerial	18	4

a) Illustrate this information on a pie chart, and make two statements that could be supported by the data.
b) If there are eight million women in employment in Spain, calculate the number in either professional or managerial employment.

Surveys (2)

Exercise 31.4

1. Below are a number of statements, some of which you may have heard or read before.
   Conduct a survey to collect data which will support or refute one of the statements. Where possible, use pie charts to illustrate your results.
   a) Women's magazines are full of adverts.
   b) If you go to a football match you are lucky to see more than one goal scored.
   c) Every other car on the road is white.
   d) Girls are not interested in sport.
   e) Children today do nothing but watch TV.
   f) Newspapers have more sport than news in them.
   g) Most girls want to be nurses, teachers or secretaries.
   h) Nobody walks to school any more.
   i) Nearly everybody has a computer at home.
   j) Most of what is on TV comes from America.

2. Overleaf are some instructions relating to a washing machine in English, French, German, Dutch and Italian.
   Analyse the data and write a report. You may wish to comment upon:
   i) the length of words in each language,
   ii) the frequency of letters of the alphabet in different languages.

ENGLISH

ATTENTION
Do not interrupt drying during the programme.
This machine incorporates a temperature safety thermostat which will cut out the heating element and **switch the programme selector to an off position** in the event of a water blockage or power failure. This can also occur if a drying programme is interrupted for **more than** 20 seconds by the opening of the door. In this event wait approximately 2 minutes for the thermostat to reset itself before selecting a further drying time.

FRENCH

ATTENTION
N'interrompez pas le séchage en cours de programme.
Une panne d'électricité ou un manque d'eau momentanés peuvent annuler le programme de séchage en cours. Dans ces cas arrêtez l'appareil, affichez de nouveau le programme et après remettez l'appareil en marche.

Pour d'ultérieures informations, rapportez-vous à la notice d'utilisation.

GERMAN

ACHTUNG
Die Trocknung soll nicht nach Anlaufen des Programms unterbrochen werden.
Ein kurzer Stromausfall bzw. Wassermangel kann das laufende Trocknungsprogramm annullieren. In diesem Falle Gerät ausschalten, Programm wieder einstellen und Gerät wieder einschalten.

Für nähere Angaben beziehen Sie sich auf die Bedienungsanleitung.

DUTCH

BELANGRIJK
Het droogprogramma niet onderbreken wanneer de machine in bedrijf is.
Door een korte stroom-of watertoevoeronderbreking kan het droogprogramma geannuleerd worden. Schakel in dit geval de machine uit, maak opnieuw uw programmakeuze en stel onmiddellijk weer in werking.

Verdere inlichtingen vindt u in de gebruiksaanwijzing.

ITALIAN

ATTENZIONE
Non interrompere l'asciugatura quando il programma è avviato.
La macchina è munita di un dispositivo di sicurezza che può annullare il programma di asciugaturea in corso quando si verifica una temporanea mancanza di acqua o di tensione.

In questi casi si dovrà spegnere la macchina, reimpostare il programma e poi riavviare la macchina.

Per ulteriori indicazioni, leggere il libretto istruzioni.

Average

'Average' is a word which in general use is taken to mean somewhere in the middle. For example, a woman may describe herself as being of average height. A student may think he or she is of average ability in maths. Mathematics is more exact and uses three principal methods to measure average.

- The **mode** is the value occurring the most often.
- The **median** is the middle value when all the data is arranged in order of size.
- The **mean** is found by adding together all the values of the data and then dividing the total by the number of data values.

Worked examples **a)** Find the mean, median and mode of the data listed below.

1, 0, 2, 4, 1, 2, 1, 1, 2, 5, 5, 0, 1, 2, 3

$$\text{Mean} = \frac{1+0+2+4+1+2+1+1+2+5+5+0+1+2+3}{15}$$

$= 2$

Arranging all the data in order and then picking out the middle number gives the median:

0, 0, 1, 1, 1, 1, 1, (2), 2, 2, 3, 4, 5, 5

The mode is the number which appeared most often. Therefore the mode is 1.

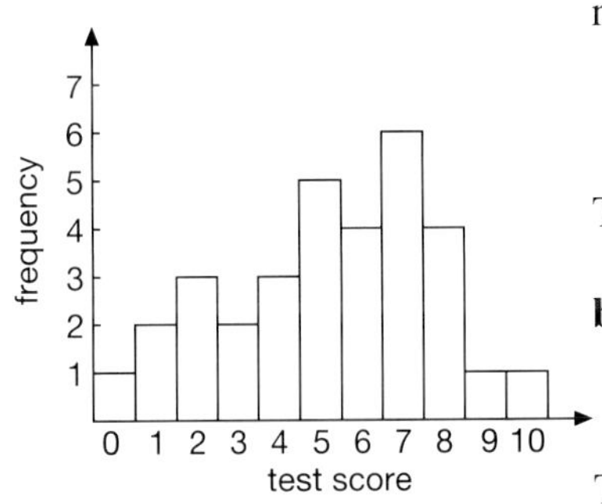

b) The frequency chart (left) shows the score out of 10 achieved by a class in a maths test.

Calculate the mean, median and mode for this data.

Transferring the results to a frequency table gives:

Test score	0	1	2	3	4	5	6	7	8	9	10	Total
Frequency	1	2	3	2	3	5	4	6	4	1	1	32
Frequency × score	0	2	6	6	12	25	24	42	32	9	10	168

By looking at the total column we can ascertain the number of students taking the test, i.e. 32, and also the total number of marks obtained by all the students, i.e. 168.

Therefore the mean score $= \frac{168}{32} = 5.25$

Arranging all the scores in order gives:

0, 1, 1, 2, 2, 2, 3, 3, 4, 4, 4, 5, 5, 5, 5, (5, 6,)
6, 6, 6, 7, 7, 7, 7, 7, 7, 8, 8, 8, 8, 9, 10

Because there is an even number of students there isn't one middle number. There is a middle pair. The

median is $\frac{(5 + 6)}{2} = 5.5$

The mode is 7 as it is the score which occurs most often.

■ The **range** of a set of numbers is found by subtracting the smallest value from the largest.

Exercise 31.5

1. Calculate the mean and range of each of the following sets of numbers:
   a) 6 7 8 10 11 12 13
   b) 4 4 6 6 6 7 8 10
   c) 36 38 40 43 47 50 55
   d) 7 6 8 9 5 4 10 11 12
   e) 12 24 36 48 60
   f) 17.5 16.3 18.6 19.1 24.5 27.8

2. Find the median and range of each of the following sets of numbers:
   a) 3 4 5 6 7
   b) 7 8 8 9 10 12 15
   c) 8 8 8 9 9 10 10 10 10
   d) 6 4 7 3 8 9 9 4 5
   e) 2 4 6 8
   f) 7 8 8 9 10 11 12 14
   g) 3.2 7.5 8.4 9.3 5.4 4.1 5.2 6.3
   h) 18 32 63 16 97 46 83

3. Find the mode and range of each of the following sets of numbers:
   a) 6 7 8 8 9 10 11
   b) 3 4 4 5 5 6 6 6 7 8 8
   c) 3 5 3 4 6 3 3 5 4 6 8
   d) 4 3 4 5 3 4 5 4
   e) 60 65 70 75 80 75
   f) 8 7 6 5 8 7 6 5 8

Exercise 31.6

In Q.1–5, find the mean, median and mode for each set of data.

1. A hockey team plays 15 matches. Below is a list of the number of goals scored in each match.

   1, 0, 2, 4, 0, 1, 1, 1, 2, 5, 3, 0, 1, 2, 2

2. The total scores when two dice are thrown 20 times are:

   7, 4, 5, 7, 3, 2, 8, 6, 8, 7, 6, 5, 11, 9, 7, 3, 8, 7, 6, 5

3. The ages of a group of girls are:

   14 years 3 months, 14 years 5 months,
   13 years 11 months, 14 years 3 months,
   14 years 7 months, 14 years 3 months, 14 years 1 month

4. The numbers of students present in a class over a three-week period are:

28, 24, 25, 28, 23, 28, 27, 26, 27, 25, 28, 28, 28, 26, 25

5. An athlete keeps a record in seconds of her training times for the 100 m race:

14.0, 14.3, 14.1, 14.3, 14.2, 14.0, 13.9, 13.8, 13.9, 13.8, 13.8, 13.7, 13.8, 13.8, 13.8

6. The mean mass of the 11 players in a football team is 80.3 kg. The mean mass of the team plus a substitute is 81.2 kg. Calculate the mass of the substitute.

7. After eight matches a basketball player had scored a mean of 27 points. After three more matches his mean was 29. Calculate the total number of points he scored in the last three games.

Exercise 31.7

1. An ordinary dice was rolled 60 times. The results are shown in the table below. Calculate the mean, median and mode of the scores.

Score	1	2	3	4	5	6
Frequency	12	11	8	12	7	10

2. Two dice were thrown 100 times. Each time their combined score was recorded. Below is a table of the results. Calculate the mean score.

Score	2	3	4	5	6	7	8	9	10	11	12
Frequency	5	6	7	9	14	16	13	11	9	7	3

3. Sixty flowering bushes are planted. At their flowering peak, the number of flowers per bush is counted and recorded. The results are shown in the table below.

Flowers per bush	0	1	2	3	4	5	6	7	8
Frequency	0	0	0	6	4	6	10	16	18

a) Calculate the mean, median and mode of the number of flowers per bush.

b) Which of the mean, median and mode would be most useful when advertising the bush to potential buyers?

Scatter graphs

Scatter graphs are particularly useful if we wish to see if there is a **correlation** (relationship) between two sets of data. The two values of data collected represent the coordinates of each point plotted. How the points lie when plotted indicates the type of relationship between the two sets of data.

Worked example The heights and weights (masses) of 20 children under the age of five were recorded. The heights were recorded in centimetres and the weights in kilograms. The data is shown below.

Height	32	34	45	46	52
Weight	5.834	3.792	9.037	4.225	10.149
Height	59	63	64	71	73
Weight	6.188	9.891	16.010	15.806	9.929
Height	86	87	95	96	96
Weight	11.132	16.443	20.895	16.181	14.000
Height	101	108	109	117	121
Weight	19.459	15.928	12.047	19.423	14.331

a) Plot a scatter graph of the above data.
b) Comment on any relationship you see.
c) If another child was measured as having a height of 80 cm, approximately what weight would you expect him or her to be?

a)

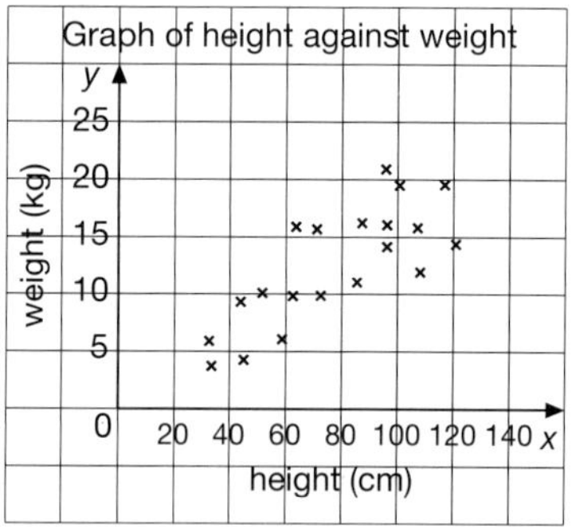

b) The points tend to lie in a diagonal direction from bottom left to top right. This suggests that as height increases then, in general, weight increases too. Therefore there is a **positive correlation** between height and weight.

c) We assume that this child will follow the trend set by the other 20 children. To deduce an approximate value for the weight, we draw a **line of best fit**. This is a solid straight line which passes through the points as closely as possible, as shown (left).

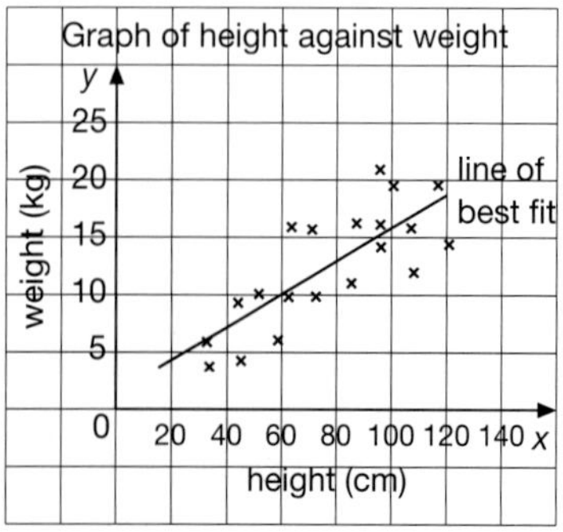

The line of best fit can now be used to give an approximate solution to the question. If a child has a height of 80 cm, you would expect his or her weight to be in the region of 13 kg.

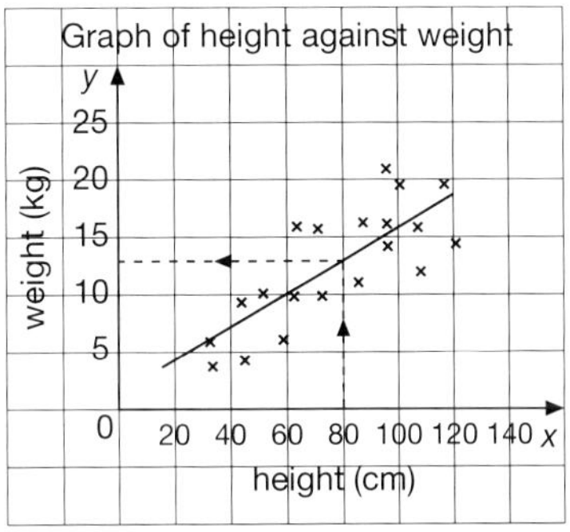

Types of correlation

There are several types of correlation, depending on the arrangement of the points plotted on the scatter graph. These are described below.

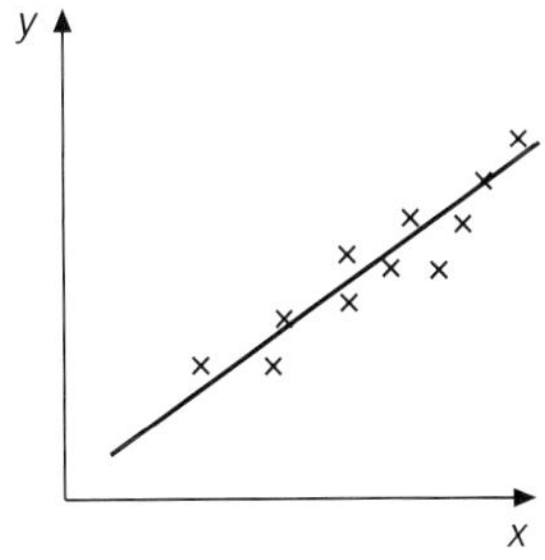

A **strong positive correlation** between the variables x and y. The points lie very close to the line of best fit.
As x increases, so does y.

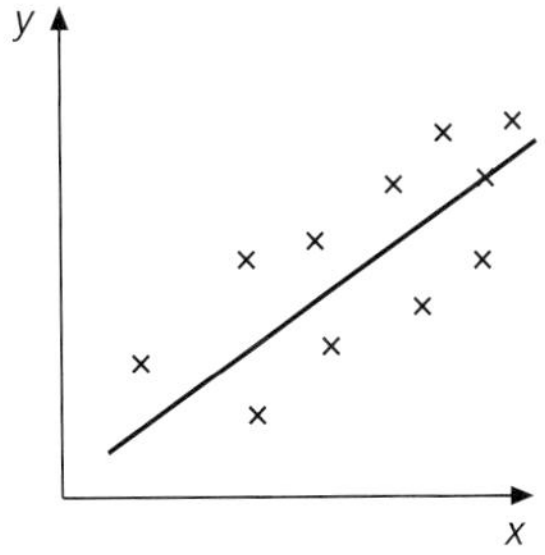

A **weak positive correlation**. Although there is direction to the way the points are lying, they are not tightly packed around the line of best fit.
As x increases, y tends to increase too.

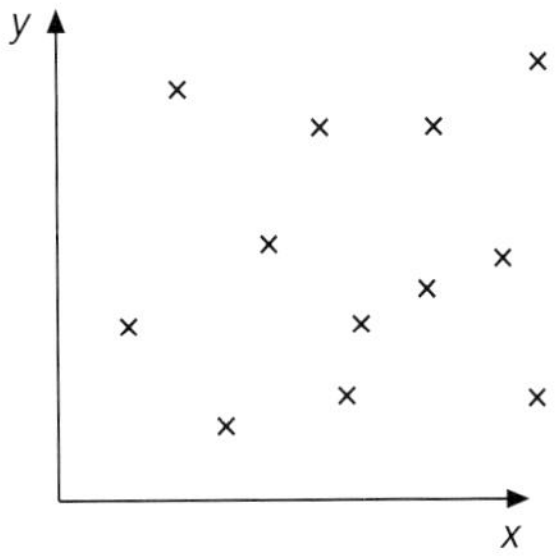

No correlation. As there is no pattern to the way in which the points are lying, there is no correlation between the variables x and y. As a result there can be no line of best fit.

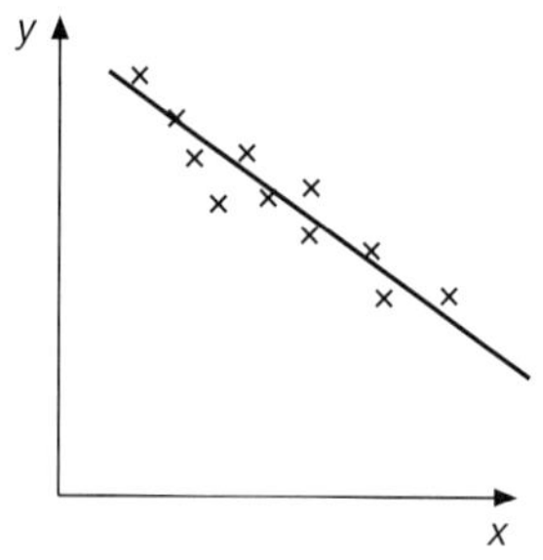

A **strong negative correlation**. The points lie close around the line of best fit.
As x increases, y decreases.

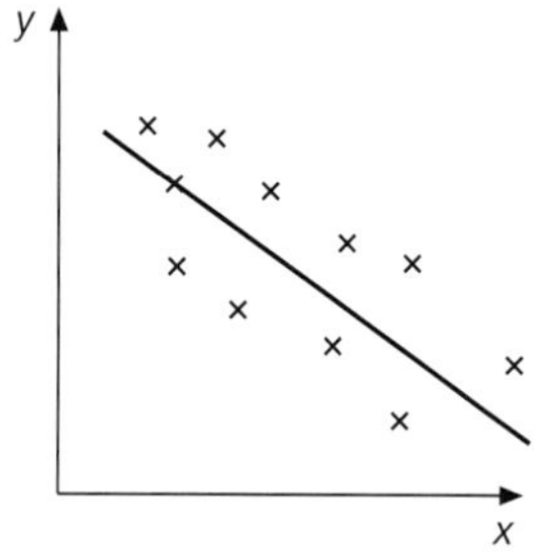

A **weak negative correlation**. The points are not tightly packed around the line of best fit.
As x increases, y tends to decrease.

Exercise 31.8

1. State what type of correlation you might expect, if any, if the following data was collected and plotted on a scatter graph. Give reasons for your answer.

a) A student's score in a maths exam and their score in a science exam.
b) A student's hair colour and the distance they have to travel to school.
c) The outdoor temperature and the number of cold drinks sold by a shop.
d) The age of a motorcycle and its second-hand selling price.
e) The number of people living in a house and the number of rooms the house has.
f) The number of goals your opponents score and the number of times you win.
g) A child's height and the child's age.
h) A car's engine size and its fuel consumption.

2. A newspaper gives daily readings for the number of hours of sunshine and the amount of rainfall in millimetres for several places in the UK. The table below is a summary.

Place	Hours of sunshine	Rainfall (mm)
Aberdeen	9.8	0.3
Aviemore	2.8	0
Belfast	4.1	6.6
Birmingham	6.4	0.8
Bournemouth	9.3	4.3
Bristol	10.2	1.8
Cardiff	10.8	2.8
Folkestone	10.5	0
Hastings	7.2	0.3
Isle of Man	9.3	2.5
Isle of Wight	10.7	0.3
London	10.1	1.5
Manchester	4.6	2.0
Margate	10.9	0
Newcastle	8.2	0.5
Newquay	4.1	2.3
Oxford	10.1	3.0
Scarborough	10.6	1.8
Skegness	12.0	3.3
Southport	8.2	7.4
Torquay	8.8	1.5

a) Plot a scatter graph of hours of sunshine against amount of rainfall. Use a spreadsheet if possible.

b) What type of correlation, if any, is there between the two variables? Comment on whether this is what you would expect.

3. The United Nations keeps an up-to-date database of statistical information on its member countries.
The table below shows some of the information available.

Country	Life expectancy at birth (years, 1990–99)		Adult illiteracy rate (%, 1995)	Infant mortality rate (per 1000 births, 1990–99)
	Female	Male		
Australia	81	76	0	6
Barbados	79	74	2.6	12
Brazil	71	63	16.8	42
Chad	49	46	51.9	112
China	72	68	18.5	41
Colombia	74	67	9.6	30
Congo	51	46	25.6	90
Cuba	78	74	4.4	9
Egypt	68	65	48.9	51
France	82	74	0	6
Germany	80	74	0	5
India	63	62	48	72
Iraq	64	61	42	95
Israel	80	76	4.9	8
Japan	83	77	0	4
Kenya	53	51	22.7	66
Mexico	76	70	10.5	31
Nepal	57	58	64.1	83
Portugal	79	72	10	9
Russian Federation	73	61	0.9	18
Saudi Arabia	73	70	27.8	23
United Kingdom	80	75	0	7
United States of America	80	73	0	7

a) By plotting a scatter graph, decide if there is a correlation between the adult illiteracy rate and the infant mortality rate.
b) Are your findings in part a) what you expected? Explain your answer.
c) Without plotting a graph, decide if you think there is likely to be a correlation between male and female life expectancy at birth. Explain your reasons.
d) Plot a scatter graph to test if your predictions for part c) were correct.

Student Assessment 1

1. A supermarket manager notes the number of delivery vans which arrive each day:

Day	Mon	Tue	Wed	Thurs	Fri
No. of vans	6	12	9	33	36

Display this information on a suitable pictogram.

2. The areas of the countries that form the United Kingdom are shown below. Illustrate this data as a bar chart.

Country	England	Scotland	Northern Ireland	Fri
Area in 1000 km^2	130	75	15	20

3. Display the data in Q.2 on a pie chart.

4. Find the mean, median, mode and range of the following sets of data:
   a) 63 72 72 84 86
   b) 6 6 6 12 18 24
   c) 5 10 5 15 5 20 5 25 15 10

5. The mean mass of the 15 players in a rugby team is 85 kg. The mean mass of the team plus a substitute is 83.5 kg. Calculate the mass of the substitute.

Student Assessment 2

1. A building company spends four months working on a building site. The number of men employed each month is as follows:

Month	July	August	September	October
No. of men employed	40	55	85	20

Display this information on a suitable pictogram.

2. The table below shows the population (in millions) of the continents:

Continent	Asia	Europe	America	Africa	Oceania
Population (millions)	3000	800	650	500	50

Display this information on a pie chart.

3. Display the data in Q.2 on a bar chart.

4. Find the mean, median, mode and range of the following sets of data:
   a) 4 5 5 6 7
   b) 3 8 12 18 18 24
   c) 4 9 3 8 7 11 3 5 3 8

5. An ordinary dice was rolled 50 times. The results are shown below. Calculate the mean, median and mode of the scores.

Score	1	2	3	4	5	6
Frequency	8	11	5	9	7	10

Student Assessment 3

1. A rugby team scores the following number of points in 12 matches:

   21, 18, 3, 12, 15, 18, 42, 18, 24, 6, 12, 3

   Calculate for the 12 matches:
   a) the mean score,
   b) the median score,
   c) the mode.

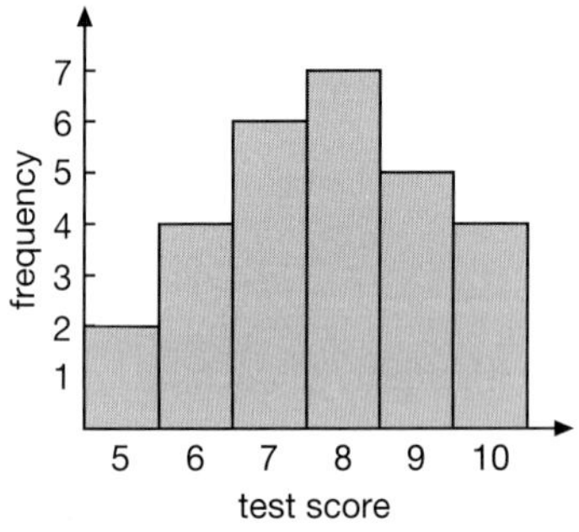

2. The bar chart (left) shows the marks out of 10 for an English test taken by a class of students.
   a) Calculate the number of students who took the test.
   b) Calculate for the class:
      i) the mean test result,
      ii) the median test result,
      iii) the mode test result.

3. Illustrate the data in Q.2 on a pie chart.

4. Fifty sacks of grain are weighed as they are unloaded from a truck. The mass of each is recorded in this grouped frequency table (below).

Mass (kg)	**Frequency**
$15 \leqslant M < 16$	0
$16 \leqslant M < 17$	3
$17 \leqslant M < 18$	6
$18 \leqslant M < 19$	14
$19 \leqslant M < 20$	18
$20 \leqslant M < 21$	8
$21 \leqslant M < 22$	1

Draw a histogram of this data.

5. Athletico football team keeps a record of attendance at their matches over a season. The attendances for the games are listed below:

18 418	16 161	15 988	13 417	12 004
11 932	8 461	10 841	19 000	19 214
16 645	14 782	17 935	8 874	19 023
19 875	16 472	14 840	18 450	16 875
13 012	17 858	19 098	6 972	8 452
7 882	11 972	16 461	11 311	19 458

a) Copy and complete the table below.

Attendance	Tally	Frequency
0–3999		
4000–7999		
8000–11 999		
12 000–15 999		
16 000–19 999		

b) Illustrate the above table using a histogram.

6. The table below gives the average time taken for 30 pupils in a class to get to school each morning, and the distance they live from the school.

Distance (km)	2	10	18	15	3	4	6	2	25	23	3	5	7	8	2
Time (min)	5	17	32	38	8	14	15	7	31	37	5	18	13	15	8
Distance (km)	19	15	11	9	2	3	4	3	14	14	4	12	12	7	1
Time (min)	27	40	23	30	10	10	8	9	15	23	9	20	27	18	4

a) Plot a scatter graph of distance travelled against time taken.
b) Describe the correlation between the two variables.
c) Explain why some pupils who live further away may get to school more quickly than some of those who live nearer.
d) Draw a line of best fit on your scatter graph.
e) A new pupil joins the class. Use your line of best fit to estimate how far away from school she might live if she takes, on average, 19 minutes to get to school each morning.

Student Assessment 4

1. A javelin thrower keeps a record of her best throws over ten competitions. These are shown in the table below.

Competition	1	2	3	4	5	6	7	8	9	10
Distance (m)	77	75	78	86	92	93	93	93	92	89

Find the mean, median and mode of her throws.

2. The bar chart below shows the marks out of 10 for a maths test taken by a class of students.
   a) Calculate the number of students who took the test.
   b) Calculate for the class:
      i) the mean test result,
      ii) the median test result,
      iii) the mode test result.

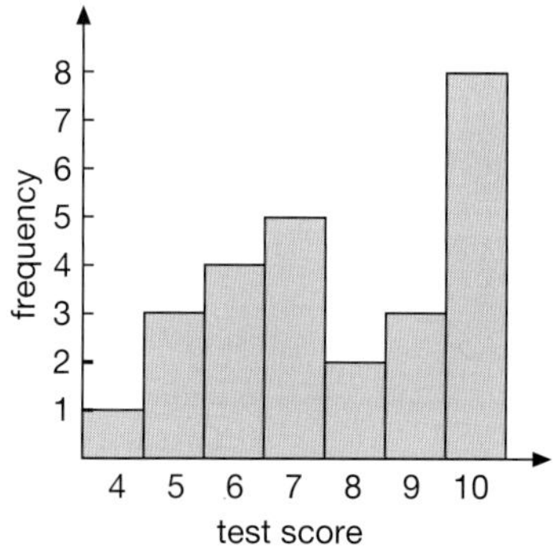

3. Illustrate on a pie chart the information shown on the bar chart in Q.2.

4. A hundred sacks of coffee with a nominal mass of 10 kg are unloaded from a train. The mass of each sack is checked and the results are presented in the table below.

Mass (kg)	Frequency
$9.8 \leqslant M < 9.9$	14
$9.9 \leqslant M < 10.0$	22
$10.0 \leqslant M < 10.1$	36
$10.1 \leqslant M < 10.2$	20
$10.2 \leqslant M < 10.3$	8

Illustrate the data on a histogram.

5. 400 students sit an IGCSE exam. Their marks (as percentages) are shown in the table below.

Mark (%)	Frequency
31–40	21
41–50	55
51–60	125
61–70	74
71–80	52
81–90	45
91–100	28

Illustrate the data on a histogram.

6. A department store decides to investigate whether there is a correlation between the number of pairs of gloves it sells and the outside temperature. Over a one-year period the store records, every two weeks, how many pairs of gloves are sold and the mean daytime temperature during the same period. The results are given in the table below.

Mean temperature (°C)	3	6	8	10	10	11	12	14	16	16	17	18	18
Number of pairs of gloves	61	52	49	54	52	48	44	40	51	39	31	43	35
Mean temperature (°C)	19	19	20	21	22	22	24	25	25	26	26	27	28
Number of pairs of gloves	26	17	36	26	46	40	30	25	11	7	3	2	0

a) Plot a scatter graph of mean temperature against number of pairs of gloves.

b) What type of correlation is there between the two variables?

c) How might this information be useful for the department store in the future?

32 PROBABILITY

Calculate the probability of a single event as a fraction or a decimal; understanding that the probability of an event occurring equals one minus the probability of it not occurring; use and understand relative frequency.

Probability is the study of chance, or the likelihood of an event happening. In this chapter we will be looking at theoretical probability. But, because probability is based on chance, what theory predicts does not necessarily happen in practice.

A **favourable outcome** refers to the event in question actually happening. The **total number of possible outcomes** refers to all the different types of outcome one can get in a particular situation. In general:

$$\text{Probability of an event} = \frac{\text{number of favourable outcomes}}{\text{total number of equally likely outcomes}}$$

If the probability = 0, the event is impossible
If the probability = 1, the event is certain to happen

Worked examples

a) An ordinary, fair dice is rolled. Calculate the probability of getting a six.

Number of favourable outcomes = 1 (i.e., getting a 6)
Total number of possible outcomes = 6
(i.e., getting a 1, 2, 3, 4, 5 or 6)
Probability of getting a six = $\frac{1}{6}$
Probability of not getting a six = $1 - \frac{1}{6} = \frac{5}{6}$

b) An ordinary, fair dice is rolled. Calculate the probability of getting an even number.

Number of favourable outcomes = 3
(i.e., getting a 2, 4 or 6)
Total number of possible outcomes = 6
(i.e., getting a 1, 2, 3, 4, 5 or 6)
Probability of getting an even number = $\frac{3}{6} = \frac{1}{2}$

Exercise 32.1

1. Draw a line in your book about 15 cm long. At the extreme left put zero, and at the extreme right put one. These represent impossible and certain events, respectively. Try to estimate the chance of the following events happening and place them on your line.

a) You will watch TV tonight.
b) You will play sport tomorrow.
c) You will miss school one day this month.
d) You will be on a plane next year.
e) You will learn a new language one day.
f) You will have a visitor at school today.

The likelihood of the events in question vary from person to person. Therefore, the probability of each event is not constant. However, the probability of some events, such as the result of throwing dice, spinning a coin or dealing cards, can be found by experiment or calculation.

Your teacher may wish you to do some probability experiments before moving on to the next section.

The line drawn in Q.1 is called a probability scale.
The scale goes from 0 to 1.
A probability of 0 means the event is impossible.
A probability of 1 means it is certain.

2. Copy the probability scale below.

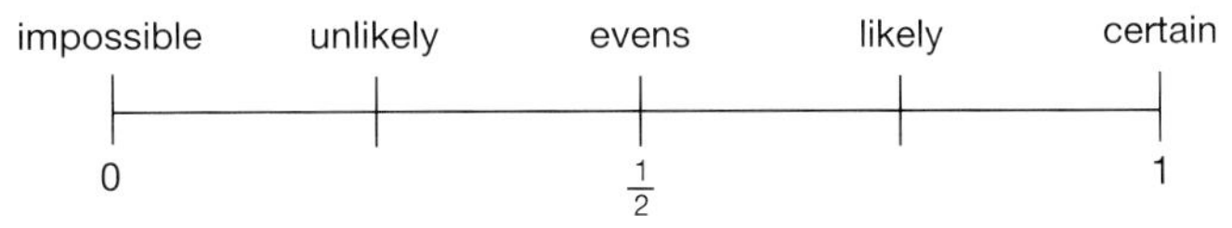

Mark on the probability scale the probability that

a) a day chosen at random is a Saturday,
b) a coin will show tails when spun,
c) the sun will rise tomorrow,
d) a woman will run a marathon in two hours,
e) the next car you see will be silver.

3. Express your chosen fractions as decimals and percentages.

Exercise 32.2

1. Calculate the theoretical probability, when rolling an ordinary, fair dice, of getting each of the following:

a) a score of 1 b) a score of 2, 3, 4, 5, 6
c) an odd number d) a score less than 6
e) a score of 7 f) a score less than 7

2. a) Calculate the probability of:
i) being born on a Wednesday,
ii) not being born on a Wednesday.
b) Explain the result of adding the answers to a) i) and ii) together.

3. 250 tickets are sold for a raffle. What is the probability of winning if you buy:

a) 1 ticket b) 5 tickets
c) 250 tickets d) 0 tickets?

4. In a class there are 25 girls and 15 boys. The teacher takes in all of their books in a random order. Calculate the probability that the teacher will
a) mark a book belonging to a girl first,
b) mark a book belonging to a boy first.

5. Tiles, each lettered with one different letter of the alphabet, are put into a bag. If one tile is drawn out at random, calculate the probability that it is:
a) an A or P
b) a vowel
c) a consonant
d) an X, Y or Z
e) a letter in your first name.

6. A boy was late for school 5 times in the previous 30 school days. If tomorrow is a school day, calculate the probability that he will arrive late.

7. a) Three red, 10 white, 5 blue and 2 green counters are put into a bag. If one is picked at random, calculate the probability that it is:
i) a green counter
ii) a blue counter.
b) If the first counter taken out is green and it is not put back into the bag, calculate the probability that the second counter picked is:
i) a green counter
ii) a red counter.

8. A roulette wheel has the numbers 0 to 36 equally spaced around its edge. Assuming that it is unbiased, calculate the probability on spinning it of getting:
a) the number 5
b) not 5
c) an odd number
d) zero
e) a number greater than 15
f) a multiple of 3
g) a multiple of 3 or 5
h) a prime number.

9. The letters R, C and A can be combined in several different ways.
a) Write the letters in as many different combinations as possible.

If a computer writes these three letters at random, calculate the probability that:
b) the letters will be written in alphabetical order,
c) that the letter R is written before both the letters A and C,
d) that the letter C is written after the letter A,
e) the computer will spell the word CART if the letter T is added.

10. A normal pack of playing cards contains 52 cards. These are made up of four suits (hearts, diamonds, clubs and spades). Each suit consists of 13 cards. These are labelled Ace, 2, 3, 4, 5, 6, 7, 8, 9, 10, Jack, Queen and King. The hearts and diamonds are red; the clubs and spades are black.

If a card is picked at random from a normal pack of cards, calculate the probability of picking:

a) a heart b) not a heart
c) a 4 d) a red King
e) a Jack, Queen or King f) the ace of spades
g) an even numbered card h) a 7 or a club.

Exercise 32.3

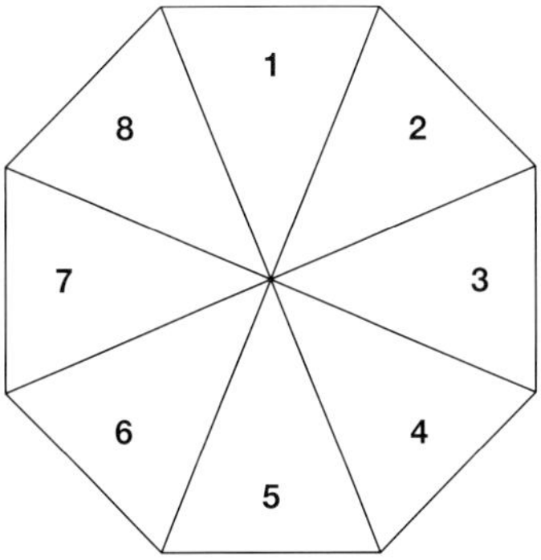

1. If you throw a fair dice, what is the probability of throwing the following?

a) a 2 b) not a 2 c) an 8
d) an even number e) a prime number f) a factor of 12

2. If you spin a coin, what is the probability of getting a head?

3. The spinner shown (left) is made in the shape of a regular octagon. What is the probability of spinning:

a) a 4 b) not a 4
c) less than 9 d) more than 3
e) an odd number f) a factor of 16
g) a prime number h) a factor of 60?

4. A normal pack of playing cards has 52 cards. There are four suits: hearts, clubs, diamonds and spades. Each suit has cards numbered 1 (ace) to 10, and three picture cards – Jack, Queen and King. Given such a pack, what is the probability of drawing each of the following cards?

a) an ace b) a 7
c) a king d) a picture card
e) a club f) a red card
g) the 9 of spades h) a club picture card
i) a card greater than 9 j) a red card less than 4
k) a red 6 or a spade l) a queen or a diamond
m) a picture card or a heart n) an ace, king or diamond
o) any odd numbered heart

5. 200 tickets are sold for a raffle. What is the probability of winning first prize if you buy the following numbers of tickets?

a) 1 b) 10 c) 15 d) 25
e) 100 f) 120 g) 150 h) 200

Relative Frequency

A football referee always used a special coin. He noticed that out of the last twenty matches the coin had come down heads far more often than tails. He wanted to know if the coin was fair, that is, if it was as likely to come down heads as tails.

He decided to do a simple experiment by spinning the coin lots of times. His results are shown below:

Number of trials	Number of heads	Relative frequency
100	40	0.4
200	90	0.45
300	142	
400	210	
500	260	
600	290	
700	345	
800	404	
900	451	
1000	499	

$$\text{The relative frequency} = \frac{\text{number of successful trials}}{\text{Total number of trials}}$$

In the "long run" that is after a large number of trials, did the coin appear to be fair?

Notice that the greater the number of trials the better the estimated probability or relative frequency is likely to be. The key idea is that increasing the number of trials gives a better estimate of the probability and the closer the result obtained by experiment will be to that obtained by calculation.

Exercise 32.4

1. Copy and complete the table above. Draw a graph with Relative frequency as the y-axis and Number of trials as the x-axis. What do you notice?
2. Conduct a similar experiment using a dice to see if the number of sixes you get is the same as the theory of probability would make you expect.
3. Make a hexagonal spinner. Conduct an experiment to see if it is fair.
4. Ask a friend to put some coloured beads in a bag. Explain how you could use relative frequency in an experiment to find out how many of each colour of beads is in the bag.

Worked examples

a) 250 tickets are sold for a raffle. A girl calculates that the tickets bought by her family give them a 0.032 probability of winning first prize. How many tickets did the family buy?

$$P = \frac{\text{no. of favourable results (F)}}{\text{no. of possible results}}$$

$$\text{so } 0.032 = \frac{F}{250}$$

$$250 \times 0.032 = F$$

$$8 = F$$

The family bought 8 tickets.

b) A man has 8 tickets for a raffle. His son knows how many tickets have been sold, and tells his father that he has a probability of 0.016 of winning the first prize. How many tickets have been sold?

$$\text{Probability} = \frac{\text{number of favourable results}}{\text{number of possible results (T)}}$$

$$0.016 = \frac{8}{T}$$

$$T = \frac{8}{0.016}$$

$$T = 500$$

So 500 tickets have been sold.

6. A boy calculates that he has a probability of 0.08 of winning the first prize in a raffle. If 500 tickets are sold, how many has he bought?

7. The probability of winning first prize in a spinner game is given as 0.04 for each number chosen. How many numbers are there on the spinner?

8. A bag contains 7 red counters, 5 blue, 3 green and 1 yellow. If one counter is drawn, what is the probability that it is:
a) yellow b) red c) blue or green
d) red, blue or green e) not blue?

9. A boy collects marbles. He has the following colours in a bag: 28 red, 14 blue, 25 yellow, 17 green and 6 purple. If he draws one marble from the bag, what is the probability that it is:
a) red b) blue c) yellow or blue
d) purple e) not purple?

10. The probability of a boy drawing a marble of one of the following colours from another bag of marbles is:

blue 0.25 red 0.2 yellow 0.15 green 0.35 white 0.05

If there are 140 green marbles, how many of each other colour does he have in his bag?

11. My 6 tickets for a raffle give me a 0.02 probability of winning first prize. How many tickets have been sold?

12. The probability of getting a bad egg in a batch of 400 is 0.035. How many bad eggs are there likely to be in a batch?

13. In a lottery for a new car, 25 000 tickets are sold. My chance of winning is 0.000 08. How many tickets do I have?

14. The probability of Juan's favourite football team winning 4–0 is 0.05. How many times are they likely to win by this score in a season of 40 matches?

Student Assessment 1

1. What is the probability of throwing the following numbers with a fair dice:
   a) a 2　　b) not a 2
   c) less than 5　　d) a 7

2. If you have a normal pack of 52 cards, what is the probability of drawing:
   a) a diamond　　b) a 6　　c) a black card
   d) a picture card　　e) a card less than 5?

3. 250 tickets are sold for a raffle. What is the probability of winning the first prize if the following number of tickets are bought?
   a) 1　　b) 5　　c) 20
   d) 75　　e) 250

4. A bag contains 11 blue, 8 red, 6 white, 5 green and 10 yellow counters. If one counter is taken from the bag, what is the probability that it is:
   a) blue　　b) green　　c) yellow　　d) not red?

5. The probability of drawing a red, blue or green marble from a bag containing 320 marbles is:

   red 0.5　blue 0.3　green 0.2

   How many marbles of each colour are there?

Student Assessment 2

1. An octagonal spinner has the numbers 1 to 8 on it as shown (below).

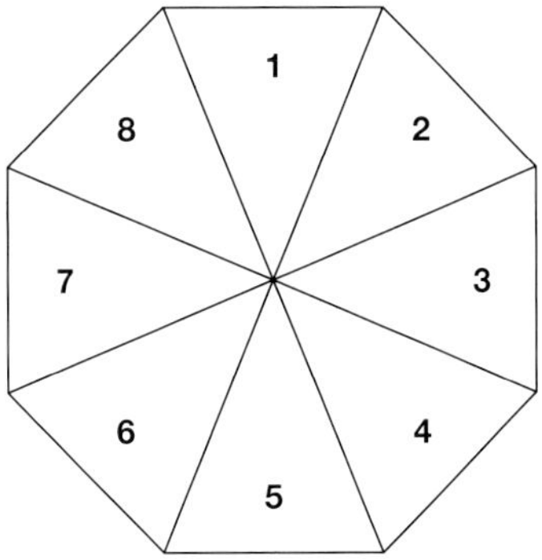

   What is the probability of spinning:
   a) a 7　　b) not a 7
   c) a factor of 12　　d) a 9?

2. A game requires the use of all the playing cards in a normal pack from 6 to King inclusive.
a) How many cards are used in the game?
b) What is the probability of drawing:
i) a 6 ii) a picture iii) a club
iv) a prime number v) an 8 or a spade?

3. 180 tickets are sold for a raffle. What is the probability of winning first prize if the following number of tickets are bought?
a) 1 b) 9 c) 15
d) 40 e) 180

4. A bag contains 11 white, 9 blue, 7 green and 5 red counters. What is the probability that a single counter drawn will be:
a) blue b) red or green c) not white?

5. The probability of drawing a red, blue or green marble from a bag containing 320 marbles is:

red 0.4 blue 0.25 green 0.35

If there are no other colours in the bag, how many marbles of each colour are there?

6. If I buy one ticket for a lottery and have a 0.000 02 probability of winning first prize, how many tickets have been sold for the lottery?

ICT SECTION

Statistics and Probability Activity 1

In this activity, you will be using the graphing facilities of a spreadsheet to compare the activities you do on a school day with the activities you do on a day at the weekend.

1. Prepare a 24-hr timeline for a weekday similar to the one shown below:

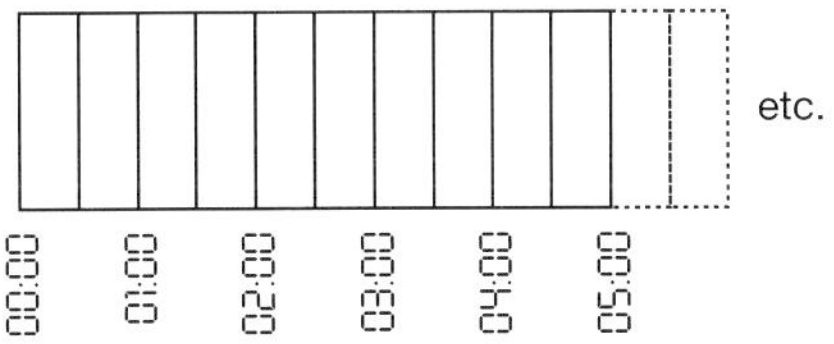

2. By using different colours for different activities, shade in the timeline to show what you did and when on a specific weekday, e.g. sleeping, eating, school, watching TV.

3. Add up the time spent on each activity and enter the results in a spreadsheet like the one below:

	A	B
1	Activity	Time spent (hrs)
2	Sleeping	
3	Eating	
4	School	
5	TV	
6		

4. Use the spreadsheet to produce a fully-labelled pie chart of this data.
5. Repeat steps 1–4 for a day at the weekend.
6. Comment on any differences and similarities between the pie charts for the two days.

Statistics and Probability Activity 2

For this activity, you will be testing the fairness of a spinner that you have constructed.

1. Using card, a pair of compasses and a ruler, construct a regular hexagon.

2. Divide your regular hexagon into six equal parts.

3. Colour the six parts using three different colours, as shown below:

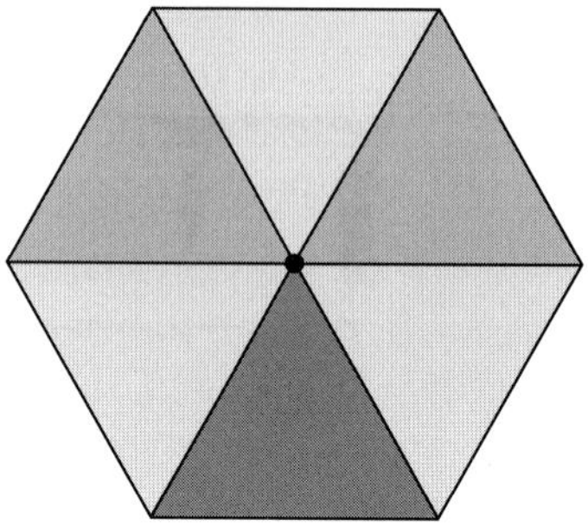

4. Calculate the theoretical probability of each colour. Record these probabilities as percentages.
5. Carefully insert a small pencil through the centre of the hexagon to form a spinner.
6. Spin the spinner 60 times, recording your results in a spreadsheet.
7. Using the spreadsheet, produce a percentage pie chart of your results.
8. Compare the actual probabilities with the theoretical ones calculated in Q.4. What conclusions can you make about the fairness of your spinner?

REVIEWS FOR CHAPTERS 31–32

Review 26

1. Find the mean, median, mode and range of the sets of numbers below:
   a) 1 4 4 5 6 7 8
   b) 4 12 7 9 3 4 8 7 16 9 4 10
   c) 6.4 8.2 7.3 6.3 8.2 8.7 5.6 8.2

2. An ordinary dice is thrown 120 times. Calculate the number of times you would expect to throw:
   a) a six b) an even number.

3. On a Saturday in November, the football results for the English Premier League showed that there were 8 home wins, 6 draws and 4 away wins. Represent this data on a pie chart.

4. The table below shows the marks out of 100 in a French test for a class of students. Draw a bar chart representing this data.

Score	Frequency
1–20	3
21–40	4
41–60	6
61–80	11
81–100	7

5. The numbers below are the numbers of words in the first forty sentences of *Macbeth*:

   5 6 10 8 3 3 5 3 2 13 5 3 11 15 3 12 15
   30 32 26 18 16 22 14 11 7 9 11 15 17 21
   28 17 15 9 7 5 23 27 34

   Record this data on a frequency table, using intervals 1–10, 11–20, etc. Draw a histogram for the data.

6. A roulette wheel has the numbers 1 to 36 and zero. When the wheel is spun, what is the probability of the ball landing on:
   a) a 4 b) an even number c) zero
   d) a factor of 36 e) a prime number?

7. A motor car manufacturer publishes this pie chart. It represents their customers' choices of colour for their new small cars in one year. If the number of new cars in the survey was 72 000, how many cars were:
a) red b) white?

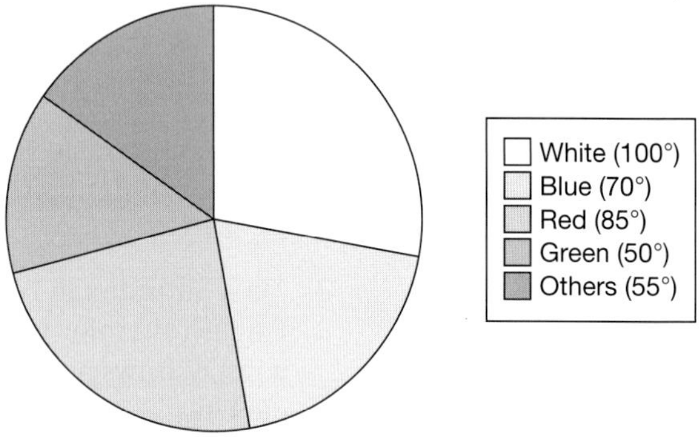

8. Illustrate the data in Q.7 in a form other than a pie chart.

9. Seats for a school concert are arranged in rows labelled A to K with 15 seats in each row numbered 1–15. Every seat is occupied, and every person is given a ticket for a raffle. Calculate the probability that the first prize is won by:
a) seat J4 b) a seat in row B
c) a person sitting in the middle of a row.

10. Explain the difference between a 'fair' game and an 'unfair' game.

Review 27

1. Find the mean shoe size of a group of students whose shoe sizes are recorded below:

Size of shoe	4	5	6	7	8	9	10
Number of students	3	4	8	6	8	4	1

2. The numbers of letters received on one day by people living in one street are recorded below. Display this information on a suitable pictogram.

Letters received	0	1	2	3	4	5
Number of houses	6	4	2	3	2	1

3. Display the information in Q.2 on a pie chart.

4. A school orchestra has the following number of players in each section.

Type of instrument	Strings	Woodwind	Brass	Percussion
Number of players	12	10	6	2

Illustrate this data on a bar chart.

5. Illustrate the data in Q.4 in another suitable way.

6. A dice in the shape of an icosahedron (20 faces) is thrown 100 times. The faces are numbered 1 to 20. How many times would you expect the top face to show:
a) a 12 b) an odd number c) a factor of 20?

7. Lettered tiles which spell the word AUSTRALIA are put into a bag. One tile is removed. Calculate the probability that it is:
a) an A b) a vowel c) a consonant.

8. 15 000 tickets are sold for a raffle to win a car. How many tickets must I buy to have a five percent chance of winning the car?

9. Explain the following terms:
a) range b) mode
c) grouped frequency d) pictogram

10. A village council is campaigning to have a new secondary school built in the village. Suggest four pieces of useful data that could be collected from people in the area by means of a questionnaire.

Review 28

1. Find the mean, median and mode of the following sets of data:
a) 6 7 8 9 10
b) 3.5 4.5 4.5 5 5.5 6 6.5
c) 8 32 16 8 24 16 8 40 32 8 24

2. A school library recorded students' choices of books under five headings as shown below. Display this data on a suitable pictogram.

Crime	28
Sci-fi	45
Romance	32
Other fiction	20
Non-fiction	55

3. Display the data in Q.2 on a pie chart.

4. 80 students were asked to record the number of minutes they spent listening to music in one week. The results are displayed below:

6	18	32	21	58	47	12	28	56	33	41	15	47	52
32	26	45	59	52	36	43	25	17	19	46	35	58	20
49	30	42	9	19	24	39	51	42	14	17	34	50	59
7	38	16	47	38	8	55	48	36	23	12	28	35	40
59	5	34	13	10	57	45	26	38	46	23	16	51	28
47	32	57	11	9	19	30	42	22	29				

Draw a grouped frequency table with intervals $0 < x \leqslant 10$; $10 < x \leqslant 20$, etc, then draw a histogram to illustrate the data.

5. 540 people were surveyed for hair colour. This information was displayed on a pie chart. If the size of the sector for black hair was 200° and for blond hair 44°, how many people had:
a) black hair b) blond hair?

6. A fair octahedral dice is thrown. The faces are numbered 1 to 8. What is the probability of throwing:
a) a 6 b) a prime number
c) not a prime number?

7. Lettered tiles which spell out the word MISSISSIPPI are put into a bag. If one tile is removed, what is the probability that it is:
a) an S b) an I c) not a vowel?

8. If I buy ten tickets for a raffle, what is the probability that I will win first prize if 400 tickets have been sold?

9. Describe briefly how you would conduct a survey to test the truth of the statement 'all boys like football'.

10. Explain briefly how you might display the data obtained in Q.9.

Review 29

1. A car park attendant records the number of cars entering the car park each hour for 12 hours. Calculate the mean and range of this data.

6 24 28 16 11 27 29 42 34 8 5 3

2. Display the data in Q.1 on a pictogram using one picture to represent 4 cars.

3. The number of matches in each of 40 boxes of matches is given below:

Number per box	37	38	39	40	41	42	43
Number of boxes	1	2	3	15	10	6	3

Calculate the mean, median, mode and range for this data.

4. Display the information in Q.3 on a bar chart.

5. Write all the ways in which the letters A, T and R can be arranged. If the letters are written at random, what is the probability that they will spell:
a) TRA b) an English word?

6. In a bag of 20 sweets, 12 are red, 6 green and 2 white. What is the probability of a sweet drawn at random being green?

7. A class of students hand in their homework books at random. If there are 12 boys and 18 girls in the class, calculate the probability that the first book in the pile will belong to:
a) a girl b) a boy.

8. A box of 26 building blocks has one of each of the letters of the alphabet on its faces. If one block is removed at random, what is the probability that it is:
a) P b) a vowel
c) a letter in the word MISSISSIPPI?

9. What is a census? Give three possible uses for the information gained from a census.

10. Give four possible questions for a questionnaire to test the statement 'young boys go to bed later than young girls'.

Review 30

1. The table (right) shows the marks out of 100 in an English test for a class of students. Draw a bar chart representing this data.

Test mark	Frequency
1–10	0
11–20	1
21–30	1
31–40	2
41–50	4
51–60	6
61–70	7
71–80	5
81–90	3
91–100	1

2. The number of apples collected from thirty trees is recorded below:

21 28 64 48 35 31 29 16 61 55 39 24 46 58
60 19 37 27 19 30 56 62 23 38 51 15 65 39
42 50

Make a grouped frequency table with interval values 0–19, 20–39, 40–59 and 60–79. Illustrate this data on a bar chart.

3. The results of a survey among 54 students to find their favourite sport are shown below. Represent this data on a pie chart.

Sport	Frequency
Football	27
Tennis	12
Hockey	6
Basketball	9

4. Two dice are thrown 60 times. The total of the two scores is recorded on a table as shown below. Calculate the mean and mode of the scores.

Score	2	3	4	5	6	7	8	9	10	11	12
Frequency	3	5	5	7	8	10	7	6	4	3	2

5. Lettered tiles which spell the word MATHEMATICS are put into a bag. If one tile is removed at random from the bag, what is the probability that it is:
a) an S b) an M c) a vowel d) not a vowel?

6. If a card is picked at random from a complete pack of 52 cards, calculate the probability that it is:
a) a spade b) a picture card c) a 7 or a heart.

7. 250 tickets are sold for a raffle. If I buy 5 tickets, what is my probability of winning first prize?

8. What is the probability of being born on:
a) a Wednesday b) a day in June
c) the 7th day of a month?

9. Devise a short questionnaire to supply data to a car manufacturer about to launch a new small car.

10. Explain briefly how you would conduct a survey to find the favourite lesson of a class of students.

GENERAL REVIEWS

General Review 1

1. Construct an equilateral triangle ABC of side 6 cm. Construct further equilateral triangles on sides AB, BC and CA.
- a) The shape produced is the net of which solid shape?
- b) Mark the lines of symmetry of the net.
- c) What is the order of rotational symmetry of the net?
- d) Calculate the surface area of the net.

2.
- a) Plot the graph of $y = x^2$ for values $-4 \leqslant x \leqslant 4$.
- b) On the same grid, plot the graph of $y = x + 6$.
- c) Use your graph to solve the quadratic equation $x^2 - x - 6 = 0$.

3. The temperature T °C at a height H metres above sea level is given by the formula $T = 20 - \dfrac{H}{150}$.
- a) Calculate the temperature at 3450 m.
- b) Rearrange the formula to make H the subject.
- c) At what height in metres is the temperature 12 °C?

4. A vet has a salary of \$57 600 per year. Her tax deductions amount to 30% of her salary, and a further 6% is deducted for her pension.
- a) Calculate:
  - i) her gross monthly salary,
  - ii) her net monthly salary.
- b) Her expenditure for one month is shown below. Illustrate this data on a pie chart.

  Tax 30%   Pension 6%   Rent \$1200   Food \$576
  Clothing and entertainment \$960
  The rest is saved.

- c) If she is awarded a pay rise of 9%, calculate her new gross annual salary.

General Review 2

1. Golf balls are sold in packs of three, either in rectangular boxes or cylindrical tubes as shown below.

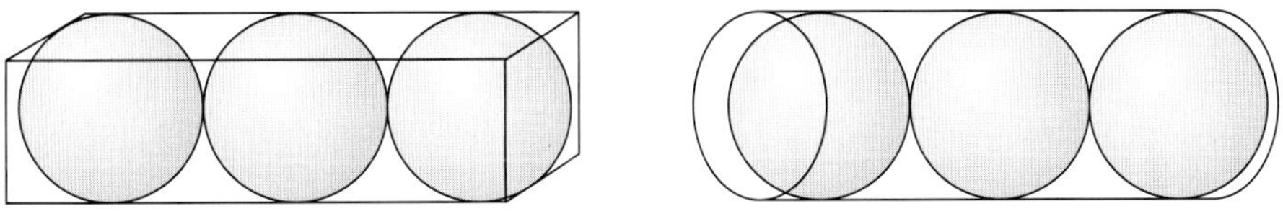

The balls exactly fit their containers. A golf ball has a diameter of 4.2 cm and a volume of 39 cm^3.

a) Write down the dimensions of the rectangular box and hence calculate its volume.
b) Calculate the percentage volume of the rectangular box occupied by the golf balls.
c) Calculate to the nearest cm^3 the volume of the cylindrical container given the formula $V = \pi r^2 h$, where V cm^3 is the volume of a cylinder of radius r cm and height h cm. π is 3.142.
d) A tennis ball has a volume of 117 cm^3. Express the volume of a golf ball and the volume of a tennis ball as a ratio in the form $1 : n$.

2. The frequency table below gives the scores out of 10 achieved by a year group of students in a maths test.

Test score	0	1	2	3	4	5	6	7	8	9	10	**Total**
Frequency	3	6	9	6	9	15	12	18	12	3	3	
Frequency × score												

a) Copy and complete the table.
b) Calculate the mean test score.
c) Find the median test score.
d) What is the modal test score?
e) Draw a frequency chart to illustrate the data.

3. Two lines AC and BD bisect each other at right angles. Construct the shape ABCD, given that AC = 12 cm and BD = 16 cm.
a) Name the shape ABCD.
b) What is the order of rotational symmetry of ABCD?
c) Measure the length of the side AB, and confirm this by calculation using Pythagoras' theorem.
d) Use trigonometry to calculate angle ABC.

4. a) Solve the equation $\dfrac{5 + 2x}{3} = \dfrac{4x - 1}{5}$.
b) Solve the simultaneous equations:

$$2x + 3y = 17$$
$$3x - 2y = 19$$

c) Rearrange this formula to make a the subject:

$$\frac{a + b}{c} = d - e$$

General Review 3

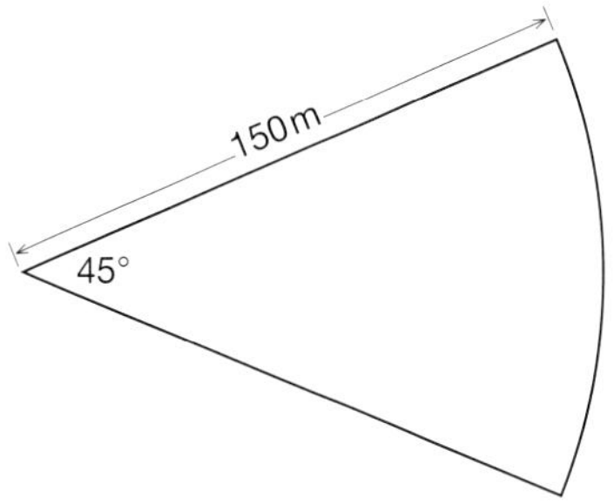

1. The sector shown (left) represents the marked area for a discus event in a school sports day:

a) Using ruler and compasses only, make a scale drawing of the sector.

b) For safety purposes, a rope is staked outside the sector so that it is always 10 m from the marked sector. Show this on your diagram.

c) The winning throw is 125 m and lands on the line of symmetry of the sector. Label this point A on your diagram. Use trigonometry to calculate the shortest distance from A to the side line of the sector.

2. a) Find the next two terms in each of the following sequences:

i) 6, 11, 16, 21 ...

ii) 3, 12, 48, 192 ...

b) The nth term of a sequence is given by the formula $(n + 1)(n - 1)$. Find:

i) the 5th term,

ii) the 99th term.

c) A sequence of numbers is 1, 8, 27, 64 ...

i) Write the next two terms.

ii) Write the 100th term.

iii) Write the nth term.

d) Write the nth term of the sequence 6, 13, 32, 69 ...

3. Draw the diagram below to scale.

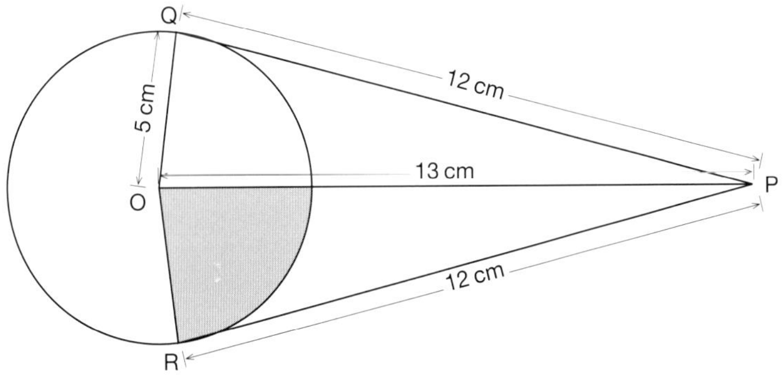

a) Prove that PQ is a tangent to the circle.

b) Use trigonometry to calculate angle POR.

c) Calculate the area of the circle.

d) Calculate the area of the shaded sector.

4. The letters A, D, R, T can be combined in various groups of four.

a) How many possible combinations are there?

b) If the letters are written by a computer at random, what is the probability that:

i) the letters will spell DART,

ii) the letter A will be written first?

5. A rugby team scores the following numbers of points in 15 matches.

21	17	6	12	42	17	13	28
17	20	23	14	23	21	42	

a) Calculate:

i) the mean score,

ii) the median score,

iii) the modal score.

b) Use the shape of a rugby goal (H) to illustrate the results on a suitable pictogram.

General Review 4

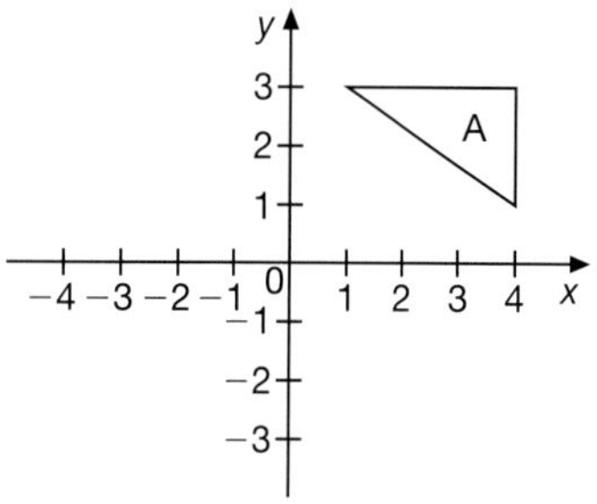

1. Copy the graph (left).

a) Reflect triangle A in the line $y = 0$. Label it B.

b) Draw the triangle formed when triangle A is rotated through 180° about the origin. Label it C.

c) Draw the translation of triangle C by the vector $\begin{pmatrix} -1 \\ +5 \end{pmatrix}$. Label it D.

d) Draw the enlargement of triangle A, centre (1, 1), scale factor 2. Label it E.

2.

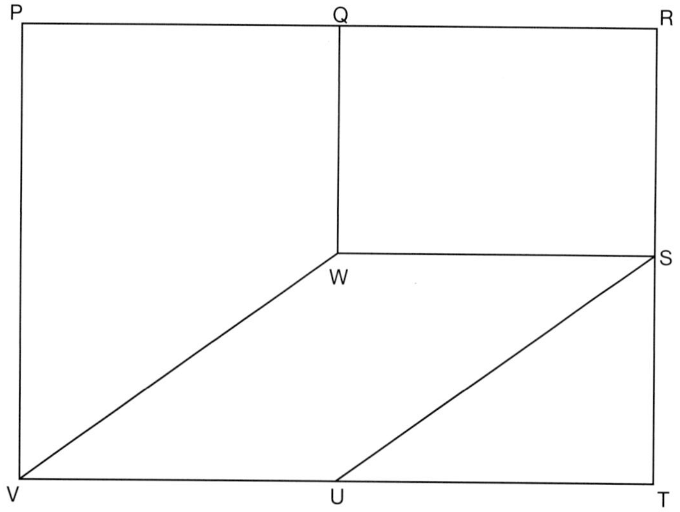

PRTV is a rectangle 100 m by 70 m. Q is the midpoint of PR. S is the midpoint of RT. U is the midpoint of TV. W is the midpoint of QU.

Calculate:
a) the area of QRSW,
b) the area of WSUV,
c) the area of PQWV,
d) the length VW,
e) angle TUS.

3. a) Expand and simplify the following expression:

$$\frac{a}{9}(18a - 27) - \frac{a}{3}(6a - 12)$$

b) Factorise the following:

$$39x^2y^2z - 26xy^2z^2 + 52x^2yz^2$$

c) If $a = 2$, $b = -2$ and $c = 5$, evaluate this expression:

$$a^2b - b^3 - c^2$$

d) Rearrange the formula below to make p the subject:

$$3m + n = r(p - q)$$

4. a) Construct a regular hexagon of side 5 cm.
b) Use your hexagon to illustrate that the sum of the internal angles of a regular hexagon is given by 180(6 − 2) degrees.
c) How many lines of symmetry does a regular pentagon have?
d) What is the order of rotational symmetry of a dodecagon?

General Review 5

1. A company has a contract to build a tunnel under a river 2 km across. The tunnel will be a cylinder 80 m wide and 2.8 km long. The volume of a cylinder is given by the formula $V = \pi r^2 h$ where r is the radius and h the height (or length) of the cylinder.
a) How many cubic metres of earth must be removed in the preparation of the tunnel?
b) The earth is removed by trucks which each carry 25 m^3 of earth. How many truck journeys are required?
c) The original cost of the tunnel was estimated at £180 million, spread evenly over three years. However, costs increased by 20% in the first year and a further 15% in each of the second and third years. Calculate the final cost of the tunnel.

2. A normal pack of playing cards contains 52 cards with four suits of thirteen cards.
a) If a card is picked at random, what is the probability of picking:
i) a 7 ii) a club iii) a picture card
iv) a 5 or a heart?

b) If all the picture cards are removed, what is the probability now of picking:
i) a 7 ii) a club iii) a picture card
iv) a 5 or a heart?

c) From the picture cards removed earlier, give an example of a choice with probability:
i) 1 ii) $\frac{1}{4}$ iii) $\frac{2}{3}$

3. ABC is an equilateral triangle of side 13 cm. ACD is a right-angled triangle in which CD is 12 cm.

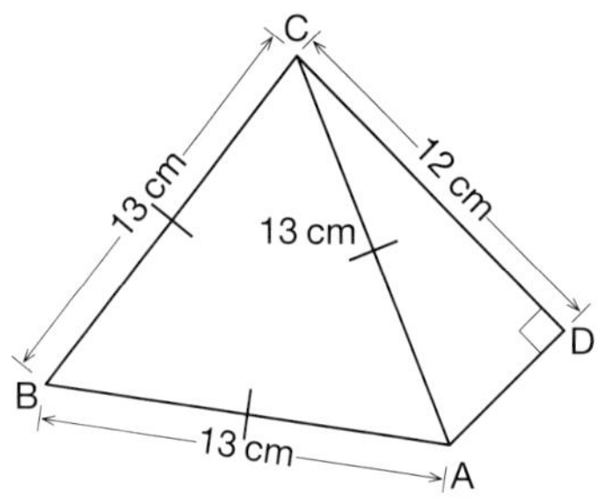

Calculate:
a) angle ABC b) the altitude of triangle ABC
c) AD d) angle ACD.
e) By considering angles ACD and BAC, prove that AB and CD are not parallel.

4. a) Sketch the graph of $y = x^2$.
b) Copy and complete the table below for the quadratic equation $y = x^2 - 9x + 18$.

x	0	1	2	3	4	5	6	7
y	18		4			–2		

c) Plot the graph of $y = x^2 - 9x + 18$ for $0 \leqslant x \leqslant 7$.
d) Use your graph to solve the following quadratic equation:

$$x^2 - 9x = -18$$

e) Use your graph to solve following quadratic equation:

$$x^2 - 9x = -20$$

SOLUTIONS

SOLUTIONS

1 Number and Language

Exercise 1.1 page 1

1. a) +3 b) 0 c) +3 d) +1 e) −2 f) −5
2. a) −2 b) −1 c) −8 d) +6 e) −6 f) +1
3. a) −6 b) −5 c) −8 d) −5 e) −6 f) −8
4. a) −1 b) −2 c) −6 d) −7 e) −6 f) +1

Exercise 1.2 page 2

1. a) 1 b) 3 c) 2 d) 3 e) 6 f) 3 g) 0 h) 2 i) 2 j) 0
2. a) −1 b) −1 c) −3 d) −5 e) −1 f) −1 g) 0 h) −2 i) −1 j) −6
3. a) −6 b) −8 c) −14 d) −7 e) −6 f) −12 g) −3 h) −6 i) −11 j) −10
4. a) −4 b) −1 c) −3 d) −1 e) −3 f) −2 g) 0 h) −1 i) −4 j) −2
5. a) 1 b) 3 c) 4 d) 1 e) 4 f) 2 g) 4 h) 0 i) 0 j) 1

Exercise 1.3 page 3

1. a) −20 b) −21 c) +16 d) −45 e) −40 f) −42
2. a) −12 b) −14 c) −16 d) −15 e) −54 f) −64
3.

	−3	−2	−1	0	+1	+2	+3
+3	−9	−6	−3	0	+3	+6	+9
+2	−6	−4	−2	0	+2	+4	+6
+1	−3	−2	−1	0	+1	+2	+3
0	0	0	0	0	0	0	0
−1	+3	+2	+1	0	−1	−2	−3
−2	+6	+4	+2	0	−2	−4	−6
−3	+9	+6	+3	0	−3	−6	−9

4. a) +6 b) −2 c) −6 d) +6 e) +3 f) +4
5. a) −21 b) −30 c) +40 d) +15 e) −36 f) +42
6. a) +5 b) −5 c) −5 d) +5 e) +4 f) −5 g) −9 h) −3 i) +9 j) −8
7. a) +10 b) +5 c) +5 d) +6 e) −5 f) −5 g) −4 h) +3 i) −6 j) −6
8. a) +3 b) +5 c) +5 d) +3 e) −8 f) −3 g) +7 h) −8 i) −1 j) +1 k) +10
9. a) 5, 6, 7, 8, 9, 10, 11, 12, 13, 14, 15

   b) 2, 3, 4, 5, 6, 7, 8, 9, 10, 11, 12

   c) −7, −6, −5, −4, −3, −2, −1, 0, 1, 2, 3

   d) 3, 4, 6, 12, −12, −6, −4, −3

   e) −4, −5, −10, −20, 20, 10, 5, 4

Exercise 1.4 page 5

Prime numbers are:
2, 3, 5, 7, 11, 13, 17, 19, 23, 29, 31, 37, 41, 43, 47, 53, 59, 61, 67, 71, 73, 79, 83, 89, 97

1. a) 9^2 b) 12^2 c) 8^2 d) 7^3 e) 4^3 f) $3^2 \times 2^3$ g) $5^3 \times 2^2$ h) $4^2 \times 3^2 \times 2^2$

Exercise 1.5 page 6

a) 1, 2, 3, 6
b) 1, 3, 9
c) 1, 7
d) 1, 3, 5, 15
e) 1, 2, 3, 4, 6, 8, 12, 24
f) 1, 2, 3, 4, 6, 9, 12, 18, 36
g) 1, 5, 7, 35
h) 1, 5, 25
i) 1, 2, 3, 6, 7, 14, 21, 42
j) 1, 2, 4, 5, 10, 20, 25, 50, 100

Exercise 1.6 page 6

a) 3, 5 b) 2, 3 c) 2, 3 d) 2 e) 2, 5 f) 13 g) 3, 11 h) 5, 7 i) 2, 5, 7 j) 2, 7

Exercise 1.7 page 7

a) $2^2 \times 3$ b) 2^5 c) $2^2 \times 3^2$ d) $2^3 \times 5$ e) $2^2 \times 11$ f) $2^3 \times 7$ g) $3^2 \times 5$ h) 3×13 i) $3 \times 7 \times 11$ j) $3^2 \times 7$

Exercise 1.8 page 7

1. a) 4 b) 5 c) 6 d) 3 e) 9 f) 22 g) 8 h) 13 i) 17 j) 12
2. a) 15 b) 12 c) 14 d) 28 e) 8 f) 30 g) 12 h) 12 i) 60 j) 60

3. **a)** 42 **b)** 60 **c)** 70 **d)** 90 **e)** 120
f) 105 **g)** 20 **h)** 231 **i)** 240 **j)** 200

Exercise 1.9 page 8

1. **a)** Rational **b)** Rational **c)** Irrational
d) Rational **e)** Rational **f)** Rational
g) Irrational **h)** Rational **i)** Rational

2. **a)** Irrational **b)** Irrational **c)** Rational
d) Rational **e)** Rational **f)** Rational

3. **a)** Rational **b)** Irrational **c)** Rational
d) Rational

Exercise 1.10 page 9

a) i) +3 from term to term ii) 30

b) i) +10 from term to term ii) 98

c) i) +22 from term to term ii) 209

d) i) −0.2 from term to term ii) −1.1

e) i) +1 to the denominator each time ii) $\frac{1}{11}$

f) i) Denominator and numerator increase by 1 each time. ii) $\frac{10}{11}$

g) i) Sequence of square numbers ii) 100

h) i) Difference between successive terms increases by 2 each time. ii) 103

i) i) Sequence of cube numbers ii) 1000

j) i) Sequence of powers of 5 ii) 5^{10} or 9 765 625

Exercise 1.11 page 10

a) 53, 71 **b)** 67, 131 **c)** 39, 63

d) 173, 275 **e)** 170, 357 **f)** 127, 221

g) 27, 29

Exercise 1.12 page 11

a) i) 20, 23 ii) $3n + 2$

b) i) 25, 29 ii) $4n + 1$

c) i) 29, 34 ii) $5n - 1$

d) i) 18, 20 ii) $2n + 6$

e) i) 36, 43 ii) $7n - 6$

f) i) 24, 28 ii) $4n - 4$

g) i) 46, 55 ii) $9n - 8$

h) i) 65, 75 ii) $10n + 5$

i) i) 64, 75 ii) $11n - 2$

j) i) 13.5, 15.5 ii) $2n - 0.5$

k) i) 5.25, 6.25 ii) $n - 0.75$

l) i) 6, 7 ii) $n - 1$

Exercise 1.13 page 11

a) i) 50, 65 ii) $n^2 + 1$

b) i) 43, 56 ii) $n^2 + 7$

c) i) 48, 63 ii) $n^2 - 1$

d) i) 216, 343 ii) n^3

e) i) 217, 344 ii) $n^3 + 1$

f) i) 226, 353 ii) $n^3 + 10$

g) i) 213, 340 ii) $n^3 - 3$

h) i) 56, 72 ii) $n^2 + n$

Student Assessment 1 page 12

1. **a)** +5 **b)** +1 **c)** +5
2. **a)** −3 **b)** −4 **c)** −10
3. **a)** −9 **b)** −17 **c)** −16
4. **a)** +1 **b)** +9 **c)** +2
5. **a)** −4 **b)** −5 **c)** −6
6. **a)** −2 **b)** −2 **c)** −8
7. **a)** +7 **b)**+7 **c)** +11
8. **a)** $2^2 \times 3 \times 5$ **b)** $3^2 \times 7$ **c)** 2^6
9. **a)** 2 **b)** 2 **c)** 24
10. **a)** 28 **b)** 30 **c)** 36
11. **a)** −32 **b)** −63 **c)** +45
12. **a)** −6 **b)** −7 **c)** +7
13. 3, 4, 5, 6, 7, 8, 9, 10, 11
14. 3, 4, 6, 12, −12, −6, −4, −3
15. 10, 5, 3, 1, −1, −2, −6

Student Assessment 2 page 12

1. **a)** +8 **b)** −8 **c)** −15
2. **a)** −4 **b)** +12 **c)** −2
3. **a)** −12 **b)** −7 **c)** −9
4. **a)** −1 **b)** +3 **c)** +1
5. **a)** −4 **b)** −4 **c)** −9
6. **a)** −1 **b)** −8 **c)** −7
7. **a)** +2 **b)** +5 **c)** 0
8. **a)** $2^3 \times 3$ **b)** $3^2 \times 5$ **c)** 2×5^2
9. **a)** 6 **b)** 12 **c)** 17
10. **a)** 42 **b)** 45 **c)** 200

11. **a)** −18 **b)** −63 **c)** +48

12. **a)** −12 **b)** −9 **c)** +3

13. −4, −3, −2, −1, 0, 1, 2

14. 2, 4, 12, −12, −6, −3, −1

15. −4, −2, −1, 3, 5, 10

Student Assessment 3 page 13

1. **a)** Rational **b)** Irrational **c)** Rational
   d) Rational **e)** Rational **f)** Irrational

2. **a)** i) 45, 54 ii) Terms increasing by 9
   b) i) 30, 24 ii) Terms decreasing by 6
   c) i) 2.25, 1.125 ii) Terms halving
   d) i) −12, −18 ii) Terms decreasing by 6
   e) i) 27, 8
   ii) Descending order of cube numbers
   f) i) 81, 243 ii) Terms multiplied by 3

3. **a)** $4n + 2$ **b)** $6n + 7$ **c)** $6n - 3$
   d) $n^2 + 3$ **e)** $10n - 10$ **f)** $n^3 - 1$

Student Assessment 4 page 14

1. **a)** $\frac{5}{8}$ **b)** 3 **c)** $\frac{11}{25}$

2. **a)** i) 30, 36 ii) Terms increasing by 6
   b) i) 12, 9 ii) Terms decreasing by 3
   c) i) −5, −10 ii) Terms decreasing by 5
   d) i) 64, 81
   ii) Ascending order of square numbers
   e) i) 1000, 10 000
   ii) Ascending order of powers of 10
   f) i) $\frac{1}{16}$, $\frac{1}{32}$ ii) Terms halving

3. **a)** $2n + 1$ **b)** $6n + 1$ **c)** $10n - 2$
   d) $8n - 7$ **e)** $8n - 12$ **f)** $n^2 + 1$

2 Squares, Square Roots and Cubes

Exercise 2.1 page 15

a) 9 **b)** 25 **c)** 64 **d)** 100 **e)** 121
f) 144 **g)** 49 **h)** 169 **i)** 225 **j)** 400

Exercise 2.2 page 15

1. **a)** 4.41 **b)** 9.61 **c)** 1.44 **d)** 4.84
   e) 6.25 **f)** 1.96

2. **a)** 5.76 **b)** 10.89 **c)** 7.84 **d)** 38.44
   e) 21.16 **f)** 53.29 **g)** 0.09 **h)** 0.64
   i) 0.01 **j)** 0.81

3. Students check their answers

4. The following are the exact answers:
   a) 6.25 **b)** 12.25 **c)** 20.25 **d)** 30.25
   e) 51.84 **f)** 40.96 **g)** 0.64 **h)** 0.04
   i) 28.09 **j)** 39.69

5. Students check their answers

Exercise 2.3 page 16

1. **a)** 5 **b)** 3 **c)** 7 **d)** 10 **e)** 11 **f)** 13
   g) 0.1 **h)** 0.2 **i)** 0.3 **j)** 0.5

2. Students check their answers

3. **a)** $\frac{1}{3}$ **b)** $\frac{1}{4}$ **c)** $\frac{1}{5}$ **d)** $\frac{1}{7}$ **e)** $\frac{1}{10}$ **f)** $\frac{2}{3}$ **g)** $\frac{3}{10}$
   h) $\frac{7}{9}$ **i)** $\frac{5}{3}$ **j)** $\frac{5}{2}$

Exercise 2.4 page 16

1. The following answers are correct to 1 d.p.:
   a) 8.4 **b)** 6.3 **c)** 7.1 **d)** 9.5 **e)** 5.9
   f) 6.7 **g)** 7.4 **h)** 7.7 **i)** 1.4 **j)** 1.7
   k) 4.5 **l)** 5.5 **m)** 3.5 **n)** 8.7 **o)** 10.7

2. Students check their answers

Exercise 2.5 page 17

1. **a)** 27 **b)** 125 **c)** 1000 **d)** 64 **e)** 729
   f) 1 million

2. 1, 8, 27, 64, 125, 216, 343, 512, 729, 1000

3. **a)** 1331 **b)** 0.125 **c)** 3.375 **d)** 15.625
   e) 8000 **f)** 27 000 **g)** 35 **h)** 125
   i) 370 **j)** 1000

4. **a)** 2 **b)** 5 **c)** 3 **d)** 0.1 **e)** 0.3 **f)** 6
   g) 10 **h)** 100 **i)** –2 **j)** –3 **k)** –10
   l) –1

Student Assessment 1 page 18

1. **a)** 49 **b)** 144 **c)** 0.01 **d)** 0.09

2. 16.81

3. **a)** 6.25 **b)** 10.89 **c)** 0.0625

4. The following are exact answers:
   a) 12.96 **b)** 42.25 **c)** 56.25 **d)** 28.09

5. **a)** 7 **b)** 12 **c)** 0.8 **d)** $\frac{5}{2}$ **e)** $\frac{9}{10}$ **f)** $\frac{10}{6}$

6. **a)** 8 **b)** 125 **c)** 1000

7. **a)** 4 **b)** 5 **c)** 10

Student Assessment 2 page 18

1. **a)** 81 **b)** 225 **c)** 0.04 **d)** 0.49

2. 6.25

3. **a)** 12.25 **b)** 16.81 **c)** 0.0225

4. **a)** 2.6 **b)** 5.5 **c)** 6.7

5. **a)** 15 **b)** 0.1 **c)** 0.9 **d)** $\frac{3}{5}$ **e)** $\frac{7}{3}$ **f)** $\frac{11}{7}$

6. **a)** 64 **b)** 0.001 **c)** $\frac{8}{27}$

7. **a)** 3 **b)** 100 **c)** $\frac{4}{5}$

3 Directed Numbers

Exercise 3.1 page 20

All answers for Q.1 and 2 are in °C.

1. **a)** 4 **b)** −4 **c)** 6 **d)** −7 **e)** −18
   f) 0 **g)** −9 **h)** 2 **i)** −6 **j)** −13

2. **a)** 2 **b)** 10 **c)** −3 **d)** −8 **e)** 5

3. 146 °C

4. 44 BC

5. 62 years

6. 2206 years

7. **a)** −£35 **b)** −£318 **c)** −£88 **d)** −£160
   e) £90

8. **a)** +75 **b)** −80 **c)** +30 **d)** −30
   e) −65 **f)** +35

9. **a)** 570 m **b)** 1080 m

10. 44 °C

11. 12 °C

12. 4 °C

13. 1700 m

14. 165 m

15. 695 m

Student Assessment 1 page 22

1. 4800

2. 2040

3. 1708

4. 2200

5. 3648 years

6. 4400 years

7. **a)** Jackson **b)** East **c)** Wilson
   d) Wilson **e)** Bridge **f)** Pear **g)** East
   h) James **i)** West **j)** Smith

8. **a)** −4 **b)** −5 **c)** −3 **d)** +9 **e)** −5
   f) +4 **g)** −5 **h)** +1 **i)** −2 **j)** −5

Student Assessment 2 page 23

1. 900 years

2. 289

3. 2170

4. 2169

5. **a)** −£84 **b)** £91 **c)** £45 **d)** £74
   e) −£43 **f)** −£15

6. **a)** Yellow **b)** Blue **c)** Red **d)** Red
   e) Orange **f)** Indigo **g)** Blue **h)** Blue
   i) Orange **j)** Blue **k)** Indigo **l)** Green

7. **a)** 30 °C **b)** −4 °C **c)** 30 °C **d)** −2 °C
   e) 30 °C **f)** 4 °C **g)** 29 °C

4 Vulgar and Decimal Fractions, and Percentages

Exercise 4.1 page 25

1. **a)** $\frac{2}{3}$ numerator / denominator **b)** $\frac{15}{22}$ num / den
   c) $\frac{4}{3}$ num / den **d)** $\frac{5}{2}$ num / den

2.

	Proper	Improper	Mixed
a)	$\frac{2}{3}$		
b)	$\frac{15}{22}$		
c)		$\frac{4}{3}$	
d)		$\frac{5}{2}$	
e)			$1\frac{1}{2}$
f)			$2\frac{3}{4}$
g)		$\frac{7}{4}$	
h)	$\frac{7}{11}$		
i)			$7\frac{1}{4}$
j)	$\frac{5}{6}$		
k)		$\frac{6}{5}$	
l)			$1\frac{1}{5}$
m)	$\frac{1}{10}$		
n)			$2\frac{7}{8}$
o)		$\frac{5}{3}$	

Exercise 4.2 page 25

1. **a)** 8 **b)** 24 **c)** 4 **d)** 20 **e)** 9 **f)** 63
   g) 5 **h)** 25 **i)** 2 **j)** 6

2. **a)** 9 **b)** 16 **c)** 20 **d)** 40 **e)** 18
   f) 72 **g)** 30 **h)** 48 **i)** 210 **j)** 52

Exercise 4.3 page 26

1. **a)** $\frac{14}{3}$ **b)** $\frac{18}{5}$ **c)** $\frac{47}{8}$ **d)** $\frac{17}{6}$ **e)** $\frac{17}{2}$ **f)** $\frac{68}{7}$
   g) $\frac{58}{9}$ **h)** $\frac{17}{4}$ **i)** $\frac{59}{11}$ **j)** $\frac{55}{7}$ **k)** $\frac{43}{10}$ **l)** $\frac{146}{13}$

2. **a)** $7\frac{1}{4}$ **b)** $6\frac{3}{5}$ **c)** $6\frac{5}{6}$ **d)** $6\frac{6}{8}$ **e)** $5\frac{4}{9}$ **f)** $1\frac{4}{12}$
   g) $9\frac{3}{7}$ **h)** $3\frac{3}{10}$ **i)** $9\frac{1}{2}$ **j)** $6\frac{1}{12}$

Exercise 4.4 page 27

1.

	Hundreds	Tens	Units	$\frac{1}{10}$	$\frac{1}{100}$	$\frac{1}{1000}$
a)			6	0	2	3
b)			5	9	4	
c)		1	8	3		
d)			0	0	7	1
e)			2	0	0	1
f)			3	5	6	

2. **a)** 4.5 **b)** 6.3 **c)** 17.8 **d)** 3.07
   e) 9.27 **f)** 11.36 **g)** 4.006 **h)** 5.027
   i) 4.356 **j)** 9.204

3. **a)** 19.14 **b)** 83.812 **c)** 6.6 **d)** 11.16
   e) 35.81 **f)** 5.32 **g)** 67.14 **h)** 6.06
   i) 1.4 **j)** 0.175

Exercise 4.5 page 27

1. **a)** 39% **b)** 42% **c)** 63% **d)** 5%

2. **a)** $\frac{58}{100} = 58\%$ **b)** $\frac{17}{25} = \frac{68}{100} = 68\%$
   c) $\frac{11}{20} = \frac{55}{100} = 55\%$ **d)** $\frac{3}{10} = \frac{30}{100} = 30\%$
   e) $\frac{23}{25} = \frac{92}{100} = 92\%$ **f)** $\frac{19}{50} = \frac{38}{100} = 38\%$
   g) $\frac{3}{4} = \frac{75}{100} = 75\%$ **h)** $\frac{2}{5} = \frac{40}{100} = 40\%$

3.

Fraction	Decimal	Percentage
$\frac{1}{10}$	0.1	10
$\frac{1}{5}$	0.2	20
$\frac{3}{10}$	0.3	30
$\frac{4}{10} = \frac{2}{5}$	0.4	40
$\frac{1}{2}$	0.5	50
$\frac{3}{5}$	0.6	60
$\frac{7}{10}$	0.7	70
$\frac{4}{5}$	0.8	80
$\frac{9}{10}$	0.9	90
$\frac{1}{4}$	0.25	25
$\frac{3}{4}$	0.75	75

Student Assessment 1 page 29

1. **a)** $\underline{3\frac{1}{2}}$ **b)** $\frac{2}{5}$ **c)** $\frac{7}{5}$ (circled) **d)** $\frac{8}{9}$ **e)** $\underline{1\frac{3}{4}}$

2. **a)** 12 **b)** 33 **c)** 6 **d)** 90

3. **a)** $\frac{5}{2}$ **b)** $\frac{27}{7}$ **c)** $\frac{71}{12}$

4. **a)** $3\frac{1}{7}$ **b)** $7\frac{1}{5}$ **c)** $7\frac{4}{9}$

5. **a)** $\frac{5}{6} = \frac{10}{12} = \frac{20}{24} = \frac{50}{60} = \frac{90}{108}$

6. **a)** $\frac{27}{100} = 0.27$ **b)** 0.105 **c)** 0.07 **d)** 0.087

7. **a)** 30% **b)** 29% **c)** 50% **d)** 70%
   e) 80% **f)** 219% **g)** 6% **h)** 75%
   i) 31% **j)** 7% **k)** 340% **l)** 200%

Student Assessment 2 page 29

1. **a)** $\frac{3}{11}$ **b)** $\underline{5\frac{3}{4}}$ **c)** $\frac{27}{8}$ (circled) **d)** $\frac{3}{7}$

2. **a)** 21 **b)** 27 **c)** 22 **d)** 39

3. **a)** $\frac{13}{5}$ **b)** $\frac{31}{9}$ **c)** $\frac{45}{8}$

4. **a)** $6\frac{3}{5}$ **b)** $5\frac{3}{9}$ **c)** $6\frac{1}{11}$

5. $\frac{2}{3} = \frac{4}{6} = \frac{8}{12} = \frac{20}{30} = \frac{18}{27}$

6. **a)** 0.35 **b)** 0.275 **c)** 6.75 **d)** 0.035

7. **a)** 60% **b)** 49% **c)** 25% **d)** 90%
   e) 150% **f)** 327% **g)** 5% **h)** 35%
   i) 77% **j)** 3% **k)** 290% **l)** 400%

5 Ordering

Exercise 5.1 page 30

1. **a)** a is less than 7
   b) b is greater than 4
   c) c is not equal to 8
   d) d is equal to or less than 3
   e) e is equal to or greater than 9
   f) f is equal to or less than 11

2. **a)** $a < 4$ **b)** $b > 7$ **c)** $c \leqslant 9$ **d)** $d \geqslant 5$
   e) $e \neq 3$ **f)** $f \leqslant 6$ **g)** $g \geqslant 9$ **h)** $h \geqslant 6$
   i) $i \neq 7$ **j)** $j \leqslant 20$

3. **a)** n is greater than 5, but less than 10
   b) n is equal to or greater than 6, but equal to or less than 15
   c) n is equal to or greater than 3, but less than 9
   d) n is greater than 8 and equal to or less than 12
   (Other wording possible)

4. **a)** $7 < p < 10$ **b)** $3 < q < 12$ **c)** $5 \leqslant r < 9$
   d) $8 < s \leqslant 15$

Exercise 5.2 page 31

1. **a)** $<$ **b)** $=$ **c)** $>$ **d)** $<$ **e)** $=$
   f) $>$

2. **a)**
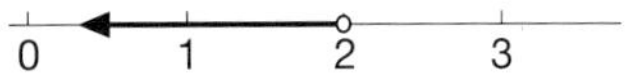

b)
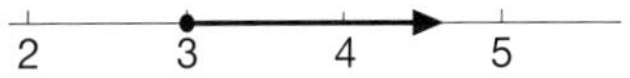

c)

d)
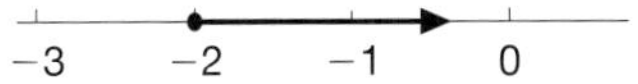

e)
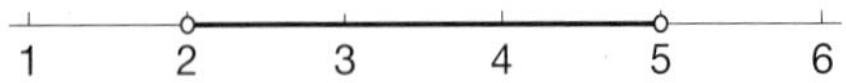

f)

g)
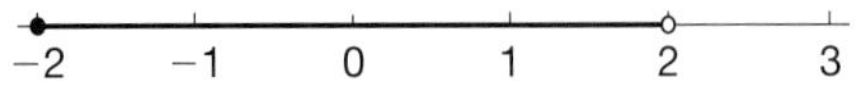

h)

3. **a)** $x > 0$ **b)** $x \leqslant 3$ **c)** $0 \leqslant x \leqslant 4$
   d) $-4 < x \leqslant -1$

4. **a)** $x \leqslant 20\,000$ **b)** $135 \leqslant x \leqslant 180$
   c) $5x + 3 < 20$ **d)** $x \leqslant 25$
   e) $350 \leqslant x \leqslant 400$ **f)** $11 \leqslant x \leqslant 28$

Exercise 5.3 page 32

1. 4.5, 4.05, 0.45, 0.405, 0.045
2. 0.06, 0.6, 0.606, 0.66, 6.0, 6.6, 6.606
3. 9.06, 0.96, 0.906, 0.690, 0.609, 0.096
4. $\frac{1}{4}, \frac{3}{10}, \frac{1}{3}, \frac{2}{5}, \frac{1}{2}, \frac{3}{4}$
5. $\frac{4}{5}, \frac{1}{2}, \frac{6}{13}, \frac{7}{18}, \frac{1}{3}, \frac{2}{19}$
6. $\frac{1}{2}, \frac{5}{9}, \frac{4}{7}, \frac{3}{5}, \frac{2}{3}, \frac{3}{4}$

Exercise 5.4 page 33

1. 15 000 cm $\frac{2}{5}$km 0.5 km 750 m 5000 m
2. 60 cm 0.75 m 800 mm 180 cm 2 m
3. 4 kg 3500 g 1 kg $\frac{3}{4}$kg 700 g
4. 150 cm^3 430 ml 800 cm^3 1 litre 120 cl

Student Assessment 1 page 34

1. **a)** m is equal to 7
   b) n is not equal to 10
   c) e is less than 4
   d) f is equal to or less than 6
   e) x is equal to or greater than 12
   f) y is greater than -1

2. **a)** $a > 5$ **b)** $b \geqslant 6$ **c)** $c \geqslant 4$ **d)** $d \neq 14$

3. **a)**
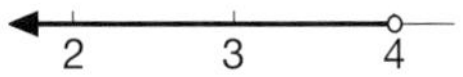

b)
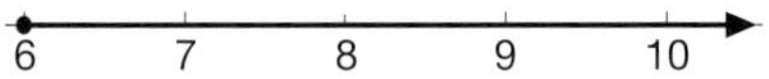

c)

d)
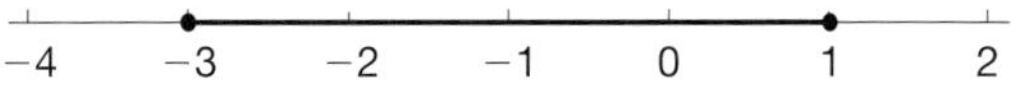

4. **a)**

b)
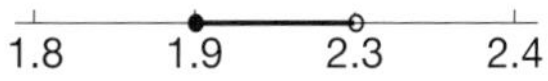

5. 0.09 0.9 0.99 9 9.09 9.99

Student Assessment 2 page 34

1. **a)** p is not equal to 2
   b) q is greater than zero
   c) r is equal to or less than 3
   d) s is equal to 2
   e) t is equal to or greater than 1
   f) u is less than -5

2. **a)** $a < 2$ **b)** $b \leqslant 4$ **c)** $c = 8$ **d)** $d \leqslant 0$

3. **a)**

b)

c)

d)
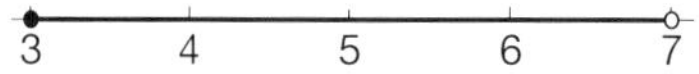

4. **a)**
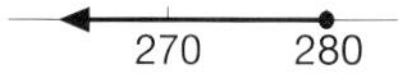

b)
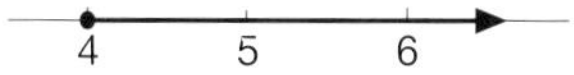

5. 1800 g, 1.09 kg, 1 kg, 900 g, 9 g

Student Assessment 3 page 35

1. **a)** = **b)** > **c)** < **d)** <

2. **a)**
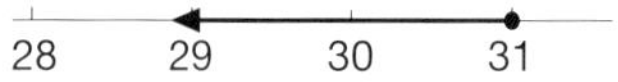

b)
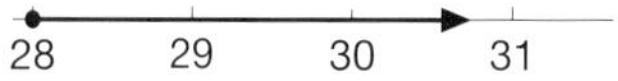

c)
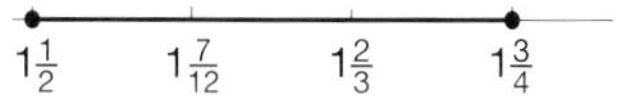

d)
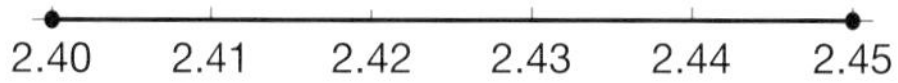

3. **a)** $x \leqslant 52$ **b)** $x \geqslant 11$ **c)** $24 \leqslant x \leqslant 38$
d) $3x + 6 < 50$ **e)** $x \geqslant 0.03$

4. **a)**
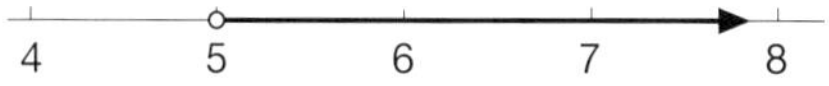

b)
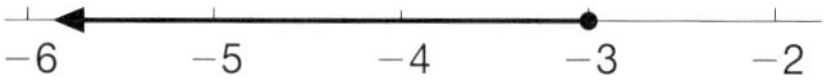

c)
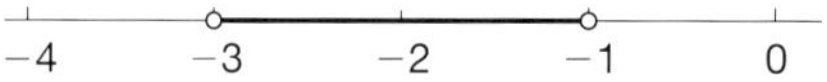

d)
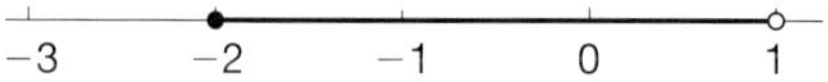

5. $\frac{6}{7}$ $\frac{13}{18}$ $\frac{2}{3}$ $\frac{7}{12}$ $\frac{1}{6}$

Student Assessment 4 page 35

1. **a)** = **b)** < **c)** > **d)** >

2. **a)**
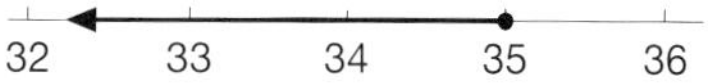

b)

c)

d)
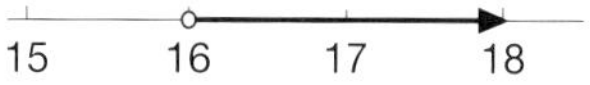

3. **a)** $x \geqslant -1$ **b)** $x < 2$ **c)** $-2 \leqslant x < 2$
d) $-1 \leqslant x \leqslant 1$

4. **a)**

b)
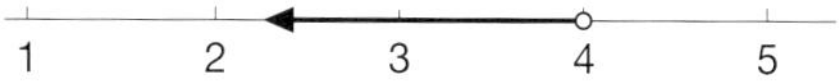

c)
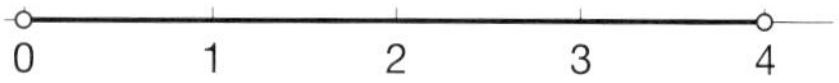

d)
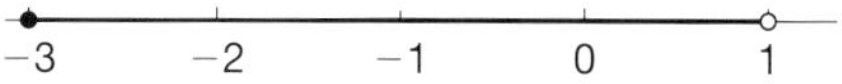

5. $\frac{3}{14}$ $\frac{2}{5}$ $\frac{1}{2}$ $\frac{4}{7}$ $\frac{9}{10}$

6 Standard Form

Exercise 6.1 page 38

1. **a)** 4 **b)** 4 **c)** 6 **d)** 6 **e)** 8 **f)** 7

2. **a)** 6.5×10^4 **b)** 4.1×10^4 **c)** 7.23×10^5
d) 1.8×10^7 **e)** 9.5×10^5 **f)** 7.6×10^8
g) 7.2×10^5 **h)** 2.5×10^5

3. 2.6×10^7 8×10^8 8.3×10^{10} 1.8×10^7
3.6×10^6 6×10^1

4. **a)** 8×10^5 **b)** 3×10^7 **c)** 1.5×10^7
d) 1.98×10^8 **e)** 2×10^9 **f)** 2.85×10^8
g) 1.5×10^7 **h)** 4.1×10^8 **i)** 2.4×10^7
j) 4.9×10^7

5. **a)** 8×10^8 **b)** 7.5×10^8 **c)** 9×10^{13}
d) 8.4×10^{11} **e)** 1.4×10^{12} **f)** 1.26×10^{10}
g) 4×10^8 **h)** 1.6×10^{17}

6. **a)** 4×10^3 **b)** 4×10^6 **c)** 1.9×10^1
d) 5×10^7 **e)** 4×10^0 (= 4) **f)** 2×10^8

7. **a)** 2×10^4 **b)** 2.5×10^8 **c)** 5×10^3
d) 7.2×10^4 **e)** 5×10^4 **f)** 5×10^3

Exercise 6.2 page 39

1. **d)** and **e)**

2. **a)** 6×10^5 **b)** 4.8×10^7 **c)** 7.84×10^{11}
d) 5.34×10^5 **e)** 7×10^6 **f)** 8.5×10^6

3. a) 6.8×10^6 b) 7.2×10^8 c) 8×10^5
d) 7.5×10^7 e) 4×10^9 f) 5×10^7

4. a) 6×10^5 b) 2.4×10^7 c) 1.4×10^8
d) 3×10^9 e) 1.2×10^{13} f) 1.8×10^7

5. 1.44×10^{11} m

6. a) 8.8×10^8 b) 2.04×10^{11}
c) 3.32×10^{11} d) 4.2×10^{22}
e) 5.1×10^{22} f) 2.5×10^{25}

7. a) 2×10^2 b) 3×10^5 c) 4×10^6
d) 2×10^4 e) 2.5×10^6 f) 4×10^4

8. a) 4.26×10^5 b) 8.48×10^9 c) 6.388×10^7
d) 3.157×10^9 e) 4.5×10^8 f) 6.01×10^7
g) 8.15×10^{10} h) 3.56×10^7

9. Mercury 5.8×10^7 km
Venus 1.08×10^8 km
Earth 1.5×10^8 km
Mars 2.28×10^8 km
Jupiter 7.78×10^8 km
Saturn 1.43×10^9 km
Uranus 2.87×10^9 km
Pluto 5.92×10^9 km

Exercise 6.3 page 41

1. a) 10^{-3} b) 10^{-3} c) 10^{-4} d) 10^{-6}
e) 10^{-7} f) 10^{-9} g) 10^{-6} h) 10^{-10}

2. a) 6×10^{-4} b) 5.3×10^{-5}
c) 8.64×10^{-4} d) 8.8×10^{-8}
e) 7×10^{-7} f) 4.145×10^{-4}

3. a) 6.8×10^{-4} b) 7.5×10^{-7}
c) 4.2×10^{-10} d) 8×10^{-9}
e) 5.7×10^{-11} f) 4×10^{-11}

4. a) −4 b) −3 c) −8 d) −5
e) −7 f) 3

5. 6.8×10^5 6.2×10^3 8.414×10^2
6.741×10^{-4} 3.2×10^{-4} 5.8×10^{-7}
5.57×10^{-9}

Student Assessment 1 page 42

1. a) 8×10^6 b) 7.2×10^{-4} c) 7.5×10^{10}
d) 4×10^{-4} e) 4.75×10^9
f) 6.4×10^{-7}

2. 7.41×10^{-9} 3.6×10^{-5} 5.5×10^{-3}
4.21×10^7 6.2×10^7 4.9×10^8

3. a) 6×10^6 8.2×10^5 4.4×10^{-3} 8×10^{-1}
5.2×10^4
b) 6×10^6 8.2×10^5 5.2×10^4 8×10^{-1}
4.4×10^{-3}

4. a) 2 b) 8 c) −4 d) 5 e) −5
f) −5

5. a) 1.2×10^8 b) 5.6×10^8 c) 2×10^5
d) 2.5×10^5

6. 6 minutes

7. 4.73×10^{15} km correct to three significant figures (3 s.f.)

Student Assessment 2 page 42

1. a) 6×10^6 b) 4.5×10^{-3} c) 3.8×10^9
d) 3.61×10^{-7} e) 4.6×10^8 f) 3×10^0

2. 4.05×10^8 3.6×10^2 9×10^1 1.5×10^{-2}
7.2×10^{-3} 2.1×10^{-3}

3. a) 1.5×10^7 4.3×10^5 4.35×10^{-4}
4.8×10^0 8.5×10^{-3}
b) 4.35×10^{-4} 8.5×10^{-3} 4.8×10^0
4.3×10^5 1.5×10^7

4. a) 3 b) 9 c) −3 d) 6 e) −1
f) 8

5. a) 1.2×10^8 b) 1.48×10^{11} c) 6.73×10^7
d) 3.88×10^6

6. 43 minutes (2 s.f.)

7. 2.84×10^{15} km (3 s.f.)

7 The Four Rules

Exercise 7.1 page 45

1. a) 9 b) 22 c) 30 d) 2 e) 7
f) 3

2. a) $6 \times (4 + 6) \div 3 = 20$
b) $6 \times (4 + 6 \div 3) = 36$
c) $8 + 2 \times (4 - 2) = 12$
d) $(8 + 2) \times (4 - 2) = 20$
e) $(9 - 3) \times 7 + 2 = 44$
f) $(9 - 3) \times (7 + 2) = 54$

3. a) 1512 b) 33 984 c) 29 830
d) 41 492 e) 20 736 f) 40 800

4. a) 5 b) 3 c) 2 d) 7 e) 1 f) 7

5. a) 7, 9 b) 100 c) 11, 16 d) 6
e) 224 f) 28

6. a) 1127.4 b) 7603.8 c) 1181.5
d) 1141.0 e) 526.1 f) 3328.2

Exercise 7.2 page 46

1. a) $\frac{1}{4} = \frac{2}{8} = \frac{4}{16} = \frac{16}{64} = \frac{3}{12}$
b) $\frac{2}{5} = \frac{4}{10} = \frac{8}{20} = \frac{20}{50} = \frac{16}{40}$
c) $\frac{3}{8} = \frac{6}{16} = \frac{9}{24} = \frac{15}{40} = \frac{18}{48}$
d) $\frac{4}{7} = \frac{8}{14} = \frac{12}{21} = \frac{32}{56} = \frac{36}{63}$
e) $\frac{5}{9} = \frac{15}{27} = \frac{20}{36} = \frac{50}{90} = \frac{55}{99}$

2. **a)** $\frac{1}{2}$ **b)** $\frac{1}{3}$ **c)** $\frac{2}{3}$ **d)** $\frac{4}{9}$ **e)** $\frac{3}{4}$ **f)** $\frac{9}{10}$

3. **a)** $4\frac{1}{4}$ **b)** $4\frac{3}{5}$ **c)** $2\frac{2}{3}$ **d)** $6\frac{1}{3}$ **e)** 4 **f)** $3\frac{7}{12}$

4. **a)** $\frac{13}{2}$ **b)** $\frac{29}{4}$ **c)** $\frac{27}{8}$ **d)** $\frac{100}{9}$ **e)** $\frac{34}{5}$ **f)** $\frac{97}{11}$

Exercise 7.3 page 47

1. **a)** $1\frac{2}{5}$ **b)** $\frac{10}{11}$ **c)** $\frac{11}{12}$ **d)** $1\frac{2}{45}$ **e)** $1\frac{1}{65}$ **f)** $1\frac{11}{12}$

2. **a)** $1\frac{1}{8}$ **b)** $1\frac{5}{7}$ **c)** $1\frac{1}{12}$ **d)** $\frac{47}{60}$ **e)** $1\frac{29}{40}$ **f)** $\frac{51}{52}$

3. **a)** $\frac{1}{7}$ **b)** $\frac{1}{10}$ **c)** $\frac{5}{9}$ **d)** $\frac{1}{12}$ **e)** $\frac{9}{40}$ **f)** $1\frac{1}{20}$

4. **a)** $\frac{17}{60}$ **b)** $\frac{45}{88}$ **c)** $\frac{17}{20}$ **d)** $\frac{44}{195}$ **e)** $\frac{9}{20}$ **f)** $\frac{1}{18}$

5. **a)** $5\frac{3}{4}$ **b)** $5\frac{3}{10}$ **c)** $3\frac{1}{10}$ **d)** $6\frac{7}{24}$ **e)** $1\frac{1}{8}$ **f)** $\frac{25}{36}$

6. **a)** $5\frac{1}{8}$ **b)** $7\frac{9}{40}$ **c)** $-\frac{3}{8}$ **d)** $\frac{7}{20}$ **e)** $-2\frac{39}{140}$ **f)** $1\frac{1}{4}$

Exercise 7.4 page 48

1. **a)** $\frac{4}{3}$ **b)** $\frac{9}{5}$ **c)** $\frac{1}{7}$ **d)** 9 **e)** $\frac{4}{11}$ **f)** $\frac{8}{37}$

2. **a)** 8 **b)** $\frac{12}{7}$ **c)** $\frac{5}{3}$ **d)** $\frac{2}{3}$ **e)** $\frac{4}{15}$ **f)** $\frac{1}{6}$

3. **a)** $\frac{1}{6}$ **b)** $\frac{3}{5}$ **c)** $\frac{4}{21}$ **d)** $\frac{2}{3}$ **e)** $\frac{1}{4}$ **f)** $\frac{7}{20}$

4. **a)** $\frac{5}{6}$ **b)** $2\frac{1}{2}$ **c)** $1\frac{1}{7}$ **d)** $4\frac{1}{6}$ **e)** $\frac{1}{5}$ **f)** $\frac{2}{3}$

5. **a)** $\frac{3}{5}$ **b)** $\frac{7}{12}$ **c)** $\frac{9}{70}$ **d)** $\frac{21}{25}$ **e)** $\frac{3}{8}$ **f)** $1\frac{25}{56}$

6. **a)** $\frac{3}{5}$ **b)** $3\frac{107}{120}$ **c)** $\frac{8}{15}$ **d)** $12\frac{1}{4}$

Exercise 7.5 page 48

1. **a)** 0.75 **b)** 0.8 **c)** 0.45 **d)** 0.34 **e)** $0.\dot{3}$ **f)** 0.375 **g)** 0.4375 **h)** $0.\dot{2}$ **i)** $0.\dot{6}\dot{3}$

2. **a)** 2.75 **b)** 3.6 **c)** 4.35 **d)** 6.22 **e)** $5.\dot{6}$ **f)** 6.875 **g)** 5.5625 **h)** $4.\dot{2}$ **i)** $5.\dot{4}\dot{2}\dot{8}\dot{5}\dot{7}\dot{1}$

Exercise 7.6 page 49

1. **a)** $\frac{1}{2}$ **b)** $\frac{7}{10}$ **c)** $\frac{3}{5}$ **d)** $\frac{3}{4}$ **e)** $\frac{33}{40}$ **f)** $\frac{1}{20}$ **g)** $\frac{1}{20}$ **h)** $\frac{201}{500}$ **i)** $\frac{1}{5000}$

2. **a)** $2\frac{2}{5}$ **b)** $6\frac{1}{2}$ **c)** $8\frac{1}{5}$ **d)** $3\frac{3}{4}$ **e)** $10\frac{11}{20}$ **f)** $9\frac{51}{250}$ **g)** $15\frac{91}{200}$ **h)** $30\frac{1}{1000}$ **i)** $1\frac{41}{2000}$

Student Assessment 1 page 50

1. **a)** 23 **b)** 18

2. 6, 15

3. 9000

4. 22 977

5. 360.2

6. $\frac{8}{18} = \frac{4}{9} = \frac{16}{36} = \frac{56}{126} = \frac{40}{90}$

7. **a)** $2\frac{1}{16}$ **b)** 9

8. **a)** 0.4 **b)** 1.75 **c)** $0.\dot{8}\dot{1}$ **d)** $1.\dot{6}$

9. **a)** $4\frac{1}{5}$ **b)** $\frac{3}{50}$ **c)** $1\frac{17}{20}$ **d)** $2\frac{1}{200}$

Student Assessment 2 page 50

1. **a)** 0 **b)** 19

2. 6, 12

3. 294

4. 18 032

5. 340.7

6. $\frac{24}{36} = \frac{8}{12} = \frac{4}{6} = \frac{20}{30} = \frac{60}{90}$

7. **a)** $1\frac{7}{10}$ **b)** 2

8. **a)** 0.875 **b)** 1.4 **c)** $0.\dot{8}$ **d)** $3.\dot{2}\dot{8}\dot{5}\dot{7}\dot{1}\dot{4}$

9. **a)** $6\frac{1}{2}$ **b)** $\frac{1}{25}$ **c)** $3\frac{13}{20}$ **d)** $3\frac{1}{125}$

8 Estimation

Exercise 8.1 page 51

1. **a)** 69 000 **b)** 74 000 **c)** 89 000 **d)** 4000 **e)** 100 000 **f)** 1 000 000

2. **a)** 78 500 **b)** 6900 **c)** 14 100 **d)** 8100 **e)** 1000 **f)** 3000

3. **a)** 490 **b)** 690 **c)** 8850 **d)** 80 **e)** 0 **f)** 1000

Exercise 8.2 page 52

1. **a)** 5.6 **b)** 0.7 **c)** 11.9 **d)** 157.4 **e)** 4.0 **f)** 15.0 **g)** 3.0 **h)** 1.0 **i)** 12.0

2. **a)** 6.47 **b)** 9.59 **c)** 16.48 **d)** 0.09 **e)** 0.01 **f)** 9.30 **g)** 100.00 **h)** 0.00 **i)** 3.00

Exercise 8.3 page 52

1. **a)** 50 000 **b)** 48 600 **c)** 7000
   d) 7500 **e)** 500 **f)** 2.57 **g)** 1000
   h) 2000 **i)** 15.0

2. **a)** 0.09 **b)** 0.6 **c)** 0.94 **d)** 1
   e) 0.95 **f)** 0.003 **g)** 0.0031
   h) 0.0097 **i)** 0.01

Exercise 8.4 page 53

1. **a)** 419.6 **b)** 5.0 **c)** 166.3 **d)** 23.8
   e) 57.8 **f)** 4427.10 **g)** 1.9 **h)** 4.1
   i) 0.6

Answers to Q.2–4 may vary slightly from those given below:

2. **a)** 1200 **b)** 3000 **c)** 3000
   d) 150 000 **e)** 0.8 **f)** 100

3. **a)** 200 **b)** 200 **c)** 30 **d)** 550
   e) 500 **f)** 3000

4. **a)** 130 **b)** 80 **c)** 9 **d)** 4 **e)** 200
   f) 250

5. **c)** and **e)** are incorrect.

Answers to Q.6 and 7 may vary slightly from those given below:

6. **a)** 120 m^2 **b)** 40 m^2 **c)** 400 cm^2

7. **a)** 200 cm^3 **b)** 4000 cm^3 **c)** 2000 cm^3

Student Assessment 1 page 55

1. **a)** 2800 **b)** 7290 **c)** 49 000 **d)** 1000

2. **a)** 3.8 **b)** 6.8 **c)** 0.85 **d)** 1.58
   e) 10.0 **f)** 0.008

3. **a)** 4 **b)** 6.8 **c)** 0.8 **d)** 10 **e)** 830
   f) 0.005

Answers to Q.4–6 may vary from those given below:

4. 18 000 yds

5. 40 m^2

6. **a)** 25 **b)** 4 **c)** 4

7. 92.3 cm^3 (1 d.p.)

Student Assessment 2 page 55

1. **a)** 6470 **b)** 88 500 **c)** 65 000 **d)** 10

2. **a)** 6.8 **b)** 4.44 **c)** 8.0 **d)** 63.08
   e) 0.057 **f)** 3.95

3. **a)** 40 **b)** 5.4 **c)** 0.06 **d)** 49 000
   e) 700 000 **f)** 687 000

Answers to Q.4–6 may vary from those given below:

4. 34 000 yds

5. 150 m^2

6. **a)** 20 **b)** 1 **c)** 3

7. 30.1 m^3 (1 d.p.)

9 Limits of Accuracy

Exercise 9.1 page 57

1. **a)** i) Lower bound = 5.5
   Upper bound = 6.5
   ii) $5.5 \leq x < 6.5$
   b) i) Lower bound = 82.5
   Upper bound = 83.5
   ii) $82.5 \leq x < 83.5$
   c) i) Lower bound = 151.5
   Upper bound = 152.5
   ii) $151.5 \leq x < 152.5$
   d) i) Lower bound = 999.5
   Upper bound = 1000.5
   ii) $999.5 \leq x < 1000.5$
   e) i) Lower bound = −0.5
   Upper bound = 0.5
   ii) $-0.5 < x < 0.5$
   f) i) Lower bound = −4.5
   Upper bound = −3.5
   ii) $-4.5 < x \leq -3.5$

2. **a)** i) Lower bound = 3.75
   Upper bound = 3.85
   ii) $3.75 \leq x < 3.85$
   b) i) Lower bound = 15.55
   Upper bound = 15.65
   ii) $15.55 \leq x < 15.65$
   c) i) Lower bound = 0.95
   Upper bound = 1.05
   ii) $0.95 \leq x < 1.05$
   d) i) Lower bound = 9.95
   Upper bound = 10.05
   ii) $9.95 \leq x < 10.05$
   e) i) Lower bound = 0.25
   Upper bound = 0.35
   ii) $0.25 \leq x < 0.35$
   f) i) Lower bound = −0.25
   Upper bound = −0.15
   ii) $-0.25 < x \leq -0.15$

3. **a)** i) Lower bound = 4.15
Upper bound = 4.25
ii) $4.15 \leqslant x < 4.25$
b) i) Lower bound = 0.835
Upper bound = 0.845
ii) $0.835 \leqslant x < 0.845$
c) i) Lower bound = 415
Upper bound = 425
ii) $415 \leqslant x < 425$
d) i) Lower bound = 4950
Upper bound = 5050
ii) $4950 \leqslant x < 5050$
e) i) Lower bound = 0.0445
Upper bound = 0.0455
ii) $0.0445 \leqslant x < 0.0455$
f) i) Lower bound = 24 500
Upper bound = 25 500
ii) $24\,500 \leqslant x < 25\,500$

4. **a)**
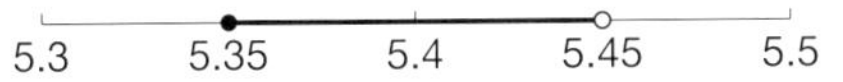

b) $5.35 \leqslant M < 5.45$

5. **a)**
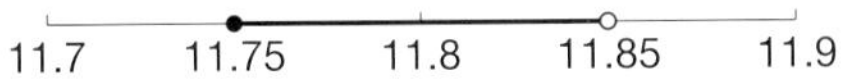

b) $11.75 \leqslant T < 11.85$

6. **a)** Lower bound = 615 m^3
Upper bound = 625 m^3
b) $615 \leqslant x < 625$

7. **a)** Lower bound = 625 m
Upper bound = 635 m
b) $395 \leqslant W < 405$

Exercise 9.2 page 58

1. **a)** i) 7.5, 8.5 ii) $7.5 \leqslant x < 8.5$
b) i) 70.5, 71.5 ii) $70.5 \leqslant x < 71.5$
c) i) 145.5, 146.5 ii) $145.5 \leqslant x < 146.5$
d) i) 199.5, 200.5 ii) $199.5 \leqslant x < 200.5$
e) i) 0.5, 1.5 ii) $0.5 \leqslant x < 1.5$
f) i) −1.5, −0.5 ii) $-1.5 < x \leqslant -0.5$

2. **a)** i) 2.45, 2.55 ii) $2.45 \leqslant x < 2.55$
b) i) 14.05, 14.15 ii) $14.05 \leqslant x < 14.15$
c) i) 1.95, 2.05 ii) $1.95 \leqslant x < 2.05$
d) i) 19.95, 20.05 ii) $19.95 \leqslant x < 20.05$
e) i) 0.45, 0.55 ii) $0.45 \leqslant x < 0.55$
f) i) −0.55, −0.45 ii) $-0.55 < x \leqslant -0.45$

3. **a)** i) 5.35, 5.45 ii) $5.35 \leqslant x < 5.45$
b) i) 0.745, 0.755 ii) $0.745 \leqslant x < 0.755$
c) i) 545, 555 ii) $545 \leqslant x < 555$
d) i) 5950, 6050 ii) $5950 \leqslant x < 6050$
e) i) 0.0115, 0.0125 ii) $0.0115 \leqslant x < 0.0125$
f) i) 9950, 10 500 ii) $9950 \leqslant x < 10\,500$

4. **a)**
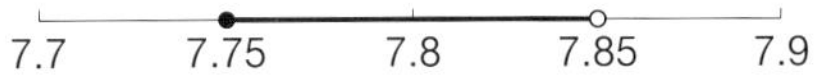

b) $7.75 \leqslant M < 7.85$

5. **a)**
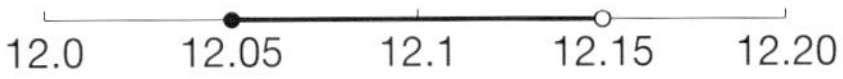

b) $12.05 \leqslant T < 12.15$

6. **a)** 735 m^3, 745 m^3 **b)** $735 \leqslant x < 745$

7. **a)** 565 m, 575 m **b)** $335 \leqslant W < 345$

Student Assessment 1 page 60

1. **a)**

b)
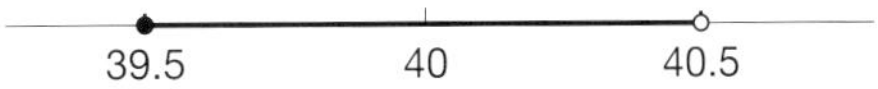

c)
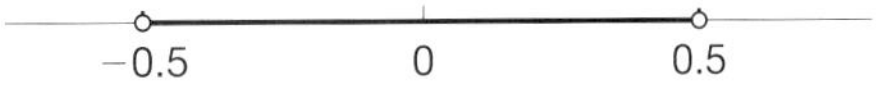

d)
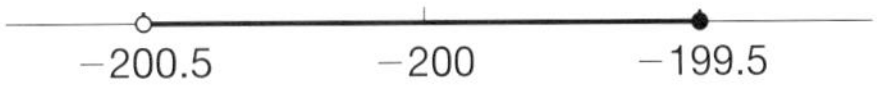

2. **a)** $205 \leqslant x < 215$ **b)** $63.5 \leqslant x < 64.5$
c) $2.95 \leqslant x < 3.05$ **d)** $0.875 \leqslant x < 0.885$

3. Length: $349.5 \leqslant L < 350.5$
Width: $199.5 \leqslant W < 200.5$

4.
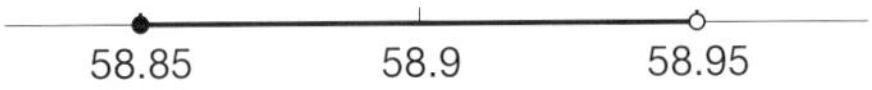

5. $0.0035 \leqslant x < 0.0045$

6. **a)** $4.825 \leqslant x < 4.835$ **b)** $5.045 \leqslant y < 5.055$
c) $9.95 \leqslant z < 10.05$
d) $-100.005 < p \leqslant -99.995$

Student Assessment 2 page 60

1. **a)**

b)

c)
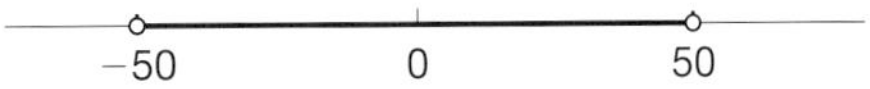

d)
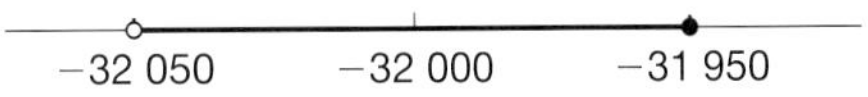

2. **a)** $253.5 \leqslant x < 254.5$ **b)** $40.45 \leqslant x < 40.55$
c) $0.4095 \leqslant x < 0.4105$
d) $99.95 \leqslant x < 100.5$

3. $20.25 \leqslant L < 20.75$ $9.75 \leqslant W < 10.25$

4. 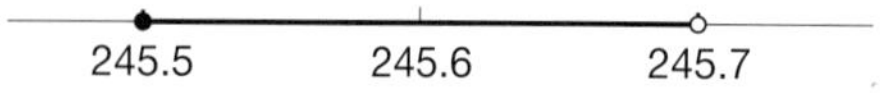

5. Lower bound = 365.245
Upper bound = 365.255

6. **a)** $10.895 \leqslant x < 10.905$
b) $2.995 \leqslant x < 3.005$ **c)** $0.45 \leqslant x < 0.55$
d) $-175.005 < x \leqslant -174.995$

10 Ratio, Proportion, and Rate

Exercise 10.1 page 62

1. 48
2. 16 h 40 min
3. 11 units
4. **a)** 7500 bricks **b)** 53 h
5. **a)** 6250 litres **b)** 128 km
6. 1111 km (4 s.f.)
7. **a)** 450 **b)** 75 **c)** 120

Exercise 10.2 page 64

1. **a)** 128 **b)** 375
2. **a)** 120 **b)** 1440 **c)** 34 560
3. **a)** 4 **b)** 30 hours
4. **a)** 22.5 **b)** 36 hours
5. 25 hours
6. 20 km
7. **a)** 270 **b)** 10 h 40 min

Exercise 10.3 page 64

1. **a)** 450 kg **b)** 1250 kg
2. **a)** Butter 600 g Flour 2 kg Sugar 200 g Currants 400 g
b) 120 cakes
3. **a)** 16.8 litres
b) Red 1.2 litres White 14.3 litres
4. **a)** 125 **b)** Red 216 Yellow 135 **c)** 20
5. **a)** 42 litres
b) Orange juice 495 litres Mango juice 110 litres

Exercise 10.4 page 65

1. 60 : 90
2. 16 : 24 : 32
3. 3.25 : 1.75
4. 18 : 27
5. 10 : 50
6. 7 : 1
7. Orange 556 ml (3 s.f.) Water 444 ml (3 s.f.)
8. **a)** 55 : 45 **b)** 440 boys 360 girls
9. $\frac{3}{5}$
10. 32 cm
11. 4 km and 3 km
12. 40°, 80°, 120°, 120°
13. 45°, 75°, 60°
14. 24 yr old \$400 000 28 yr old \$466 667
32 yr old \$533 333
15. Alex £2000 Maria £3500 Ahmet £2500

Exercise 10.5 page 67

1. 4
2.

Speed (km/h)	60	40	30	120	90	50	10
Time (h)	2	3	4	1	$1\frac{1}{3}$	$2\frac{2}{5}$	12

3. 30
4. **a)** i) 12 h ii) 4 h iii) 48 h
b) i) 16 ii) 3 iii) 48
5. **a)** 30 rows **b)** 42 chairs
6. 6 h 40 min
7. 4
8. 18 h

Student Assessment 1 page 68

1. 16 cm and 14 cm
2. 1200 g
3. 200 g
4. **a)** 2 km **b)** 48 cm
5. **a)** 26 litres of petrol and 4 litres of oil
   b) 3250 ml of oil
6. **a)** 1 : 40 **b)** 13.75 cm
7. Girl £1040 Boy £960
8. 24°, 60°, 96°
9. **a)** 15 s **b)** 8 copiers
10. 6 h

Student Assessment 2 page 68

1. **a)** $\frac{7}{10}$ **b)** 45 cm
2. **a)** 375 g **b)** 625 g
3. **a)** 450 m **b)** 80 cm
4. **a)** 1 : 25 **b)** 1.75 m
5. 300 : 750 : 1950
6. 60°, 90°, 90°, 120°
7. 150°
8. **a)** 13.5 h **b)** 12 pumps
9.

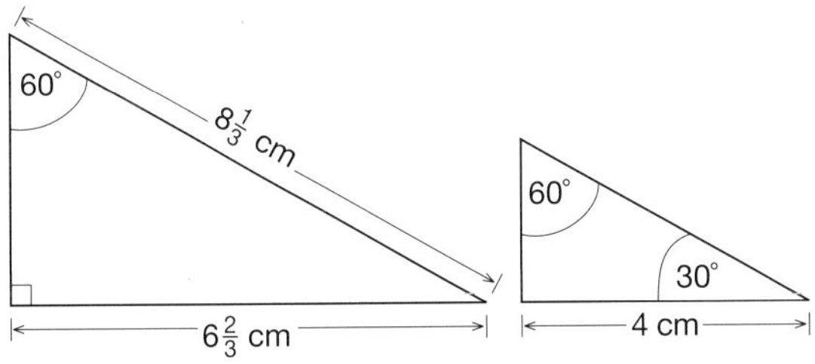

10. **a)** 4 min 48 s **b)** 1.6 litres/min

11 Percentages

Exercise 11.1 page 71

1. White = 47% Blue = 23% Red = 30%
2. 70%
3. **a)** 60% **b)** 40%
4. **a)** $\frac{73}{100}$ **b)** $\frac{28}{100}$ **c)** $\frac{10}{100}$ **d)** $\frac{25}{100}$
5. **a)** 27% **b)** 30% **c)** 14% **d)** 25%
6. **a)** 0.39 **b)** 0.47 **c)** 0.83 **d)** 0.07
   e) 0.02 **f)** 0.2
7. **a)** 31% **b)** 67% **c)** 9% **d)** 5%
   e) 20% **f)** 75%

Exercise 11.2 page 71

1. **a)** 25% **b)** $66.\dot{6}\%$ **c)** 62.5%
   d) 180% **e)** 490% **f)** 387.5%
2. **a)** 0.75 **b)** 0.8 **c)** 0.2 **d)** 0.07
   e) 1.875 **f)** $0.1\dot{6}$
3. **a)** 20 **b)** 100 **c)** 50 **d)** 36
   e) 4.5 **f)** 7.5
4. **a)** 8.5 **b)** 8.5 **c)** 52 **d)** 52
   e) 17.5 **f)** 17.5
5. **a)** Black 6 **b)** Blonde 3 **c)** Brown 21
6. Beef 66 Chicken 24 Pork 12 Lamb 18
7. English 143 Pakistani 44 Greek 11 Other 22
8. Newspapers 69 Pens 36 Books 18 Other 27

Exercise 11.3 page 72

1. **a)** 48% **b)** 36.8% **c)** 35% **d)** 50%
   e) 45% **f)** 40% **g)** $33\frac{1}{3}\%$
   h) 57% (2 s.f.)
2. Win 50% Lose $33\frac{1}{3}\%$ Draw $16\frac{2}{3}\%$
3. A = 34.5% (1 d.p.) B = 25.6% (1 d.p.)
   C = 23.0% (1 d.p.) D = 16.9% (1 d.p.)
4. Red 35.5% Blue 31.0% White 17.7%
   Silver 6.6% Green 6.0% Black 3.2%

Exercise 11.4 page 74

1. **a)** 187.5 **b)** 322 **c)** 7140 **d)** 245
   e) 90 **f)** 121.5
2. **a)** 90 **b)** 38 **c)** 9 **d)** 900 **e)** 50
   f) 43.5
3. **a)** 20% **b)** 80% **c)** 110% **d)** 5%
   e) 85% **f)** 225%
4. **a)** 50% **b)** 30% **c)** 5% **d)** 100%
   e) 36% **f)** 5%
5. 7475 tonnes
6. $6825
7. **a)** £75 **b)** £93.75
8. **a)** 43 **b)** 17.2%
9. 1100

Student Assessment 1 page 75

1.

Fraction	Decimal	Percentage
$\frac{3}{4}$	0.75	75%
$\frac{4}{5}$	0.8	80%
$\frac{1}{3}$	$0.\dot{3}$	$33\frac{1}{3}$%
$\frac{5}{8}$	0.625	62.5%
$\frac{3}{2}$	1.5	150%

2. 640 m
3. £345.60
4. $10 125
5. **a)** 20% **b)** 41.7% (1 d.p.) **c)** 22.5% **d)** 85.7% (1 d.p.) **e)** 7% **f)** 30%
6. 16% profit
7. **a)** £36 **b)** 25%
8. 2001

Student Assessment 2 page 75

1.

Fraction	Decimal	Percentage
$\frac{1}{4}$	0.25	25%
$\frac{3}{5}$	0.6	60%
$\frac{5}{8}$	0.625	62.5%
$\frac{2}{3}$	$0.\dot{6}$	$66\frac{2}{3}$%
$2\frac{1}{4}$	2.25	225%

2. 750 m
3. $2100
4. £97 200
5. **a)** 29.2% (1 d.p.) **b)** 21.7% (1 d.p.) **c)** 125% **d)** 8.3% (1 d.p.) **e)** 20% **f)** 10%
6. 8.3%
7. **a)** 325 fr **b)** 61.8% (1 d.p.)
8. 8 days

12 Use of an Electronic Calculator

Exercise 12.1 page 77

1. **a)** 25 **b)** 22.68 **c)** 17.82 **d)** 121.24 **e)** 134.33 **f)** 156.1
2. **a)** 12.3 **b)** 11.9 **c)** −1.24 **d)** −0.01 **e)** −30.3 **f)** −8.4 **g)** −13 **h)** 0
3. **a)** 80.04 **b)** 118.26 **c)** 91.02 **d)** 3 **e)** −42.6 **f)** −3.1 **g)** 4.84 **h)** 4.4

Exercise 12.2 page 78

1. **a)** 26 **b)** 10 **c)** 42 **d)** 16 **e)** 8 **f)** −6
2. **a)** 20 **b)** 34 **c)** 32 **d)** 31 **e)** 20 **f)** 23
3. **a)** 27 **b)** 64 **c)** 30 **d)** 3 **e)** 144 **f)** 1.6

Exercise 12.3 page 78

1. **a)** $6 \times (2 + 1) = 18$ **b)** $1 + 3 \times 5 = 16$
   c) $(8 + 6) \div 2 = 7$ **d)** $(9 + 2) \times 4 = 44$
   e) $9 \div 3 \times 4 + 1 = 13$
   f) $(3 + 2) \times (4 - 1) = 15$
2. **a)** $12 \div 4 - 2 + 6 = 7$
   b) $12 \div (4 - 2) + 6 = 12$
   c) $12 \div 4 - (2 + 6) = -5$
   d) $12 \div (4 - 2 + 6) = 1.5$
   e) $4 + 5 \times 6 - 1 = 33$
   f) $4 + 5 \times (6 - 1) = 29$
   g) $(4 + 5) \times 6 - 1 = 53$
   h) $(4 + 5) \times (6 - 1) = 45$

Exercise 12.4 page 79

1. **a)** 2 **b)** 3 **c)** 7 **d)** 4 **e)** 23 **f)** 0
2. **a)** 1 **b)** 5 **c)** 2 **d)** 50 **e)** 7 **f)** −1.5

Student Assessment 1 page 80

1. **a)** 25.1 **b)** 7.3 **c)** 4.9 **d)** 13 **e)** −13.86 **f)** 3.75
2. **a)** 44 **b)** 13 **c)** 25 **d)** 19 **e)** 49 **f)** 3
3. **a)** $(7 - 4) \times 2 = 6$
   b) $12 + 3 \times (3 + 4) = 33$
   c) $(5 + 5) \times (6 - 4) = 20$
   d) $(5 + 5) \times 6 - 4 = 56$
4. **a)** 3.5 **b)** 10

Student Assessment 2 page 80

1. **a)** 15.12 **b)** −3.6 **c)** −2.1 **d)** 9.4 **e)** −14.76 **f)** 7.2
2. **a)** 20 **b)** 15 **c)** 11 **d)** 13 **e)** 1 **f)** 4

3. a) $(7 - 5) \times 3 = 6$
   b) $16 + 4 \times (2 + 4) = 40$
   c) $(4 + 5) \times (6 - 1) = 45$
   d) $(1 + 5) \times 6 - 6 = 30$

4. a) 5.5 b) 9

13 Measures

Exercise 13.1 page 81

1. a) 100 b) 1000 c) $\frac{1}{1000}$ d) 1000
   e) 1000 f) $\frac{1}{1000}$ g) Tonne
   h) 1 milligram i) 1 litre j) Millilitre

2. a) m, cm b) cm c) g d) ml
   e) cm, m f) tonne g) litres h) km
   i) tonne j) litres

3. Students draw lines

4. Students draw, estimate length of and measure lines

5. a) m b) km c) m d) kg e) l
   f) m g) tonne h) m i) g j) l

Exercise 13.2 page 83

1. a) mm b) mm c) km d) m e) mm
   f) m g) mm h) m i) km j) m

2. a) 20 b) 85 c) 230 d) 1200 e) 830
   f) 50 g) 625 h) 87 i) 4 j) 2000

3. a) 3000 b) 4.7 c) 5.6 d) 6400 e) 800
   f) 0.96 g) 0.625 h) 87 i) 4 j) 0.012

4. a) 5 b) 6.3 c) 1.15 d) 2.535 e) 250
   f) 0.5 g) 0.07 h) 0.008 i) 1000
   j) 700 000

Exercise 13.3 page 84

1. a) 3800 b) 28.5 c) 4280 d) 0.32
   e) 500

2. 1.35

3. a) 3530 kg b) 3.77 g c) 9.3 kg
   d) 5.15 kg e) 14 kg

Exercise 13.4 page 84

1. a) 8.4 b) 0.65 c) 87.5 d) 0.05
   e) 2.5

2. a) 3200 b) 750 c) 87 d) 8000 e) 8
   f) 300

3. a) 4500 ml b) 1530 ml c) 7050 ml
   d) 1000 ml

4. a) 1.2 litres b) 1.34 litres c) 1.4 litres
   d) 1.4 litres

Exercise 13.5 page 85

1. a) 160 cm b) 22.5 cm c) 71.2 cm
   d) 1.52 m

2. a) 4, 1 b) 4, 1 c) 8, 4 d) 2, 0.25
   e) 1, 0.0625 f) 3, 0.5625 g) 5, 1
   h) 4.5, 0.5

3. a) 400, 10 000 b) 400, 10 000
   c) 800, 40 000 d) 200, 2500 e) 100, 625
   f) 300, 5625 g) 500, 10 000 h) 450, 5000

4. a) 1 b) 0.125 c) 0.015 625 d) 0.001

5. a) 0.5 b) 0.031 25 c) 0.25 d) 0.08

Student Assessment 1 page 87

1. a) 26 b) 625 c) 88 d) 7 e) 4.8
   f) 7810 g) 6.8 h) 875 i) 2000 j) 85

2. a) 4200 b) 0.75 c) 3.94 d) 4100
   e) 720 f) 4.1 g) 6.28 h) 830 000
   i) 47 000 j) 1 000 000

3. 8330

4. a) 1.8 b) 3200 c) 83 d) 250

5. a) 180 b) 52 c) 340 d) 180

6. a) 720 cm^2 b) 1.05 m^2

7. a) 1200 cm^3 b) 0.012 m^2

Student Assessment 2 page 88

1. a) 47 b) 3 c) 310 d) 6400 e) 49
   f) 4000 g) 4 h) 34 i) 46 j) 50

2. a) 0.0036 b) 0.55 c) 6.5 d) 6700
   e) 370 f) 1.51 g) 380 h) 77 i) 6
   j) 2 million

3. 430 kg

4. a) 3.4 b) 6700 c) 730 d) 300

5. a) 180 b) 190 c) 240

6. a) 128 cm^2 b) 0.75 m^2

7. a) 0.375 m^3 b) 0.04 m^3

14 Time

Exercise 14.1 page 90

1. a) 14.30 b) 21.00 c) 08.45 d) 06.00
   e) 12.00 f) 22.55 g) 07.30 h) 19.30
   i) 01.00 j) 00.00

2. **a)** 07.15 **b)** 20.00 **c)** 09.10 **d)** 08.45
e) 14.45 **f)** 19.40

3. **a)** 7.20 a.m. **b)** 9.00 a.m. **c)** 2.30 p.m.
d) 6.25 p.m. **e)** 11.40 p.m. **f)** 1.15 a.m.
g) 12.05 a.m. **h)** 11.35 a.m. **i)** 5.50 p.m.
j) 11.59 p.m. **k)** 4.10 a.m. **l)** 5.45 a.m.

4. **a)** 08.05 **b)** 08.40 **c)** 09.05 **d)** 09.30

5. **a)** 18.20 **b)** 18.45 **c)** 19.00 **d)** 19.15

6. **a)** 08.25 **b)** 09.35 **c)** 09.50 **d)** 10.05

7.

Depart	Arrive
06.15	**07.55**
06.30	08.10
09.25	**11.05**
10.20	12.00
13.18	**14.58**
14.48	16.28
18.54	**20.34**
19.25	21.05

8.

Depart	Arrive
06.00	**08.05**
06.45	08.50
08.55	**11.00**
09.09	11.14
13.48	**15.53**
14.17	16.22
21.25	**23.30**
22.05	00.10

9.

Cambridge	04.00	08.35	12.50	19.45	21.10
Stansted	05.15	**09.50**	**14.05**	**21.00**	**22.25**
Gatwick	06.50	**11.25**	**15.40**	**22.35**	**00.00**
Heathrow	07.35	**12.10**	**16.25**	**23.20**	**00.45**

Heathrow	06.25	09.40	14.35	18.10	22.15
Gatwick	08.12	**11.27**	**16.22**	**19.57**	**00.02**
Stansted	10.03	**13.18**	**18.13**	**21.48**	**01.53**
Cambridge	11.00	**14.15**	**19.10**	**22.45**	**02.50**

10.

	London	Jo'burg	London	Jo'burg
Sunday	06.15	**17.35**	14.20	**01.40**
Monday	**07.25**	18.45	**18.05**	05.25
Tuesday	07.20	**18.40**	15.13	**02.33**
Wednesday	**07.52**	19.12	**20.10**	07.30
Thursday	06.10	**17.30**	16.27	**03.47**
Friday	**06.05**	17.25	**20.55**	08.15
Saturday	09.55	**21.15**	18.50	**06.10**

11.

	London	Kuala Lumpur	London	Kuala Lumpur	London	Kuala Lumpur
Sunday	08.28	**22.13**	14.00	**03.45**	18.30	**08.15**
Monday	**08.15**	22.00	**13.30**	03.15	**20.05**	09.50
Tuesday	09.15	**23.00**	15.25	**05.10**	17.55	**07.40**
Wednesday	**07.50**	21.35	**14.15**	04.00	**18.37**	08.22
Thursday	07.00	**20.45**	13.45	**03.30**	18.40	**08.25**
Friday	**10.25**	00.10	**15.00**	04.45	**17.53**	07.38
Saturday	10.12	**23.57**	14.20	**04.05**	19.08	**08.53**

Exercise 14.2 page 92

1. **a)** $T = d/s = \frac{240}{60} = 4$ h **b)** $\frac{340}{40} = 8$ h 30 min
c) $\frac{270}{80} = 3$ h 22.5 min **d)** $\frac{100}{60} = 1$ h 40 min
e) $\frac{70}{30} = 2$ h 20 min **f)** $\frac{560}{90} = 6$ h 13.3 min
g) $\frac{230}{100} = 2$ h 18 min **h)** $\frac{70}{50} = 1$ h 24 min
i) $\frac{4500}{750} = 6$ h **j)** $\frac{6000}{800} = 7$ h 30 min

2. (1) 30 min (2) 32 min 44 s (3) 34 min 17 s
(4) 35 min 7 s (5) 36 min

3. $\frac{1500}{100} = 15$ h. Therefore, leave
09.30 + 15 h = 00.30

4. $\frac{3200}{600} = 5$ h 20 min. It will land in Athens at
16.25 + 5 h 20 min = 21.45

5. $\frac{5200}{800} = 6.5$ h. Leaves 09.45 arrives 16.15, less 5 h
time difference = 11.15

Student Assessment 1 page 94

1. **a)** 04.35 **b)** 18.30 **c)** 07.45 **d)** 19.30

2. **a)** 8.45 a.m. **b)** 6.35 p.m. **c)** 9.12 p.m.
d) 12.15 a.m.

3. **a)** 08.10 **b)** 08.40 **c)** 09.03

4.

Alphaville	07.50	**11.01**	**15.53**
Betatown	**08.27**	11.38	**16.30**
Gammatown	**08.45**	**11.56**	16.48
Deltatown	**09.27**	**12.38**	**17.30**

5. **a)** $\frac{250}{50} = 5$ h **b)** $\frac{375}{100} = 3$ h 45 min
c) $\frac{80}{60} = 1$ h 20 min **d)** $\frac{200}{120} = 1$ h 40 min
e) $\frac{70}{30} = 2$ h 20 min **f)** $\frac{300}{80} = 3$ h 45 min

Student Assessment 2 page 94

1. **a)** 05.20 **b)** 20.15 **c)** 08.50 **d)** 23.30

2. **a)** 7.15 a.m. **b)** 4.43 p.m. **c)** 7.30 p.m.
d) 12.35 a.m.

3. **a)** 08.10 **b)** 08.39 **c)** 9.00

4.

Apple	10.14	**15.42**	**19.33**
Peach	**11.52**	17.20	**21.11**
Pear	**13.56**	**19.24**	23.15
Plum	**15.49**	**21.17**	**01.08**

5. **a)** $\frac{350}{70} = 5$ h **b)** $\frac{425}{100} = 4$ h 15 min
c) $\frac{160}{60} = 2$ h 40 min **d)** $\frac{450}{120} = 3$ h 45 min
e) $\frac{600}{160} = 3$ h 45 min

15 Money

Exercise 15.1 page 96

1. **a)** \$A32 **b)** \$A48 **c)** \$A8 **d)** \$A40
e) \$A56 **f)** \$A24

2. **a)** €12.50 **b)** €18.75 **c)** €25
d) €21.88 **e)** €15.63 **f)** €30

3. Student's own graph
a) 500 rupees **b)** 2000 rupees
c) 2500 rupees **d)** 4500 rupees
e) 50 000 rupees **f)** 36 000 rupees

4. **a)** €2 **b)** €14.40 **c)** €12.60 **d)** €9
e) €108 **f)** €54

5. Student's own graph
a) 1000 \$ **b)** 2500 \$ **c)** 3750 \$
d) 1500 \$ **e)** 12 500 \$

6. **a)** €21.05 **b)** €57.89 **c)** €70.53
d) €1052.63

7. Student's own graph
L120 m = €85.71 L400 m = €285.71
L880 m = €628.57

8. Student's own graph
a) i) Y37 895 ii) Y94 373 iii) Y126 316
b) i) US\$39.58 ii) US\$67.29 iii) US\$7917

9. Student's own graph
a) \$6250 **b)** \$25 000 **c)** \$15 000
d) \$75 000

10. Student's own graph
a) L14 million **b)** L19.6 million
c) L33.6 million

Student Assessment 1 page 98

1. Student's own graph
a) 240 rupees **b)** 1920 rupees
c) 40 000 rupees

2. **a)** HK\$15 **b)** HK\$56.25 **c)** HK\$125

3. Student's own graph
a) €164.10 **b)** €36.92 **c)** €492.30

4. **a)** NZ\$195 **b)** NZ\$68.25 **c)** NZ\$877.50

5. **a)** 1200 rupees **b)** 6000 rupees
c) 2000 rupees

Student Assessment 2 page 98

1. Student's own graph
a) \$A84.38 **b)** \$A50.63 **c)** \$A118.13

2. **a)** €32 **b)** €48 **c)** €59.26

3. Student's own graph
a) HK\$296 **b)** HK\$5920 **c)** HK\$740

4. **a)** NAB\$139.73 **b)** NAB\$838.38
c) NAB\$1270.27

5. **a)** C\$52.66 **b)** C\$78.79 **c)** C\$11.97

16 Personal and Household Finance

Exercise 16.1 page 100

1. **a)** £138.50 **b)** £164.14 **c)** £154.82
d) £165.06

2. **a)** 229 **b)** £25 **c)** £123 **d)** £55
e) £182

3. **a)** £131 **b)** £306 **c)** £72 **d)** £15

4. **a)** £126 **b)** £214.72 **c)** £190.76
d) £290.50 **e)** £348

5.

Basic hours worked	Rate per hour (£)	Basic pay (£)	Overtime hours worked	Overtime pay (£)	Total gross pay (£)
a) 40	3.60	**144**	8	**43.20**	**187.20**
b) 35	**5.80**	203	4	**34.80**	**237.80**
c) 38	4.15	**157.70**	6	**37.35**	**195.05**
d) **42**	6.10	256.20	5	**45.75**	**301.95**
e) 44	5.25	**231.00**	4	**31.50**	**262.50**
f) **37**	4.87	180.19	3	**21.92**	**202.11**
g) 36	6.68	**240.48**	6	**60.12**	**300.60**
h) 45	7.10	319.50	7	**74.55**	**394.05**

6. **a)** £127.30 **b)** £161.70 **c)** £132.63
d) £205.33 **e)** £178.50 **f)** £137.43
g) £204.41 **h)** £267.95

Exercise 16.2 page 101

1.

	Mon	Tue	Wed	Thur	Fri	Total	Gross pay
Pepe	4	5	7	6	6	**28**	**154**
Felicia	3	4	4	5	5	**21**	**115.50**
Delores	5	6	6	5	6	**28**	**154**
Juan	3	4	6	6	6	**25**	**137.50**

2.

	Mon	Tue	Wed	Thur	Fri	Total	Gross pay
Maria	240	360	288	192	180	**1260**	**525**
Bea	168	192	312	180	168	**1020**	**425**
Joanna	288	156	192	204	180	**1020**	**425**
Bianca	228	144	108	180	120	**780**	**325**
Selina	192	204	156	228	144	**924**	**385**

3.

	Gross pay (rand)	Net pay (rand)
Smith	308	261.80
Jones	395	335.75
White	338	287.30
Green	431	366.35
Brown	453	385.05

4. **a)**

km walked	Amount per km in euro	Total raised
10	0.80	8
14	0.65	9.10
18	0.38	**6.84**
10.15	0.72	7.31
12	**0.66**	7.92
13	1.20	15.60
15	1	**15**
18	0.88	15.84
18	0.58	10.44
17	0.95	16.15

b) 112.20 euro
c) Total of 10 parts, therefore 1 part = 11.22 euro
Charity 1 receives €22.44
Charity 2 receives €33.66
Charity 3 receives €56.10

Exercise 16.3 page 103

1. **a)** £72 **b)** £420 **c)** £102 **d)** £252
e) £3250 **f)** £369.60

2. **a)** $t = 5$ **b)** $t = 0.41$ **c)** $t = 5$ **d)** $t = 8$
e) $t = 6$ **f)** $t = 7$

3. **a)** $r = 7$ **b)** $r = 4$ **c)** $r = 3.5$ **d)** $r = 7.5$
e) $r = 8$ **f)** $r = 11$

4. **a)** $P = 400$ **b)** $P = 200$ **c)** $P = 850$
d) $P = 1200$ **e)** $P = 4000$ **f)** $P = 1200$

5. $r = 4$

6. $t = 2$

7. $r = 4.5$

8. $r = 9.5$

9. £315

10. $r = 6$

11. £11 033 750

12. 52 087.50 euro

13. £10 368

14. 1331 students

15. 3 276 800 tonnes
16. 2 years
17. 5 years
18. 3 years
19. 4 years
20. 5 years

Exercise 16.4 page 107

1. £1.80 loss
2. £2.88 profit
3. £70.38 profit per seat
4. £240 extra
5. £250 loss
6. **a)** £4.50 profit **b)** £16 profit **c)** £80 profit **d)** £13 profit **e)** £2.88 profit **f)** £23 loss

Exercise 16.5 page 108

1. **a)** 11% **b)** 25% **c)** 50% **d)** 20% **e)** 30% **f)** 66.7%
2. **a)** 25% **b)** 30% **c)** 20% **d)** 50% **e)** 30.8% **f)** 30%
3. Type A = 30% Type B = 15.4%
   Type C = 33.3% Type D = 30.9%
   Type E = 48.5%
   Type E makes most profit.
4. 80%
5. Mirror = 30% Wayfarer = 40%
   Laser = 45% Fireball = 37.6%

Student Assessment 1 page 109

1. £225 (£3.75 × 3 h × 5 days × 4 weeks)
2. £64.90 (£16.80 + £48.10)
3. **a)** £54 **b)** 4 years **c)** 6.5% **d)** £840
4. 12.5%
5. 11.1%

Student Assessment 2 page 109

1. £122.20
2. £26.16 (£12.96 + £13.20)
3. **a)** £72 **b)** 8% **c)** 4 years **d)** 7.5% **e)** £1250
4. £6000
5. 30%

ICT Section

Number Activity 1 page 110

1.

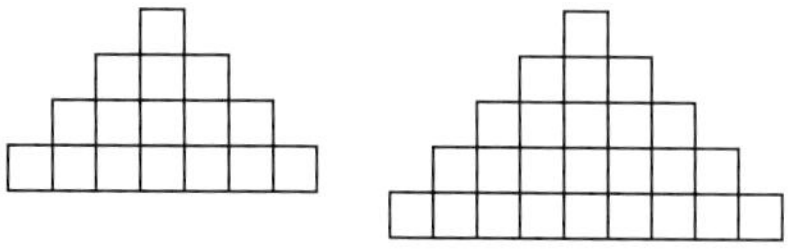

2.

	A	B
1	Pattern	Number of squares
2	1	1
3	2	4
4	3	9
5	4	16
6	5	25
7	10	
8	20	
9	50	

3. Number of squares is the pattern number squared.

4.

	A	B
1	Pattern	Number of squares
2	1	1
3	2	4
4	3	9
5	4	16
6	5	25
7	10	100
8	20	400
9	50	2500

The formula in cell B7 is =A7^2 or =A7*A7

5. **a)** i)

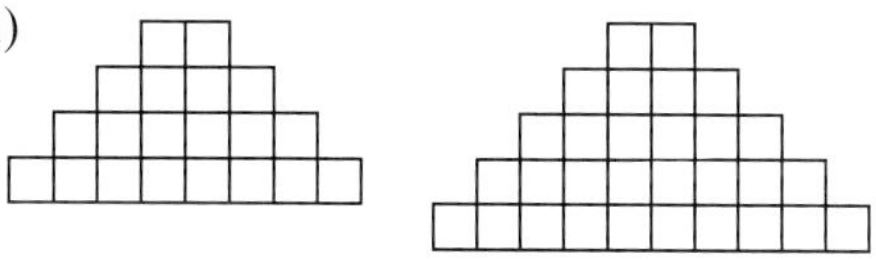

ii)

	A	B
1	Pattern	Number of squares
2	1	2
3	2	6
4	3	12
5	4	20
6	5	30
7	10	
8	20	
9	50	

iii) The number of squares is the pattern number squared + the pattern number.

iv)

	A	B
1	Pattern	Number of squares
2	1	2
3	2	6
4	3	12
5	4	20
6	5	30
7	10	110
8	20	420
9	50	2550

The formula in cell B7 is =A7^2+A7

b) i)

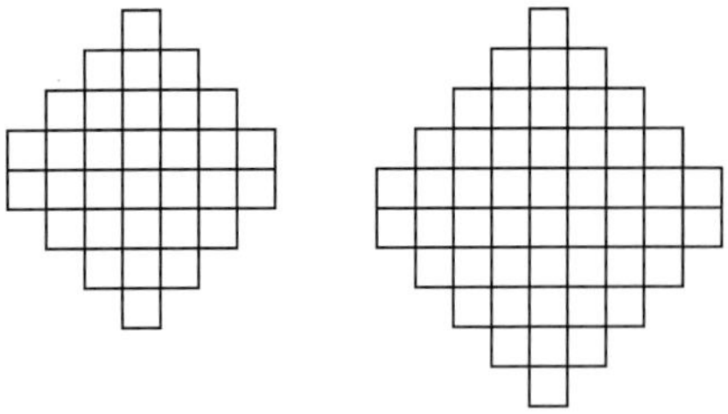

ii)

	A	B
1	Pattern	Number of squares
2	1	2
3	2	8
4	3	18
5	4	32
6	5	50
7	10	
8	20	
9	50	

iii) The number of squares is the pattern number squared and then multiplied by two.

iv)

	A	B
1	Pattern	Number of squares
2	1	2
3	2	8
4	3	18
5	4	32
6	5	50
7	10	200
8	20	800
9	50	5000

The formula in cell B7 is =2*A7^2

Number Activity 2 page 111

3. The example below shows a possible screen depicting an exchange rate between euro and pounds sterling of £1 = €1.40

	A	B	C	D	E
1	Currency Converter				
2	£		€		
3		£1=€1.40	=A3*1.40		
4					
5					
6	Write the exchange rate in this cell		Write a formula in this cell to convert one currency to the other		
7					
8					

Reviews for Chapters 1 to 16

Review 1 page 112

1. $2^3 \times 3 \times 5$

2. Student's drawing. 10.24 units^2

3. -12 °C

4. a) 0.6 **b)** 0.75 **c)** 0.375

5. 0.05 m 18 cm 236 mm $\frac{1}{4}$m

6. 8.5×10^7

7. a) $\frac{3}{7}$ **b)** $\frac{27}{63}$

8. 26 400 yds

9. 31.2 cm

10. area = 1.21 cm^2 perimeter = 4.4 cm

Review 2 page 112

1. $3\sqrt{8}$ $\sqrt{7}$

2. a) $175 \leqslant x < 185$ **b)** $0.745 \leqslant x < 0.755$

3. 42

4. 196

5. a) 0745 **b)** 2030

6. Student's graph
a) 57 rand **b)** £6.32

7. a) 1.6 m **b)** 2800 m **c)** 0.12 m

8. £146.25

9. 40.5 cm

10. 5400 cm^3

Review 3 page 113

1. 36

2.

x	0	1	2	3	4	5	6	7	8
y	0	1	4	9	16	25	36	49	64

a) ≈ 10.2 **b)** ≈ 56.3

3. 56 °C

4. a) $\frac{1}{4}$ **b)** $\frac{5}{8}$ **c)** $\frac{21}{20}$

5. a)

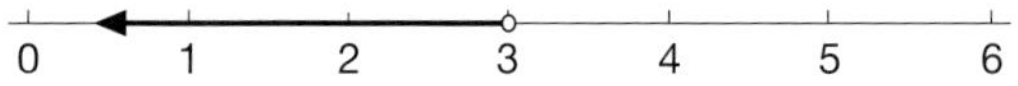

b)

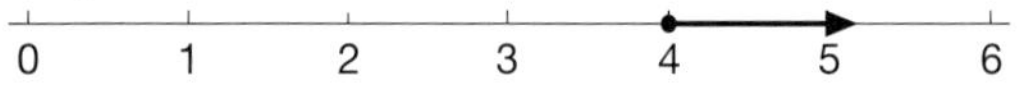

c)

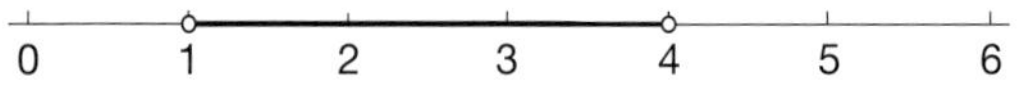

d)

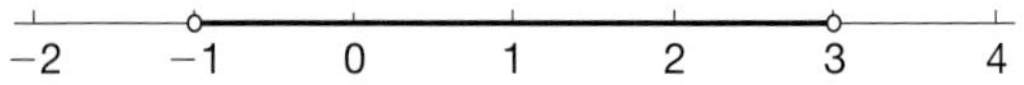

6. 7.83×10^{-6}

7. $\frac{19}{10}$

8. **a)** 70 **b)** 480 000 **c)** 600 **d)** 645

9. 140 cm

10. area = 56.25 cm^2 perimeter = 30 cm

Review 4 page 113

1. **a)** 60 48 **b)** 34 55

2. **a)**

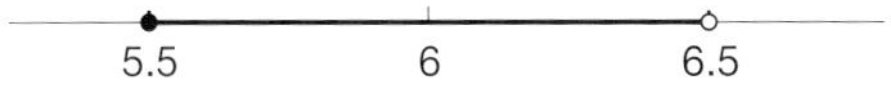

b)

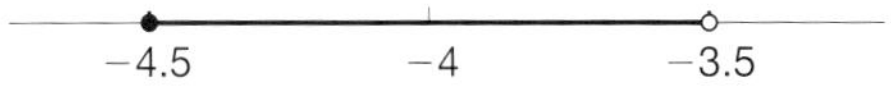

c)

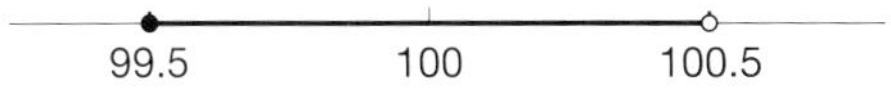

3. 6000

4. 486 euro

5. 10 : 35

6. Student's conversion graph
   a) ≈ \$146 **b)** ≈ 135 rand

7. **a)** 28 kg **b)** 2.4 kg **c)** 0.15 kg

8. 4 years

9. 19 cm

10. 3.375 cm^3

Review 5 page 114

1. 168

2. **a)** 7 **b)** 0.9 **c)** 0.2

3. 2130 m

4. 5865

5. $28 \leqslant n \leqslant 31$

6. 1.3×10^{12}

7. 6

8. ≈ 10

9. 43 cm

10. 1.331 cm^3

Review 6 page 114

1. **a)** and **c)** are rational

2.

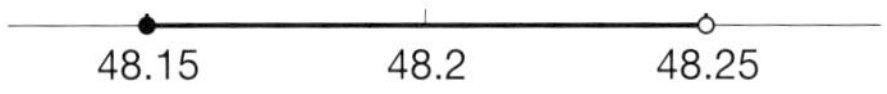

3. 2.25 litres

4. £196

5. 1545

6. Student's conversion graph
   a) ≈ \$22 **b)** ≈ 2860 taka

7. **a)** 8.5 litres **b)** 0.4 litre **c)** 15.5 litres

8. 6%

9. area = 42.25 cm^2 perimeter = 26 cm

10. 768 cm^3

Review 7 page 115

1. **a)** −10 **b)** 24 **c)** −6

2. **a)** $\frac{6}{7}$ **b)** $\frac{10}{9}$ **c)** $\frac{5}{2}$

3. 380 + present year

4. 210.4

5. $\frac{1}{2}$ $\frac{5}{9}$ $\frac{4}{7}$ $\frac{3}{5}$ $\frac{2}{3}$

6. 3.8×10^{13}

7. $\frac{45}{7}$

8. 480 m^2

9. 44.1 cm

10. 13.824 cm^3

Review 8 page 116

1. **a)** $6n + 1$ **b)** $5n - 5$

2. **a)** $4.545 \leqslant x < 4.555$ **b)** $-10.05 < x \leqslant -9.95$

3. 12 : 16

4. 480

5. 2.5 h

6. Student's conversion graph
   a) 1200 rupees **b)** ≈ 14.2 euro

7. **a)** 6.5 m **b)** 0.0085 m **c)** 0.15 m

8. £51.00

9. perimeter = 370 m area = 8250 m^2

10. 42.875 cm^3

Review 9 page 116

1. $-3, -2, -1, -\frac{1}{7}, 0.04, 1, 1.5$

2. **a)** 6.25 **b)** 0.81 **c)** 1.0201

3. 35 BC

4. **a)** 80% **b)** 87.5% **c)** 225%

5. **a)** = **b)** < **c)** < **d)** <

6. 6.66×10^7

7. $\frac{25}{24}$

8. 120 km

9. 20 : 25

10. 224

Review 10 page 117

1. a) +3 to previous term
b) Add two previous terms together to get the next term.

2. a)

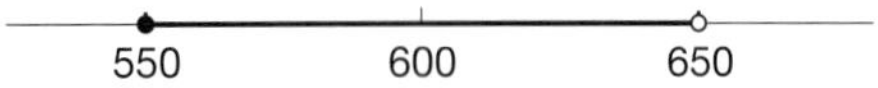

b)

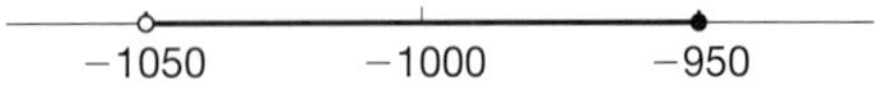

3. 12 : 18 : 30

4. \$207

5. 510 km

6. Student's conversion graph
a) 4 125 000 lira **b)** \$5.82

7. a) 85 g **b)** 3 g **c)** 500 g

8. 281.25 euro

9. $\frac{128}{25}$

10. a) 60% **b)** 75% **c)** 37.5%

Review 11 page 117

1. a) LCM = 12 HCF = 2
b) LCM = 117 HCF = 1

2. a) 8 **b)** 15 **c)** 0.09

3. −4.5 °C

4.

Fraction	Decimal	Percentage
$\frac{1}{8}$	**0.125**	**12.5%**
$\frac{3}{8}$	0.375	**37.5%**
$\frac{7}{5}$	**1.4**	140%

5. a)

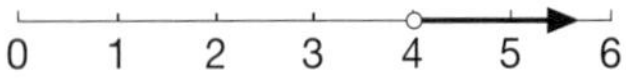

b)

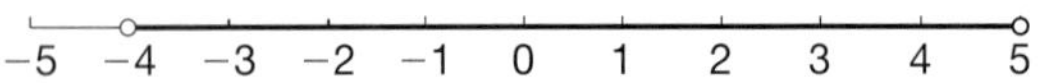

c)

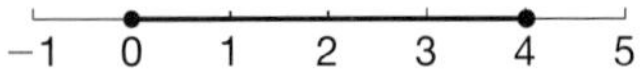

d)

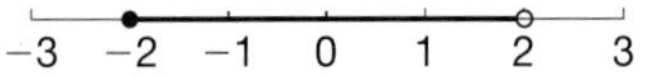

6. 2600 s

7. 3

8. ≈ 800 cm^3

9. $\frac{45}{8}$

10. £168

Review 12 page 118

1. Student's reply

2. a) $265.5 \leqslant x < 266.5$ **b)** $30.65 \leqslant x < 30.75$

3. 7.5 h

4. 30%

5. 133 km/h (3 s.f.)

6. Student's conversion graph
a) 2100 rupees **b)** 50 dollars

7. a) 2500 ml **b)** 30 ml **c)** 7.5 ml

8. £128.70

9. area = 30 cm^2 perimeter = 30 cm

10. 7.75×10^6

17 Indices

Exercise 17.1 page 120

1. a) 3^3 **b)** 2^5 **c)** 4^2 **d)** 6^4 **e)** 8^6
f) 5^1

2. a) $2^3 \times 3^2$ **b)** $4^5 \times 5^2$ **c)** $3^2 \times 4^3 \times 5^2$
d) 2×7^4 **e)** 6^2 **f)** $3^3 \times 4^2 \times 6^5$

3. a) 4×4 **b)** $5 \times 5 \times 5 \times 5 \times 5 \times 5 \times 5$
c) $3 \times 3 \times 3 \times 3 \times 3$
d) $4 \times 4 \times 4 \times 6 \times 6 \times 6$
e) $7 \times 7 \times 2 \times 2 \times 2 \times 2 \times 2 \times 2 \times 2$
f) $3 \times 3 \times 4 \times 4 \times 4 \times 2 \times 2 \times 2 \times 2$

4. a) 32 **b)** 81 **c)** 64 **d)** 216
e) 1 000 000 **f)** 256 **g)** 72
h) 125 000

Exercise 17.2 page 121

1. a) 3^6 **b)** 8^7 **c)** 5^9 **d)** 4^{10} **e)** 2^4
f) $3^5 \times 6^6$ **g)** $4^8 \times 5^9 \times 6^2$
h) $2^4 \times 5^{10} \times 6^8$

2. a) 4^4 **b)** 5^3 **c)** 2 **d)** 6^3 **e)** 6^3
f) 8 **g)** 4^3 **h)** 3^7

3. a) 5^4 **b)** 4^{12} **c)** 10^{10} **d)** 3^{15}
e) 6^8 **f)** 8^6

4. **a)** 2^3 **b)** 3 **c)** 5^3 **d)** 4^5 **e)** 2^8
f) $6^4 \times 8^5$ **g)** $4^3 \times 5^2$ **h)** 4×6^7

5. **a)** c^8 **b)** m^2 **c)** b^9 **d)** m^3n^6
e) $2a^4b$ **f)** $3x^3y^2$ **g)** $\frac{uv^3}{2}$ **h)** $\frac{x^2y^3z^2}{3}$

6. **a)** $12a^5$ **b)** $8a^5b^3$ **c)** $8p^6$ **d)** $16m^4n^6$
e) $200p^{13}$ **f)** $32m^5n^{11}$ **g)** $24xy^3$
h) $a^{(d+e)}b^{(d+e)}$

Exercise 17.3 page 123

1. **a)** 8 **b)** 25 **c)** 1 **d)** 1 **e)** 1
f) 0.25

2. **a)** $\frac{1}{4}$ **b)** $\frac{1}{9}$ **c)** $\frac{3}{50}$ **d)** $\frac{1}{200}$ **e)** 1
f) $\frac{1}{1000}$

3. **a)** 1 **b)** 2 **c)** 4 **d)** $\frac{1}{2}$ **e)** $\frac{1}{6}$
f) 10

4. **a)** 12 **b)** 32 **c)** 225 **d)** 80 **e)** 7
f) 64

Student Assessment 1 page 124

1. **a)** $2^3 \times 5^2$ **b)** $2^2 \times 3^5$

2. **a)** $4 \times 4 \times 4$ **b)** $6 \times 6 \times 6 \times 6$

3. **a)** 800 **b)** 27

4. **a)** 3^7 **b)** $6^5 \times 3^9$ **c)** 2^7 **d)** 6
e) $3^2 \times 4^2$ **f)** 1

5. **a)** 4 **b)** 9 **c)** 5 **d)** 1

Student Assessment 2 page 124

1. **a)** $2^2 \times 3^5$ **b)** 2^{14}

2. **a)** $6 \times 6 \times 6 \times 6 \times 6$
b) $\frac{1}{2 \times 2 \times 2 \times 2 \times 2}$

3. **a)** 27 000 **b)** 125

4. **a)** 2^7 **b)** $7^7 \times 3^{12}$ **c)** 2^6 **d)** 3^3
e) 4^{-1} **f)** 2^2

5. **a)** 5 **b)** 16 **c)** 49 **d)** 48

18 Algebraic Representation and Manipulation

Exercise 18.1 page 125

1. **a)** $2a + 6$ **b)** $4b + 28$ **c)** $10c + 40$
d) $21d + 63$ **e)** $72e - 63$ **f)** $24f - 18$

2. **a)** $3a^2 + 6ab$ **b)** $8ab + 12b^2$
c) $2ac + 2bc + 2c^2$ **d)** $6bd + 9cd + 12d^2$
e) $3ce - 3de - e^2$ **f)** $3df - ef - 2f^2$

3. **a)** $4a^2 + 6b^2$ **b)** $12a^2 + 16b^2$ **c)** $-6c - 9d$
d) $-2c - 3d$ **e)** $-4c^2 + 8d^2 - 12e^2$
f) $-10e + 15f^2$

4. **a)** $2a^2 + 2ab$ **b)** $3ab - 3b^2$ **c)** $4b^2c - 4c^3$
d) $3a^2d^2 - 6b^2d^2 + 3c^2d^2$ **e)** $-12de^2 + 3e^3$
f) $-4df + 6e^2f + 4f^2$

Exercise 18.2 page 125

1. **a)** $4x - 12$ **b)** $10p - 20$ **c)** $-42x + 24y$
d) $6a - 9b - 12c$ **e)** $-14m + 21n$
f) $-16x + 6y$

2. **a)** $3x^2 - 9xy$ **b)** $a^2 + ab + ac$
c) $8m^2 - 4mn$ **d)** $-15a^2 + 20ab$
e) $4x^2 - 4xy$ **f)** $24p^2 - 8pq$

3. **a)** $-2x^2 + 3y^2$ **b)** $a - b$ **c)** $7p - 2q$
d) $3x - 4y + 2z$ **e)** $3x - \frac{3}{2}y$
f) $2x^2 - 3xy$

4. **a)** $12r^3 - 15rs + 6rt$ **b)** $a^3 + a^2b + a^2c$
c) $6a^3 - 9a^2b$ **d)** $p^2q + pq^2 - p^2q^2$
e) $m^3 - m^2n + m^3n$ **f)** $a^6 + a^5b$

Exercise 18.3 page 126

1. **a)** $8a + 4$ **b)** $7b - 8$ **c)** $5c - 6$
d) $d - 8$ **e)** $-2e - 1$ **g)** $3f + 7$

2. **a)** $2a + 3b + 8$ **b)** $4a - 6b + 2$ **c)** $5c - 7$
d) $d + 2$ **e)** $-3e + 7$ **f)** $-2f - 2$

3. **a)** $2a^2 + 6a + 2b^2 - 2b$
b) $3a^2 - 2b^2 - 12a + 6b$
c) $2a^2 - 2b^2 + 2ac + 2bc$ **d)** $a^2d^2 - c^2d^2$
e) $ac + bc$ **f)** $2ad + ae + 2ce$

Exercise 18.4 page 126

1. **a)** $-a - 8$ **b)** $4x - 20$ **c)** $3p - 16$
d) $21m - 6n$ **e)** 3 **f)** $-p - 3p^2$

2. **a)** $8m^2 + 28m + 2$ **b)** $x - 4$ **c)** $2p + 22$
d) $m - 12$ **e)** $a^2 + 6a + 2$
f) $7ab - 16ac + 3c$

3. **a)** $4x + 4$ **b)** $5x - \frac{3}{2}y$ **c)** $\frac{9}{4}x - \frac{5}{2}y$
d) $\frac{9}{2}x - \frac{1}{2}y$ **e)** $7x - 4y$ **f)** 0

Exercise 18.5 page 127

1. **a)** $2(2x - 3)$ **b)** $6(3 - 2p)$ **c)** $3(2y - 1)$
d) $2(2a - 3b)$ **e)** $3(p - q)$
f) $4(2m + 3n + 4r)$

2. **a)** $a(3b + 4c - 5d)$ **b)** $2p(4q + 3r - 2s)$
c) $a(a - b)$ **d)** $2x(2x - 3y)$
e) $ab(c + d + f)$ **f)** $3m(m + 3)$

3. **a)** $3pq(r - 3s)$ **b)** $5m(m - 2n)$
c) $4xy(2x - y)$ **d)** $b^2(2a^2 - 3c^2)$
e) $12(p - 3)$ **f)** $6(7x - 9)$

4. **a)** $6(3 + 2y)$ **b)** $7(2a - 3b)$
c) $11x(1 + y)$ **d)** $4(s - 4t + 5r)$
e) $5q(p - 2r + 3s)$ **f)** $4y(x + 2y)$

5. **a)** $m(m + n)$ **b)** $3p(p - 2q)$
c) $qr(p + s)$ **d)** $ab(1 + a + b)$
e) $p^3(3 - 4p)$ **f)** $b^2c(7b + c)$

6. **a)** $m(m^2 - mn + n^2)$ **b)** $2r^2(2r - 3 + 4s)$
c) $7xy(8x - 4y)$ **d)** $18mn(4m + 2n - mn)$

7. **a)** $a(3a - 2b + 4c)$ **b)** $b(2a - 3b + 4c)$
c) $2c(a^2 - 2b^2 + 3bc)$ **d)** $13cd(3d + 4c)$

8. **a)** $4ac(3 - 2c + a)$ **b)** $17ab(2a - 3b)$
c) $11c^2(3a + 11c - b^2)$
d) $19c^2d^2(2c - 3d + 5)$

9. **a)** $\frac{5}{c}(3a - 5b + 2d)$ **b)** $\frac{23}{c^2}(2a - b)$
c) $\frac{1}{2a}(1 - \frac{1}{2})$ **d)** $\frac{1}{5d}(3 - \frac{1}{2} + \frac{4}{3})$

10. **a)** $\frac{1}{a}\left(\frac{5}{a} - 3\right)$ **b)** $\frac{3}{b}\left(\frac{2}{b} - 1\right)$
c) $\frac{1}{3a}\left(2 - \frac{3}{a}\right)$ **d)** $\frac{1}{5d}\left(\frac{3}{d} - 4\right)$

Exercise 18.6 page 128

1. **a)** 12 **b)** −1 **c)** −6 **d)** 5

2. **a)** −15 **b)** 10 **c)** 78 **d)** 24

3. **a)** 13 **b)** 34 **c)** −19 **d)** 57

4. **a)** −4 **b)** 4 **c)** −27 **d)** −27

5. **a)** −125 **b)** −125 **c)** 100 **d)** −100

Exercise 18.7 page 128

1. **a)** 0 **b)** 30 **c)** 14 **d)** 20 **e)** −13
f) −4

2. **a)** −3 **b)** −30 **c)** 20 **d)** −16
e) −40 **f)** 42

3. **a)** −160 **b)** −23 **c)** 42 **d)** −17
e) −189 **f)** 113

4. **a)** 48 **b)** −8 **c)** 15 **d)** 16 **e)** −5
f) 9

5. **a)** 12 **b)** −5 **c)** −5 **d)** 7 **e)** 7
f) 36

Exercise 18.8 page 129

1. **a)** $b = c - a$ **b)** $b = d - 2c$
c) $c = 4a - 2b$ **d)** $b = 2a - 3d$

2. **a)** $b = \frac{c}{a}$ **b)** $c = \frac{bd}{a}$
c) $a = \frac{c + 3}{b}$ **d)** $c = \frac{b - 4}{a}$

3. **a)** $n = r - m$ **b)** $m = p - n$
c) $n = 3p - 2m$ **d)** $q = 3x - 2p$
e) $a = \frac{cd}{b}$ **f)** $d = \frac{ab}{c}$

4. **a)** $x = \frac{4m}{3y}$ **b)** $r = \frac{7pq}{5}$ **c)** $x = \frac{c}{3}$
d) $x = \frac{y - 7}{3}$ **e)** $y = \frac{3r + 9}{5}$
f) $x = \frac{5y - 9}{3}$

5. **a)** $b = \frac{2a - 5}{6}$ **b)** $a = \frac{6b + 5}{2}$
c) $z = \frac{3x - 7y}{4}$ **d)** $x = \frac{4x + 7y}{3}$
e) $y = \frac{3x - 4z}{7}$ **f)** $p = \frac{8 + q}{2r}$

6. **a)** $p = 4r$ **b)** $p = \frac{4}{3r}$ **c)** $p = \frac{n}{10}$
d) $n = 10p$ **e)** $p = \frac{2t}{q + r}$ **f)** $q = \frac{2t}{p} - r$

7. **a)** $r = \frac{3m - n}{t(p + q)}$ **b)** $t = \frac{3m - n}{r(p + q)}$
c) $m = \frac{rt(p + q) + n}{3}$
d) $n = 3m - rt(p + q)$ **e)** $p = \frac{3m - n}{rt} - q$

f) $q = \frac{3m - n}{rt} - p$

8. a) $d = \frac{ab}{ce}$ b) $a = \frac{dec}{b}$ c) $c = \frac{ab}{de}$
d) $a = cd - b$ e) $b = d - \frac{a}{c}$
f) $c = \frac{a}{d - b}$

Student Assessment 1 page 130

1. a) $3a + 12$ b) $28b - 14$ c) $3c^2 + 12cd$
d) $15cd - 10d^2$ e) $-12e + 18f$
f) $-2f + 3g$

2. a) $6a + 12$ b) $b - 4$ c) $-4c + 1$
d) $5d + 1$ e) $e^2 - 3e + 6$
f) $-2e^2 + 2f^2 - 2de + 2df$

3. a) $3(a + 4)$ b) $5b(3b + 5)$
c) $2f(2c - 3d + 4e)$ d) $2d^3(2 - d)$

4. a) 12 b) 0 c) 50 d) −16

5. a) $a = c - b$ b) $b = c + d$
c) $c = \frac{bd}{a}$ d) $d = \frac{2c + e}{3}$
e) $e = \frac{3a}{f - g}$ f) $f = \frac{3a}{e} + g$

Student Assessment 2 page 130

1. a) $4a + 8$ b) $10b - 15$ c) $2c^2 + 4cd$
d) $6cd - 12d^2$ e) $-15e + 5f$ f) $f - 2g$

2. a) $7a + 10$ b) $5b - 9$ c) $-2c - 4$
d) $d - 2$ e) $e^2 - 3e + 6$
f) $df - 2ef - e^2 - f^2$

3. a) $7(a + 2)$ b) $13b(2b + 3)$
c) $3f(c - 2d + 3g)$ d) $5d^2(1 - 2d)$

4. a) −6 b) 1 c) 20 d) 9

5. a) $a = c + b$ b) $2c + 3d = b$
c) $\frac{ad}{b} = c$ d) $\frac{e + 3c}{5} = d$
e) $\frac{4a}{f + g} = e$ f) $\frac{4a}{e} - g = f$

Student Assessment 3 page 131

1. a) $10a - 30b + 15c$ b) $15x^2 - 27x$
c) $-15xy^2 - 5y^3$ d) $15x^3y + 9x^2y^2 - 3x^5$
e) $12 - p$ f) $14m^2 - 14m$ g) $4x + 3$
h) $\frac{13}{2}m^2 - 3m$

2. a) $4(3a - b)$ b) $x(x - 4y)$
c) $4p^2(2p - q)$ d) $8xy(3 - 2x + y)$

3. a) −21 b) 26 c) 43 d) 7 e) 12
f) 12

4. a) $q = x - 3p$ b) $n = \frac{3m - 8r}{5}$
c) $y = \frac{2mt}{3}$ d) $w = \frac{2y}{x} - y$ e) $p = \frac{xyt}{2rs}$
f) $x = w(m + n) - y$

Student Assessment 4 page 131

1. a) $6x - 9y + 15z$ b) $8pm - 28p$
c) $-8m^2n + 4mn^2$ d) $20p^3q - 8p^2q^2 - 8p^3$
e) $-2x - 2$ f) $22x^2 - 14x$ g) 2
h) $\frac{5}{2}x^2 - x$

2. a) $8(2p - q)$ b) $p(p - 6q)$
c) $5pq(p - 2q)$ d) $3pq(3 - 2p + 4q)$

3. a) 0 b) −7 c) 29 d) 7 e) 7
f) 35

4. a) $n = p - 4m$ b) $y = \frac{4x - 5z}{3}$
c) $y = \frac{10px}{3}$ d) $y = \frac{3w}{m} - x$
e) $r = \frac{pqt}{4mn}$ f) $q = r(m - n) - p$

19 Equations

Exercise 19.1 page 132

1. a) $a = 4$ b) $b = 2$ c) $c = 8$ d) $d = 8$
e) $e = 10$ f) $f = 6$

2. a) $a = 7$ b) $b = -9$ c) $c = 5$
d) $d = -8$

3. a) $a = 11$ b) $b = -12$ c) $c = 13$
d) $d = 3$

4. a) $a = 5$ b) $b = 5$ c) $c = 3$
d) $d = -1$

5. a) $a = 12$ b) $b = 8$ c) $c = 10$
d) $d = 15$ e) $e = 12$ f) $f = -16$

6. a) $a = 9$ b) $b = 20$ c) $c = 18$
d) $d = -35$ e) $e = 6$ f) $f = -4$

7. **a)** $a = 4.5$ **b)** $b = \frac{10}{3}$ **c)** $c = \frac{5}{2}$
d) $d = 11\frac{1}{5}$ **e)** $e = -16$ **f)** $f = 14$

8. **a)** $a = 5$ **b)** $b = 1$ **c)** $c = 17$ **d)** $d = 11$
e) $e = 8$ **f)** $f = 8$

9. **a)** $a = 2$ **b)** $b = 7$ **c)** $c = 7$
d) $d = -0.5$ **e)** $e = -2$ **f)** $f = 0.5$

10. **a)** $a = 13$ **b)** $b = -23$ **c)** $c = 27$
d) $d = -1$ **e)** $e = 7$ **f)** $f = -1$

Exercise 19.2 page 133

1. **a)** $x = -4$ **b)** $y = 5$ **c)** $y = -5$
d) $p = -4$ **e)** $y = 8$ **f)** $x = -5.5$

2. **a)** $x = 4\frac{1}{3}$ **b)** $x = 5$ **c)** $x = 6$
d) $y = -8$ **e)** $y = 4$ **f)** $m = 10$

3. **a)** $m = 1$ **b)** $p = 3$ **c)** $k = -1$
d) $x = -21$ **e)** $x = 2$ **f)** $y = 3$

4. **a)** $x = 6$ **b)** $y = 14$ **c)** $x = 4$
d) $m = 12$

5. **a)** $x = 15$ **b)** $x = -5$ **c)** $x = 7.5$
d) $x = 8$ **e)** $x = 2.5$ **f)** $x = 10$

6. **a)** $x = 5$ **b)** $x = 14$ **c)** $x = 22$
d) $x = 5$ **e)** $x = 8$ **f)** $x = 2$

7. **a)** $y = 10$ **b)** $x = 17$ **c)** $x = 13$
d) $y = -5$ **e)** $x = 4$ **f)** $x = 6.5$

Exercise 19.3 page 135

1. **a)** i) $3x + 60 = 180$ ii) $x = 40°$
iii) 40°, 60°, 80°
b) i) $3x = 180$ ii) $x = 60°$ iii) 20°, 80°, 80°
c) i) $18x = 180$ ii) $x = 10°$
iii) 20°, 50°, 110°
d) i) $6x = 180$ ii) $x = 30°$ iii) 30°, 60°, 90°
e) i) $7x - 30 = 180$ ii) $x = 30°$
iii) 10°, 40°, 130°
f) i) $9x - 45 = 180$ ii) $x = 25°$
iii) 25°, 55°, 100°

2. **a)** i) $12x = 360$ ii) $x = 30°$
iii) 90°, 120°, 150°
b) i) $11x + 30 = 360$ ii) $x = 30°$
iii) 90°, 135°, 135°
c) i) $12x + 60 = 360$ ii) $x = 25°$
iii) 35°, 80°, 90°, 155°
d) i) $10x + 30 = 360$ ii) $x = 33°$
iii) 33°, 94°, 114°, 119°

3. **a)** i) $11x - 80 = 360$ ii) $x = 40°$
iii) 40°, 80°, 80°, 160°
b) i) $10x + 60 = 360$ ii) $x = 30°$
iii) 45°, 90°, 95°, 130°
c) i) $9x + 45 = 360$ ii) $x = 35°$
iii) 35°, 55°, 120°, 150°
d) i) $16x + 8 = 360$ ii) $x = 22°$
iii) 44°, 96°, 100°, 120°
e) i) $6x + 18 = 180$ ii) $x = 27°$
iii) 50°, 50°, 130°, 130°
f) i) $14x - 6 = 540$ ii) $x = 39°$
iii) 77°, 86°, 105°, 119°, 153°

4. **a)** 20° **b)** 25° **c)** 14 **d)** 25°
e) 31 **f)** 40°

5. **a)** 50° **b)** 13° **c)** 40° **d)** 40°

6. **a)** 5 **b)** 2 **c)** 7 **d)** 1.1 **e)** 25
f) 15

7. **a)** $n = 3$ **b)** $n = 7$

8. **a)** $n = 10$ **b)** $n = 4$

9. **a)** $n = 8$ **b)** $n = 14$

10. **a)** $n = 2$ **b)** $n = 3$

Exercise 19.4 page 139

1. **a)** $x = 4$ $y = 2$ **b)** $x = 6$ $y = 5$
c) $x = 6$ $y = -1$ **d)** $x = 5$ $y = 2$
e) $x = 5$ $y = 2$ **f)** $x = 4$ $y = 9$

2. **a)** $x = 3$ $y = 2$ **b)** $x = 7$ $y = 4$
c) $x = 1$ $y = 1$ **d)** $x = 1$ $y = 5$
e) $x = 1$ $y = 10$ **f)** $x = 8$ $y = 2$

3. **a)** $x = 5$ $y = 4$ **b)** $x = 4$ $y = 3$
c) $x = 10$ $y = 5$ **d)** $x = 6$ $y = 4$
e) $x = 4$ $y = 4$ **f)** $x = 10$ $y = -2$

4. **a)** $x = 5$ $y = 4$ **b)** $x = 4$ $y = 2$
c) $x = 5$ $y = 3$ **d)** $x = 5$ $y = -2$
e) $x = 1$ $y = 5$ **f)** $x = -3$ $y = -3$

5. **a)** $x = -5$ $y = -2$ **b)** $x = -3$ $y = -4$
c) $x = 4$ $y = 3\frac{2}{3}$ **d)** $x = 2$ $y = 7$
e) $x = 1$ $y = 1$ **f)** $x = 2$ $y = 9$

6. **a)** $x = 2$ $y = 3$ **b)** $x = 5$ $y = 10$
c) $x = 4$ $y = 6$ **d)** $x = 4$ $y = 4$
e) $x = 5$ $y = 1$ **f)** $x = -3$ $y = -3$

7. **a)** $x = 1$ $y = -1$ **b)** $x = 11\frac{2}{3}$ $y = 8$
c) $x = 4$ $y = 0$ **d)** $x = 3$ $y = 4$
e) $x = 2$ $y = 8$ **f)** $x = 1$ $y = 1$

Exercise 19.5 page 141

1. **a)** $a = 2, b = 1$ **b)** 4, 3 **c)** 5, 2 **d)** 7, 2
e) 3, 4 **f)** 2, 1

2. **a)** 2, 3 **b)** 5, 1 **c)** 1, 2 **d)** 3, 3 **e)** 4, 5
f) 2, 1

3. **a)** $-1, 1$ **b)** $-1, 2$ **c)** $3, -1$ **d)** $-2, 2$
e) $-3, 2$ **f)** $-1, -1$

4. **a)** $x = 2$ $y = 3$ **b)** $x = 1$ $y = 4$
c) $x = 5$ $y = 2$ **d)** $x = 3$ $y = 3$
e) $x = 4$ $y = 2$ **f)** $x = 6$ $y = 1$

5. **a)** $x = 1$ $y = 4$ **b)** $x = 5$ $y = 2$
c) $x = 3$ $y = 3$ **d)** $x = 6$ $y = 1$
e) $x = 2$ $y = 3$ **f)** $x = 2$ $y = 3$

6. **a)** $x = 0$ $y = 3$ **b)** $x = 5$ $y = 2$
c) $x = 1$ $y = 7$ **d)** $x = 6$ $y = 4$
e) $x = 2$ $y = 5$ **f)** $x = 3$ $y = 0$

7. **a)** $x = 1$ $y = 0.5$ **b)** $x = 2.5$ $y = 4$
c) $x = \frac{1}{5}$ $y = 4$ **d)** $x = \frac{3}{4}$ $y = \frac{1}{2}$
e) $x = 11$ $y = -\frac{5}{3}$ **f)** $x = \frac{1}{2}$ $y = 1$

Exercise 19.6 page 142

1. 10 and 7
2. 16 and 9
3. $x = 1$ $y = 4$
4. $x = 2$ $y = 5$
5. 60 and 20 years old
6. 60 and 6 years old

Student Assessment 1 page 143

1. **a)** $a = 6$ **b)** $b = -7$ **c)** $c = -3$
d) $d = 4$

2. **a)** $a = -10$ **b)** $b = 4$ **c)** $c = -0.5$
d) $d = 4$

3. **a)** $a = 10$ **b)** $b = 21$ **c)** $c = 6$
d) $d = 18$

4. **a)** $a = 8$ **b)** $b = 12$ **c)** $c = 24$ **d)** $d = 9$

5. **a)** $a = -10$ **b)** $b = 7$ **c)** $c = 23$
d) $d = 11$

6. **a)** $2, 1$ **b)** $4, 3$ **c)** $-1, +1$ **d)** $-5, +4$

Student Assessment 2 page 143

1. **a)** $a = 2$ **b)** $b = 4$ **c)** $c = -3$
d) $d = -5$

2. **a)** $a = -4$ **b)** $b = 2$ **c)** $c = 1$
d) $d = 1.2$

3. **a)** $a = 21$ **b)** $b = -8$ **c)** $c = -12$
d) $d = 15$

4. **a)** $a = 10$ **b)** $b = 15$ **c)** $c = 2$
d) $d = -\frac{7}{3}$

5. **a)** $a = -20$ **b)** $b = -3.5$ **c)** $c = \frac{1}{4}$
d) $d = 2$

6. **a)** $5, 1$ **b)** $1, -2$ **c)** $2, 0.5$ **d)** $-2, -3$

Student Assessment 3 page 144

1. **a)** 9 **b)** 11 **c)** -4 **d)** 6
2. **a)** 1.5 **b)** 7 **c)** 4 **d)** 3
3. **a)** -10 **b)** 12 **c)** 10 **d)** $11\frac{1}{4}$
4. **a)** 16 **b)** $-8\frac{2}{3}$ **c)** 2 **d)** 3.5
5. **a)** $x = 5$ $y = 2$ **b)** $x = 3\frac{1}{3}$ $y = 4\frac{1}{3}$
c) $x = 5$ $y = 4$ **d)** $x = 5$ $y = 1$

Student Assessment 4 page 144

1. **a)** -6 **b)** 6 **c)** 4 **d)** 2.4
2. **a)** 0.5 **b)** 4 **c)** 9.5 **d)** 5
3. **a)** 6 **b)** 15 **c)** 22 **d)** 6
4. **a)** 8.5 **b)** $4\frac{1}{3}$ **c)** 8.5 **d)** 12
5. **a)** $x = 7$ $y = 4$ **b)** $p = 1$ $q = 2$
c) $x = 7$ $y = 1$ **d)** $m = 3$ $n = 5$

20 Graphs in Practical Situations

Exercise 20.1 page 146

1.

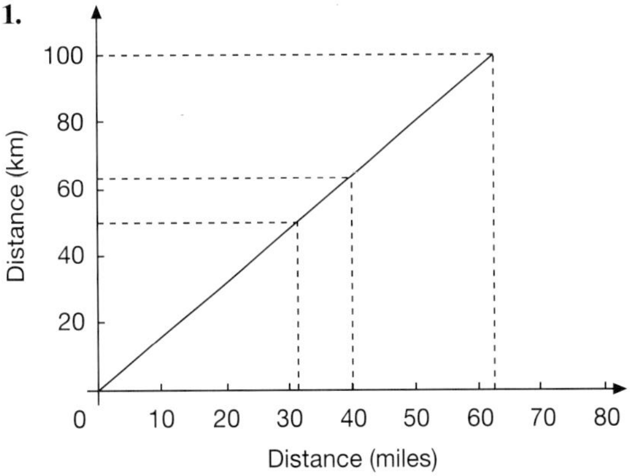

a) 50 km = 31 miles **b)** 40 miles = 64 km, therefore 80 miles = 128 km
c) 100 km/h = 62 mph
d) 40 mph = 64 km/h

2.

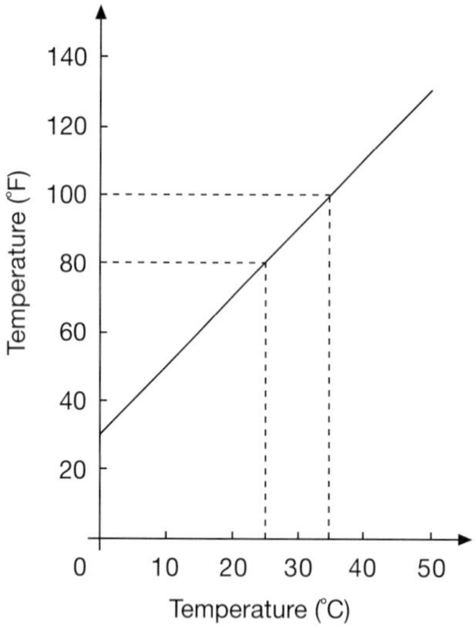

a) 25 °C = 80 °F
b) 100 °F = 35 °C
c) 0 °C = 30 °F
d) 100 °F = 35 °C, therefore 200 °F = 70 °C

3.

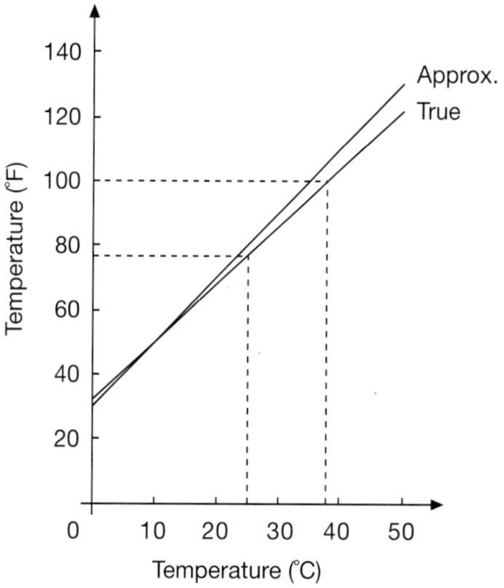

i) a) 25 °C = 77 °F
b) 100 °F = 38 °C
c) 0 °C = 32 °F
d) 100 °F = 38 °C, therefore 200 °F = 76 °C

ii) The rough conversion is most useful at lower temperatures (i.e. between 0 and 20 °C).

4. **a)**

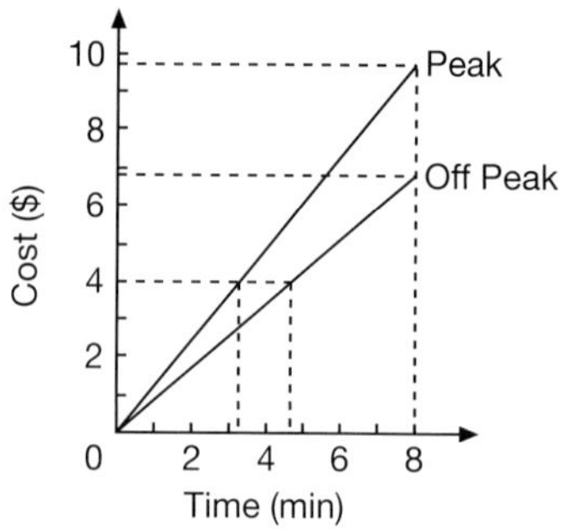

b) 8 min = $6.80
c) 8 min = $9.60
d) Extra time = 1 min 20 s

5.

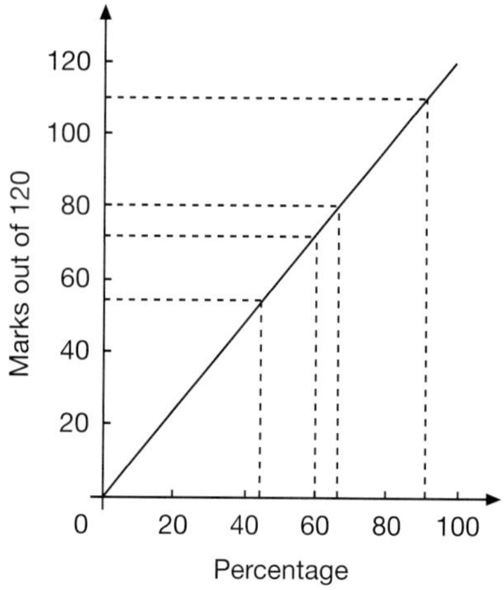

a) 80 = 67%
b) 110 = 92%
c) 54 = 45%
d) 72 = 60%

Exercise 20.2 page 147

1. **a)** 6 m/s **b)** 4 m/s **c)** 39 km/h **d)** 20 km/h **e)** 160 km/h **f)** 50 km/h

2. **a)** 400 m **b)** 182 m **c)** 210 km **d)** 255 km **e)** 10 km **f)** 79.2 km

3. **a)** 5 s **b)** 50 s **c)** 4 min **d)** 71 min 26 s (nearest second) **e)** 5 s **f)** 4 min

Exercise 20.3 page 148

1.

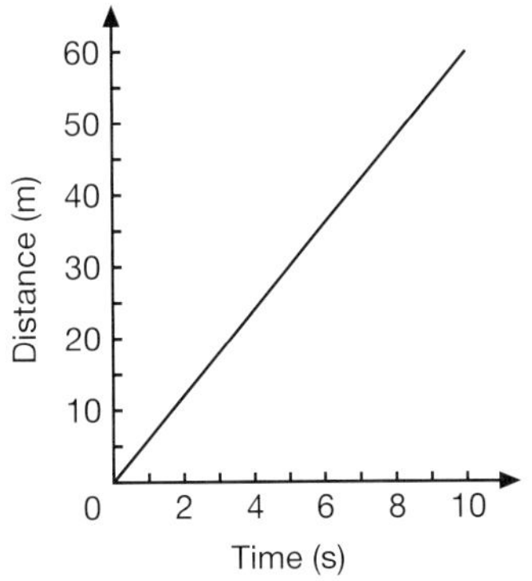

2.

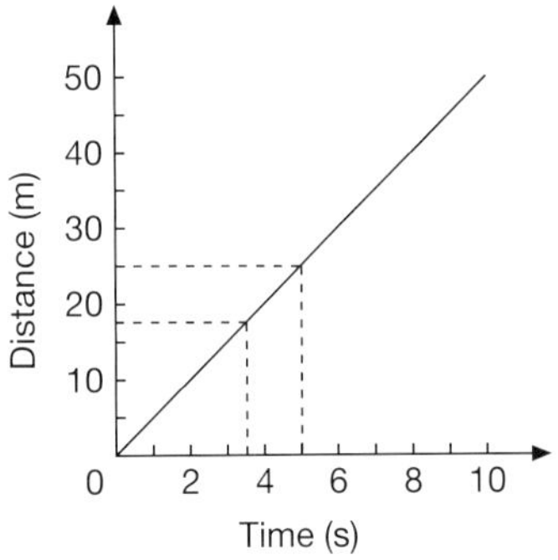

a) 25 m = 5 s
b) 3.5 s = 17 m

3. **a)** Speed A = 40 m/s Speed B = $13\frac{1}{3}$m/s
b) Distance apart = $133\frac{1}{3}$m

4. **a)** $\frac{2}{3}$m/s **b)** 6 m/s, $\frac{2}{3}$m/s **c)** 1 m/s
d) $\frac{1}{2}$m **e)** $7\frac{1}{3}$m

Exercise 20.4 page 149

1. **a)** 45 km/h **b)** 20 km/h
c) Paul has arrived at the restaurant and is waiting for Helena.

2.

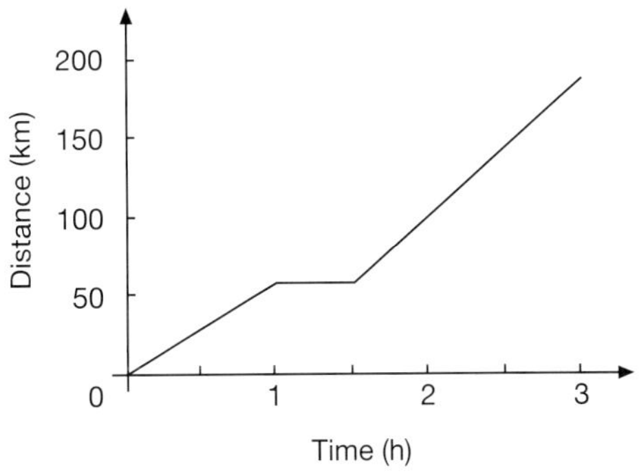

3.

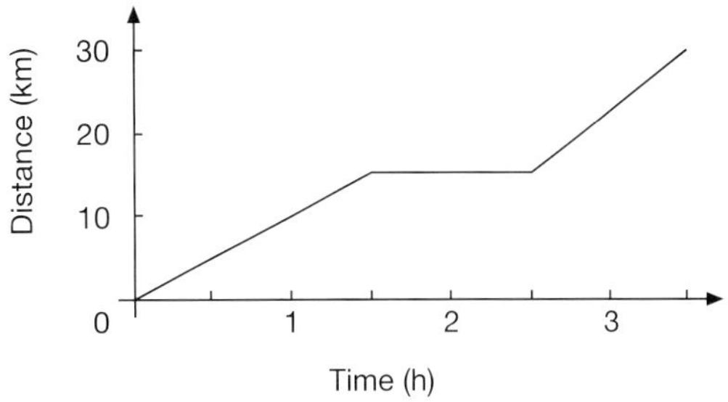

4. **a)**

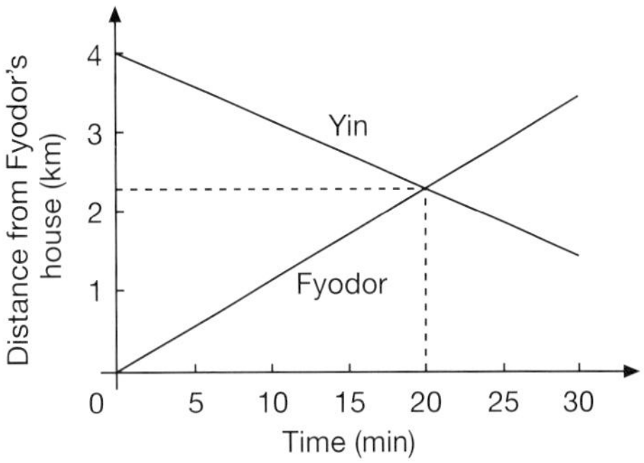

b) After 20 min **c)** Distance = $2\frac{1}{3}$km

5. **a)**

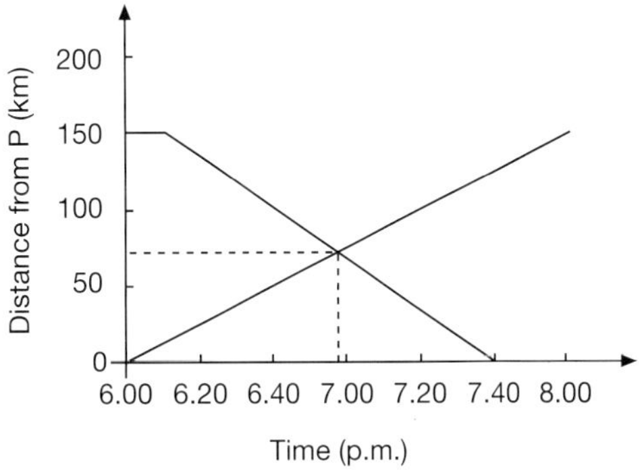

b) Time ≈ 6.57 p.m.
c) Distance from Q ≈ 79 km
d) The 6.10 p.m. train from station Q arrives first.

6. **a)** a: $133\frac{1}{3}$km/h b: 0 km/h c: 200 km/h
b) d: 100 km/h e: 200 km/h

Student Assessment 1 page 150

1. a)

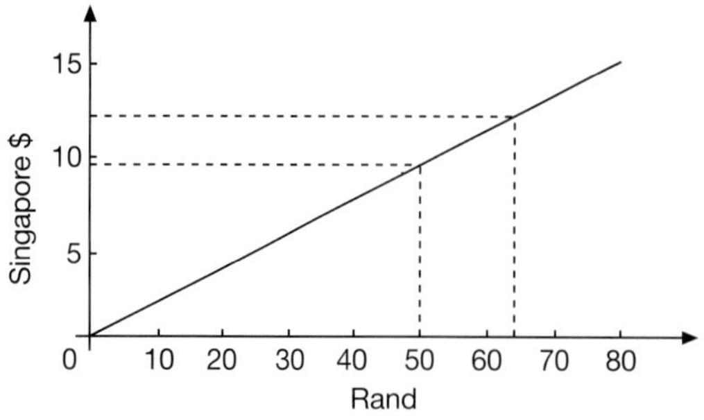

b) 50 rand ≈ 9.00 dollars
c) 11.5 dollars ≈ 64 rand

2. a)

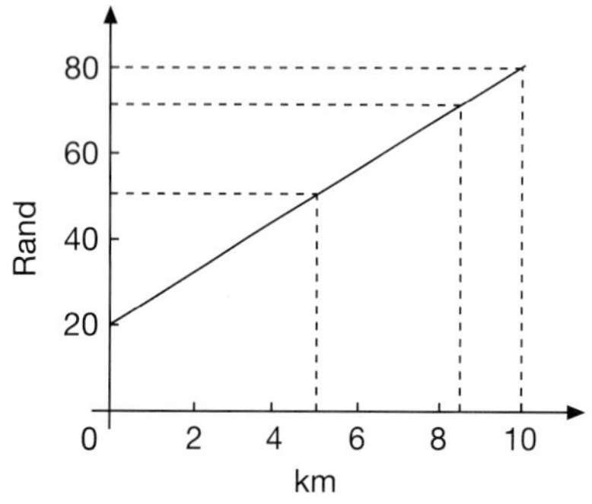

i) 5 km: 50 rand ii) 8.5 km: 71 rand
b) 80 rand: 10 km

3. a)

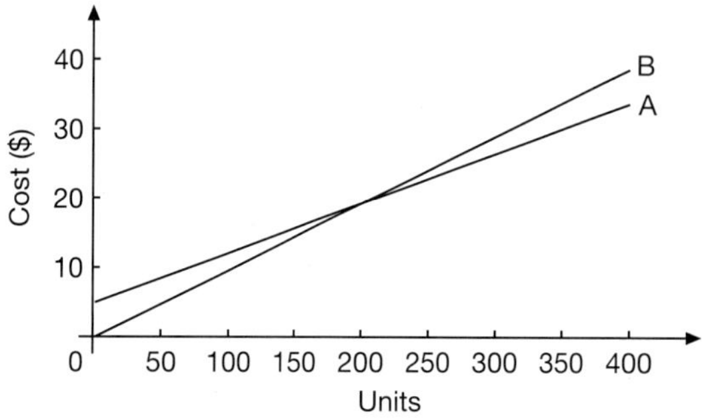

b) If the customer uses under 200 units/quarter then he should use account type B.

4. a)

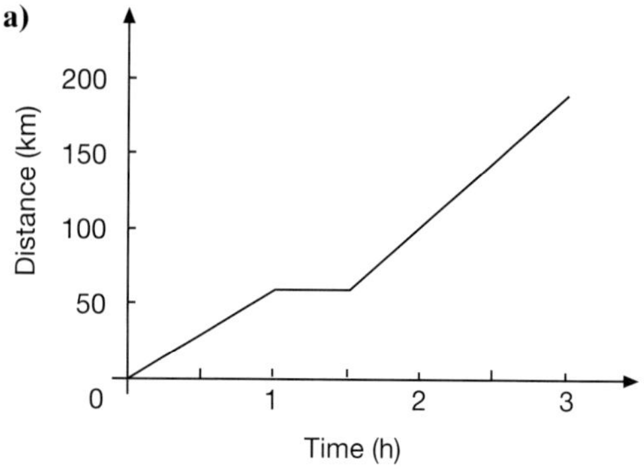

b) 180 km

5. a)

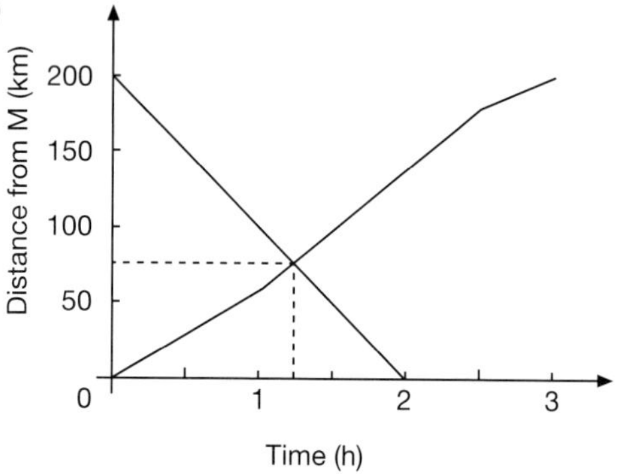

b) Distance ≈ 122 km
c) Time ≈ 1 h 13 min after start

Student Assessment 2 page 151

1.

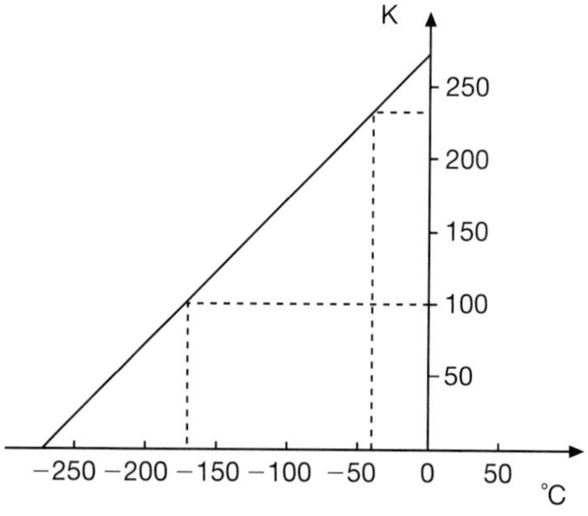

a) −40 °C = 233 K **b)** 100 K = −173 °C

2. a)

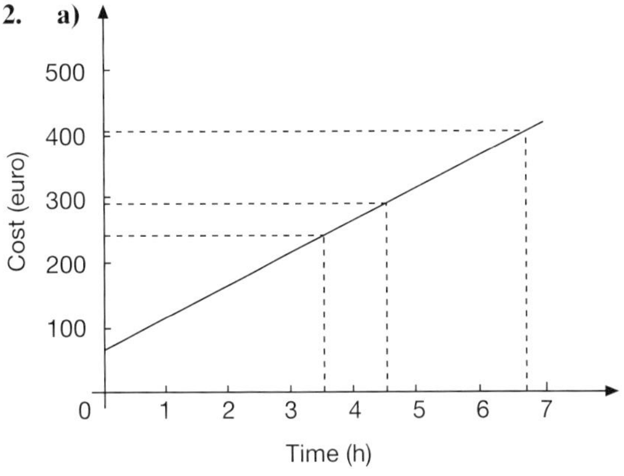

i) 295 euro   ii) ≈408 euro   **b)** $3\frac{1}{2}$h

3. a)

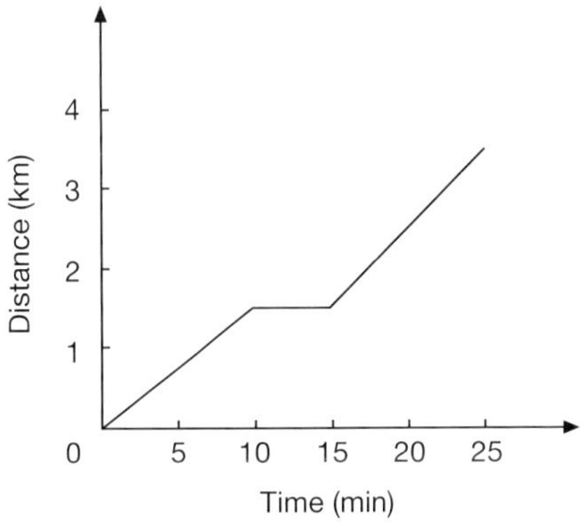

b) 25 min

4. a) B, C
b) B because it illustrates going back in time, C because it illustrates infinite speed

21 Graphs of Functions

Exercise 21.1 page 153

1. a) 1 **b)** 1 **c)** 1 **d)** 2 **e)** 3 **f)** 2 **g)** 4 **h)** $\frac{1}{2}$ **i)** 0 **j)** Infinite **k)** $\frac{1}{4}$ **l)** $\frac{3}{2}$

2. a) -1 **b)** -1 **c)** -2 **d)** $-\frac{1}{2}$ **e)** -1 **f)** -2 **g)** $-\frac{3}{2}$ **h)** $\frac{2}{3}$ **i)** $-\frac{1}{4}$ **j)** -1 **k)** 0 **l)** -4

Exercise 21.2 page 154

a) $y = 7$ **b)** $y = 2$ **c)** $x = 7$ **d)** $x = 3$

e) $y = x$ **f)** $y = \frac{1}{2}x$ **g)** $y = -x$

h) $y = -2x$

Exercise 21.3 page 155

1. a) $y = x + 1$ **b)** $y = x + 3$ **c)** $y = x - 2$ **d)** $y = 2x + 2$ **e)** $y = \frac{1}{2}x + 5$ **f)** $y = \frac{1}{2}x - 1$

2. a) $y = -x + 4$ **b)** $y = -x - 2$ **c)** $y = -2x - 2$ **d)** $y = -\frac{1}{2}x + 3$ **e)** $y = -\frac{3}{2}x + 2$ **f)** $y = -4x + 1$

3. a) 1) a) 1 **b)** 1 **c)** 1 **d)** 2 **e)** $\frac{1}{2}$ **f)** $\frac{1}{2}$
2) a) -1 **b)** -1 **c)** -2 **d)** $-\frac{1}{2}$ **e)** $-\frac{3}{2}$ **f)** -4
b) The gradient is equal to the coefficient of x.
c) The constant being added/subtracted indicates where the line intersects the y-axis.

Exercise 21.4 page 158

1. a) +3, +1 **a)** +3, +3 **a)** +3, –2
These lines are parallel to each other.
d) –2, +4 **e)** –2, +1 **f)** –2, –3
These lines are parallel to each other.
g) –1, 7 **h)** –1, 3 **i)** –1, –3
These lines are parallel to each other.

2. a) $m = 2$ $c = 1$ **b)** $m = 3$ $c = 5$ **c)** $m = 1$ $c = -2$ **d)** $m = \frac{1}{2}$ $c = 4$ **e)** $m = -3$ $c = 6$ **f)** $m = -\frac{2}{3}$ $c = 1$ **g)** $m = -1$ $c = 0$ **h)** $m = -1$ $c = -2$ **i)** $m = -2$ $c = 2$

3. a) $m = 3$ $c = 1$ **b)** $m = -\frac{1}{2}$ $c = 2$ **c)** $m = -2$ $c = -3$ **d)** $m = -2$ $c = -4$ **e)** $m = \frac{1}{4}$ $c = 6$ **f)** $m = 3$ $c = 2$ **g)** $m = 1$ $c = -2$ **h)** $m = -8$ $c = 6$ **i)** $m = 3$ $c = 1$

4. a) $m = 2$ $c = -3$ **b)** $m = \frac{1}{2}$ $c = 4$ **c)** $m = 2$ $c = -4$ **d)** $m = -8$ $c = 12$ **e)** $m = 2$ $c = 0$ **f)** $m = -3$ $c = 3$ **g)** $m = 2$ $c = 1$ **h)** $m = -\frac{1}{2}$ $c = 2$ **i)** $m = 2$ $c = -\frac{1}{2}$

Exercise 21.5 page 159

1. a) $y = 2x - 1$ **b)** $y = 3x + 1$ **c)** $y = 2x + 3$ **d)** $y = x - 4$ **e)** $y = 4x + 2$ **f)** $y = -x + 4$ **g)** $y = -2x + 2$ **h)** $y = -3x - 1$ **i)** $y = \frac{1}{2}x$

2. a) $y = \frac{1}{7}x + \frac{26}{7}$ **b)** $y = \frac{6}{7}x + \frac{4}{7}$ **c)** $y = \frac{3}{2}x + \frac{15}{2}$ **d)** $y = 9x - 13$ **e)** $y = -\frac{1}{2}x + \frac{5}{2}$ **f)** $y = -\frac{3}{13}x + \frac{70}{13}$ **g)** $y = 2$ **h)** $y = -3x$ **i)** $x = 6$

Exercise 21.6 page 160

1. a)

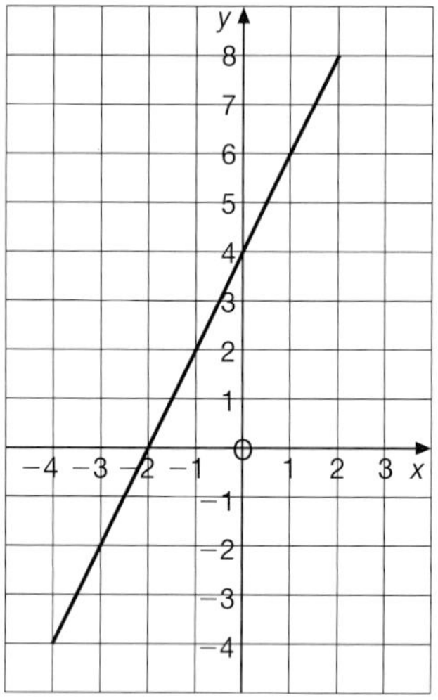

b)

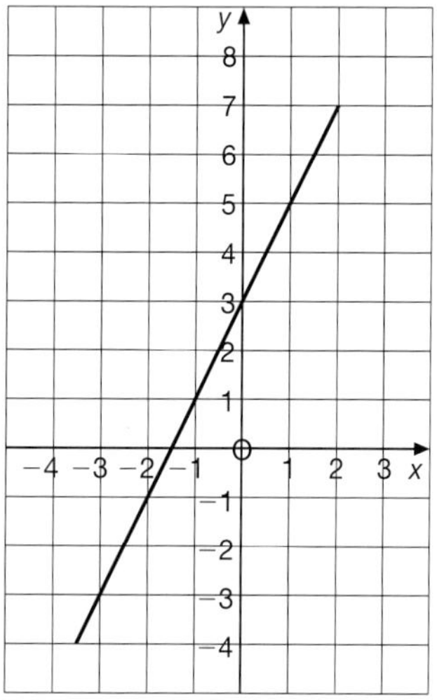

c)

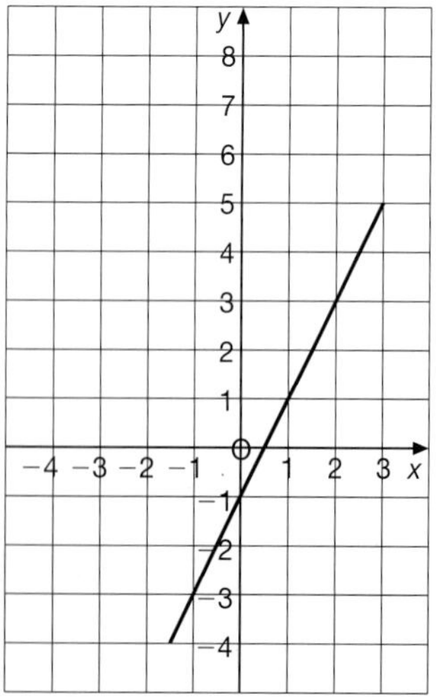

d)

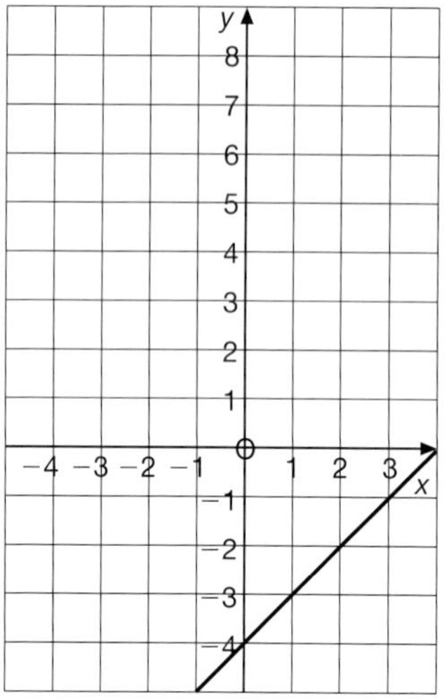

e)

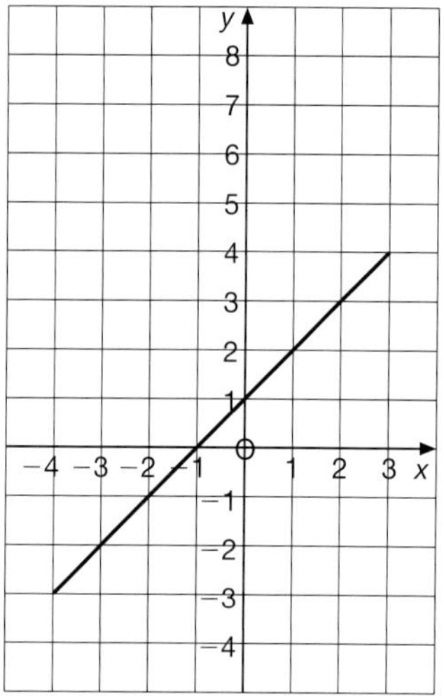

f)

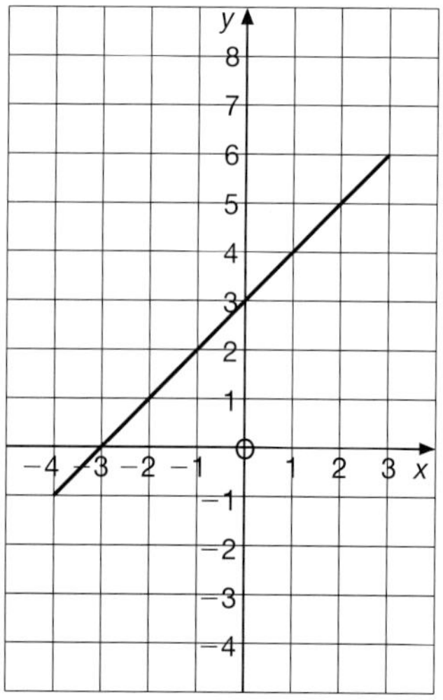

g)

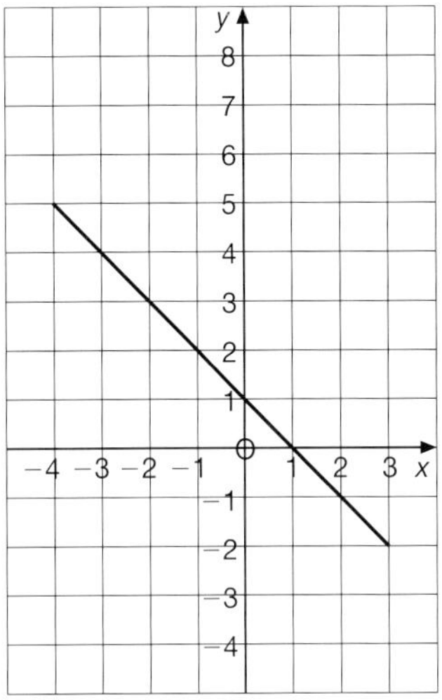
y
8
7
6
5
4
3
2
1
O
−4 −3 −2 −1 1 2 3 x
−1
−2
−3
−4

h)

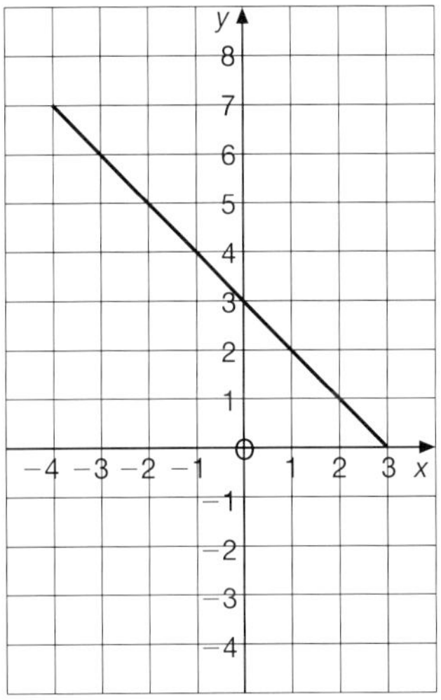
y
8
7
6
5
4
3
2
1
O
−4 −3 −2 −1 1 2 3 x
−1
−2
−3
−4

i)

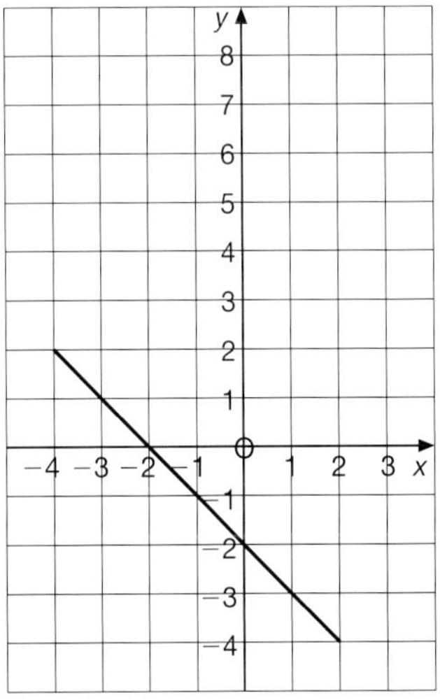
y
8
7
6
5
4
3
2
1
O
−4 −3 −2 −1 1 2 3 x
−1
−2
−3
−4

2. a)

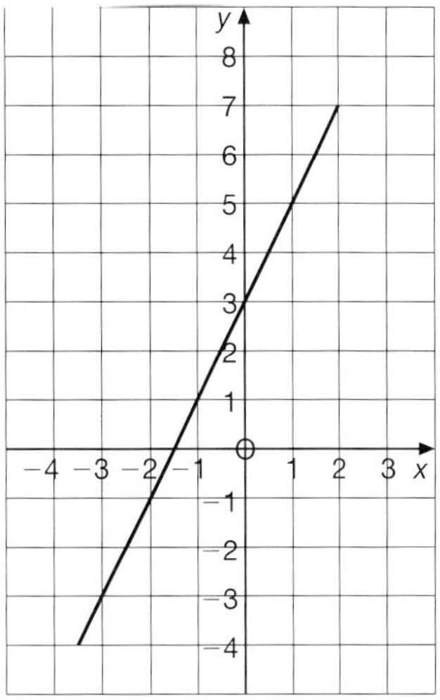
y
8
7
6
5
4
3
2
1
O
−4 −3 −2 −1 1 2 3 x
−1
−2
−3
−4

b)

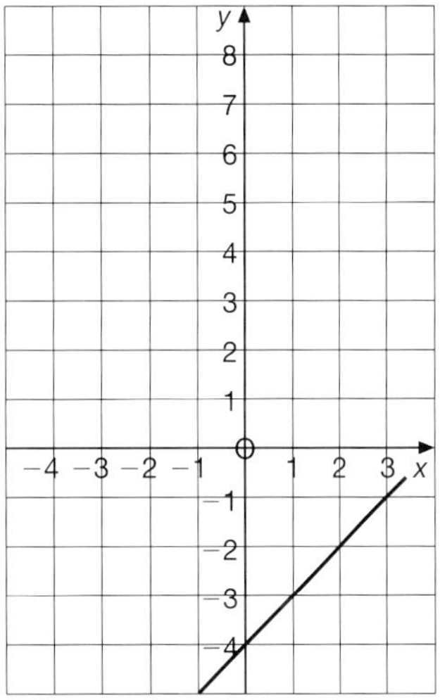
y
8
7
6
5
4
3
2
1
O
−4 −3 −2 −1 1 2 3 x
−1
−2
−3
−4

c)

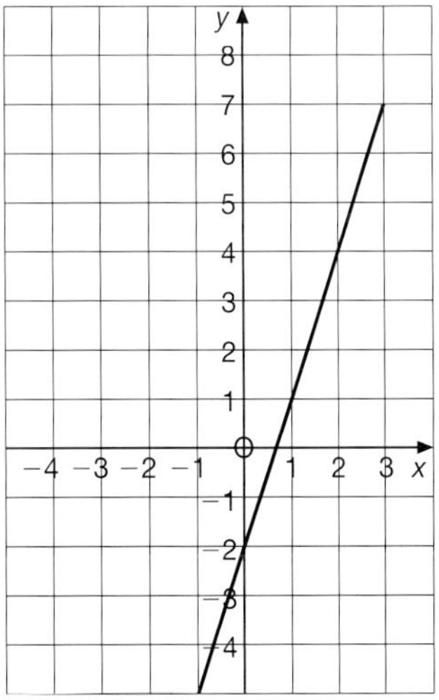
y
8
7
6
5
4
3
2
1
O
−4 −3 −2 −1 1 2 3 x
−1
−2
−3
−4

d)

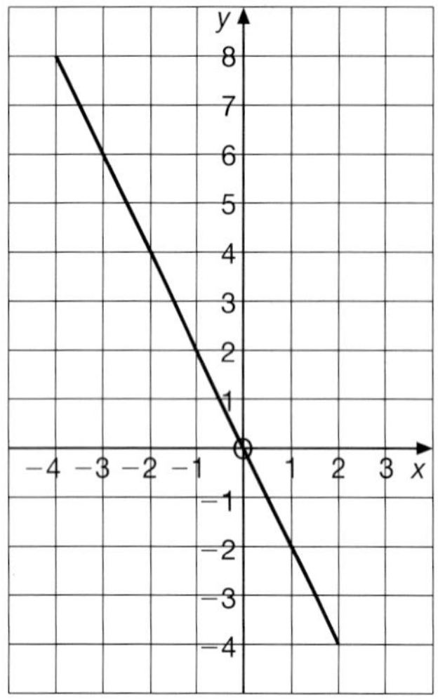
y
x
−4 −3 −2 −1 O 1 2 3
8 7 6 5 4 3 2 1 −1 −2 −3 −4

e)

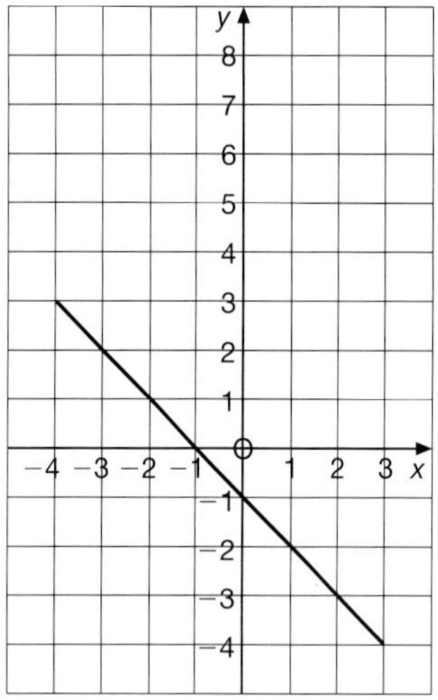
y
x
−4 −3 −2 −1 O 1 2 3
8 7 6 5 4 3 2 1 −1 −2 −3 −4

f)

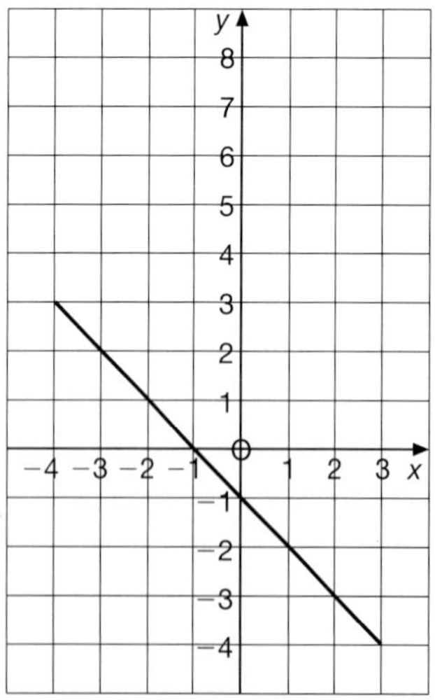
y
x
−4 −3 −2 −1 O 1 2 3
8 7 6 5 4 3 2 1 −1 −2 −3 −4

g)

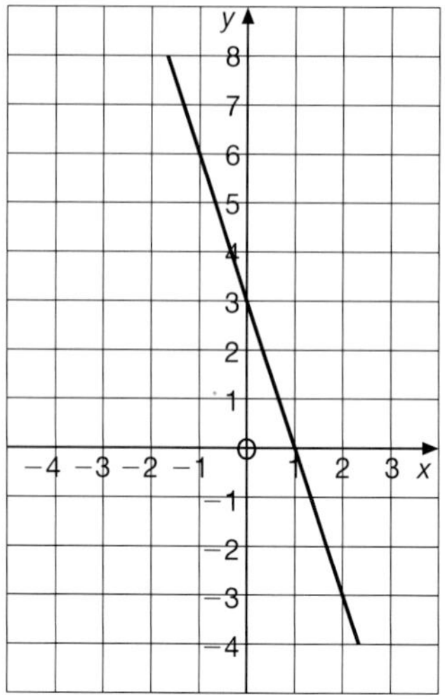
y
x
−4 −3 −2 −1 O 1 2 3
8 7 6 5 4 3 2 1 −1 −2 −3 −4

h)

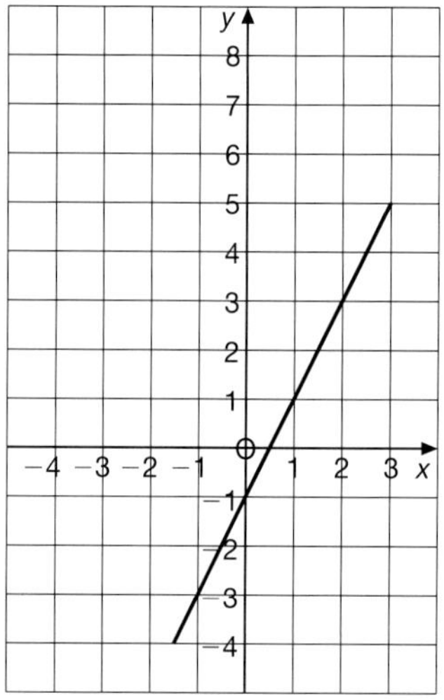
y
x
−4 −3 −2 −1 O 1 2 3
8 7 6 5 4 3 2 1 −1 −2 −3 −4

i)

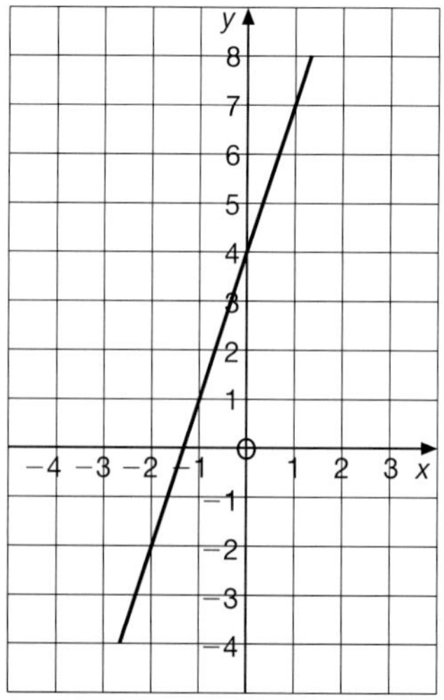
y
x
−4 −3 −2 −1 O 1 2 3
8 7 6 5 4 3 2 1 −1 −2 −3 −4

3. a)

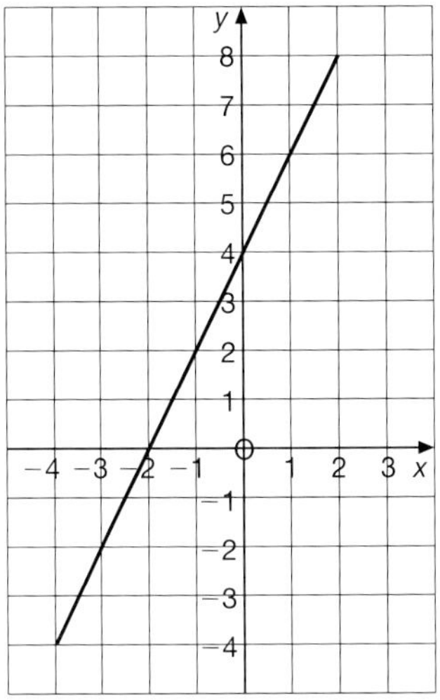
y
8 7 6 5 4 3 2 1 −1 −2 −3 −4
−4 −3 −2 −1 1 2 3 x

d)

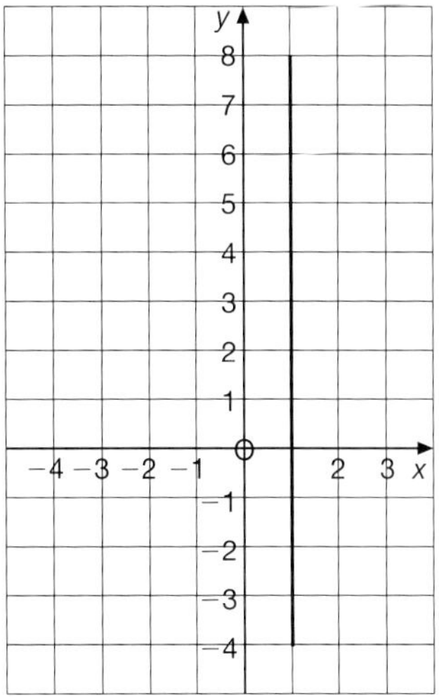
y
8 7 6 5 4 3 2 1 −1 −2 −3 −4
−4 −3 −2 −1 2 3 x

b)

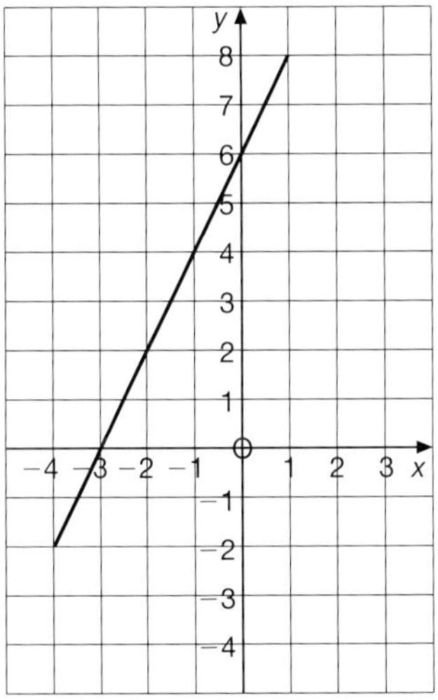
y
8 7 6 5 4 3 2 1 −1 −2 −3 −4
−4 −3 −2 −1 1 2 3 x

e)

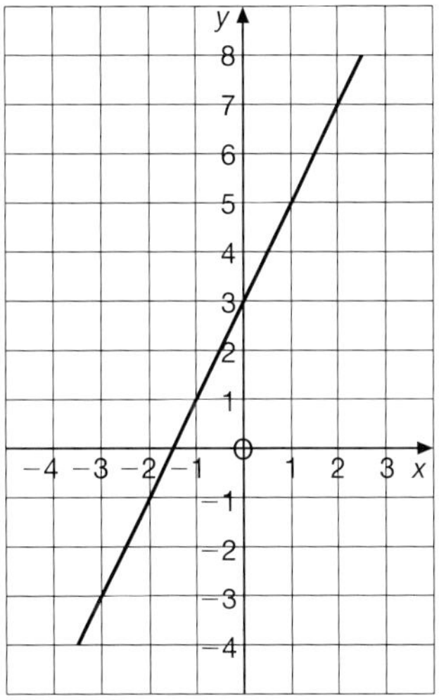
y
8 7 6 5 4 3 2 1 −1 −2 −3 −4
−4 −3 −2 −1 1 2 3 x

c)

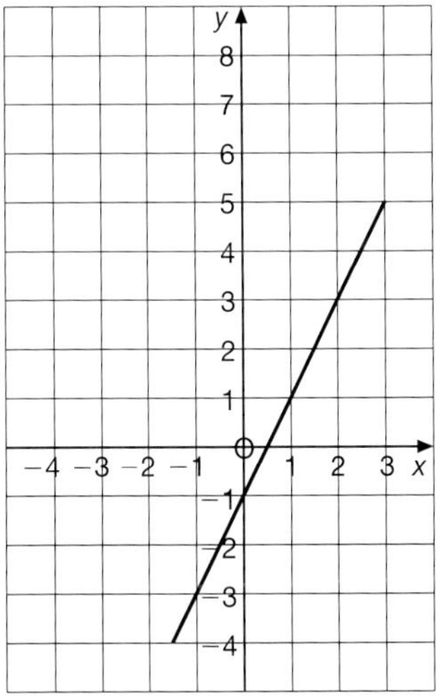
y
8 7 6 5 4 3 2 1 −1 −2 −3 −4
−4 −3 −2 −1 1 2 3 x

f)

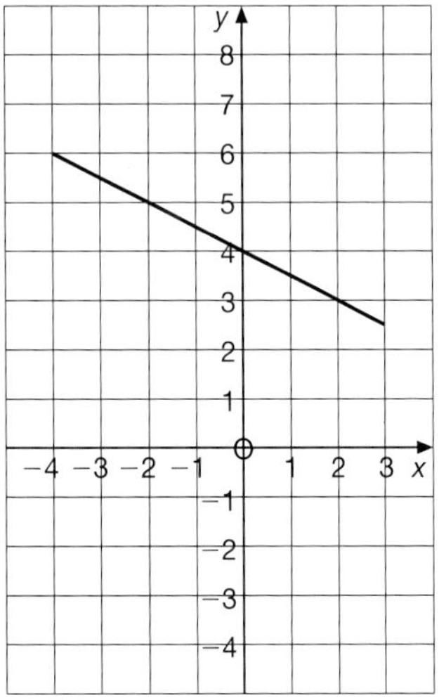
y
8 7 6 5 4 3 2 1 −1 −2 −3 −4
−4 −3 −2 −1 1 2 3 x

g)

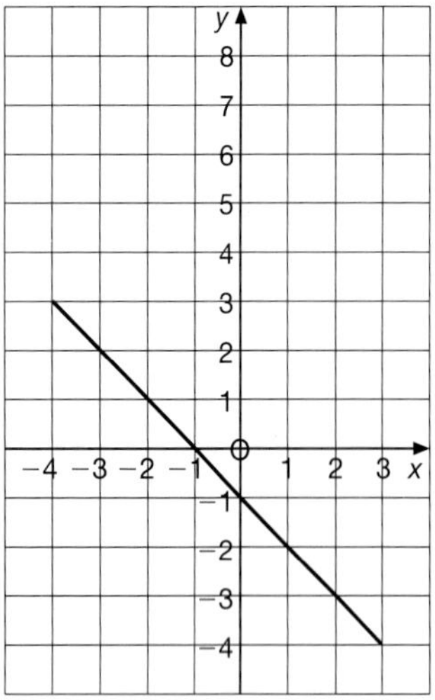

h)

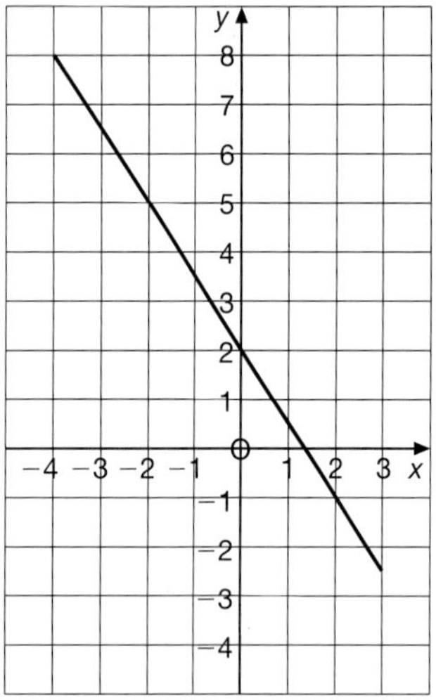

i)

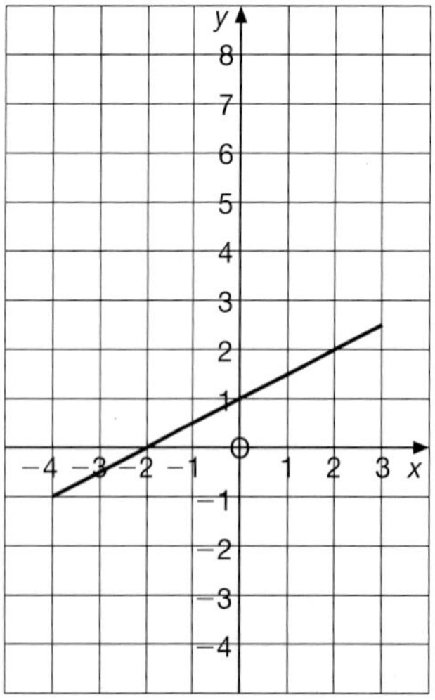

Exercise 21.7 page 162

1.

x	−4	−3	−2	−1	0	1	2	3
y	10	4	0	−2	−2	0	4	10

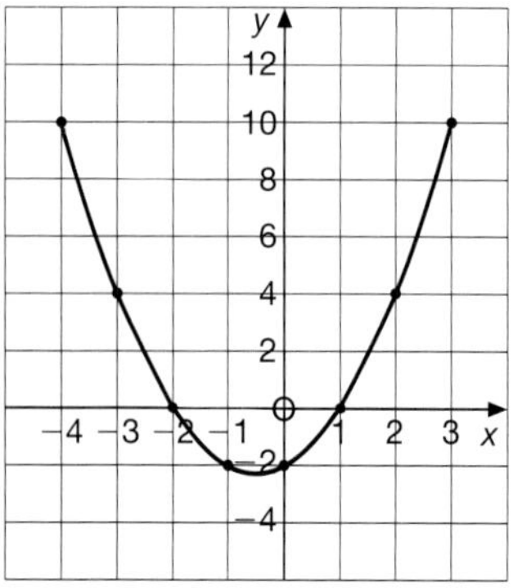

2.

x	−3	−2	−1	0	1	2	3	4	5
y	−12	−5	0	3	4	3	0	−5	−12

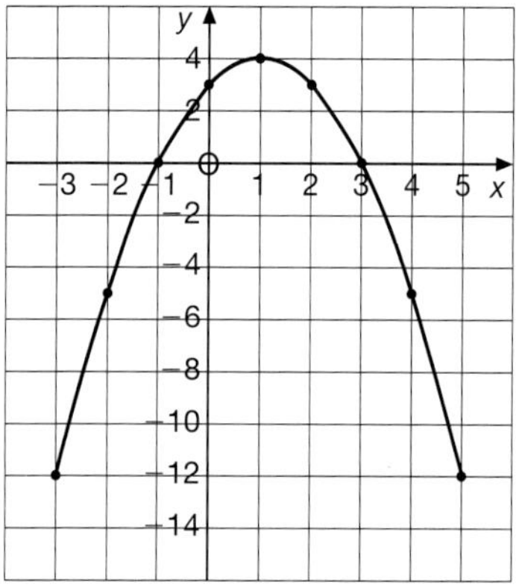

3.

x	−1	0	1	2	3	4	5
y	9	4	1	0	1	4	9

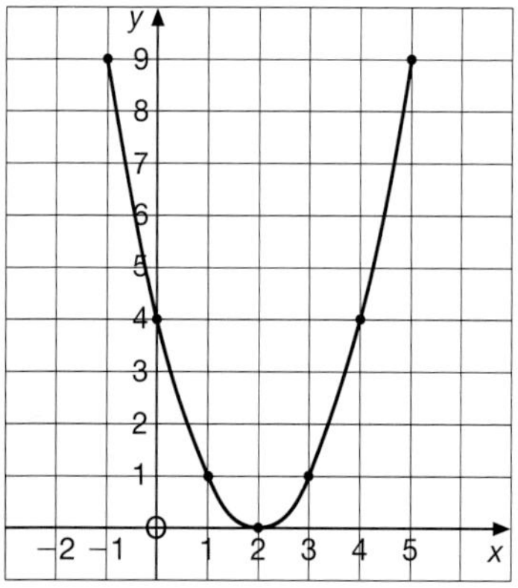

4.

x	−4	−3	2	−1	0	1	2
y	−9	−4	−1	0	−1	−4	−9

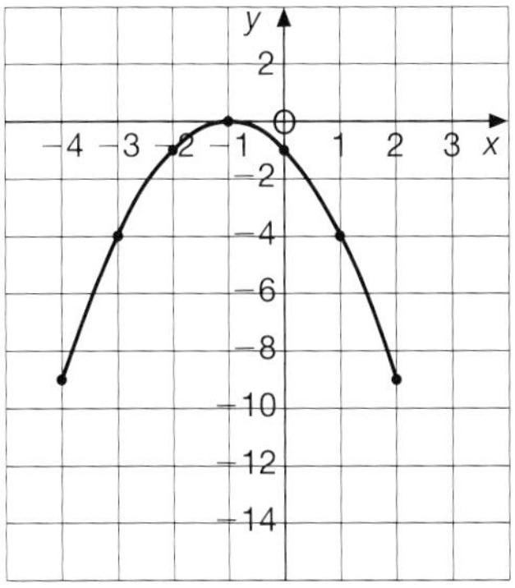

5.

x	−4	−3	−2	−1	0	1	2	3	4	5	6
y	9	0	−7	−12	−15	−16	−15	−12	−7	0	9

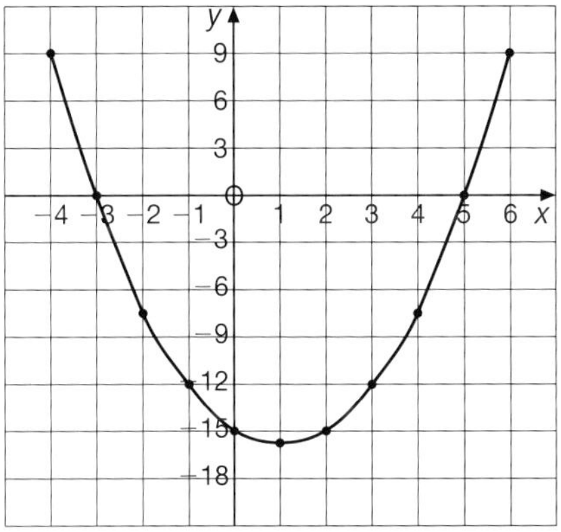

Exercise 21.8 page 162

1. −2 and 3
2. −1 and 1
3. 3
4. −4 and 3
5. 2

Exercise 21.9 page 163

1. −1.6 and 2.6
2. No solution
3. 2 and 4
4. −3.5 and 2.5
5. 0.3 and 3.7

Exercise 21.10 page 164

1.

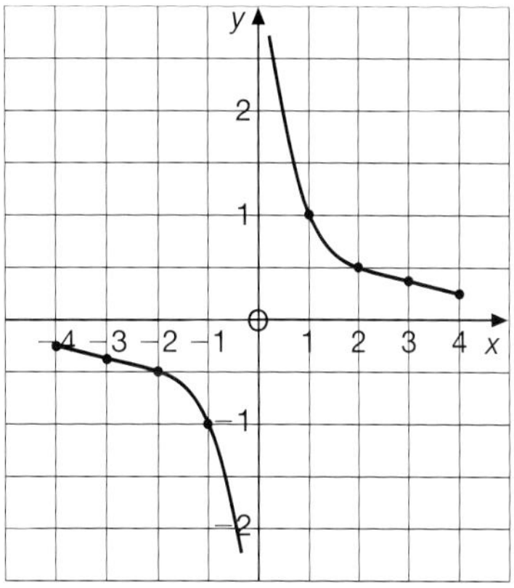

2.

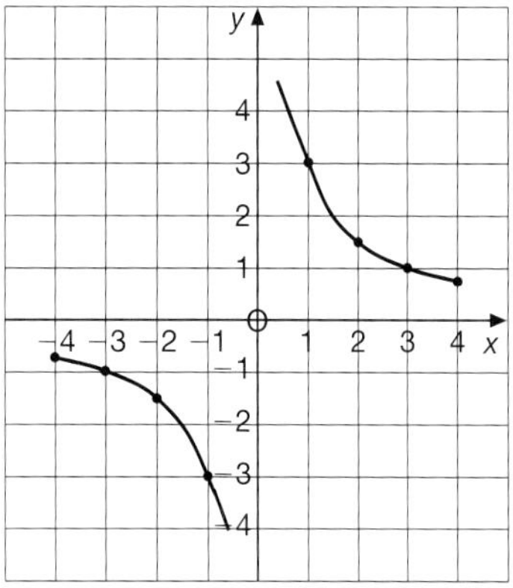

3.

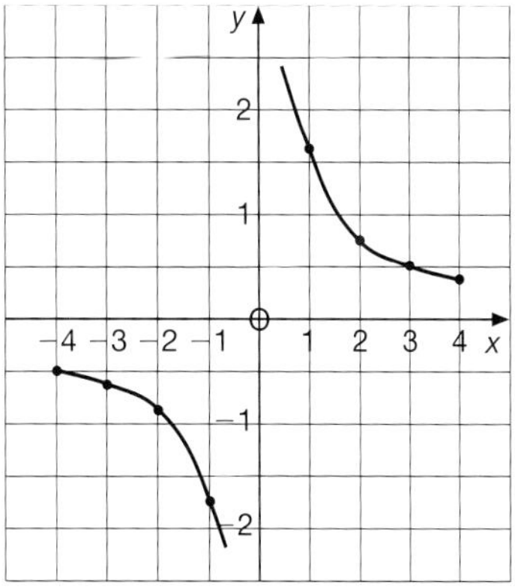

3. a)

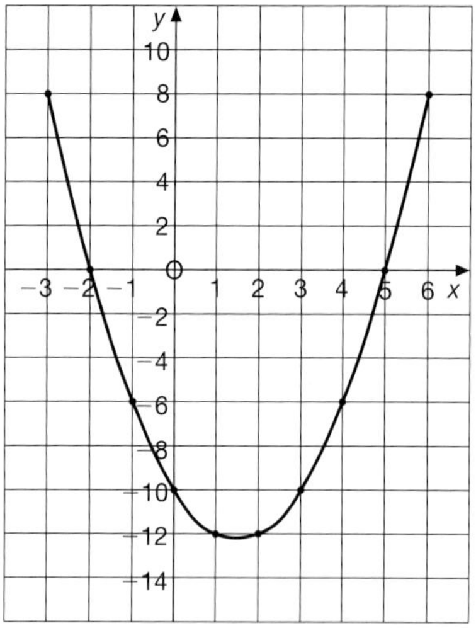

b)

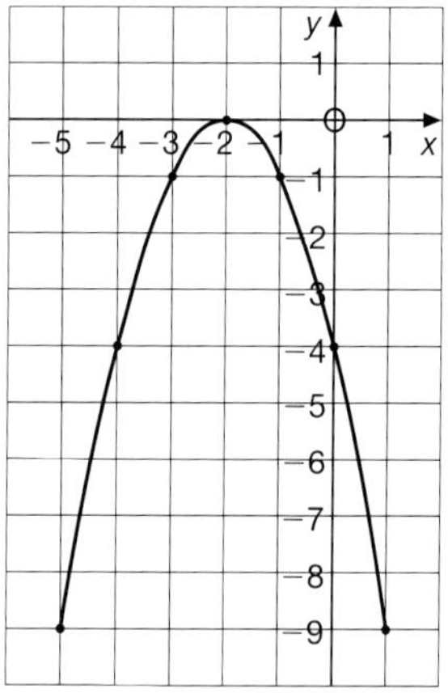

4. a)

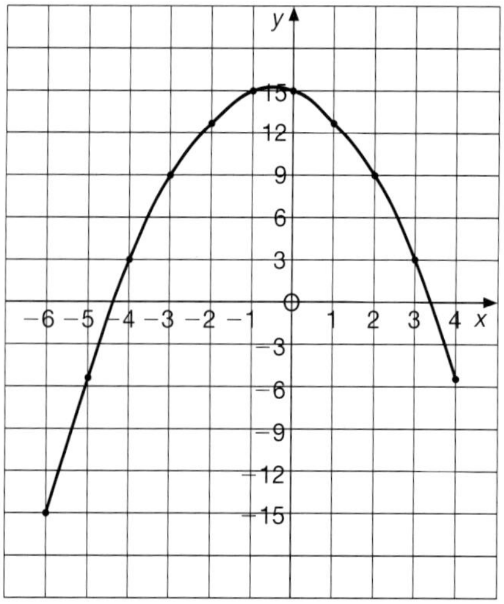

b) i) $x = -3.7$ and 2.7 ii) $x = -1.8$ and 2.8

5. a)

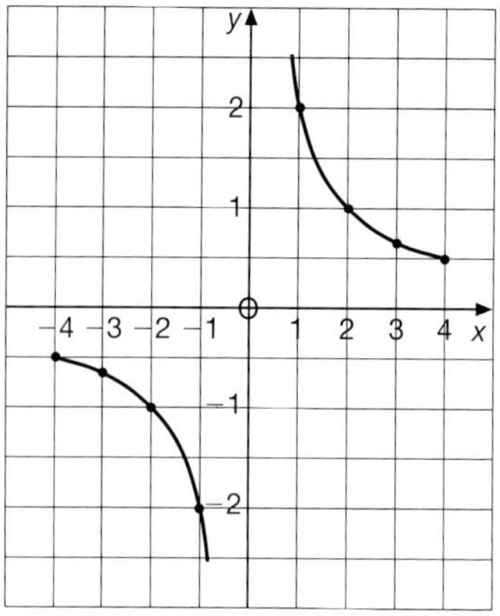

b) $x = -2$ and 1

ICT Section

Algebra Activity 1 page 167

1. **a)** (2, 2)
b) (1, 2)
c) $(1\frac{1}{2}, 3)$
d) (2, 4)
e) (4, 1)

2. Student's solutions to the simultaneous equations, giving the same answers as in Q.1 above.

Algebra Activity 2 page 167

1. **a)** $x = -2$ and $x = 1$
b) $x = 1$ and $x = 6$
c) $x = -4$ and $x = 3$
d) $x = -3$ and $x = 0.5$
e) $x = 1$
f) $x = -2$

Reviews for Chapters 17 to 21

Review 13 page 169

1. **a)** 8 **b)** 81 **c)** 144

2. **a)** $-2m - 12$ **b)** $2x^2 - 2y^2$

3. **a)** $5b(3a - 2c)$ **b)** $3pq(2p + 3q + 4)$

4. **a)** 4 **b)** −30

5. **a)** $a = 3c - 2b$ **b)** $b = \frac{12a - c}{2}$

6. **a)** $a = 5$ **b)** $b = -1$

7. $a = 2, b = 3$

8. Student's graph
a) 100 km **b)** 44 miles

9. Student's graphs

10. Student's graphs. $\sqrt{6.25} = 2.5$

Review 14 page 169

1. **a)** 36 **b)** 125 **c)** 2

2. **a)** $-7a - 2b$ **b)** $-4a + 5b$

3. **a)** $12xy(x - 4y)$ **b)** $3bc(a - 2b + 3c)$

4. **a)** 4 **b)** 12

5. **a)** $r = 3q - 2p$ **b)** $z = 3y - 2x$

6. **a)** $a = 2$ **b)** $b = 9$

7. $a = 1, b = -1$

8. Student's graph
a) 130 °F **b)** 35 °C

9. Student's graphs

10. Student's graphs. $-\sqrt{5.29} = -2.3$

Review 15 page 170

1. **a)** 49 **b)** 64 **c)** 216

2. **a)** $5x^2 - 5xy + 5xz$ **b)** $6b - 2ab - a$

3. **a)** $13y(3x - 4z)$ **b)** $4pqr(r + 2q - 4p)$

4. **a)** 0 **b)** −9

5. **a)** $\frac{3b - c + e}{2} = d$ **b)** $\frac{p}{rs} = q$

6. **a)** $a = 1.5$ **b)** $c = 6.5$

7. $a = 2, b = -3$

8. **a)** 43 **b)** 71 **c)** 89

9. Student's graph

10. **a)** 6, 1, −2, −3, −2, 1, 6
b) Student's graph

Review 16 page 171

1. **a)** 81 **b)** 10 000 **c)** 1

2. **a)** $17x(x - 3y)$ **b)** $9bc(2a - 3c + 4b)$

3. **a)** $9b - 9a$ **b)** 4

4. **a)** 38 **b)** −105

5. **a)** $d = \frac{c - 4a + 3b}{5}$ **b)** $s = \frac{2nr}{3m}$

6. **a)** $a = 2$ **b)** $b = 7$

7. $a = 5, b = 4$

8. Student's graph
a) 196 reals **b)** 107 euro

9. Student's graph

10. **a)** 6, 2, 0, 0, 2, 6, 12
b) Student's graph

Review 17 page 172

1. **a)** $\frac{1}{9}$ **b)** $\frac{1}{32}$ **c)** 2

2. **a)** $12mn(2n - 3m)$ **b)** $17bc(2a - 3d + 4c)$

3. **a)** 21 **b)** $3ab + 18b + 3a$

4. **a)** 2 **b)** 2

5. **a)** $c = -\frac{ad}{b}$ **b)** $a = cd - b$

6. **a)** $m = 14$ **b)** $b = 1\frac{5}{6}$

7. $a = 4, b = 1$

8. Student's graph
a) 63 **b)** 35 **c)** 90

9. Student's graph

10. **a)** 10, 4, 0, −2, −2, 0, 4, 10
b) Student's graph
c) $x = -2$ or $x = +1$

Review 18 page 173

1. **a)** 3^2 **b)** 3^1

2. **a)** $xy + 6x + 6y$ **b)** $2a^2 + 2b^2 - 2c^2$

3. **a)** $11xy(11x - 7y)$ **b)** $15pq(3r - 2p + 4q)$

4. **a)** 18 **b)** 18

5. **a)** $q = -\frac{np}{m}$ **b)** $a - cd = b$

6. **a)** $a = -1$ **b)** $b = 1.5$

7. $m = 6, n = -5$

8. Student's graph
   a) 223 K **b)** −73 K

9. Student's graph

10. **a)** 12, 6, 2, 0, 0, 2, 6
    b) Student's graph
    c) $x = +1$ or $x = +2$

22 Geometrical Terms and Relationships

Exercise 22.1 page 174

1.

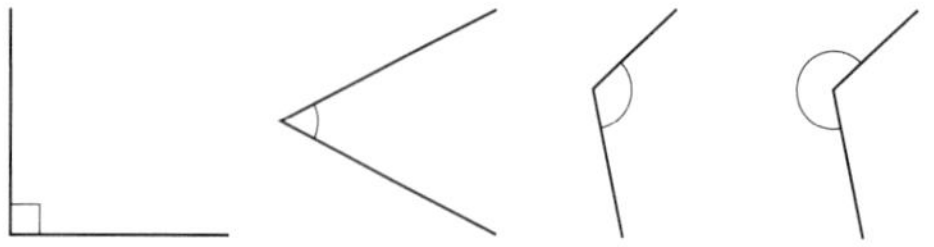

2. **a)** Acute **b)** Right angle **c)** Obtuse
   d) Reflex **e)** Acute **f)** Obtuse
   g) Reflex **h)** Right angle

3. **a)** Complementary **b)** Supplementary
   c) Complementary **d)** Neither
   e) Complementary **f)** Supplementary
   g) Neither **h)** Neither
   i) Complementary **j)** Supplementary

4. 125°, 65°, 276°, 84°, 31°, 149°, 46°, 44°, 320°, 40°

5. Student's own diagram

6. **a)** BAC **b)** ABC **c)** ECD **d)** CDE
   e) CED **f)** IFG **g)** FGH **h)** GHI
   i) FIH

7. Student's own measurements

8. **a)** RQ and PU, ST and PU
   b) QT and AB, AT and BP, BT and AR

Exercise 22.2 page 177

1. **a)** Obtuse-angled isosceles
   b) Right-angled isosceles
   c) Obtuse-angled scalene
   d) Acute-angled scalene
   e) Right-angled scalene
   f) Acute-angled triangle

2. Student's own diagram

3. **a)** and **d)** **c)** and **e)** **b)** and **f)**

4. Student's own diagram

5. **a)** and **f)** – hypotenuse right-angle
   b) and **c)** – ASA
   d) and **e)** – SAS
   g) and **h)** – SSS

Exercise 22.3 page 180

1. **a)** Rectangle **b)** Parallelogram
   c) Square **d)** Isosceles trapezium
   e) Trapezium **f)** Quadrilateral **g)** Kite
   h) Rhombus

2.

	Rectangle	Square	Parallelogram	Kite	Rhombus	Equilateral triangle
Opposite sides equal in length	Yes	Yes	Yes		Yes	
All sides equal in length		Yes			Yes	Yes
All angles right angles	Yes	Yes				
Both pairs of opposite sides parallel	Yes	Yes	Yes		Yes	
Diagonals equal in length	Yes	Yes				
Diagonals intersect at right angles		Yes		Yes	Yes	
All angles equal	Yes	Yes				Yes

3. Student's own diagram

4.

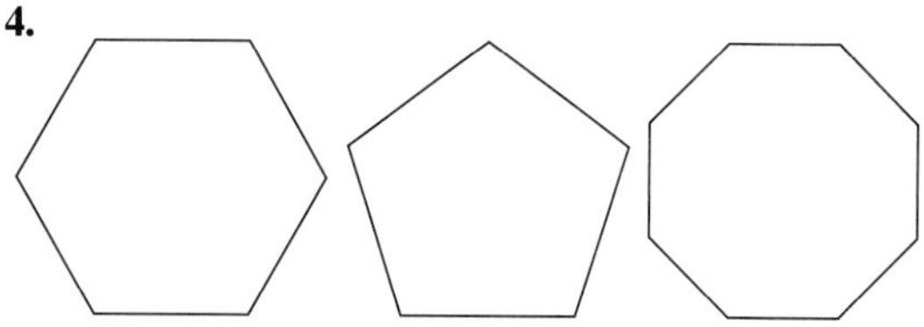

Exercise 22.4 page 182

Student's own diagrams

Student Assessment 1 page 183

1. **a)** Acute **b)** Obtuse **c)** Reflex
   d) Right angle

2. Student's own diagram
3. **a)** Acute scalene triangle
   b) Right-angled scalene
4. 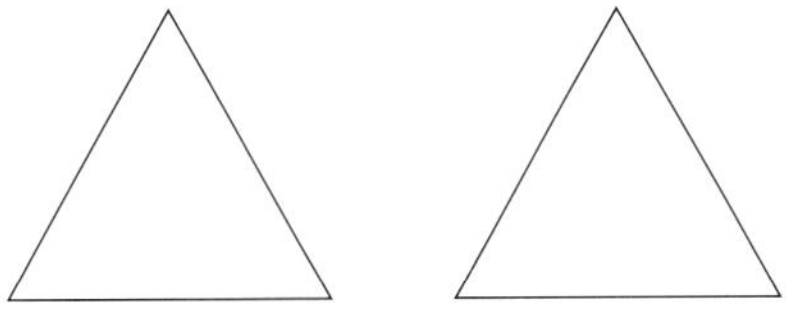
5. Student's own diagram
6. All sides equal in length
   Both pairs of opposite sides are parallel
   Diagonals are equal in length
7. **a)** Yes. Student's own explanation

Student Assessment 2 page 184

1. Student's own diagram
2. Student's own diagram
3. Congruent and right-angled
4. 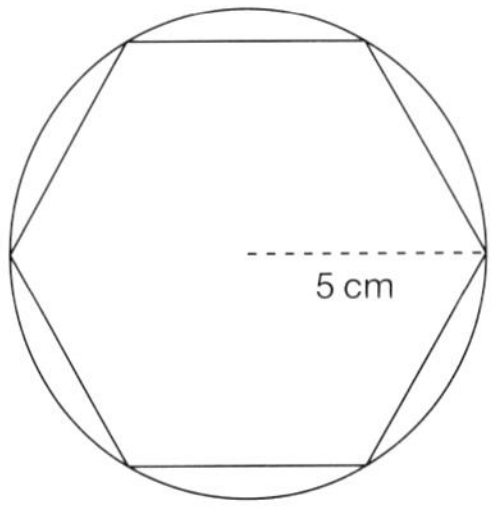

5. Student's own diagram

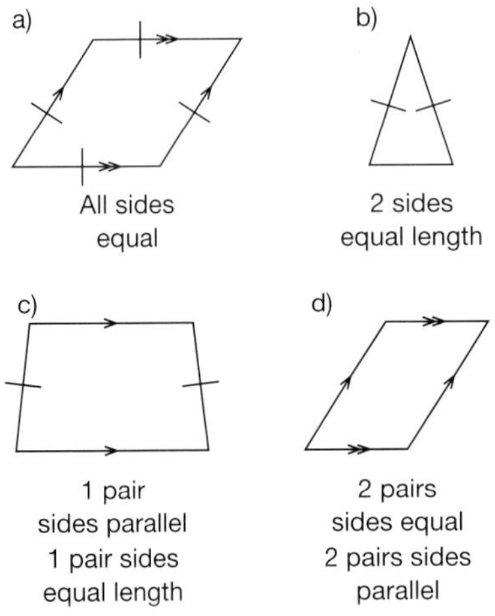

6. Student's own diagram
7. A and C

23 Geometrical Constructions

Exercise 23.1 page 185

1. **a)** 60 mm **b)** 20 mm **c)** 83 mm
   d) 91 mm **e)** 58 mm **f)** 99 mm
2. Student's own lines

Exercise 23.2 page 187

1. **a)** 47° **b)** 24° **c)** 100° **d)** 150°
   e) 110° **f)** 158°
2. **a)** $a = 90°$ $b = 166°$ $c = 104°$
   b) $d = 24°$ $e = 55°$ $f = 125°$ $g = 156°$
   c) $h = 44°$ $i = 316°$
   d) $j = 35°$ $k = 85°$ $l = 240°$
   e) $m = 30°$ $n = 141°$ $o = 60°$ $p = 330°$
   f) $q = 112°$ $r = 339°$ $s = 336°$ $t = 45°$
3. Student's own angles

Exercise 23.3 page 189

1. Student's own constructions
2. Student's own constructions
3. Student's own constructions
4. **a)** Student's own construction attempt
   b) It is not possible as AC + BC < AB

Exercise 23.4 page 189

1. Student's own circles
2. Student's own patterns
3. Student's own patterns

Exercise 23.5 page 191

1–9. Student's own constructions

Exercise 23.6 page 194

1. Student's own constructions
2. Student's own constructions
3. **a)** Student's own construction
   b) Student's own construction
   c) It is equidistant from all three points.
4. Student's own construction
5. Student's own construction

Exercise 23.7 page 197

1. **a)** 300 m **b)** 250 m **c)** 300 m **d)** 416 m
2. **a)** 10 cm **b)** 8 cm **c)** 6 cm **d)** 6.8 cm
3. **a)** Student's own construction
   b)

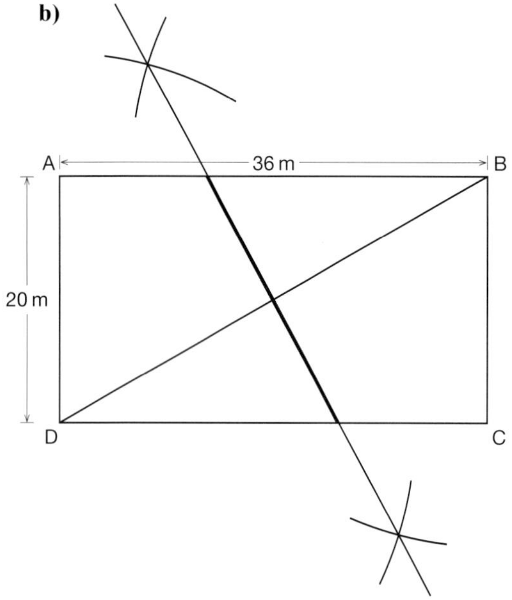

   c) Approx. 23 m
4. **a)** Student's own construction
   b) Approx. 23 cm^2
5. **a)** Student's own construction
   b) Approx. 12 km
6. **a)** Student's own construction
   b)

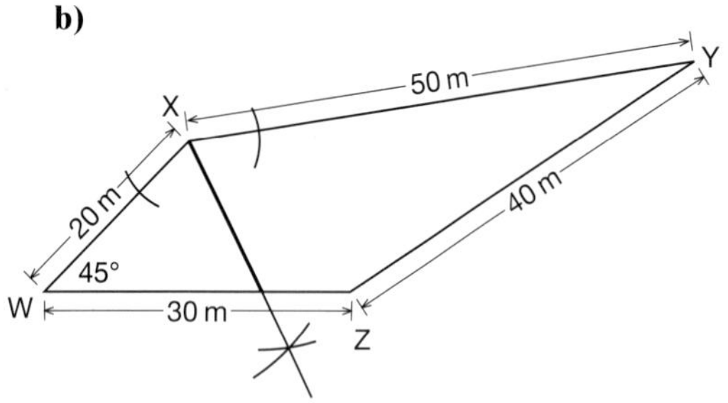

   c) Approx. 16 m

Student Assessment 1 page 199

1. **a)** 7.2 cm **b)** Student's own line
2. **a)** 145° **b)** Student's own angle
3. Student's own construction
4. Student's own construction
5. Student's own construction
6. **a)** Student's own construction
   b) Approx. 53 m^2

Student Assessment 2 page 200

1. **a)** 48° 130° 60° 24° 277°
2. Student's own drawing
3. Student's own constructions
4. Student's own construction
5. **a)** 22 cm **b)** 24.4 cm
6. **a)** Student's own construction
   b) Approx. 23 cm^2

24 Angle Properties

Exercise 24.1 page 203

1. **a)** 75° **b)** 96° **c)** 96° **d)** 146°
2. **a)** 111° **b)** 37° **c)** 22° **d)** 62°
3. **a)** 40° **b)** 45° **c)** 42° **d)** 30°

Exercise 24.2 page 205

1. Student's own diagram and measurements
2. Student's own diagram and measurements
3. Student's own diagram and measurements
4. Student's own observations

Exercise 24.3 page 206

1. Student's own diagram and measurements
2. Student's own diagram and measurements
3. Student's own diagram and measurements
4. Student's own observations

Exercise 24.4 page 207

1. $p = 54°$ $q = 63°$
2. $a = 55°$ $b = 80°$ $c = 100°$
3. $v = 120°$ $w = 60°$ $x = 120°$ $y = 60°$
   $z = 60°$
4. $a = 50°$ $b = 130°$ $c = 45°$ $d = 135°$
5. $p = 45°$ $q = 135°$ $r = 45°$ $s = 45°$
   $t = 135°$
6. $d = 70°$ $e = 30°$

7. $a = 37°$

8. $a = 36°$

Exercise 24.5 page 209

1. **a)** Student's own diagrams and measurements
   b) Student's own observations

2. **b)** Student's own observations

Exercise 24.6 page 210

1. **a)** Equilateral **b)** Isosceles
   c) Equilateral **d)** Scalene **e)** Scalene
   f) Isosceles **g)** Isosceles **h)** Scalene

2. **a)** 70° **b)** 55° **c)** 60° **d)** 73° **e)** 45°
   f) 110°

3. **a)** $a = 30°$ $b = 45°$
   b) $x = 50°$ $y = 80°$ $z = 70°$
   c) $p = 130°$ $q = 15°$ $r = 60°$
   d) $d = 35°$ $e = 55°$ $f = 55°$
   e) $a = 27.5°$ $b = 27.5°$ $c = 55°$
   $d = 27.5°$ $e = 97.5°$
   f) $p = 45°$ $q = 45°$ $r = 67.5°$ $s = 112.5°$

Exercise 24.7 page 212

1. Kite
2. Parallelogram
3. Rhombus
4. Square
5. Trapezium
6. Trapezium
7. Rectangle
8. Parallelogram
9. Kite

Exercise 24.8 page 214

1. $a = 115°$
2. $x = 40°$ $y = 140°$ $z = 140°$
3. $m = 75°$ $n = 75°$
4. $s = 65°$ $t = 115°$ $u = 115°$
5. $h = 120°$ $i = 60°$ $j = 120°$ $k = 60°$
6. $a = 80°$ $b = 20°$ $c = 20°$ $d = 20°$
   $e = 140°$
7. $p = 40°$ $q = 130$ $r = 50°$
8. $p = 75°$ $q = 30°$ $r = 50°$ $s = 80°$
   $t = 70°$ $u = 70°$ $v = 40°$

Exercise 24.9 page 216

1. **a)** 720° **b)** 1260° **c)** 900°

2. **a)** 135° **b)** 90° **c)** 144° **d)** 150°

3. **a)** 72° **b)** 30° **c)** 51.4° (1 d.p.)

4. **a)** 18 **b)** 10 **c)** 36 **d)** 8 **e)** 20
   f) 120

5. **a)** 5 **b)** 12 **c)** 20 **d)** 15 **e)** 40
   f) 360

6. 12

7.

Number of sides	Name	Sum of exterior angles	Size of an exterior angle	Sum of interior angles	Size of an interior angle
3	Equilateral triangle	360°	120°	180°	60°
4	Square	360°	90°	360°	90°
5	Pentagon	360°	72°	540°	108°
6	Hexagon	360°	60°	720°	120°
7	Heptagon	360°	51.4°	900°	128.6°
8	Octagon	360°	45°	1080°	135°
9	Nonagon	360°	40°	1260°	140°
10	Decagon	360°	36°	1440°	144°
12	Dodecagon	360°	30°	1800°	150°

Exercise 24.10 page 218

1. 60° 2. 135°
3. 20° 4. 32°
5. 110° 6. 22.5°

Exercise 24.11 page 218

1. 35° 2. 60°
3. 40° 4. 45°
5. 24° 6. 26°
7. 13 cm 8. 8 cm
9. 17.7 cm (1 d.p.)

Student Assessment 1 page 220

1. **a)** $a = 80°$ $b = 100°$
   b) $x = 65°$ $y = 65°$ $z = 20°$
   c) $a = 45°$ **d)** $p = 73°$

2. **a)** $p = 135°$ $q = 135°$ $r = 45°$
   b) $a = 120°$ $b = 60°$ $c = 60°$ $d = 60°$

3. a) $m = 50°$ $n = 90°$ $p = 40°$ $q = 140°$
b) $w = 55°$ $x = 70°$ $y = 55°$ $z = 55°$
c) $a = 70°$ $b = 110°$ $c = 110°$ $d = 70°$ $e = 30°$
d) $a = 30°$ $b = 120°$ $c = 80°$ $d = 80°$ $e = 80°$

Student Assessment 2 page 221

1. a) $a = 70°$ $b = 110°$
b) $x = 95°$ $y = 55°$ $z = 55°$
c) $a = 30°$ d) $p = 40°$

2. a) $p = 130°$ $q = 130°$ $r = 50°$
b) $a = 50°$ $b = 50°$ $c = 50°$ $d = 50°$

3. a) $m = 50°$ $n = 70°$ $p = 70°$ $q = 50°$
b) $w = 85°$ $x = 50°$ $y = 50°$ $z = 45°$
c) $a = 130°$ $b = 25°$ $c = 25°$ $d = 50°$
d) $a = 40°$ $b = 100°$ $c = 40°$ $d = 70°$ $e = 70°$

Student Assessment 3 page 222

1.
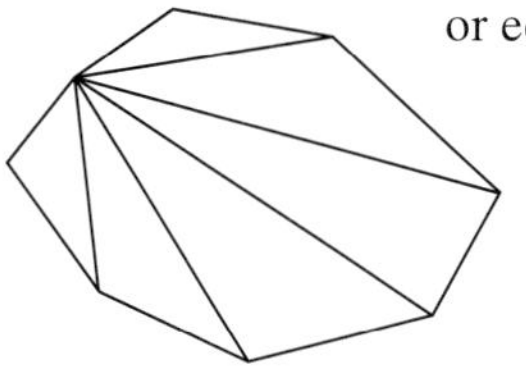
or equivalent

2. 162°
3. 1260°
4. 360°
5. 72°
6. a) 90° b) 6.5 cm
7. 58°
8. 30°
9. 25°
10. 152°

Student Assessment 4 page 223

1.
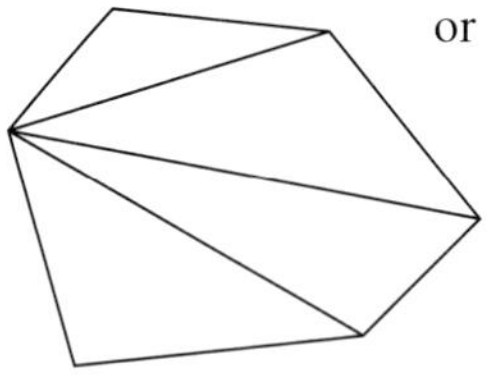
or equivalent

2. 165°
3. 1800°
4. 30°
5. a) 90° b) 13 cm

6. 28°
7. 125°
8. 45°
9. 42°
10. 38°

25 Loci

Exercise 25.1 page 227

1.
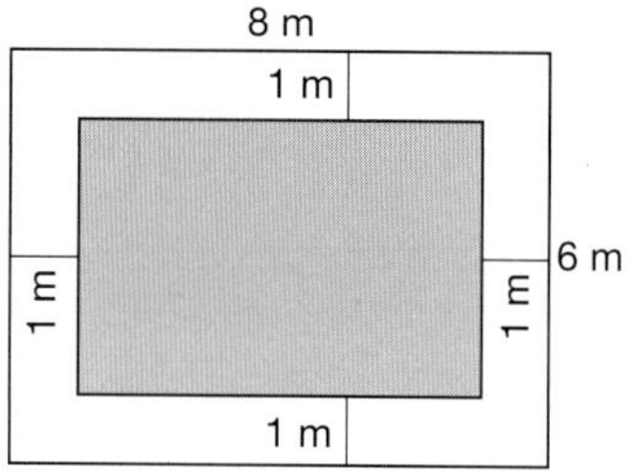

2.
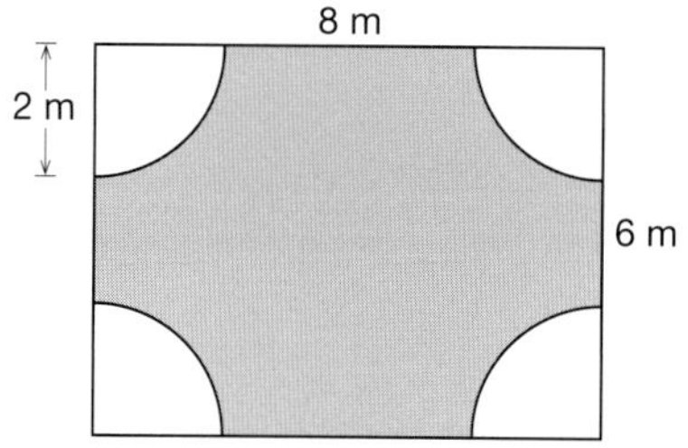

3.
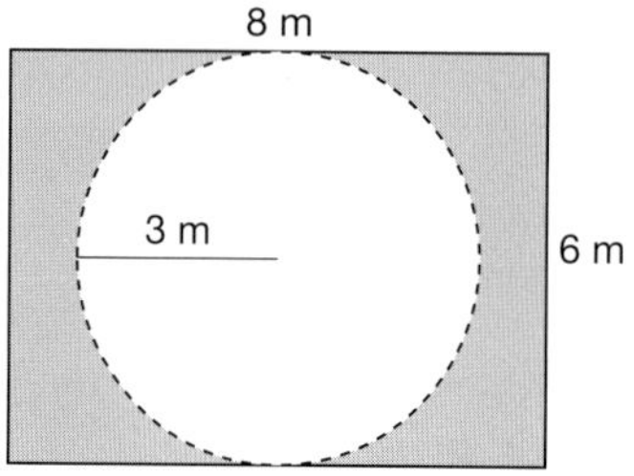

4.
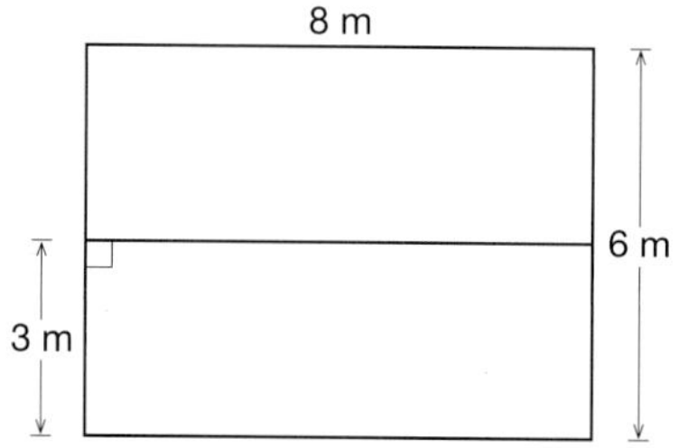

5.

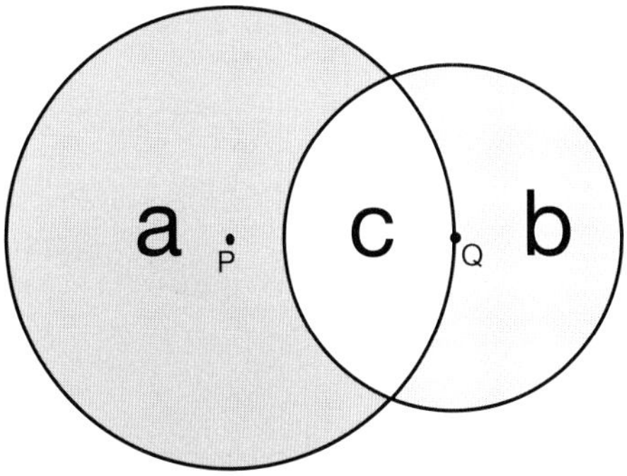

6.

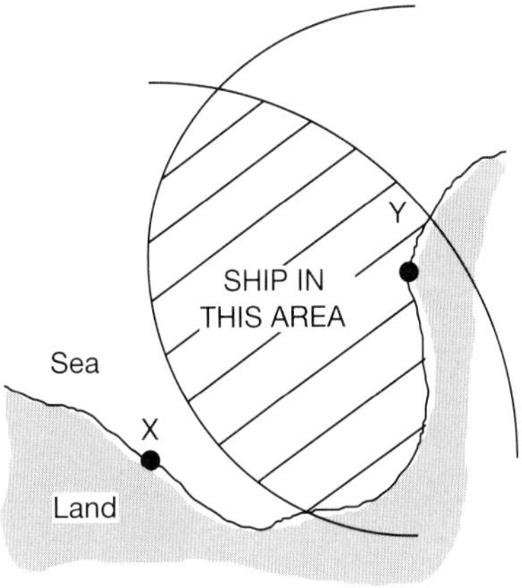

7. a) Student's own diagrams. L, M and N will all lie on the circumference of a circle, the centre of the circle being the point equidistant from L, M and N.

b) There would be no point equidistant from all three (except in the infinite!).

8. C is on the circumference of a circle with AB as its diameter.

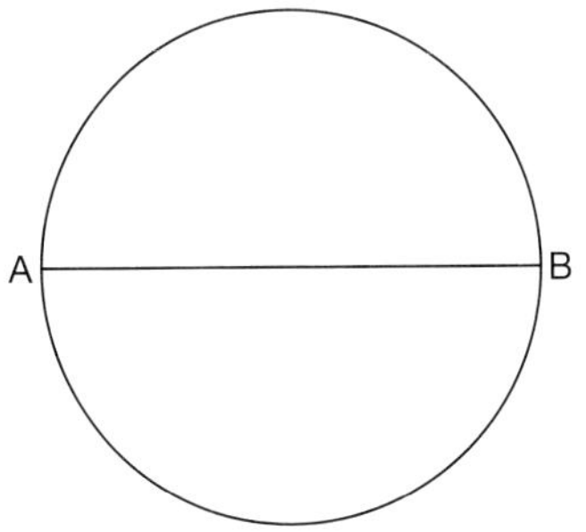

9. Student's own construction

10.

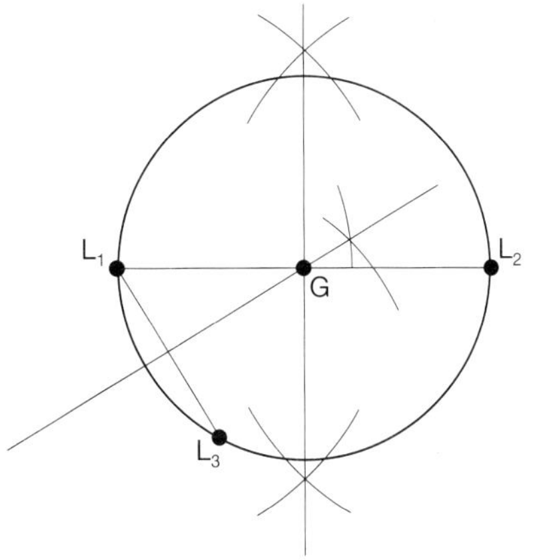

Exercise 25.2 page 228

1. a)

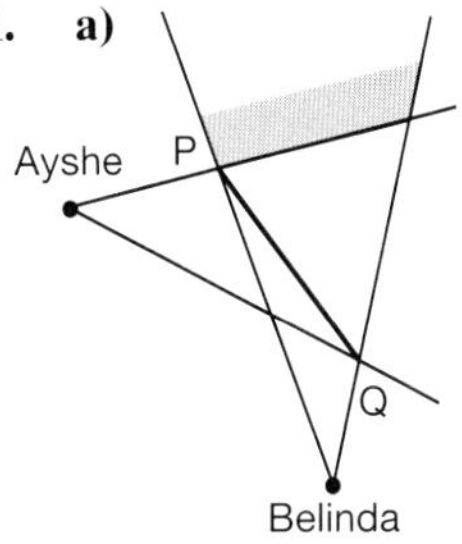

b)

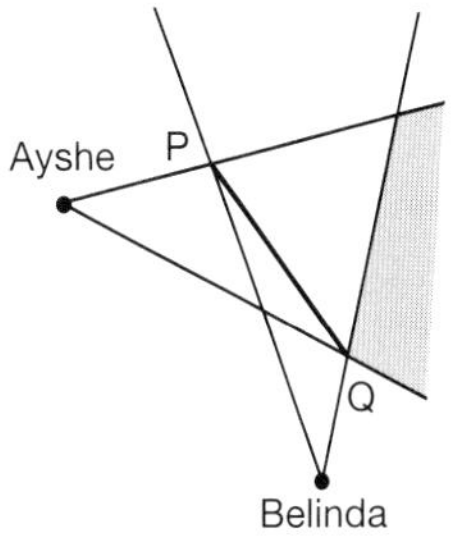

c)

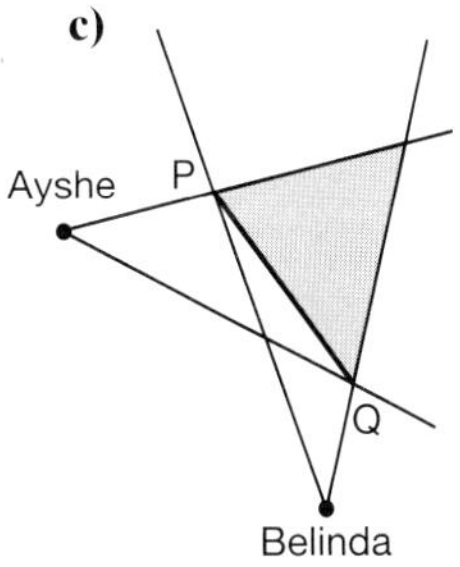

d)

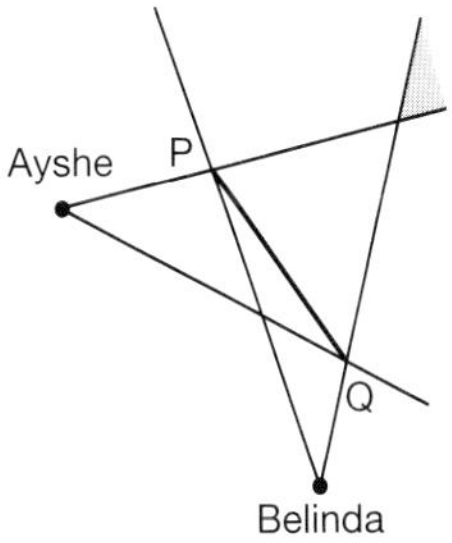

2.

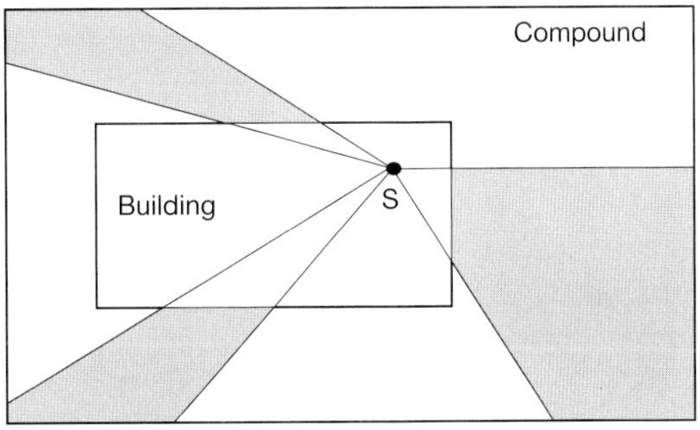

3.

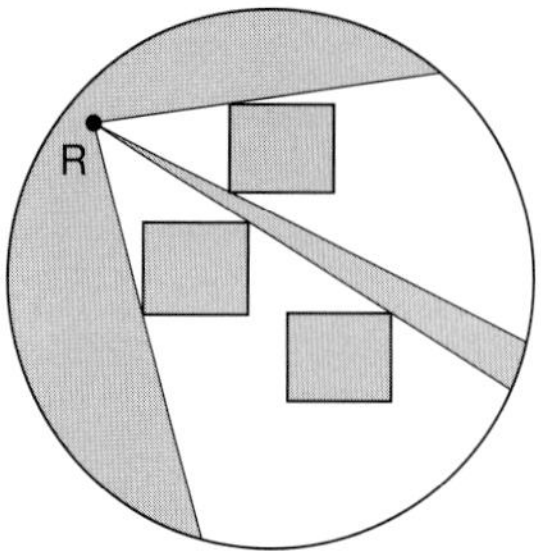

Exercise 25.3 page 229

1.

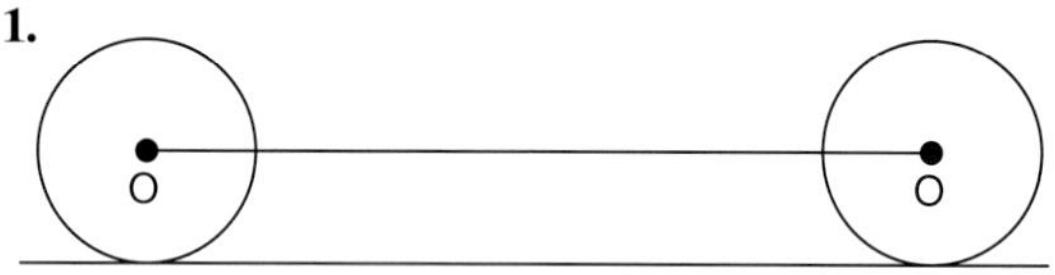

2.

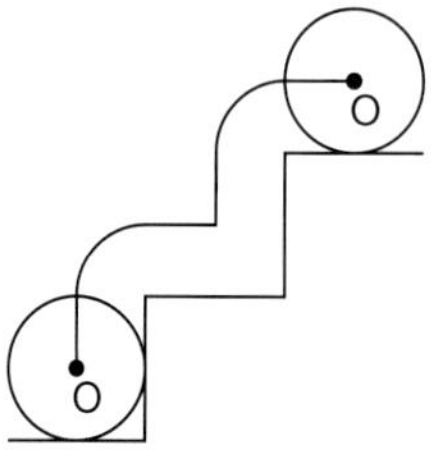

3.

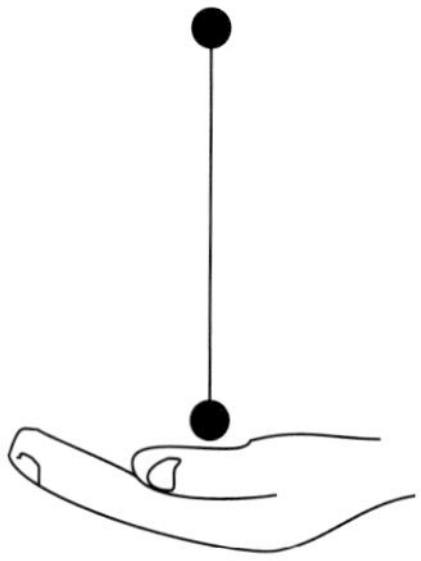

4

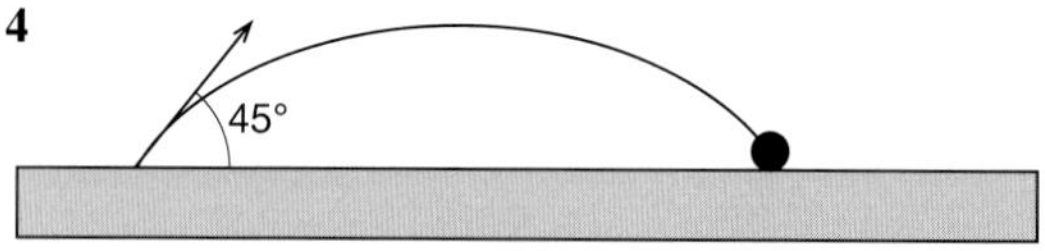

5.

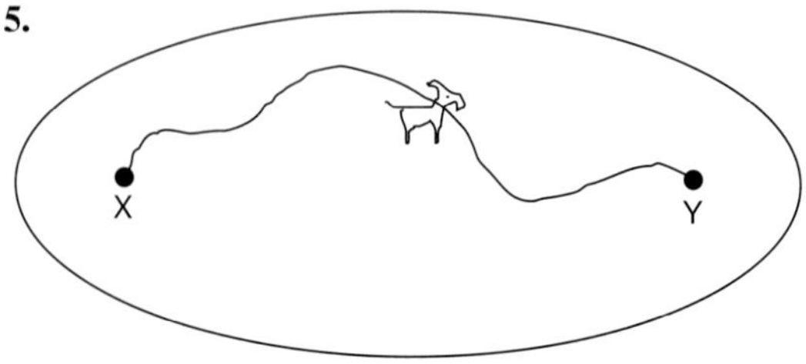

Student Assessment I page 230

1.

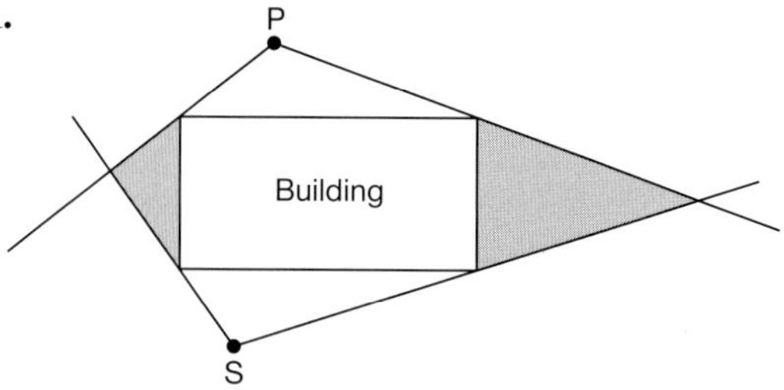

2.

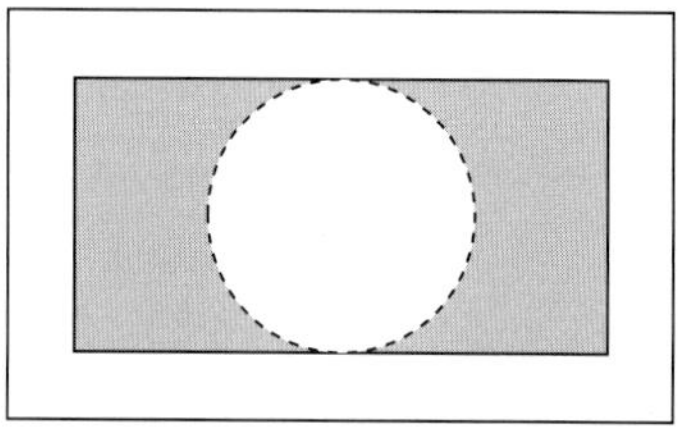

3.

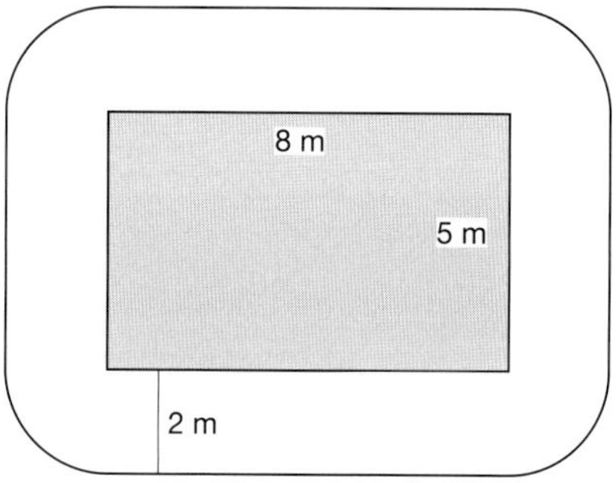

4.

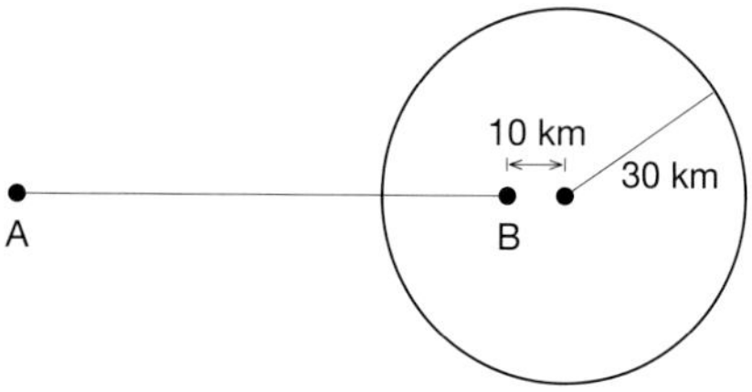

5.

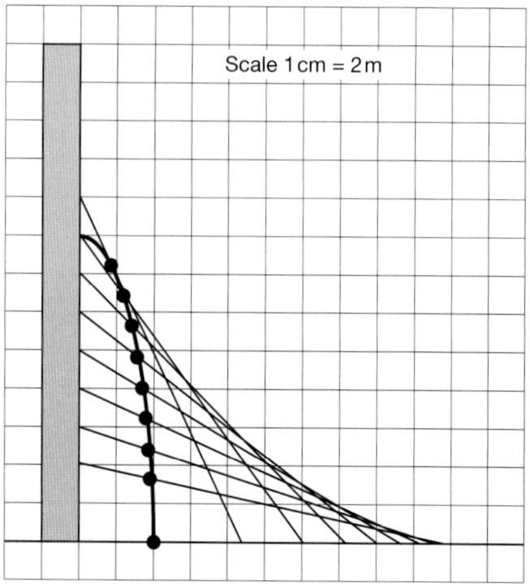

6.

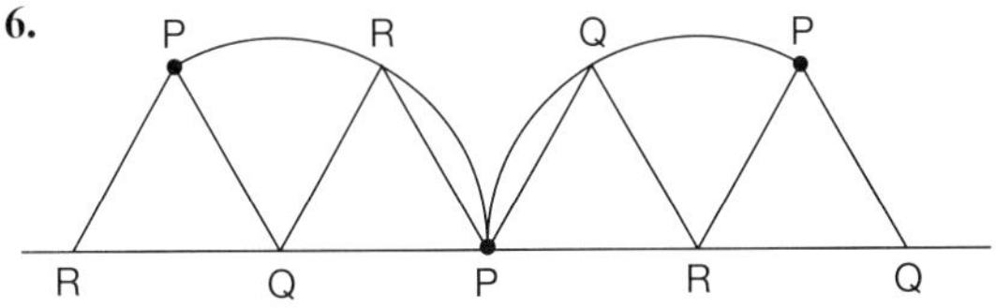

Student Assessment 2 page 231

1.

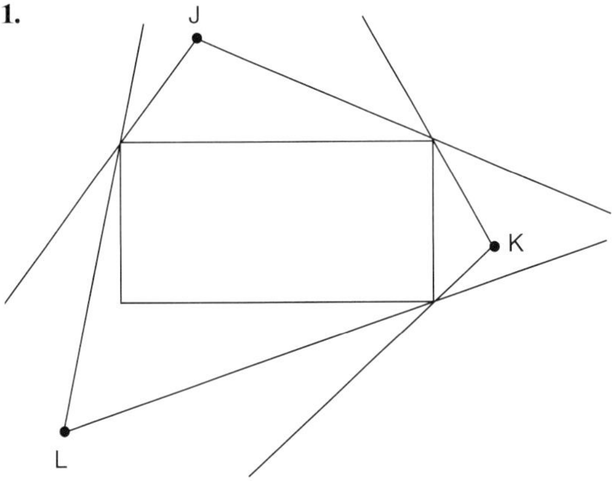

None of the friends can see each other, as shown above.

2.

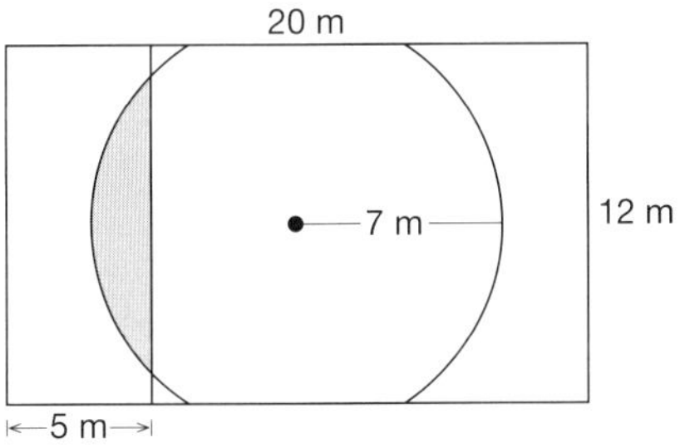

3.

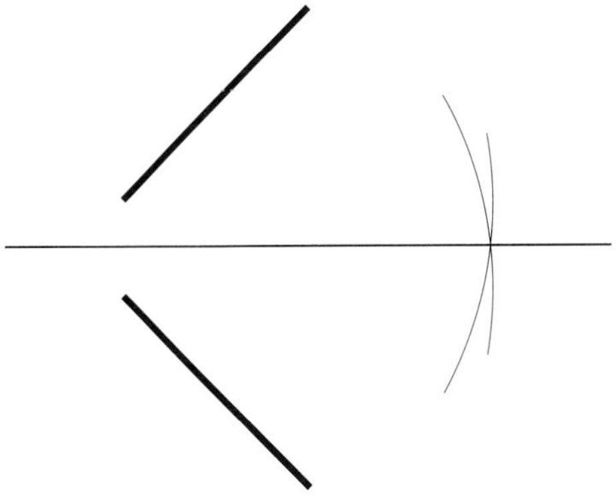

4.

5.

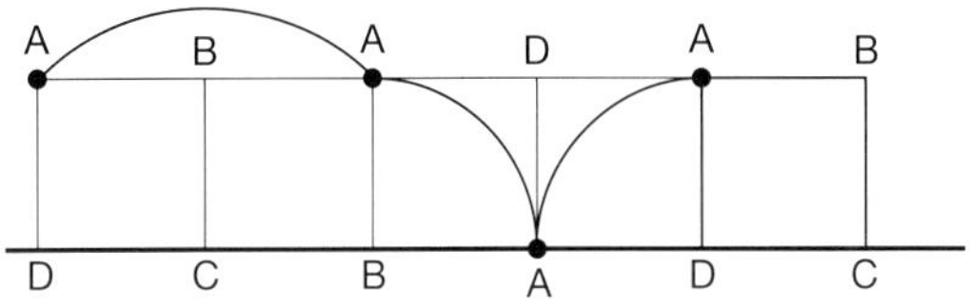

26 Symmetry

Exercise 26.1 page 233

1. a)

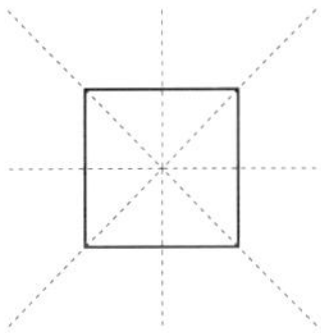

b)

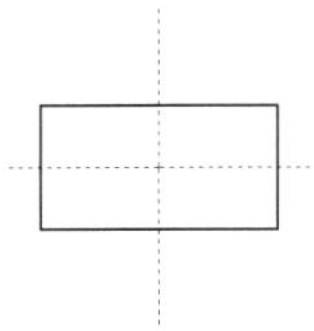

c)

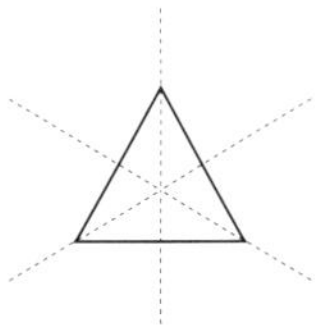

d)

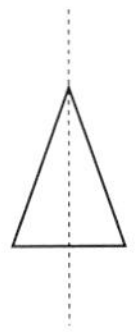

e)

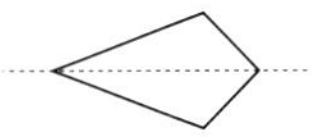

f)

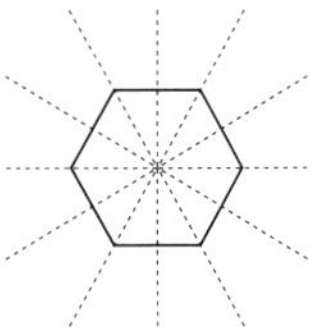

g)

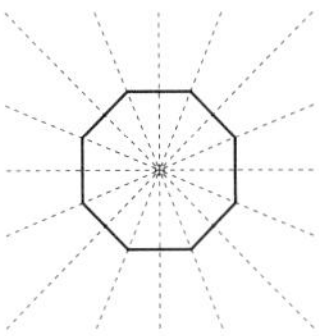

h)

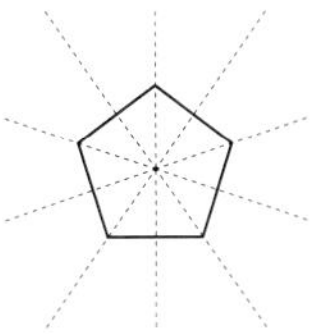

i)

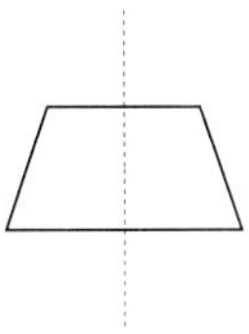

2. **a)**

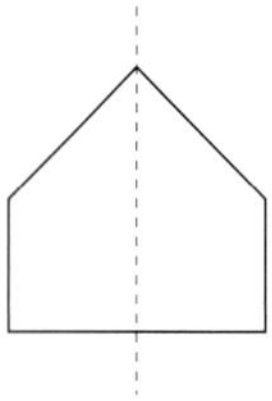

b)

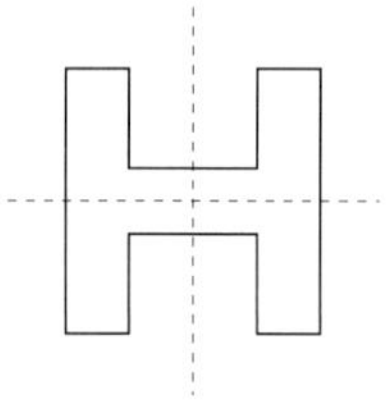

c)

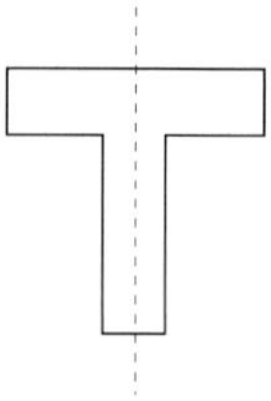

d)

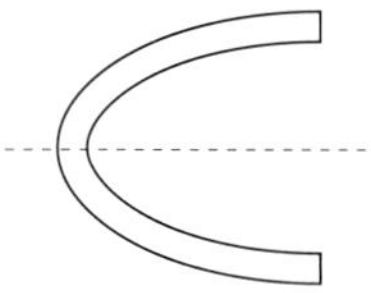

e)

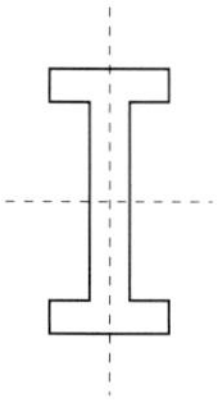

f)

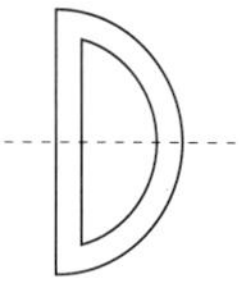

g)

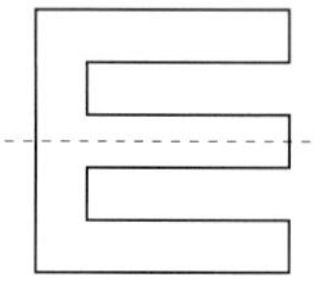

h)

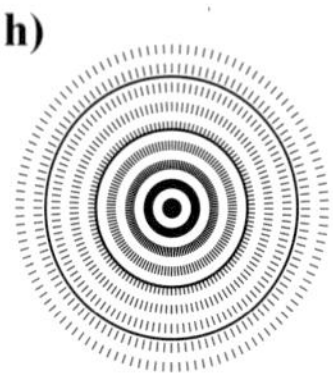

infinite number

i)

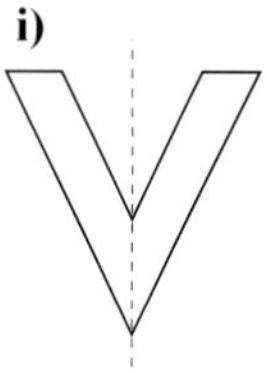

3. **a)**

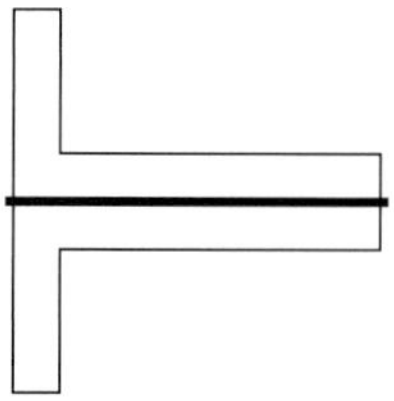

b)

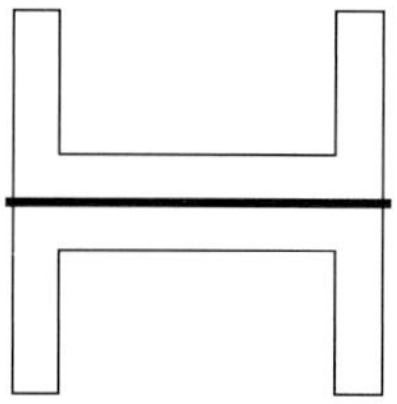

c)

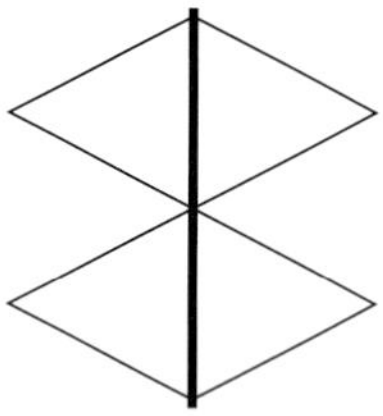

d)

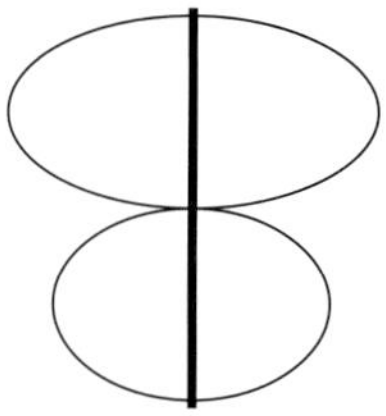

e)

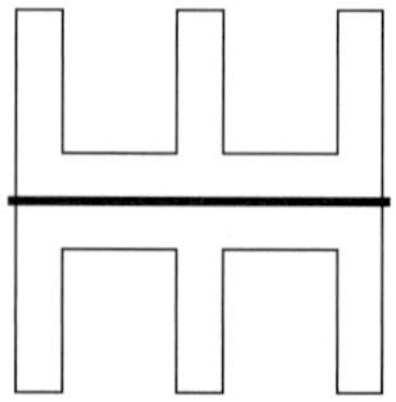

f)

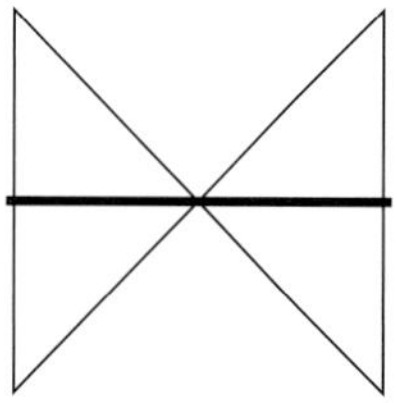

4. **a)**

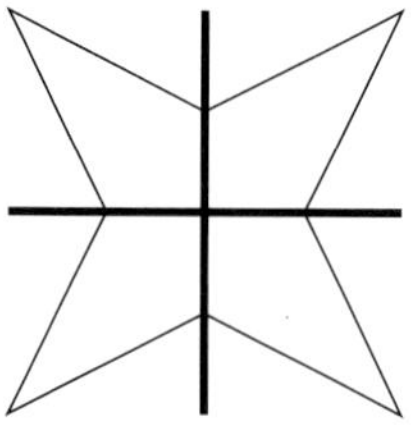

b)

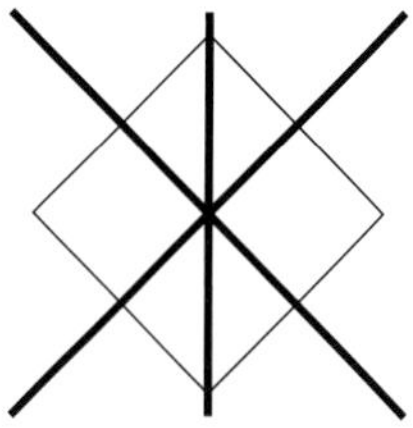

c)

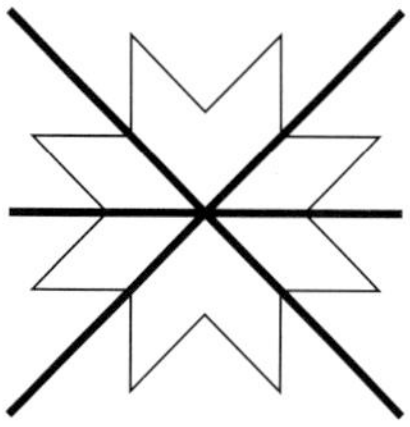

d)

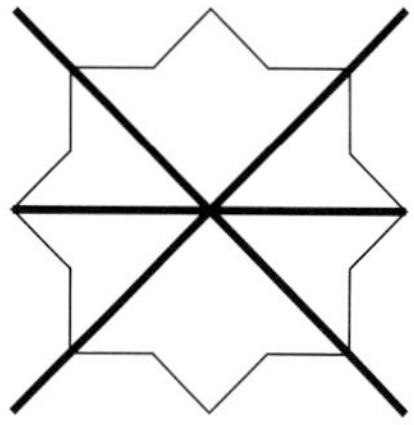

Exercise 26.2 page 234

1. **a)**

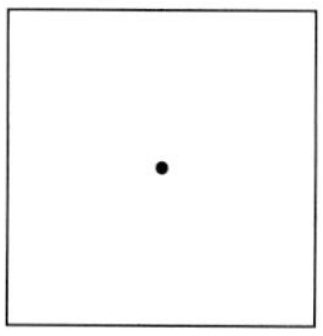

order 4

b)

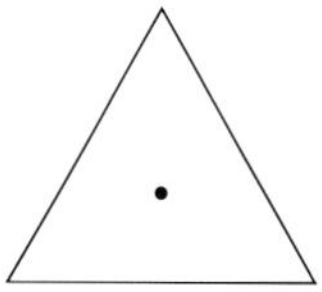

order 3

c)

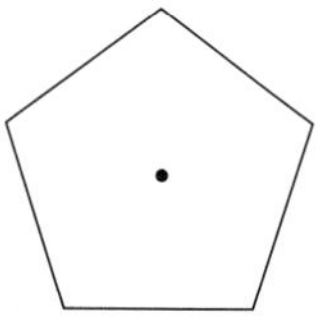

order 5

d)

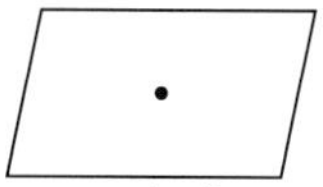

order 2

e)

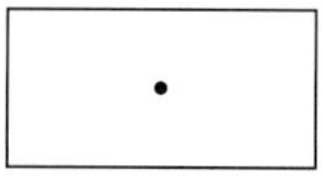

order 2

f)

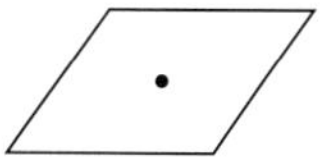

order 2

g)

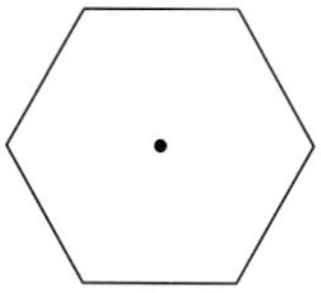

order 6

h)

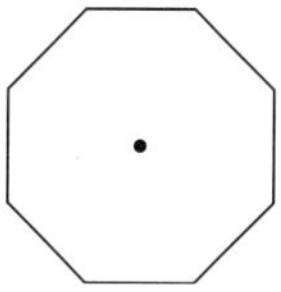

order 8

2. **a)**

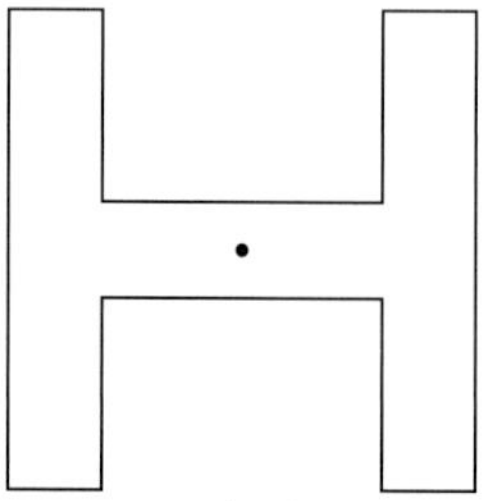

order 2

b)

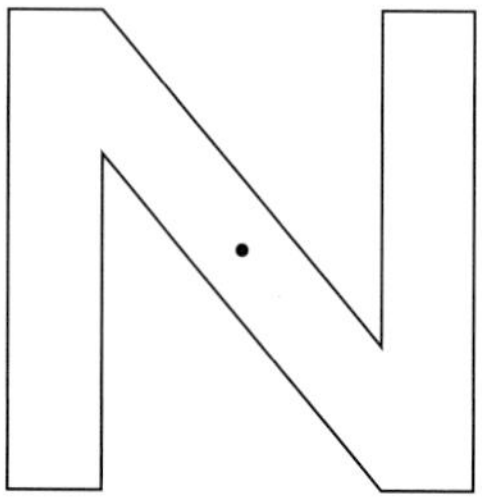

order 2

c)

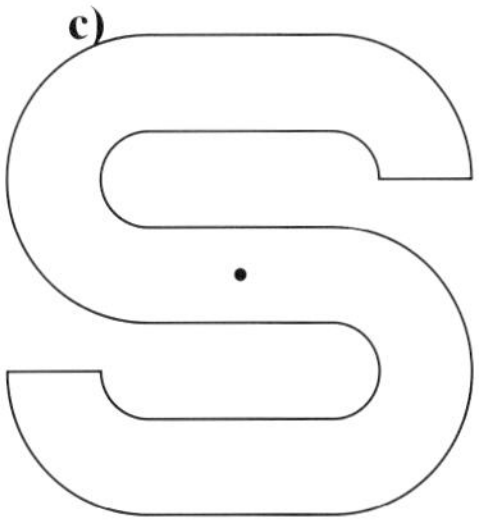

order 2

d)

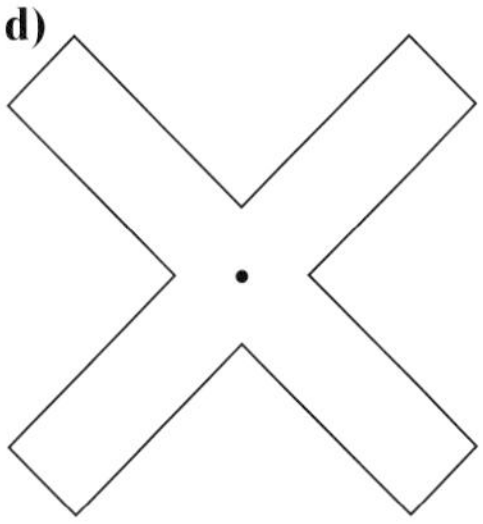

order 4

e)

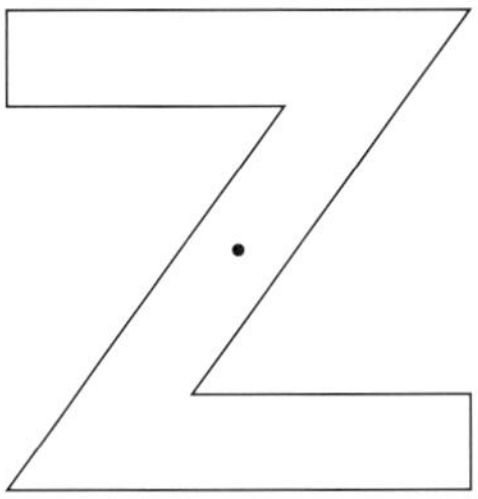

order 2

3. **a)** i) Student's planes ii) 3
b) i) Student's planes ii) 2
c) i) Student's planes ii) 4
d) i) Student's planes ii) 4
e) i) Student's planes ii) Infinite
f) i) Student's planes ii) Infinite
g) i) Student's planes ii) Infinite
h) i) Student's planes ii) 9

4. **a)** 2 **b)** 2 **c)** 3 **d)** 4 **e)** Infinite
f) Infinite **g)** Infinite **h)** 4

Student Assessment 1 page 236

1. **a)** Student's own diagram (e.g. isosceles triangle)
b) Student's own diagram (e.g. rectangle)
c) Student's own diagram (e.g. equilateral triangle)

2. **a)** Student's own diagram (e.g. parallelogram)
b) Student's own diagram (e.g. equilateral triangle)

3. **a)**

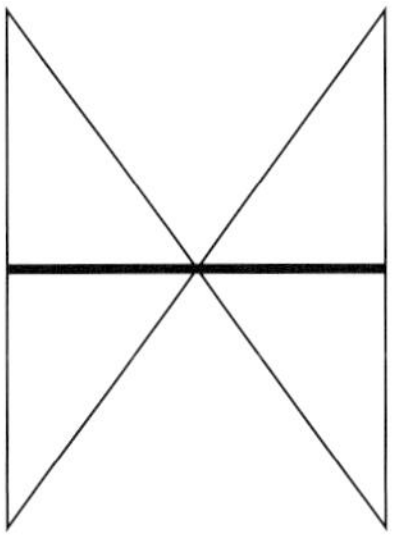

b)

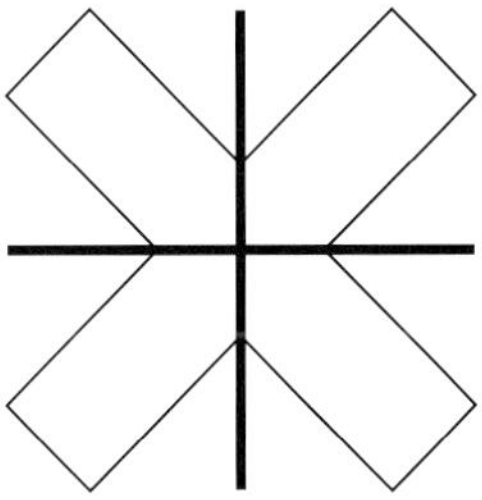

c)

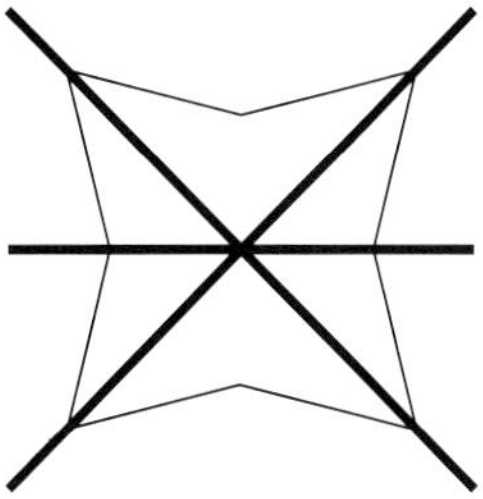

d)

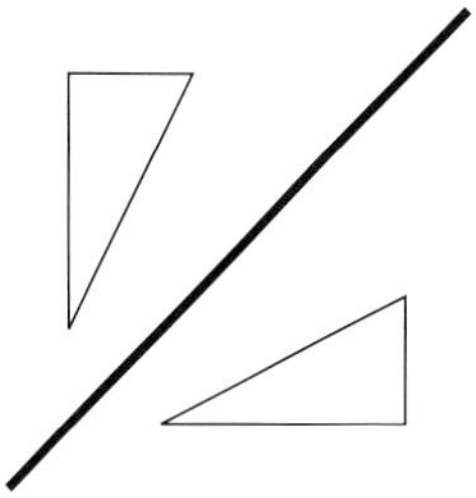

4. **a)** Order 2 **b)** Order 4
c) Order 4 **d)** Order 1

Student Assessment 2 page 237

1. **a)** Student's own diagram (e.g. square)
b) Student's own diagram (e.g. regular octagon)

2. **a)** Student's own diagram (e.g. cuboid)
b) Student's own diagram (e.g. equilateral triangular prism)

3. **a)**

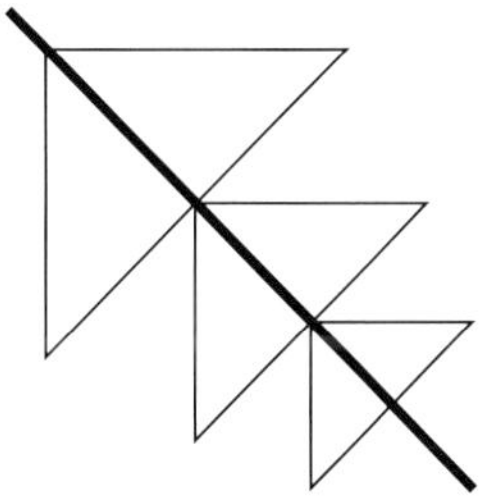

b)

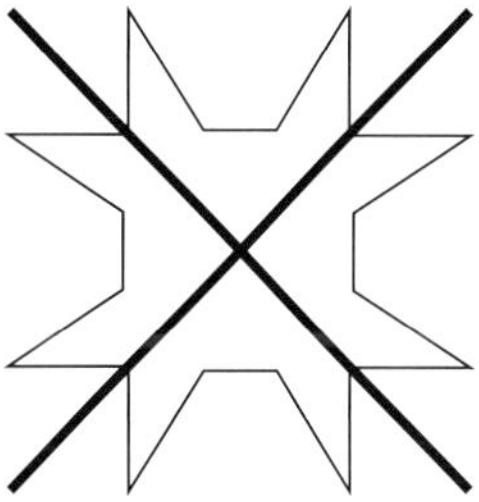

c)

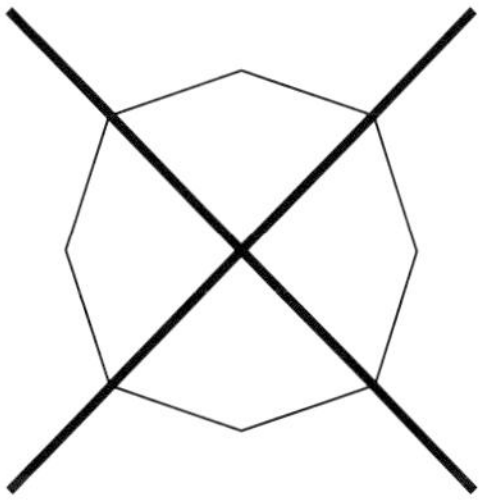

d)

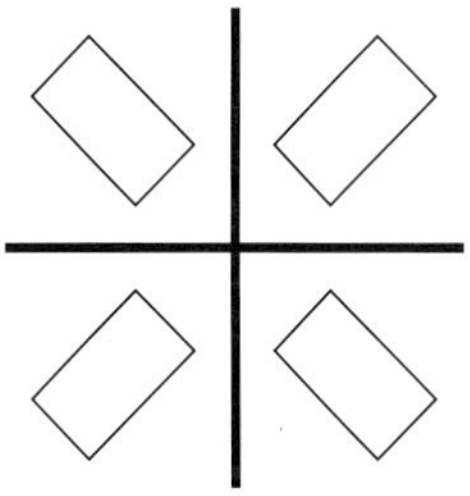

4. **a)** Order 4 **b)** Order 4
c) Order 4 **d)** Order 4

27 Transformations

Exercise 27.1 page 238

1.

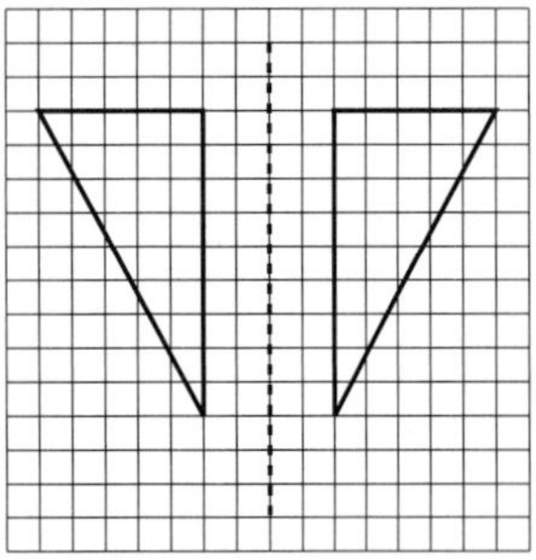

2.

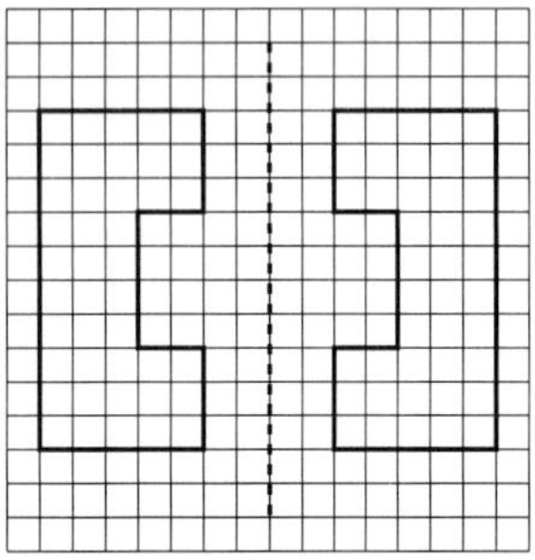

3.

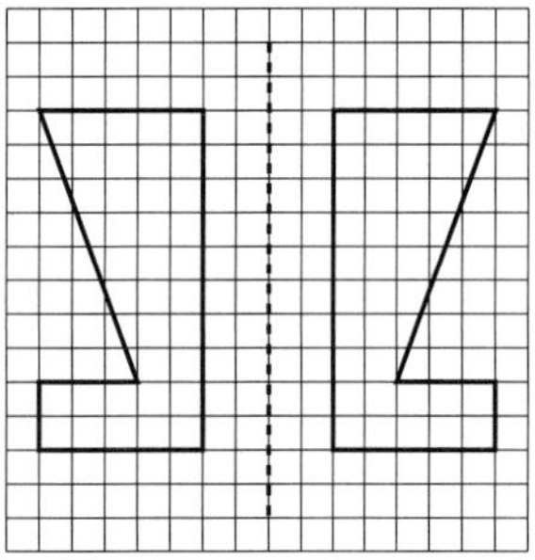

4.

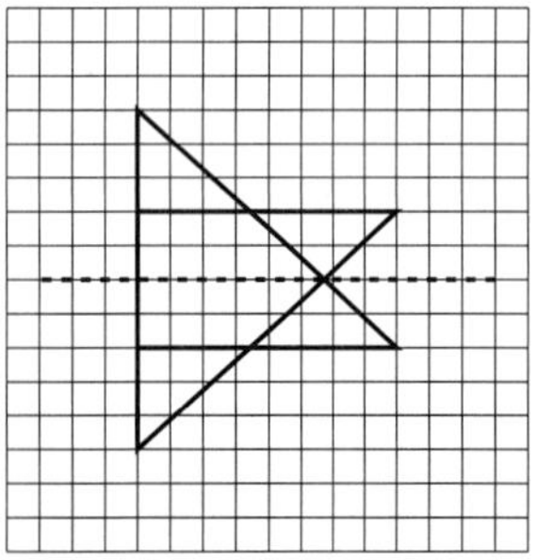

5.

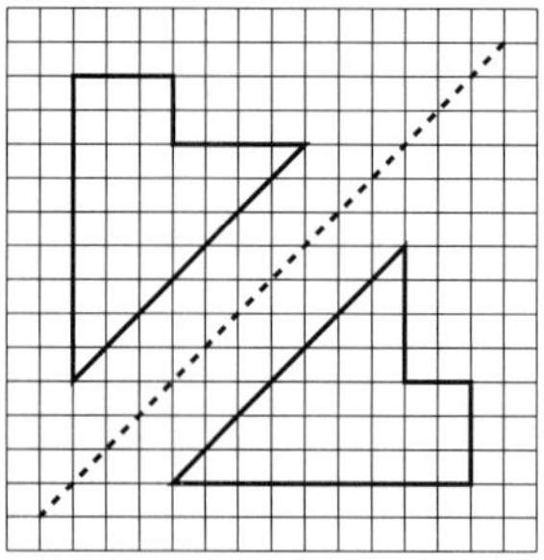

6.

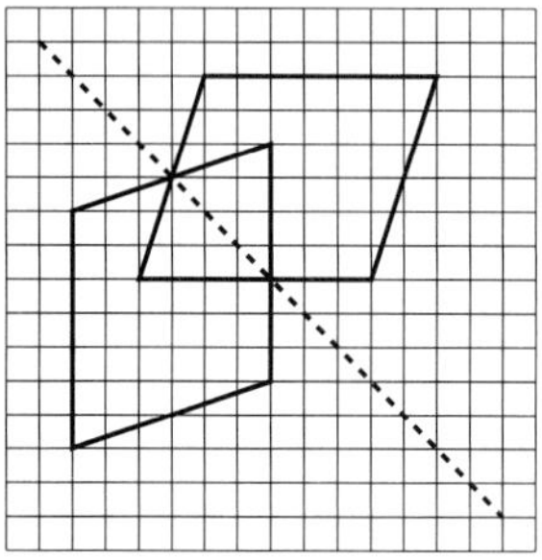

7.

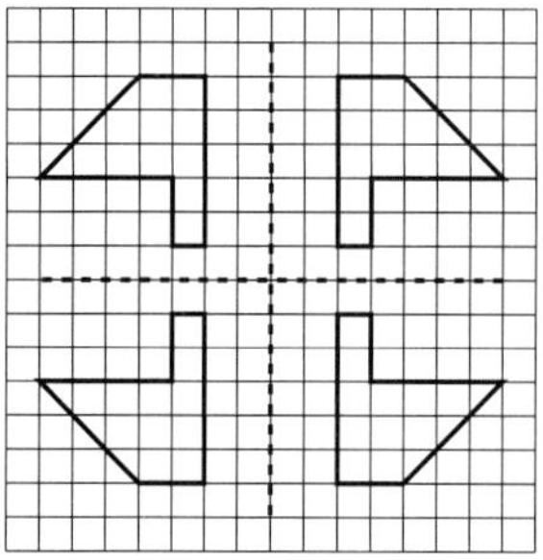

8.

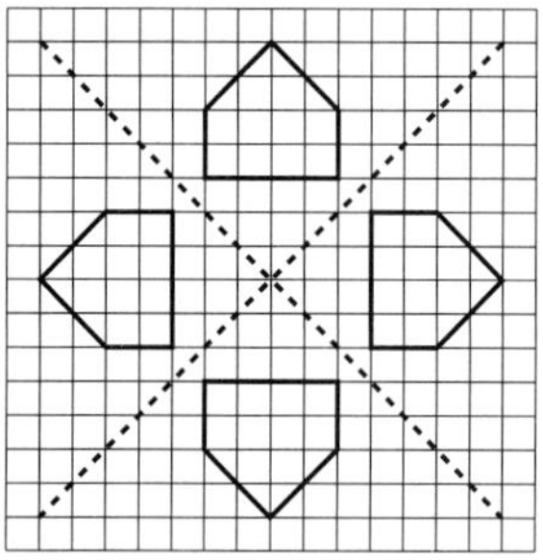

Exercise 27.2 page 239

1.

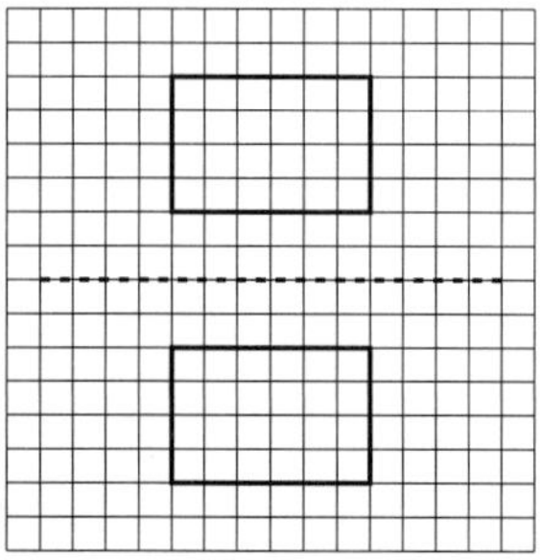

2.

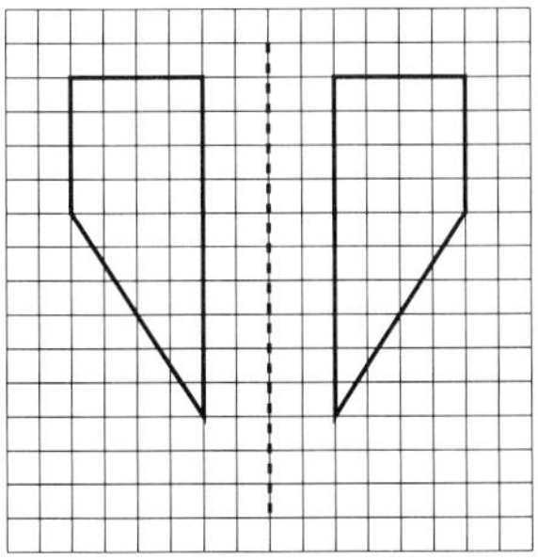

3.

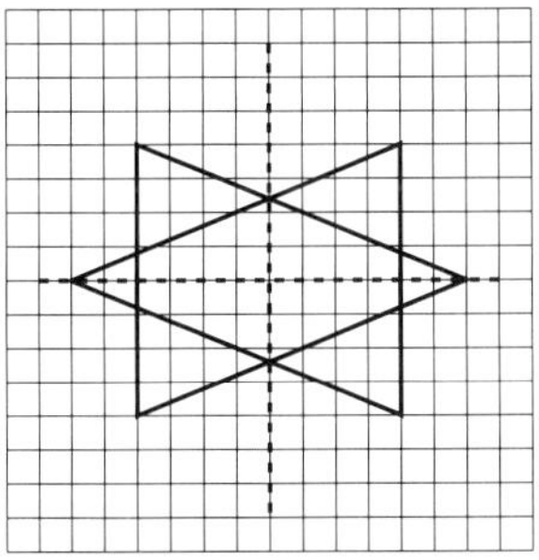

4.

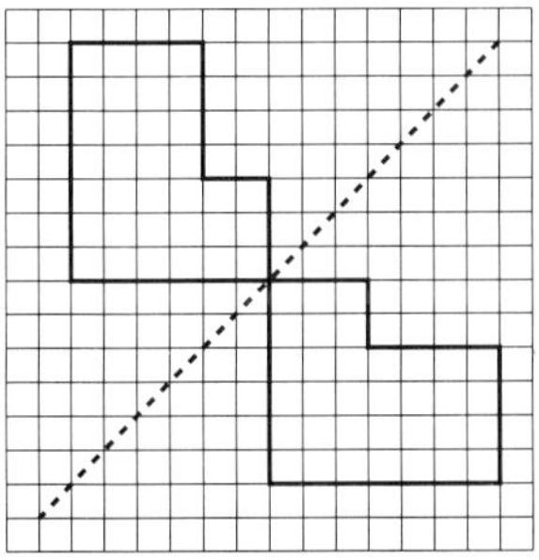

5.

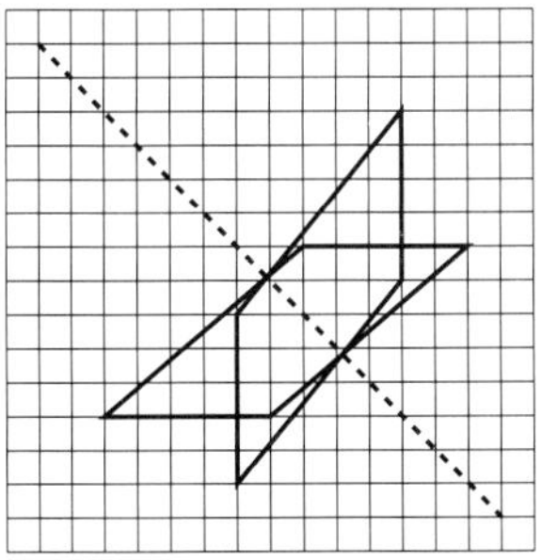

6.

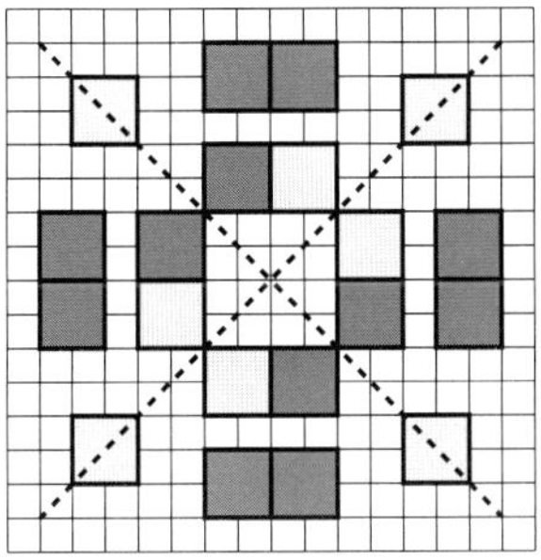

Exercise 27.3 page 239

1.

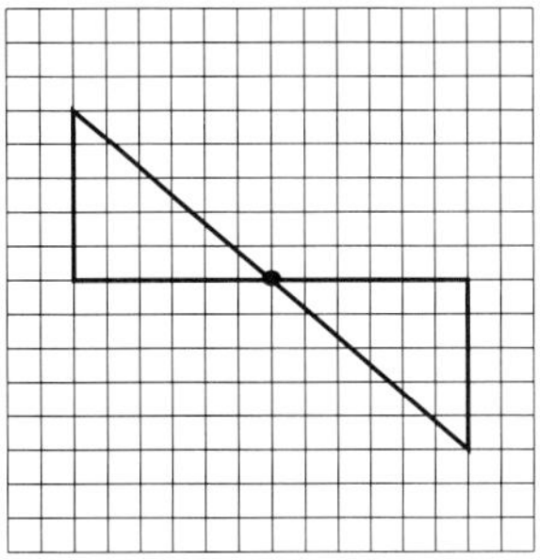

2.

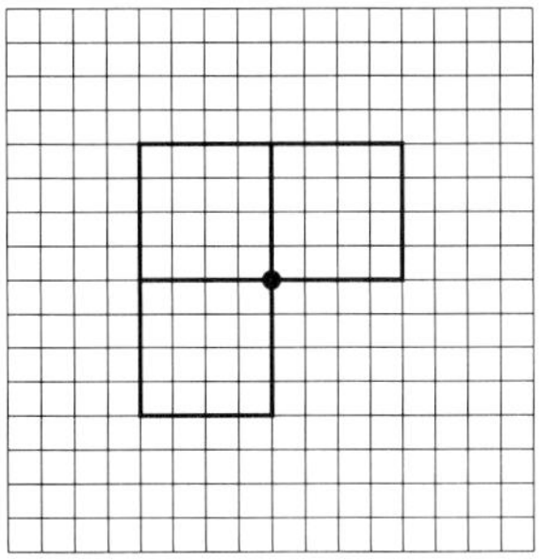

3.

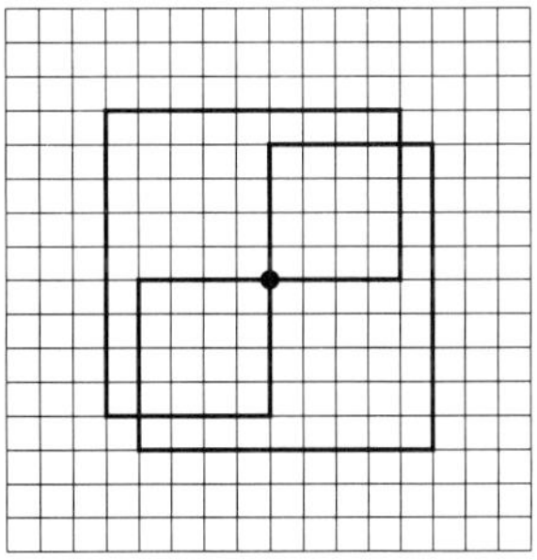

4.

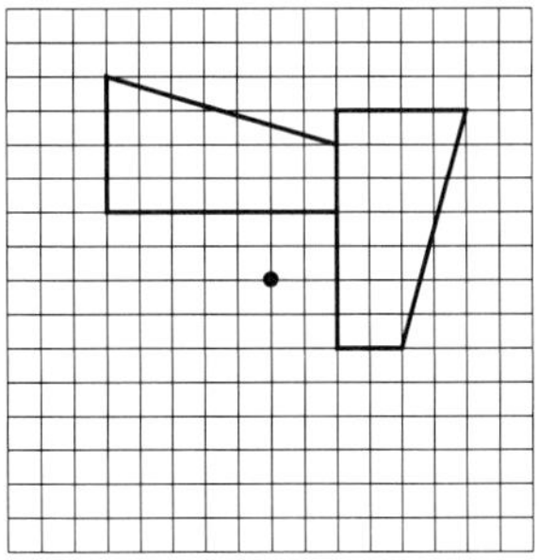

5.

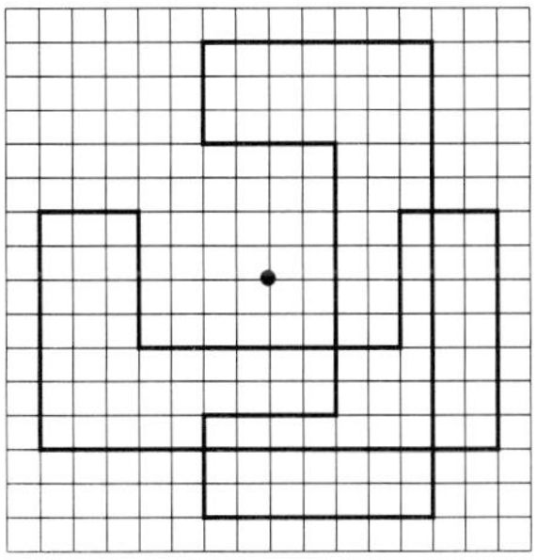

6.

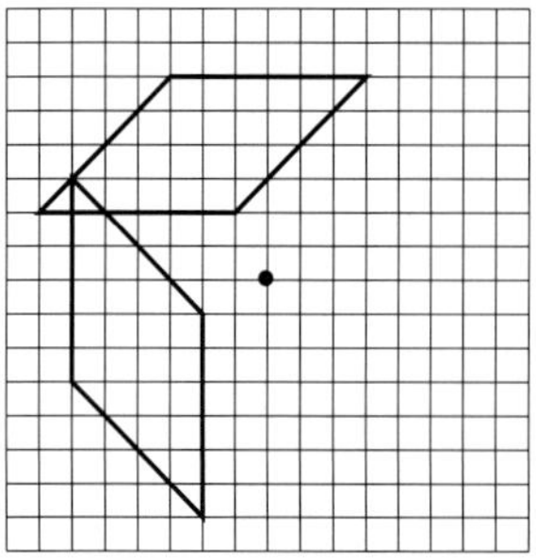

Exercise 27.4 page 240

1. a)

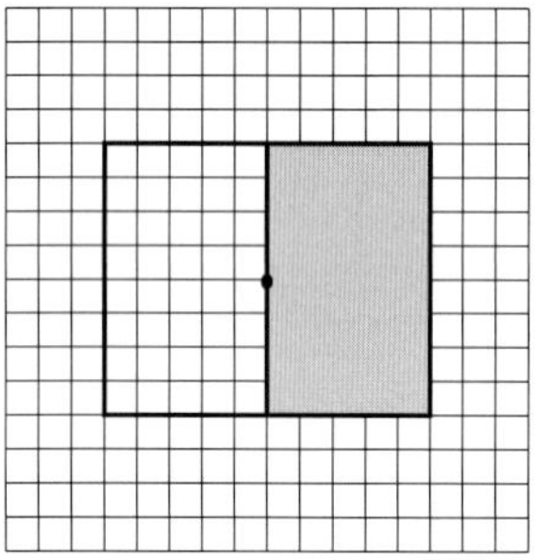

b) 180° clockwise/anti-clockwise

2. a)

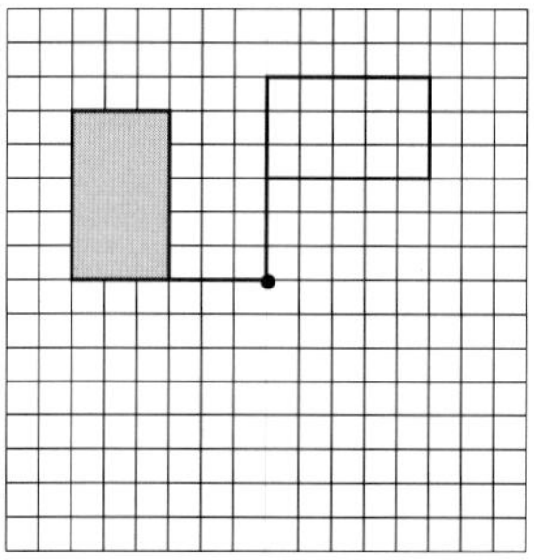

b) 90° anti-clockwise

3. a)

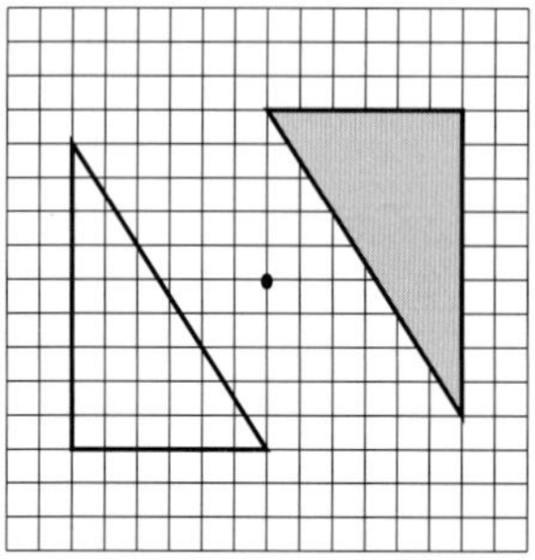

b) 180° clockwise/anti-clockwise

4. a)

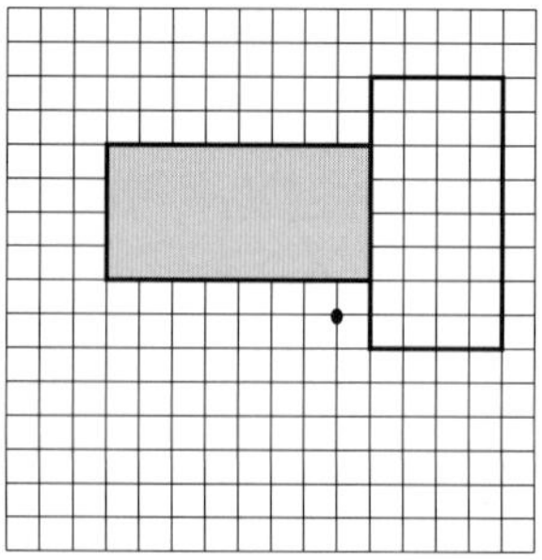

b) 90° anti-clockwise

5. a)

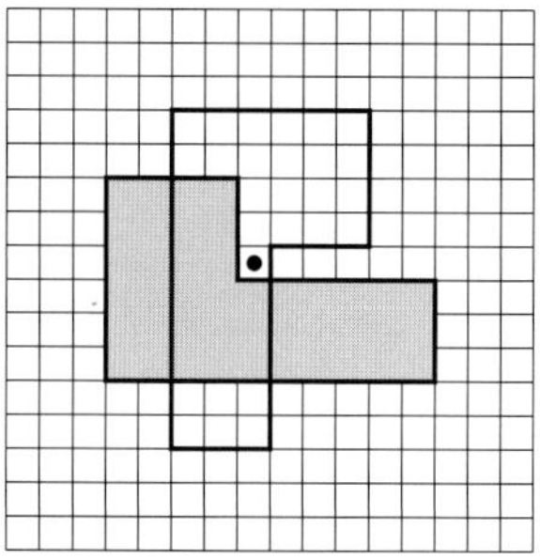

b) 90° anti-clockwise

6. a)

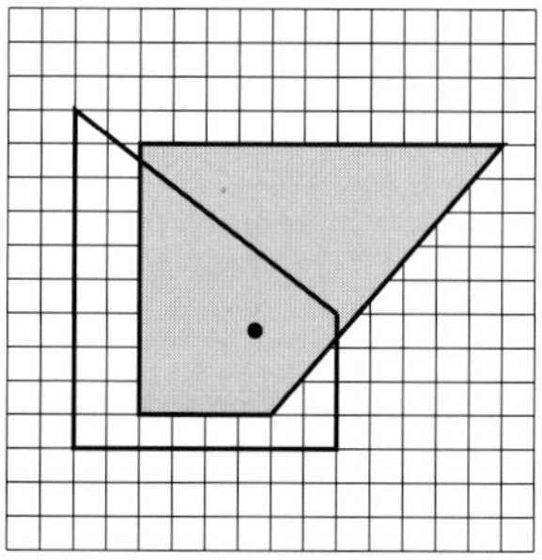

b) 90° clockwise

Exercise 27.5 page 241

1. $A \to B = \begin{pmatrix} -6 \\ 0 \end{pmatrix}$ $A \to C = \begin{pmatrix} 3 \\ 6 \end{pmatrix}$

2. $A \to B = \begin{pmatrix} 0 \\ -7 \end{pmatrix}$ $A \to C = \begin{pmatrix} -6 \\ 1 \end{pmatrix}$

3. $A \to B = \begin{pmatrix} 0 \\ 6 \end{pmatrix}$ $A \to C = \begin{pmatrix} 6 \\ -3 \end{pmatrix}$

4. $A \to B = \begin{pmatrix} 5 \\ 0 \end{pmatrix}$ $A \to C = \begin{pmatrix} -3 \\ -6 \end{pmatrix}$

Exercise 27.6 page 241

1.

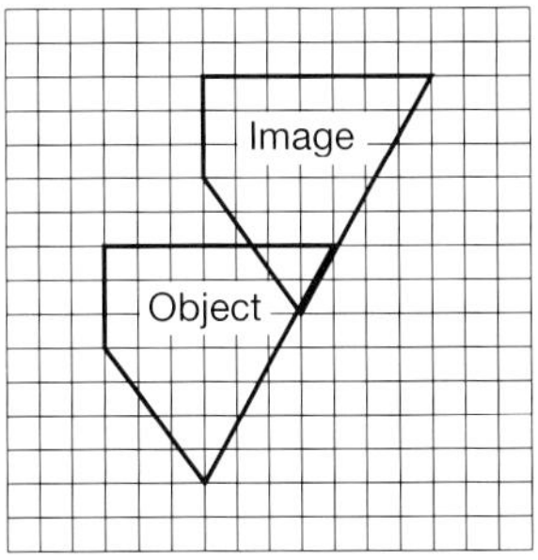

2.

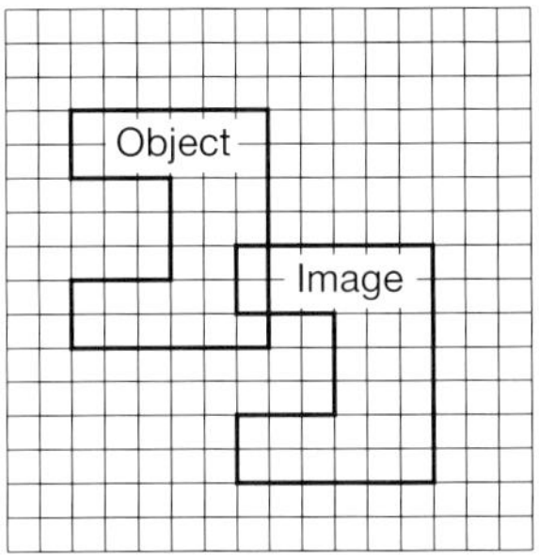

3.

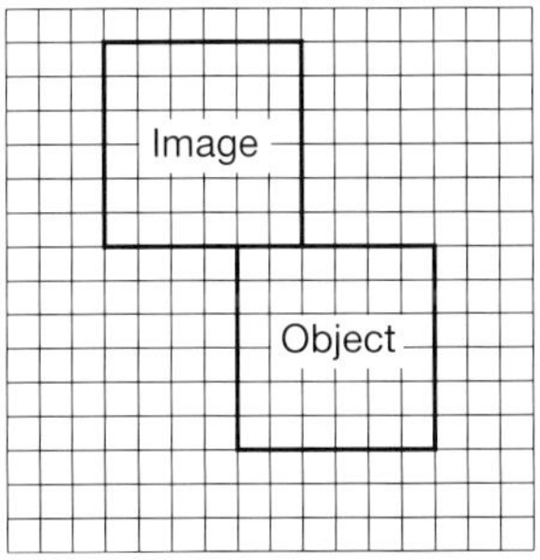

4.

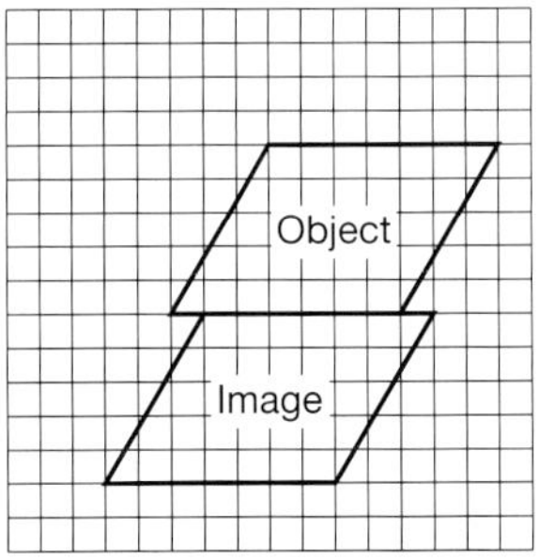

5.

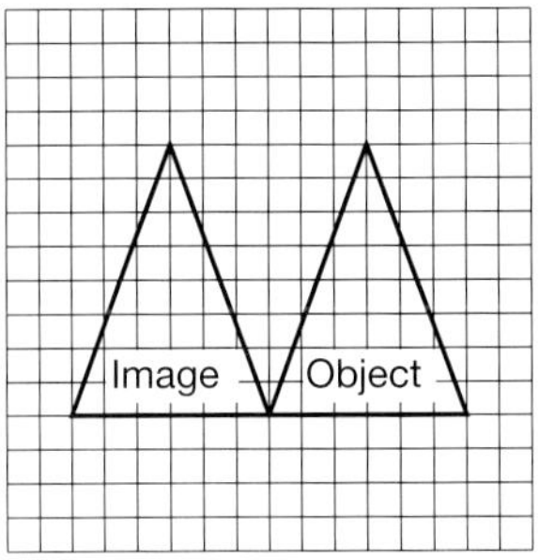

6.

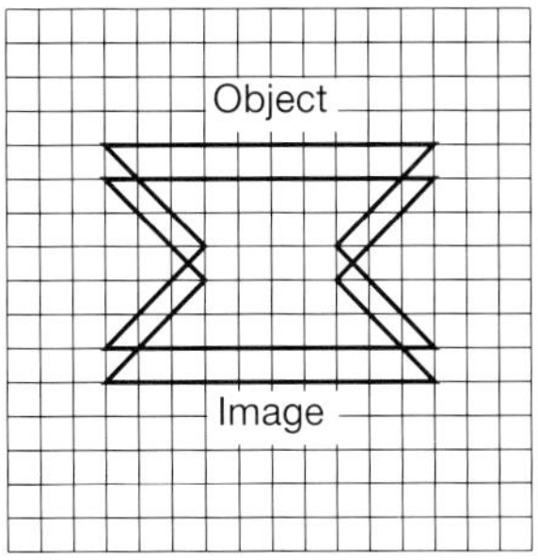

Exercise 27.7 page 244

1.

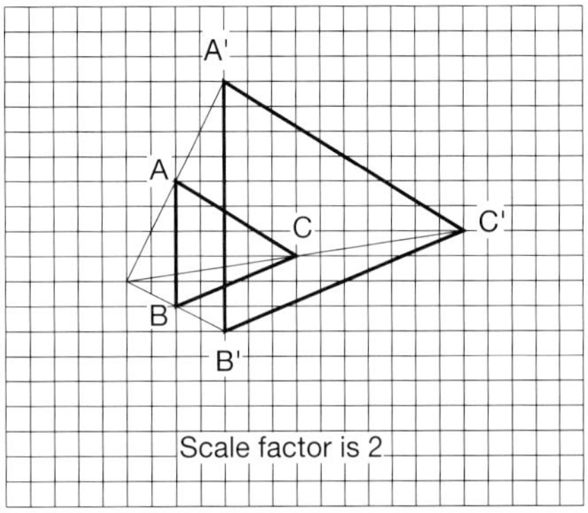

2.

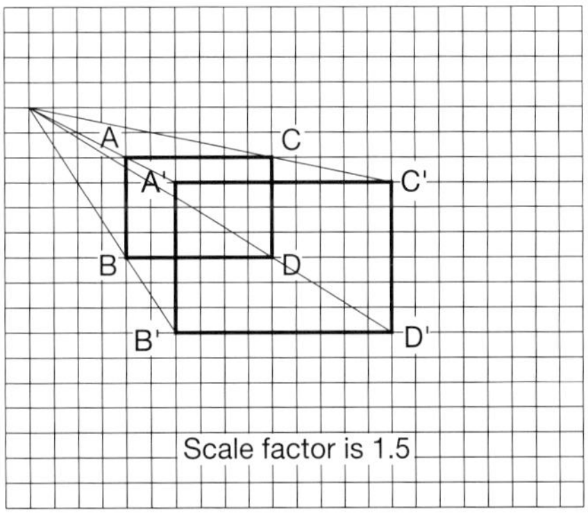

3.

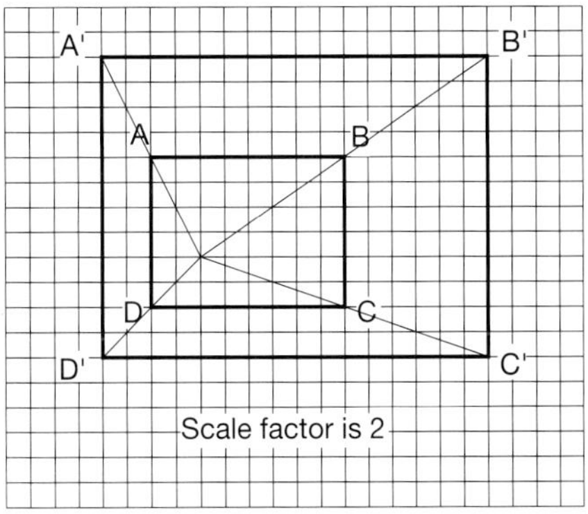

4.

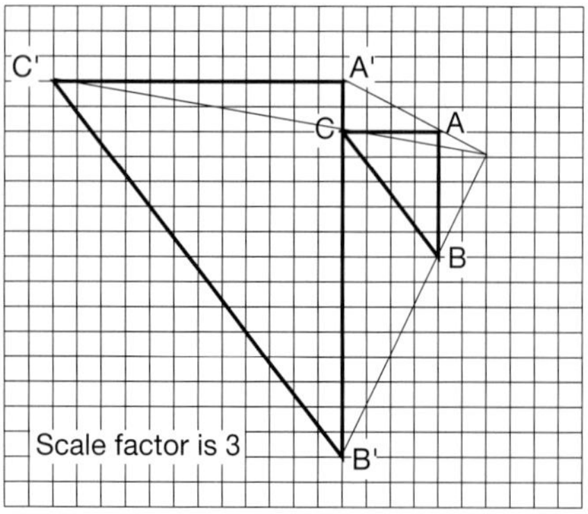

5.

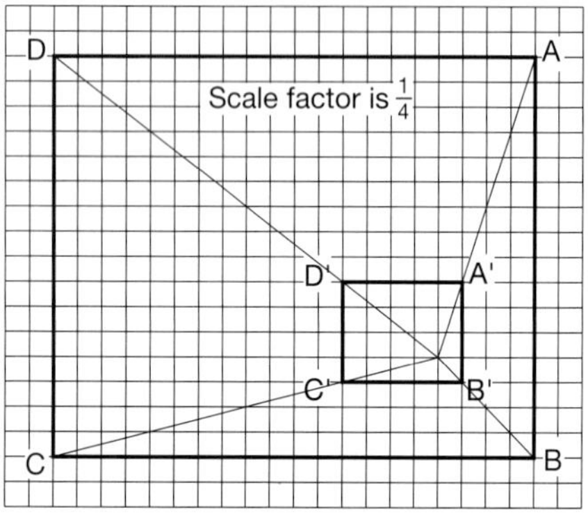

Exercise 27.8 page 245

1.

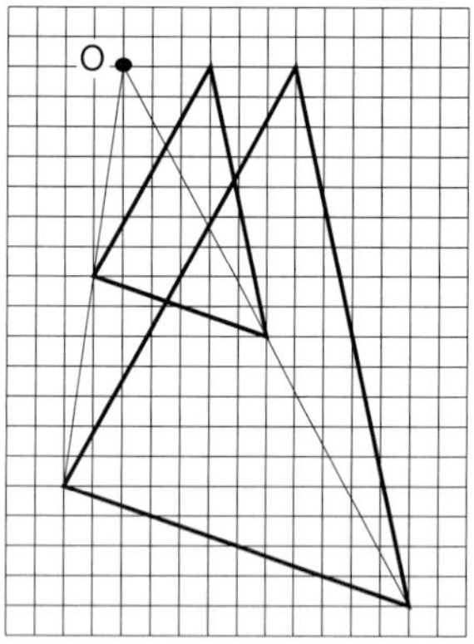

2.

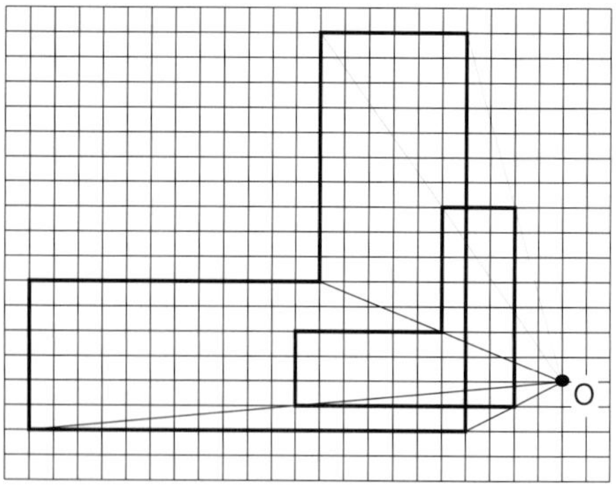

3.

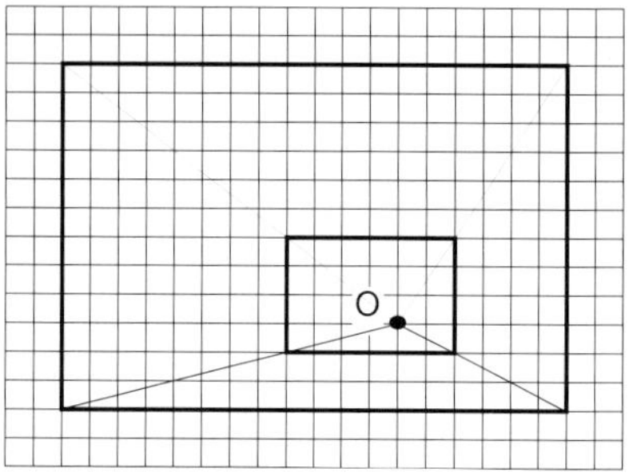

4.

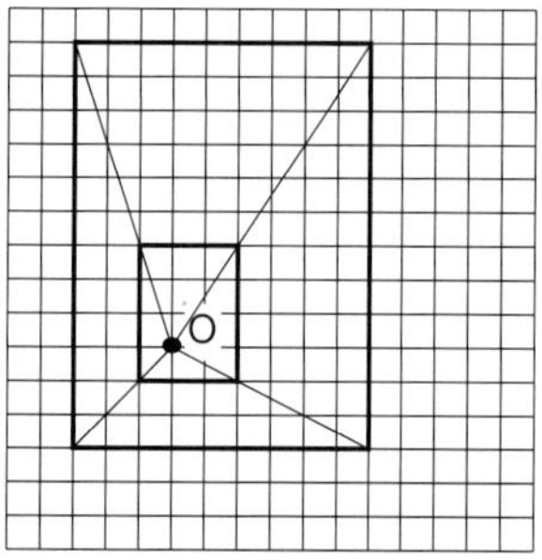

Exercise 27.9 page 246

1.

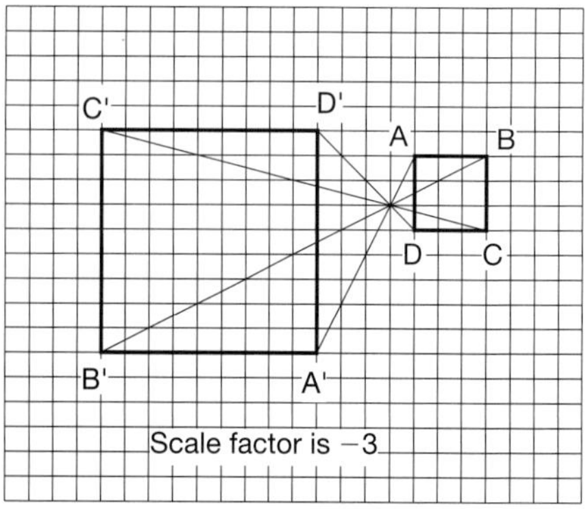

2.

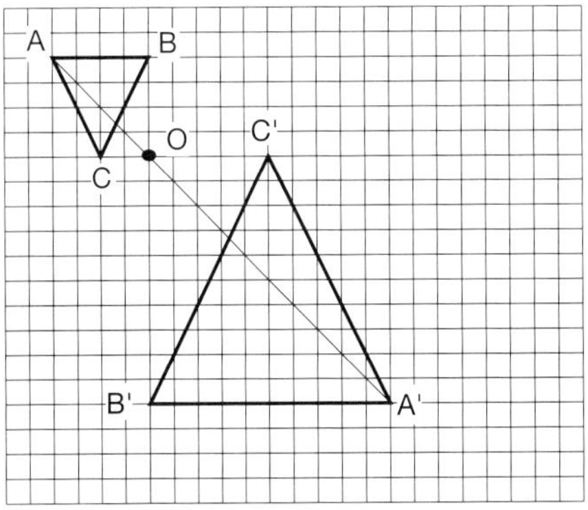

3.

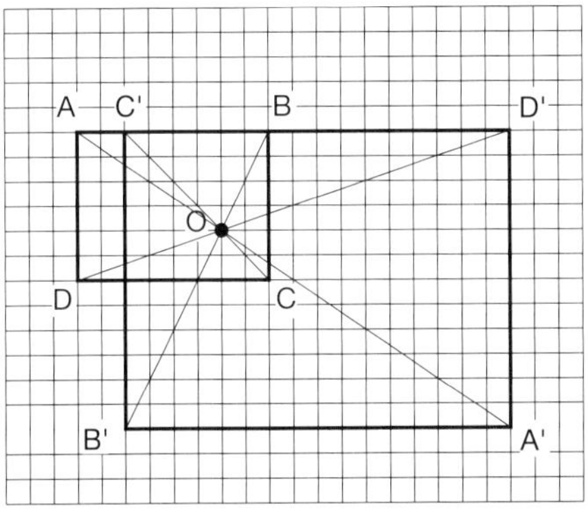

4.

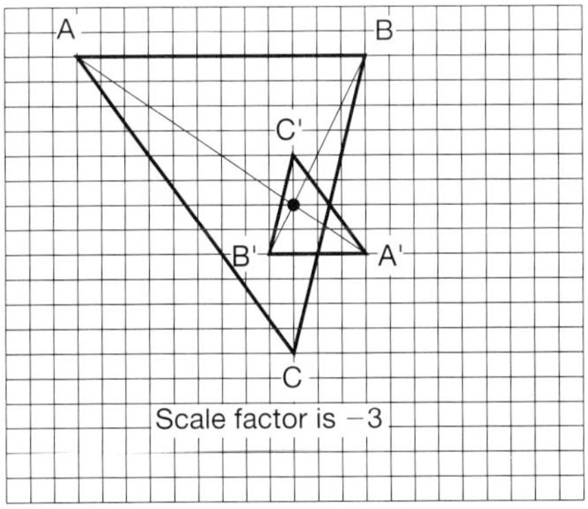

5.

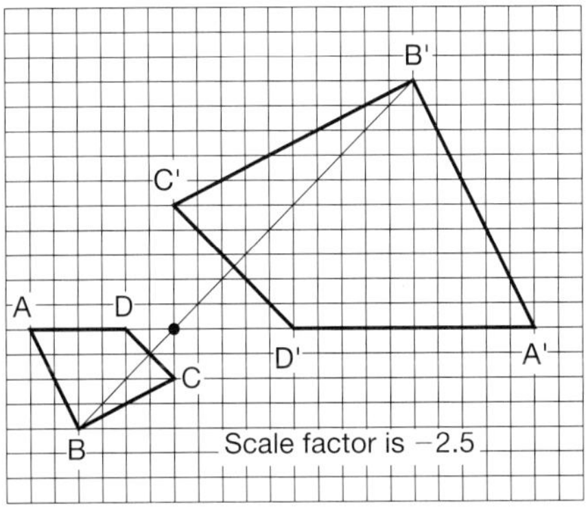

6.

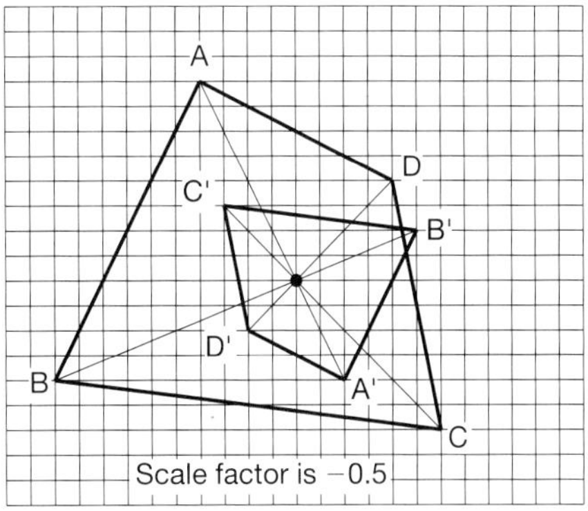

Student Assessment 1 page 249

1.

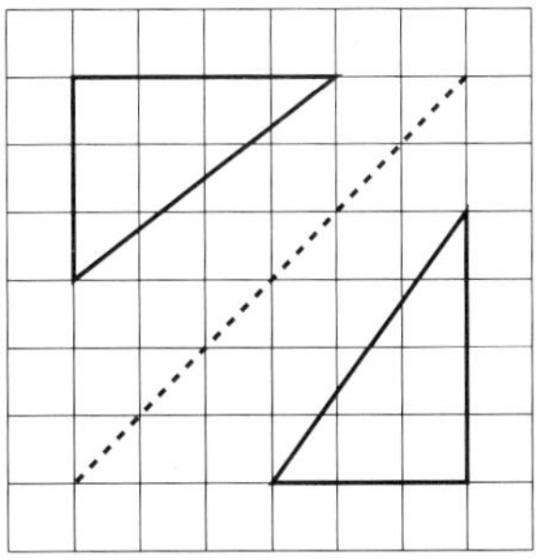

2.

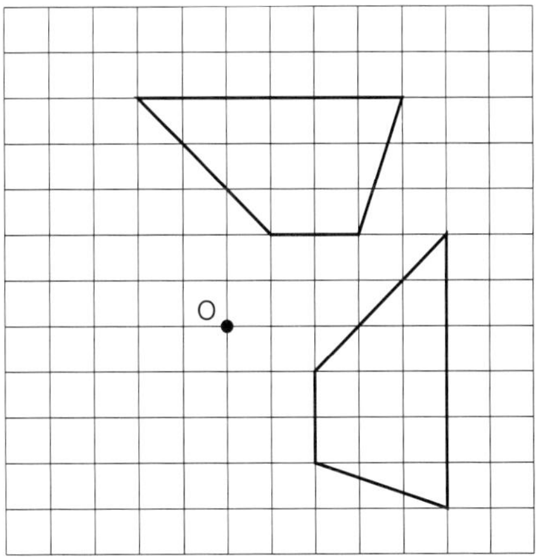

3. **a)** $\begin{pmatrix} 6 \\ 0 \end{pmatrix}$ **b)** $\begin{pmatrix} -3 \\ -5 \end{pmatrix}$

4.

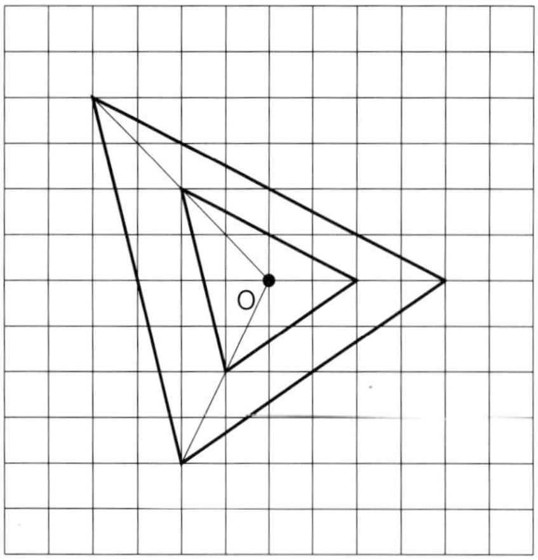

5. a)

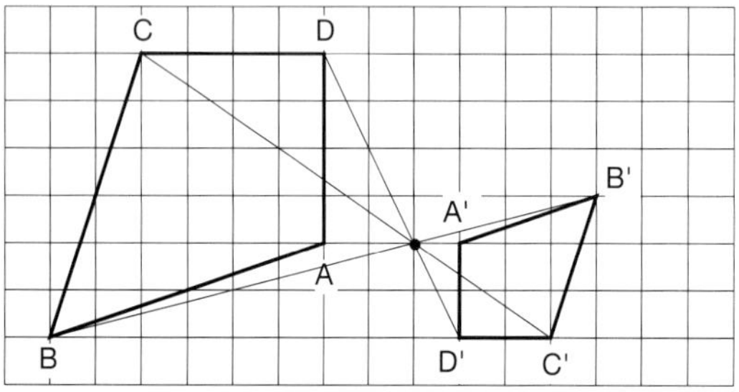

b) The scale factor of enlargement is -0.5.

Student Assessment 2 page 250

1.

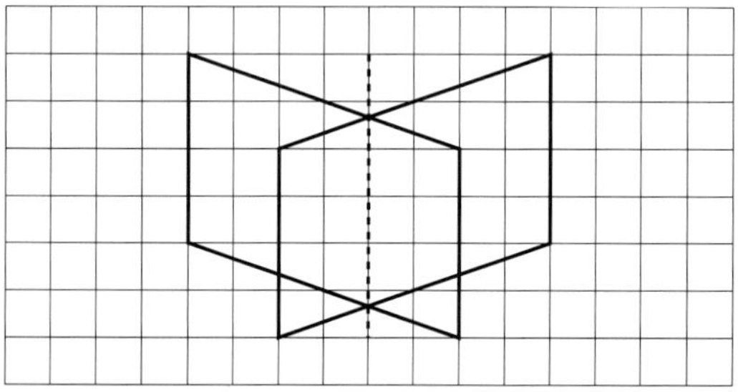

2.

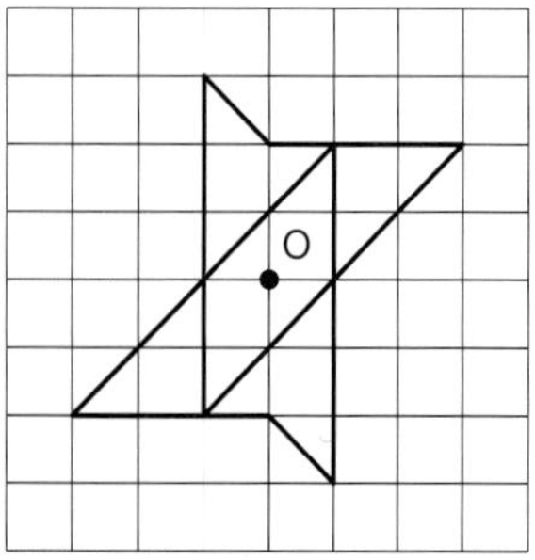

3. **a)** $\begin{pmatrix} 0 \\ -3 \end{pmatrix}$ **b)** $\begin{pmatrix} -8 \\ 2 \end{pmatrix}$

4.

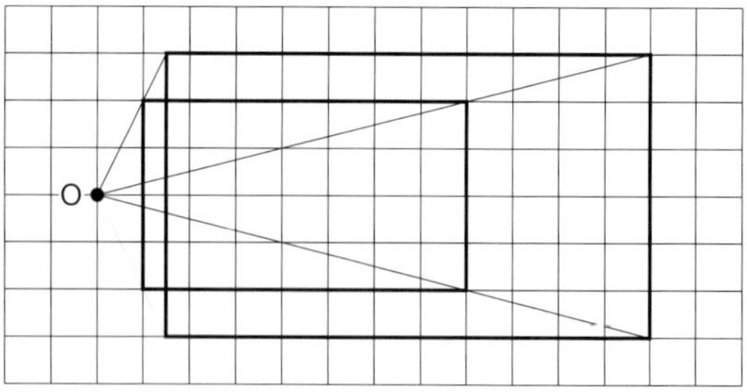

5. a)

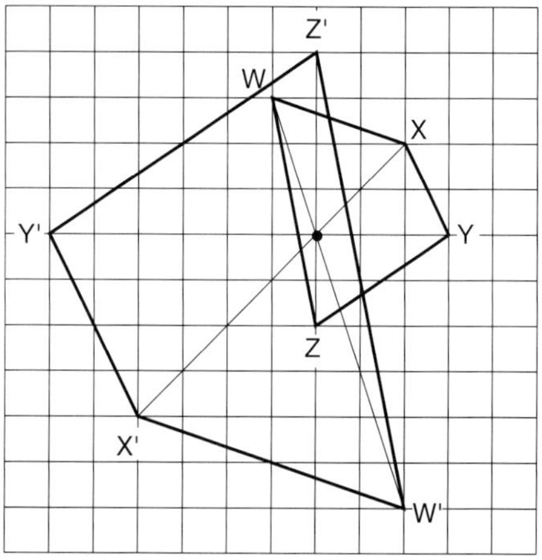

b) The scale factor of enlargement is -2.

28 Mensuration

Exercise 28.1 page 253

1. **a)** $A = 24\text{ cm}^2$ $P = 20\text{ cm}$
   b) $A = 45\text{ cm}^2$ $P = 28\text{ cm}$
   c) $A = 27\text{ cm}^2$ $P = 21\text{ cm}$
   d) $A = 38\text{ m}^2$ $P = 27.6\text{ m}$
   e) $A = 21\text{ m}^2$ $P = 18.4\text{ m}$
   f) $A = 22.5\text{ cm}^2$ $P = 19.5\text{ cm}$
   g) $A = 15.04\text{ cm}^2$ $P = 15.8\text{ cm}$
   h) $A = 102.85\text{ m}^2$ $P = 48.4\text{ m}$
   i) $A = 1.02\text{ m}^2$ $P = 4.1\text{ m}$
   j) $A = 2.475\text{ m}^2$ $P = 8.1\text{ m}$

2. **a)** $A = 61.2\text{ cm}^2$ **b)** $b = 10\text{ cm}$
   c) $l = 16\text{ cm}$ **d)** $b = 10\text{ cm}$ **e)** $l = 12\text{ cm}$
   f) $b = 15\text{ cm}$ **g)** $l = 25\text{ cm}$

3. a) $A = 60\ m^2$ $P = 31\ m$
b) $A = 875\ mm^2$ $P = 120\ mm$
c) $A = 20.5\ m^2$ $P = 21.4\ m$
d) $A = 11\ 050\ m^2$ $P = 430\ m$
e) $A = 2025\ cm^2$ $P = 180\ cm$
f) $A = 875\ cm^2$ $P = 120\ m$
g) $A = 2.8\ km^2$ $P = 8.6\ km$
h) $A = 0.03\ km^2$ $P = 2.45\ km$
i) $A = 2925\ m^2$ $P = 220\ m$
j) $A = 15\ 625\ mm^2$ $P = 500\ mm$

Exercise 28.2 page 254

1. a) 6 cm^2 b) 32.5 cm^2 c) 20 cm^2
d) 60 cm^2 e) 108 cm^2 f) 55 cm^2
2. a) 90 cm^2 b) 104 cm^2 c) 128 cm^2
d) 1168 mm^2 e) 300 cm^2 f) 937.5 mm^2

Exercise 28.3 page 257

1. a) 25.13 cm b) 21.99 cm c) 28.90 cm
d) 1.57 m
2. a) 50.27 cm^2 b) 38.48 cm^2 c) 66.48 cm^2
d) 0.20 m^2
3. a) 2.4 cm (1 d.p.) b) 0.5 cm (exact)
c) 0.6 m (1 d.p.) d) 1.3 mm (1 d.p.)
4. a) 4.5 cm (1 d.p.) b) 6.0 cm (1 d.p.)
c) 3.2 m (1 d.p.) d) 4.3 mm (1 d.p.)

Exercise 28.4 page 257

1. i) 1.57 m (2 d.p.) ii) 637 times (3 s.f.)
2. 188.5 m (1 d.p.)
3. 264 mm^2 (3 s.f.)
4. a) 144 cm^2 b) 28.3 cm^2 (1 d.p.)
c) 30.9 cm^2 (1 d.p.)
5. a) 57.1 m (1 d.p.) b) 178.3 m^2 (1 d.p.)

Exercise 28.5 page 258

1. 58.5 cm^2
2. 84 cm^2
3. 153.3 cm^2 (1 d.p.)
4. 157.5 cm^2

Exercise 28.6 page 259

1. 4
2. 3
3. 23.0 m^2 (1 d.p.)
4. a) 16 m^2, 24 m^2 b) 100 m^2 c) 15

Exercise 28.7 page 260

1. a) 460 cm^2 b) 208 cm^2 c) 147.78 cm^2
d) 33.52 cm^2
2. a) 2 cm b) 4 cm c) 6 cm d) 5 cm
3. a) 100.5 cm^2 (1 d.p.) b) 276.5 cm^2 (1 d.p.)
c) 279.3 cm^2 (1 d.p.) d) 25.6 cm^2 (1 d.p.)
4. a) 1.2 cm b) 0.5 cm c) 1.7 cm
d) 7.0 cm

Exercise 28.8 page 261

1. a) 24 cm^2 b) 2 cm
2. a) 216 cm^2 b) 15.2 cm (1 d.p.)
3. a) 94.2 cm^2 (1 d.p.) b) 14 cm
4. 4.4 cm

Exercise 28.9 page 262

1. a) 24 cm^3 b) 18 cm^3 c) 27.6 cm^3
d) 8.82 cm^3
2. a) 452.4 cm^3 (1 d.p.) b) 277.1 cm^3 (1 d.p.)
c) 196.3 cm^3 (1 d.p.) d) 0.5 cm^3 (1 d.p.)
3. a) 108 cm^3 b) 140 cm^3 c) 42 cm^3
d) 6.2 cm^3
4. a) 70 cm^3 b) 96 cm^3 c) 380 cm^3
d) 137.5 cm^3

Exercise 28.10 page 263

1. a) 16 cm b) 4096 cm^3 c) 3216 cm^3
d) 21.5% (1 d.p.)
2. a) 42 cm^2 b) 840 cm^3
3. 5.7 cm
4. 2.90 m^3 (2 d.p.)

Student Assessment 1 page 264

1. a) $A = 1000\ m^2$ b) $P = 140\ m$
c) Therefore, 560 tiles (or 564 if 4 corners added).
2. 192 tiles
3. $A = 9.12\ m^2$ $P = 12.4\ m$
4. 500 cm^2
5. a) 3700 cm^2 b) 148 m^2

Student Assessment 2 page 264

1. a) Circumference = 34.6 cm Area = 95.0 cm^2
   b) Circumference = 50.3 mm
   Area = 201.1 mm^2
2. 9.9 cm^2
3. a) 39.3 cm^2 b) 34 cm^2 c) 101.3 cm^2
4. a) 10.2 cm^2 b) 282.7 cm^2 c) 633.0 cm^2
5. a) 339.3 mm^3 b) 9.8 cm^3

Student Assessment 3 page 265

1. 1.25 km^2
2. a) 1200 tiles b) £3500
3. Student's own diagram
   $A = 6\text{ m}^2$ $P = 12\text{ m}$
4. $A = 600\text{ cm}^2$
5. 3600 cm^2

Student Assessment 4 page 265

1. a) Circumference = 27.0 cm Area = 58.1 cm^2
   b) Circumference = 47.1 mm
   Area = 176.7 mm^2
2. 325.8 cm^2
3. a) 56.5 cm^2 b) 108 cm^2 c) 254.5 cm^2
4. 418 cm^3
5. a) 1011.6 cm^2 b) 523.1 cm^2

29 Trigonometry

Exercise 29.1 page 267

1–2. Student's own diagrams

3. Student's own diagrams leading to:
   a) 310° b) 325° c) 040° d) 020°
   e) 332° f) 054° g) 343° h) 034°
4. Student's own diagrams leading to:
   a) 120° b) 140° c) 110° d) 102°

Exercise 29.2 page 270

1. a) 1.8 cm b) 4.0 cm c) 19.2 cm
   d) 4.9 cm e) 37.3 cm f) 13.9 cm
2. a) 14.3 cm b) 9.0 cm c) 9.3 cm
   d) 4.1 cm e) 13.9 cm f) 6.2 cm
3. a) 49.4° b) 51.1° c) 51.3° d) 63.4°
   e) 50.4° f) 71.6°

Exercise 29.3 page 272

1. a) 2.4 cm b) 18.5 cm c) 6.2 cm
   d) 2.4 cm e) 43.8 cm f) 31.8 cm
2. a) 38.7° b) 48.6° c) 38.1° d) 49.8°
   e) 32.6° f) 14.5°

Exercise 29.4 page 274

a) 36.0 cm b) 15.1 cm c) 48.2° d) 81.1°
e) 6.7 cm f) 16.8 cm g) 70.5° h) 2.1 cm

Exercise 29.5 page 275

1. a) 5 cm b) 11.4 mm (1 d.p.) c) 12 cm
   d) 13.2 cm (1 d.p.)
2. a) 11.0 cm (1 d.p.) b) 14.8 cm (1 d.p.)
   c) 7.9 cm (1 d.p.) d) 7.3 cm (1 d.p.)
   e) 3 cm f) 13.9 cm (1 d.p.)
3. 71.6 km
4. 66.9 km
5. a) 225° − 135° = 90° b) 73.8 km
6. 57 009 m
7. a) 8.5 km b) 15.5 km (1 d.p.)
8. a) 13.3 m (1 d.p.) b) 15.0 m (1 d.p.)

Exercise 29.6 page 276

1. a) 43.6° (1 d.p.) b) 19.5 cm (1 d.p.)
   c) 16.7 cm (1 d.p.) d) 42.5° (1 d.p.)
2. a) i) 20.8 km (1 d.p.) ii) 215.2° (1 d.p.)
   b) i) 228.4 km (1 d.p.) ii) 101.7 km (1 d.p.)
   iii) 103.2 km (1 d.p.) iv) 147.4 km (1 d.p.)
   v) 414.8 km (1 d.p.) vi) 216.9° (1 d.p.)
   c) i) 6.7 m (1 d.p.) ii) 19.6 m (1 d.p.)
   iii) 15.3 m (1 d.p.)
   d) i) 48.2° (1 d.p.) ii) 41.8° (1 d.p.)
   iii) 8 cm iv) 8.9 cm (1 d.p.)
   v) 76.0 cm^2 (1 d.p.)

Student Assessment 1 page 278

1. a) Student's own diagram
   b) Student's own diagram
2. Student's own diagram
3. a) 10 cm b) 6.5 cm c) 2 cm
   d) 6.49 cm e) 16 cm
4. 52.2 cm

Student Assessment 2 page 279

1. Student's own diagram
2. Student's own diagram
3. **a)** 20 cm **b)** 32.5 cm **c)** 0.9 cm **d)** 18 cm **e)** 13 cm
4. 4.6 m

Student Assessment 3 page 280

1. **a)** 4.0 cm **b)** 43.9 cm **c)** 20.8 cm **d)** 3.9 cm
2. **a)** 37° **b)** 56° **c)** 31° **d)** 34°
3. **a)** 5.0 cm **b)** 6.6 cm **c)** 9.3 cm **d)** 28.5 cm

Student Assessment 4 page 281

1. **a)** 160.8 km **b)** 177.5 km
2. **a)** $\tan\theta = \frac{5}{x}$ **b)** $\tan\theta = \frac{7.5}{(x+16)}$
   c) $\frac{5}{x} = \frac{7.5}{(x+16)}$ **d)** 32 m
   e) 8.9° (1 d.p.)
3. **a)** 285 m (3 s.f.) **b)** 117° (3 s.f.) **c)** 297° (3 s.f.)
4. **a)** 1.96 km (2 d.p.) **b)** 3.42 km (2 d.p.) **c)** 3.57 km (2 d.p.)

30 Vectors

Exercise 30.1 page 283

1. **a)** $\begin{pmatrix}2\\4\end{pmatrix}$ **b)** $\begin{pmatrix}3\\1\end{pmatrix}$ **c)** $\begin{pmatrix}-2\\-4\end{pmatrix}$ **d)** $\begin{pmatrix}3\\-2\end{pmatrix}$
   e) $\begin{pmatrix}-6\\1\end{pmatrix}$ **f)** $\begin{pmatrix}6\\-1\end{pmatrix}$ **g)** $\begin{pmatrix}-3\\-1\end{pmatrix}$
   h) $\begin{pmatrix}-5\\-5\end{pmatrix}$ **i)** $\begin{pmatrix}-1\\3\end{pmatrix}$
2. **a)** $\begin{pmatrix}4\\4\end{pmatrix}$ **b)** $\begin{pmatrix}0\\4\end{pmatrix}$ **c)** $\begin{pmatrix}-3\\0\end{pmatrix}$ **d)** $\begin{pmatrix}-2\\6\end{pmatrix}$
   e) $\begin{pmatrix}-4\\-2\end{pmatrix}$ **f)** $\begin{pmatrix}0\\-4\end{pmatrix}$ **g)** $\begin{pmatrix}3\\0\end{pmatrix}$ **h)** $\begin{pmatrix}2\\-6\end{pmatrix}$
   i) $\begin{pmatrix}-4\\-4\end{pmatrix}$
3.

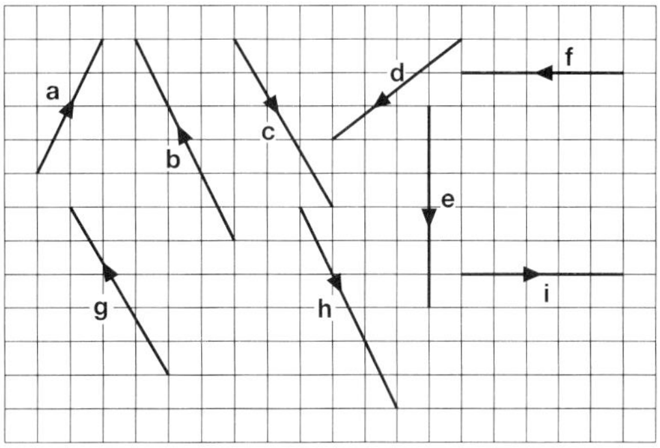

Exercise 30.2 page 284

1. **a) – f)**

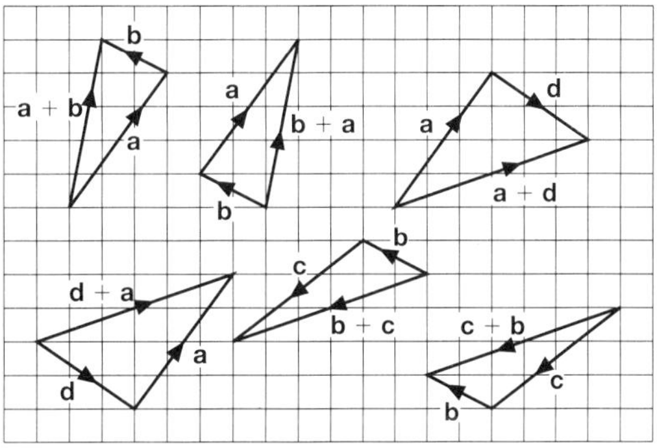

2. $\mathbf{a} + \mathbf{b} = \mathbf{b} + \mathbf{a}$, $\mathbf{a} + \mathbf{d} = \mathbf{d} + \mathbf{a}$, $\mathbf{b} + \mathbf{c} = \mathbf{c} + \mathbf{b}$

Exercise 30.3 page 285

1. $\mathbf{d} = -\mathbf{c}$ $\mathbf{e} = -\mathbf{a}$ $\mathbf{f} = 2\mathbf{a}$ $\mathbf{g} = \frac{1}{2}\mathbf{c}$ $\mathbf{h} = \frac{1}{2}\mathbf{b}$
   $\mathbf{i} = -\frac{1}{2}\mathbf{b}$ $\mathbf{j} = \frac{3}{2}\mathbf{b}$ $\mathbf{k} = -\frac{3}{2}\mathbf{a}$
2. **a)** $\begin{pmatrix}4\\6\end{pmatrix}$ **b)** $\begin{pmatrix}-12\\-3\end{pmatrix}$ **c)** $\begin{pmatrix}2\\-4\end{pmatrix}$
   d) $\begin{pmatrix}-2\\2\end{pmatrix}$ **e)** $\begin{pmatrix}-6\\3\end{pmatrix}$ **f)** $\begin{pmatrix}-4\\15\end{pmatrix}$
   g) $\begin{pmatrix}-6\\1\end{pmatrix}$ **h)** $\begin{pmatrix}-1\\1\end{pmatrix}$ **i)** $\begin{pmatrix}-2\\18\end{pmatrix}$
3. **a)** $2\mathbf{a}$ **b)** $-\mathbf{b}$ **c)** $\mathbf{b} + \mathbf{c}$

Student Assessment 1 page 286

1. **a)** $\begin{pmatrix}-2\\3\end{pmatrix}$ **b)** $\begin{pmatrix}7\\2\end{pmatrix}$ **c)** $\begin{pmatrix}6\\-2\end{pmatrix}$
2. $\mathbf{a} = \begin{pmatrix}2\\4\end{pmatrix}$ $\mathbf{b} = \begin{pmatrix}4\\0\end{pmatrix}$ $\mathbf{c} = \begin{pmatrix}0\\-5\end{pmatrix}$ $\mathbf{d} = \begin{pmatrix}-4\\-2\end{pmatrix}$
   $\mathbf{e} = \begin{pmatrix}-2\\1\end{pmatrix}$

3.

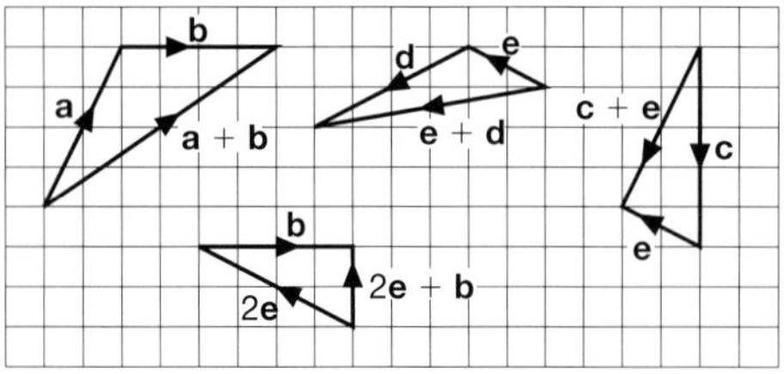

4. **a)** $\begin{pmatrix} -1 \\ 5 \end{pmatrix}$ **b)** $\begin{pmatrix} -5 \\ 3 \end{pmatrix}$ **c)** $\begin{pmatrix} 1 \\ 11 \end{pmatrix}$ **d)** $\begin{pmatrix} -12 \\ 10 \end{pmatrix}$

Student Assessment 2 page 286

1. **a)** $\begin{pmatrix} -2 \\ 7 \end{pmatrix}$ **b)** $\begin{pmatrix} -5 \\ -4 \end{pmatrix}$ **c)** $\begin{pmatrix} -1 \\ -3 \end{pmatrix}$

2. **a)** $\begin{pmatrix} 0 \\ 3 \end{pmatrix}$ **b)** $\begin{pmatrix} -6 \\ 0 \end{pmatrix}$ **c)** $\begin{pmatrix} -1 \\ -5 \end{pmatrix}$ **d)** $\begin{pmatrix} 3 \\ 3 \end{pmatrix}$

   e) $\begin{pmatrix} 4 \\ -1 \end{pmatrix}$

3.

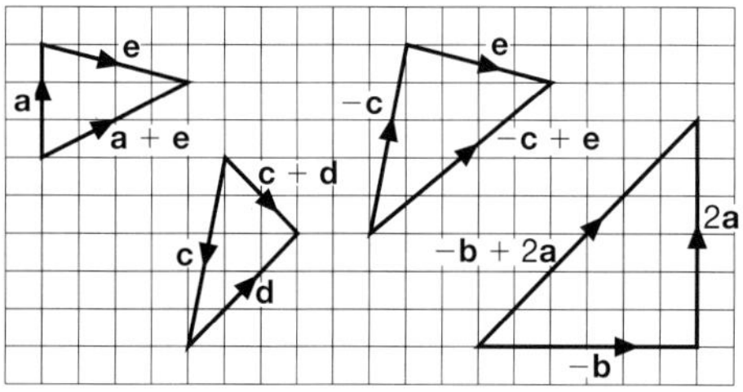

4. **a)** $\begin{pmatrix} -1 \\ 1 \end{pmatrix}$ **b)** $\begin{pmatrix} 3 \\ -1 \end{pmatrix}$ **c)** $\begin{pmatrix} 6 \\ -6 \end{pmatrix}$ **d)** $\begin{pmatrix} -6 \\ 8 \end{pmatrix}$

ICT Section

Shape and Space Activity 1 page 287

1.

	A	B	C	D	E	F
1		Regular Polygons				
2	Number of sides	Name	Sum of exterior angles	Size of an exterior angle	Size of an interior angle	Sum of interior angles
3	3	Equilateral triangle	360	120	60	180
4	4	Square	360	90	90	360
5	5	Pentagon	360	72	108	540
6	6	Hexagon	360	60	120	720
7	7	Heptagon	360	51.4285714	128.571429	900
8	8	Octagon	360	45	135	1080
9	9	Nonagon	360	40	140	1260
10	10	Decagon	360	36	144	1440
11	12	Dodecagon	360	30	150	1800
12	20	Icosagon	360	18	162	3240
13						
14		Formulae in row 3	=360	=C3/A3	=180-D3	=E3*A3

2. Size of an exterior angle = $360/n$
   Size of an interior angle = $180 - 360/n$

Shape and Space Activity 2 page 287

2. **a)**

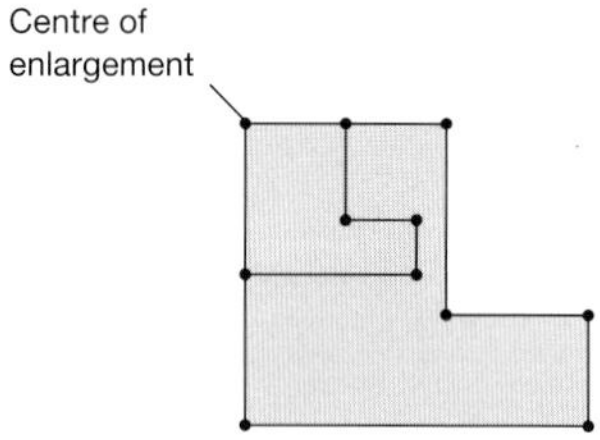

b)

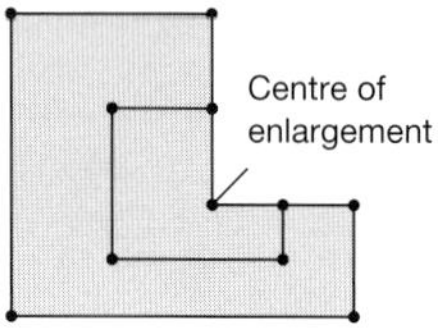

c)

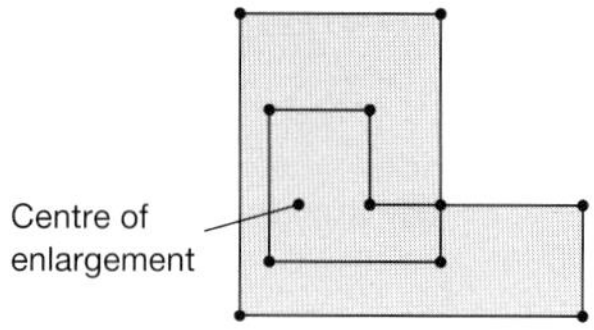

d)

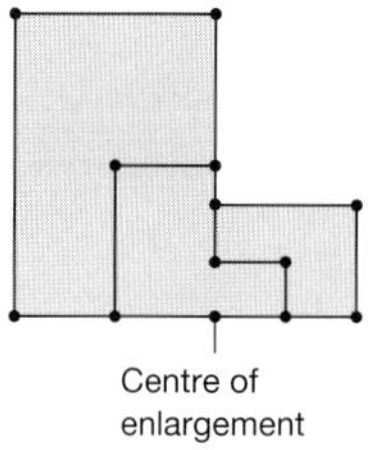

4. Scale factor of enlargement is –2
5. A negative scale factor inverts the image

Reviews for Chapters 22 to 30

Review 19 page 289

1. Student's own angle and measurement
2. Student's own construction. Base angles ≈ 71°
3. 72°

4. 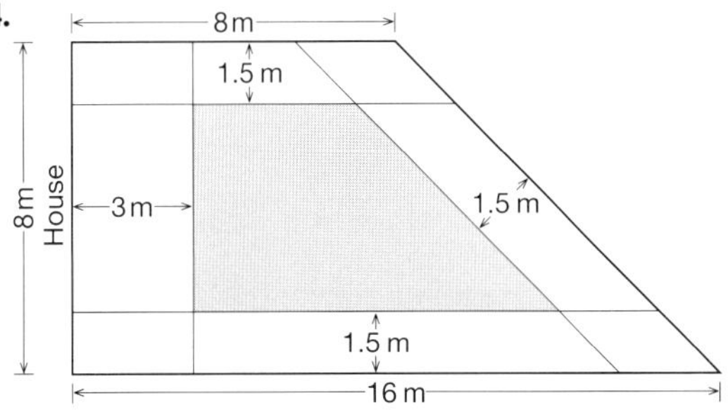

5. 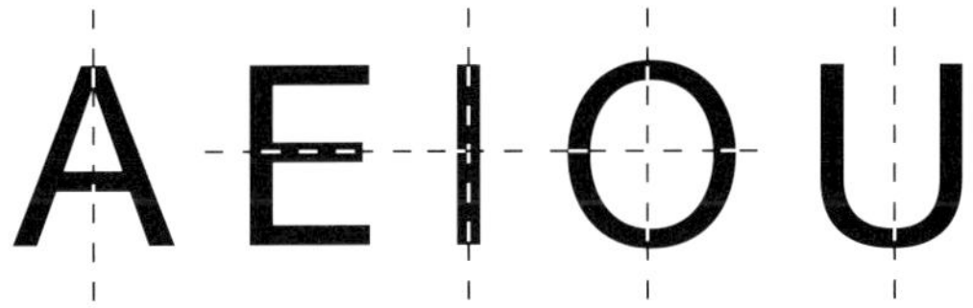

Note that if the 'O' is drawn as a circle it will have an infinite number of lines of symmetry.

6. 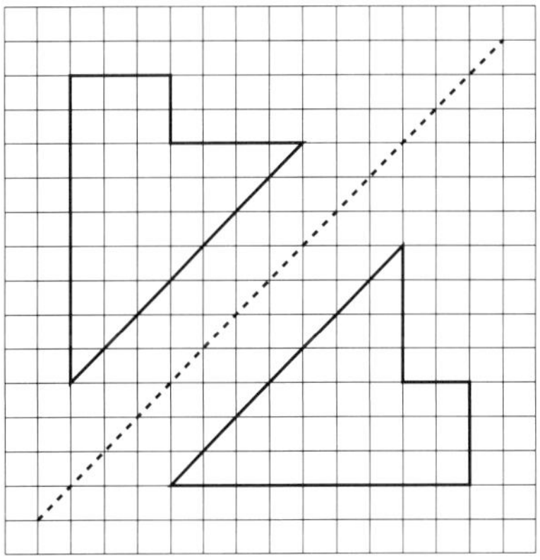

Review 20 page 290

1. 262.5 cm^2
2. 4.0 km
3. Student's own diagram. Distance ≈ 5.7 km. Bearing ≈ 047°
4. 46°
5. **a)** $\begin{pmatrix} 2 \\ 4 \end{pmatrix}$ **b)** $\begin{pmatrix} -2 \\ -4 \end{pmatrix}$ **c)** $\begin{pmatrix} 3 \\ -2 \end{pmatrix}$ **d)** $\begin{pmatrix} -6 \\ 1 \end{pmatrix}$

Review 21 page 291

1. **a)** 120° **b)** 30°
2. Student's own construction
3. Student's sketch of a dodecagon
4. **a)** Student's construction in which O is the centre of OA, OB and OC are all radii
   b) Student's own explanation
5. Student's own drawing (possible solution is a regular hexagon)
6. **a)**

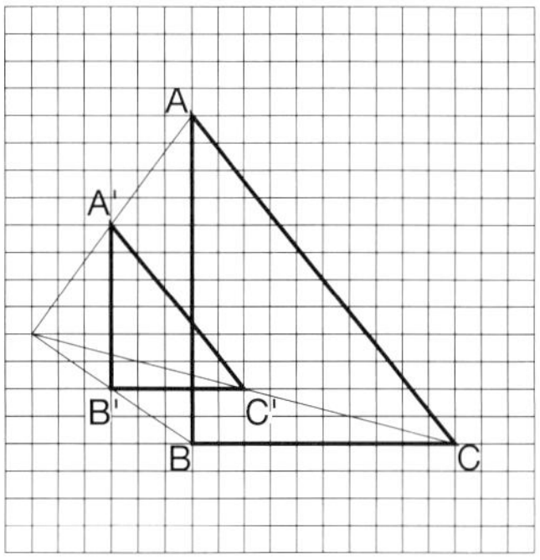

b) Scale factor of enlargement is 0.5

Review 22 page 292

1. 4 cm
2. 603 cm^3
3. 14.8 cm
4. 18°
5. **a)**

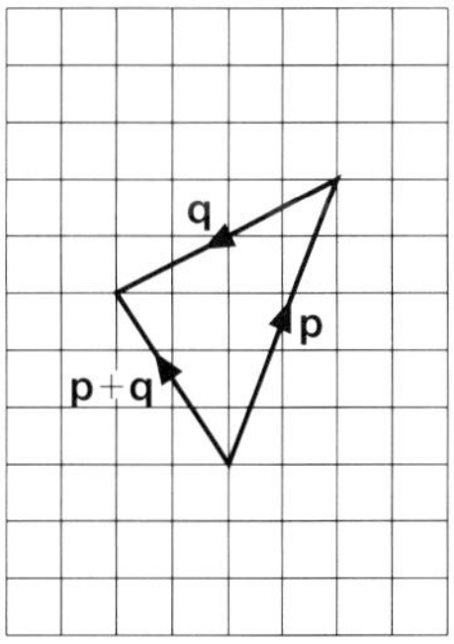

b) $\begin{pmatrix} -2 \\ 3 \end{pmatrix}$

Review 23 page 292

1. Student's own diagram
   ∠ABC = 60°. BC ≈ 5.5 cm, AC ≈ 7.4 cm
2. **a)** Parallelogram **b)** Rhombus
3. 22.5°

4. 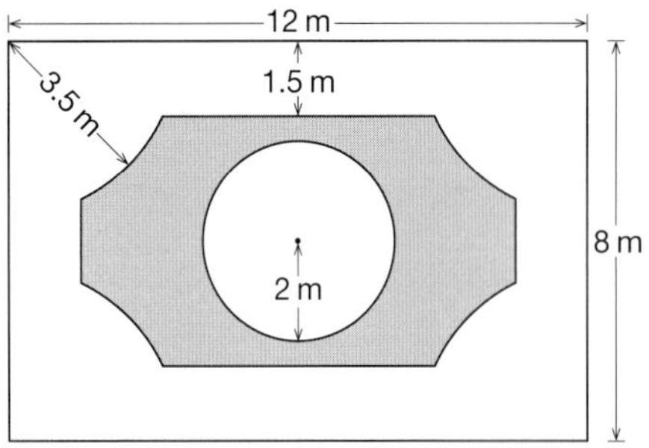

5. Student's own diagram

6. **a)** 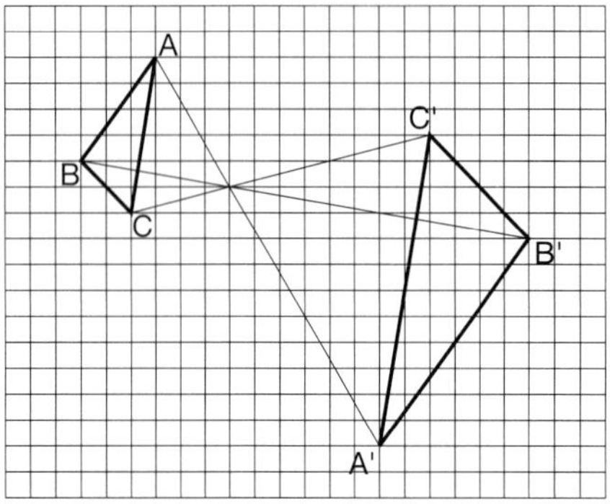

b) Scale factor of enlargement is -2

Review 24 page 294

1. 707 mm^2
2. 550 cm^2
3. 130 m
4. 113° (3 s.f.)
5. **a)** $\begin{pmatrix} -4 \\ 0 \end{pmatrix}$ **b)** $\begin{pmatrix} -1 \\ 1 \end{pmatrix}$ **c)** $\begin{pmatrix} -11 \\ -7 \end{pmatrix}$

Review 25 page 294

1. Student's own diagram
2. **a)** 98 cm^3 (2 s.f.) **b)** 78.5 cm (1 d.p.)
3. 1 cm
4. **a)** Several possible solutions, e.g. equilateral triangle
   b) Kite or isosceles trapezium
   c) Parallelogram

31 Handling Data

Exercise 31.1 page 297

1. Dog = 8 Cat = 10 Fish = 3
   Hamster = 4 Horse = 2 Snake = 1

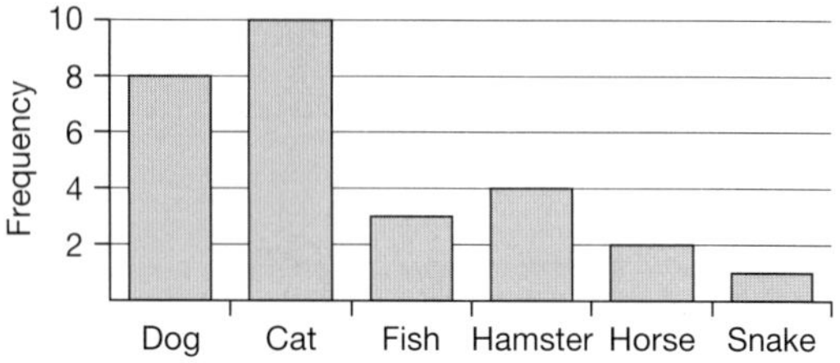

2. Red = 11 Blue = 6 Green = 3
   White = 12 Black = 1
   Student's own pictogram

3. Art = 6 Maths = 5 Science = 5
   English = 5

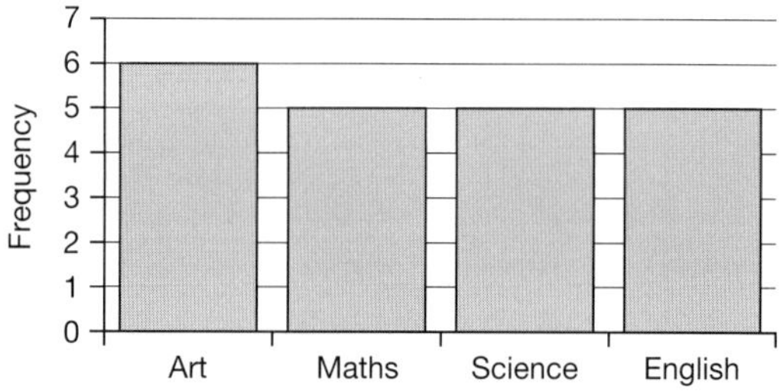

Exercise 31.2 page 299

1.

Score	30–39	40–49	50–59	60–69	70–79	80–89	90–99
Frequency	1	6	12	16	6	2	1

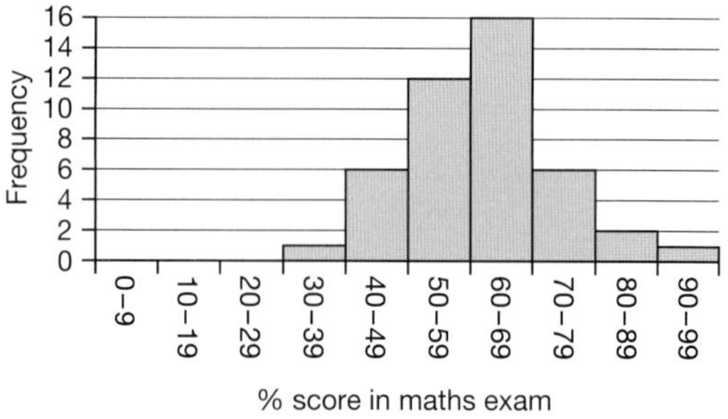

2.

Score	0–9	10–19	20–29	30–39	40–49	50–59	60–69	70–79	80–89	90–99
Frequency	3	6	3	3	3	4	6	4	5	3
Score	100–109	110–119	120–129	130–139	140–149					
Frequency	1	3	2	2	2					

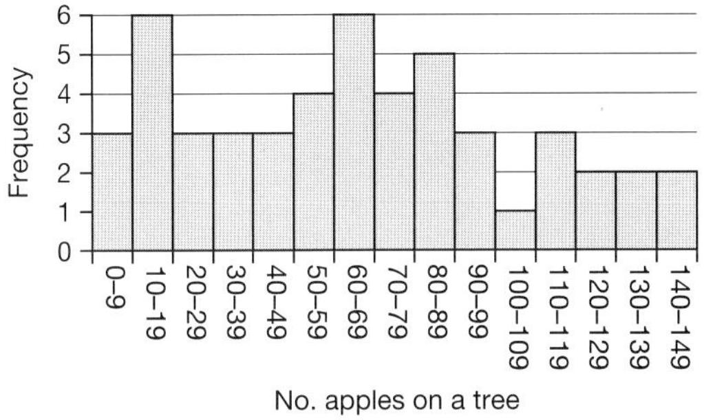

3.

Score	0–9	10–19	20–29	30–39	40–49	50–59	60–69
Frequency	3	4	5	7	4	3	4

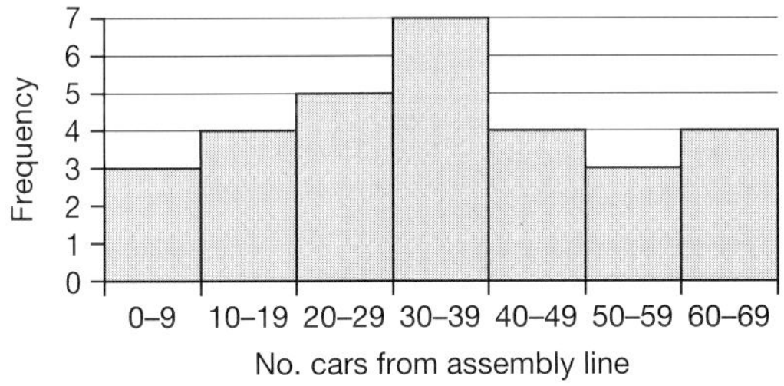

4.

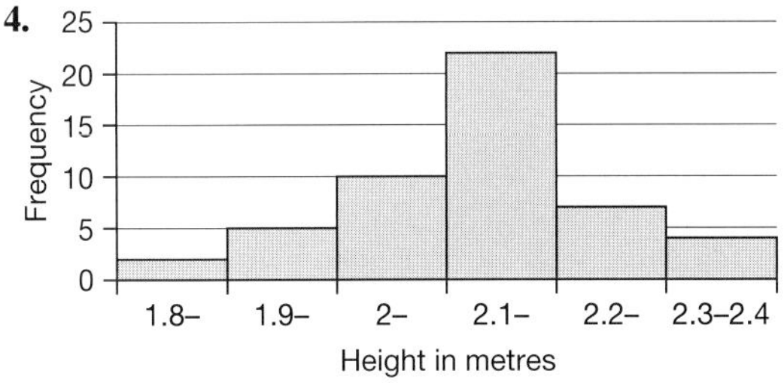

5.

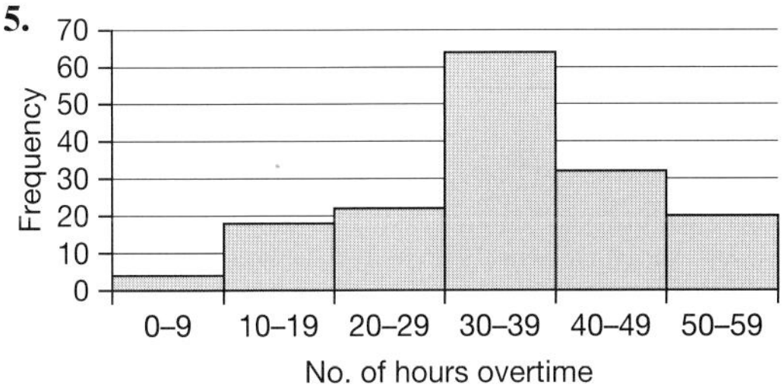

6.

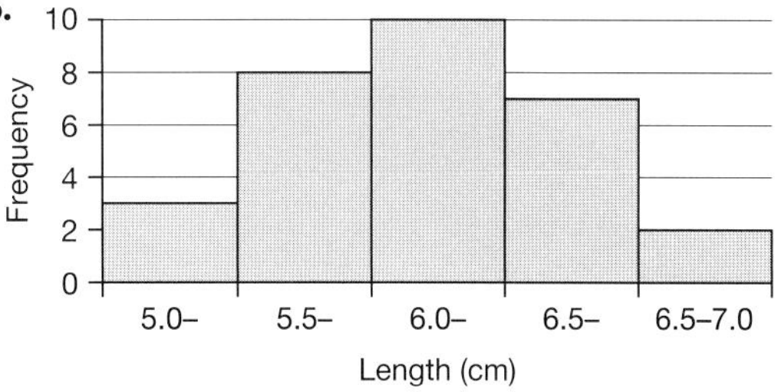

Exercise 31.3 page 303

1.

	Sleep	Meals	Music	Sport	TV	School
Ayse	8 h 20	2 h	5 h	n/a	2 h	6 h 40
Ahmet	8 h 40	2 h	n/a	5 h 20	2 h	6 h

2.

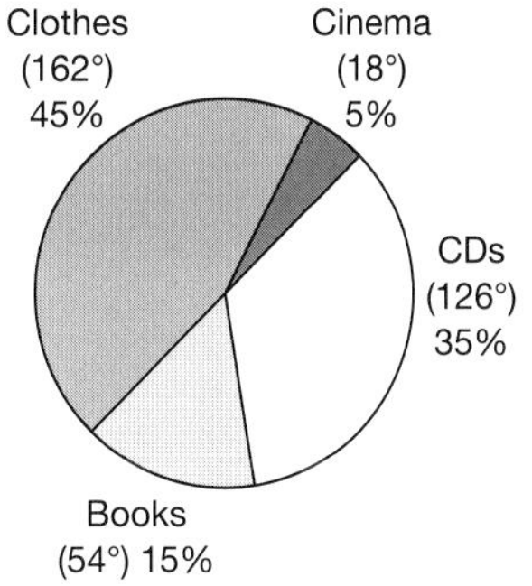

3.

	Fraction	£	Degrees
Clothes	$\frac{1}{3}$	800	120
Transport	$\frac{1}{5}$	480	72
Ents.	$\frac{1}{4}$	600	90
Saved	$\frac{13}{60}$	520	78

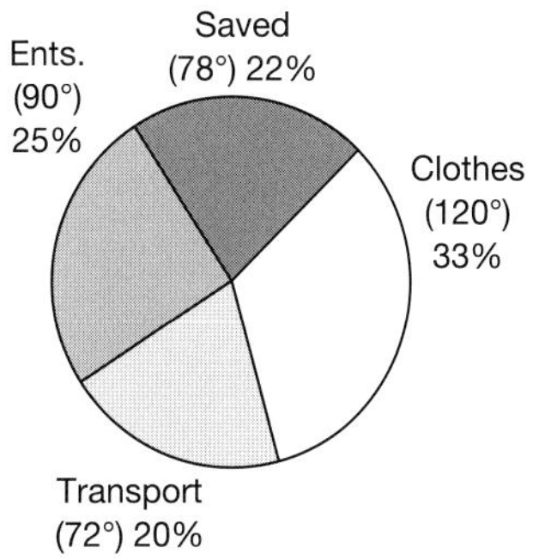

4.

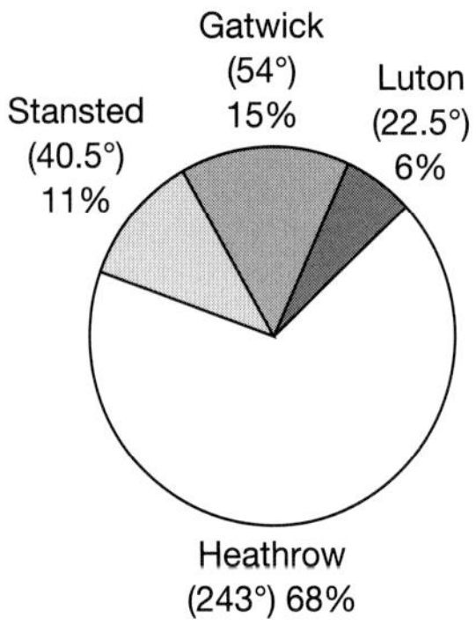

5. a)

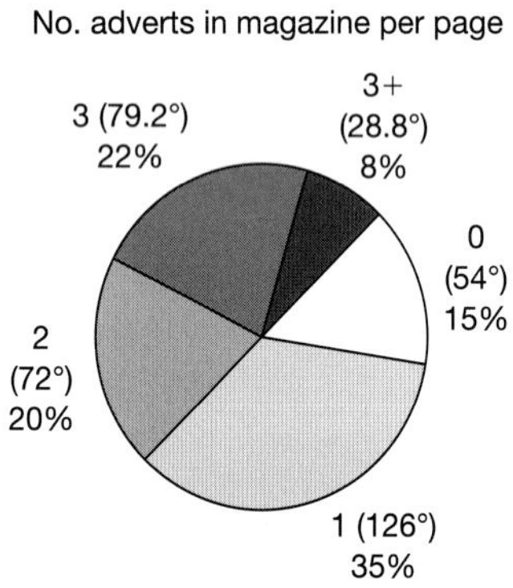

b) 240 pages × 15% = 36 pages

6. a)

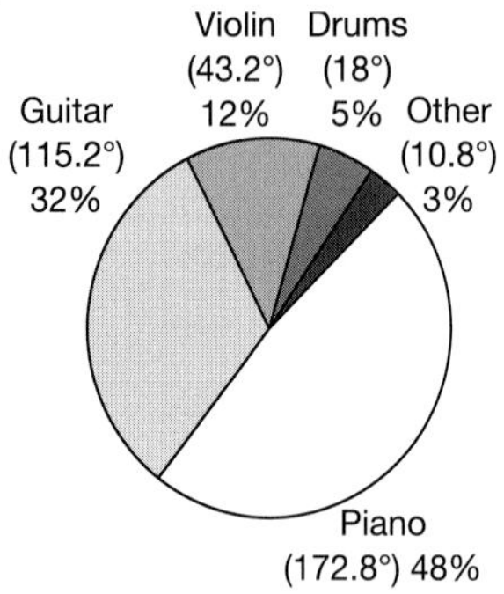

b) 96 play guitar

c) 144 play piano. 14.4% of the whole school

7.

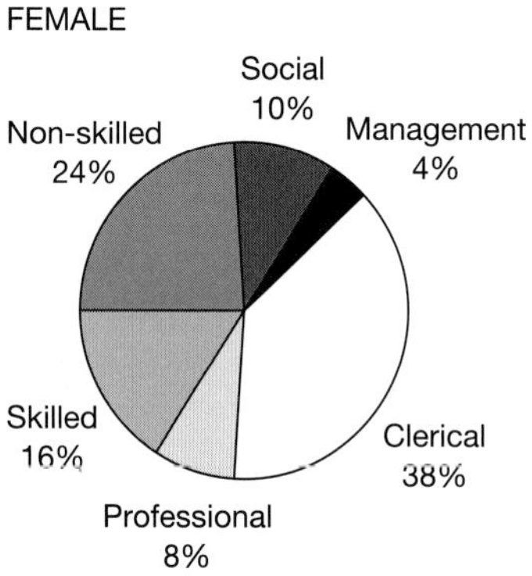

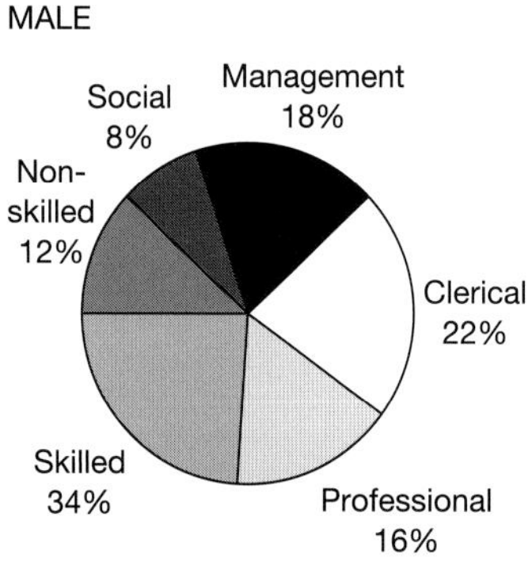

Student's own two statements

b) Professional = 8 million × 8% = 640 000
Management = 8 million × 4% = 320 000

Exercise 31.4 page 305

1. Student's own survey results and pie charts

2. Student's report

Exercise 31.5 page 308

1.
a) Mean = 9.6 Range = 7
b) Mean = 6.4 Range = 6
c) Mean = 44.1 Range = 19
d) Mean = 8 Range = 8
e) Mean = 36 Range = 48
f) Mean = 20.7 Range = 11.5

2.
a) Median = 5 Range = 4
b) Median = 9 Range = 8
c) Median = 9 Range = 2
d) Median = 6 Range = 6
e) Median = 5 Range = 6
f) Median = 9.5 Range = 7
g) Median = 5.85 Range = 6.1
h) Median = 46 Range = 81

3. **a)** Mode = 8 Range = 5
b) Mode = 6 Range = 5
c) Mode = 3 Range = 5
d) Mode = 4 Range = 2
e) Mode = 75 Range = 20
f) Mode = 8 Range = 3

Exercise 31.6 page 308

1. Mean = 1.7 (1 d.p.)
Median = 1
Mode = 1

2. Mean = 6.2
Median = 6.5
Mode = 7

3. Mean = 14 yrs 3 mths
Median = 14 yrs 3 mths
Mode = 14 yrs 3 mths

4. Mean = 26.4
Median = 27
Mode = 28

5. Mean = 13.9 s (1 d.p.)
Median = 13.9 s
Mode = 13.8 s

6. 91.1 kg

7. 103 points

Exercise 31.7 page 309

1. Mean = 3.35
Median = 3
Mode = 1 and 4

2. Mean = 7.03

3. **a)** Mean = 6.3 (1 d.p.)
Median = 7
Mode = 8
b) The mode, as it gives the highest number of flowers per bush

Exercise 31.8 page 312

1. Student's answers may differ from those given below.
a) Possible positive correlation (strength depending on topics tested)
b) No correlation
c) Positive correlation (likely to be quite strong)
d) Negative correlation (likely to be strong). Assume that motorcycles are not rare/vintage.
e) Factors such as social class, religion and income are likely to affect results. Therefore little correlation is likely.
f) Negative correlation (likely to be strong)
g) 0–16 years likely to be a positive correlation
h) Strong positive correlation

2. **a)**

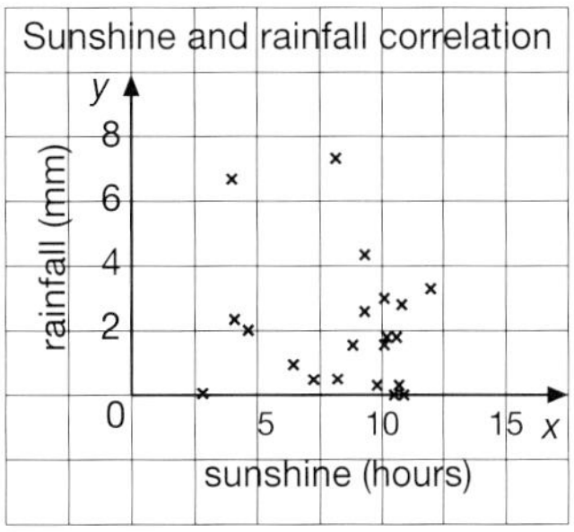

b) Graph shows a very weak negative correlation.

3. **a)**

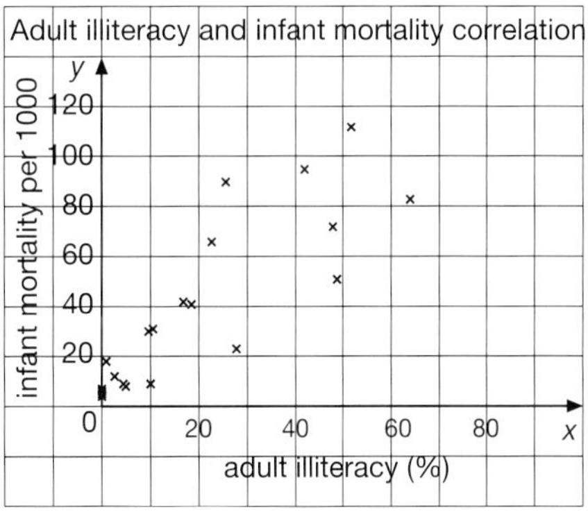

b) Positive correlation
c) Student's answer. However, although there is a correlation, it doesn't imply that one variable affects the other.

d)

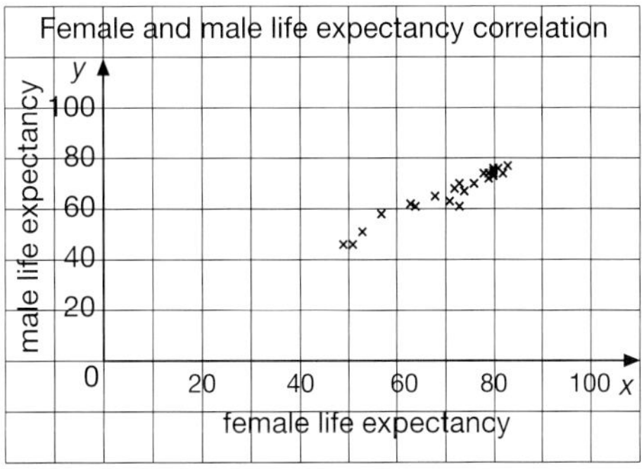

Student Assessment 1 page 315

1. Student's own pictogram

2.

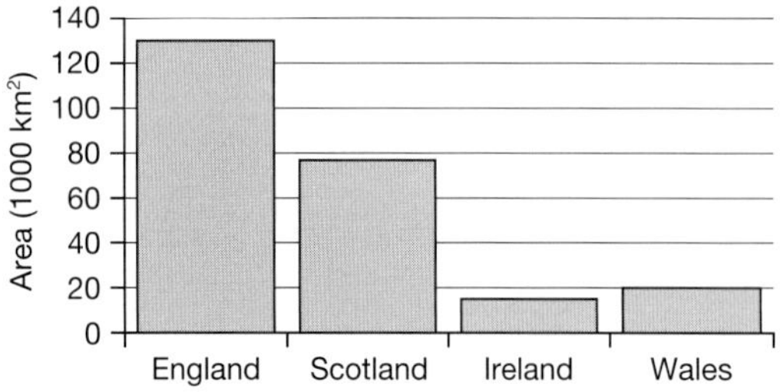

3.

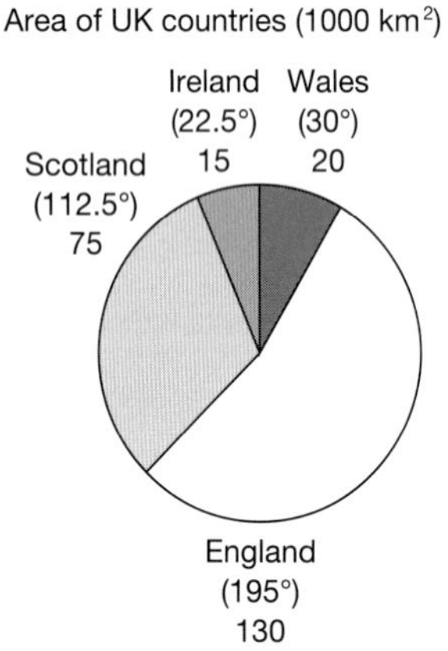

4. **a)** Mean = 75.4 Median = 72 Mode = 72
Range = 23
b) Mean = 12 Median = 9 Mode = 6
Range = 18
c) Mean = 11.5 Median = 10 Mode = 5
Range = 20

5. 61 kg

Student Assessment 2 page 315

1. Student's own pictogram

2.

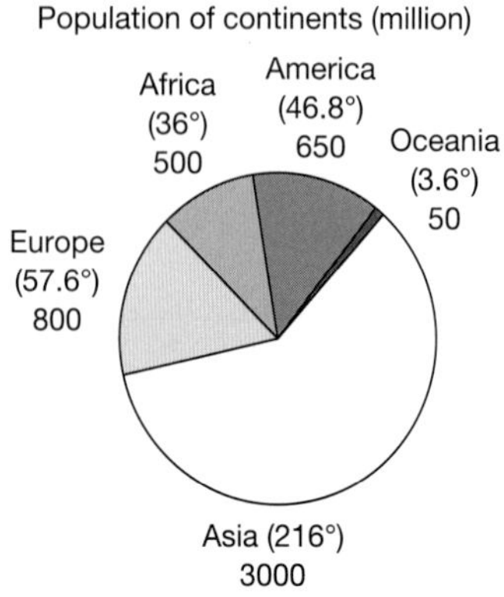

3.

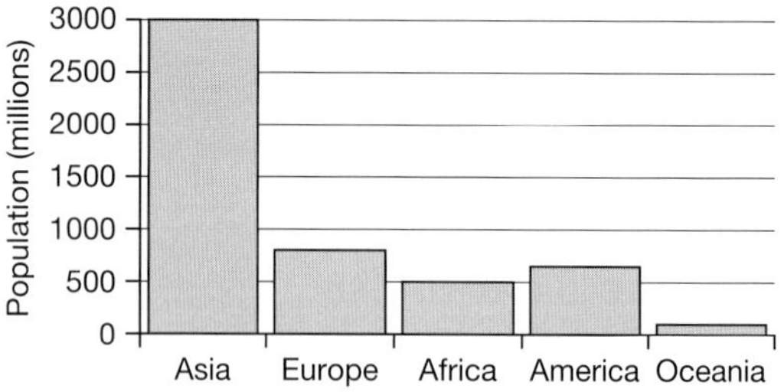

4. **a)** Mean = 5.4 Median = 5 Mode = 5
Range = 3
b) Mean = 13.8 Median = 15 Mode = 18
Range = 21
c) Mean = 6.1 Median = 6 Mode = 3
Range = 8

5. Mean = 3.52 Median = 4 Mode = 2

Student Assessment 3 page 316

1. **a)** Mean = 16 **b)** Median = 16.5
c) Mode = 18

2. **a)** 28
b) i) Mean = 7.8 (1 d.p.) ii) Median = 8
iii) Mode = 8

3. Student's own pie chart

4. Student's own histogram

5. **a)**

Attendance	Tally	Frequency
0–3999		0
4000–7999	II	2
8000–11 999	卌 II	7
12 000–15 999	卌 I	6
16 000–19 999	卌 卌 卌	15

b)

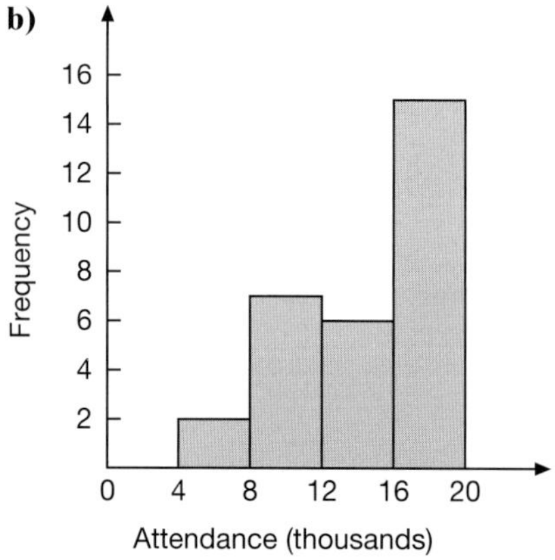

6. a)

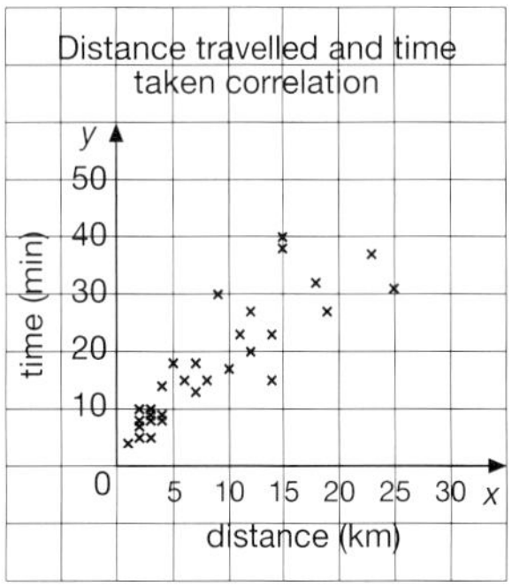

b) (Strong) positive correlation

c) It depends on their mode of transport.

d)

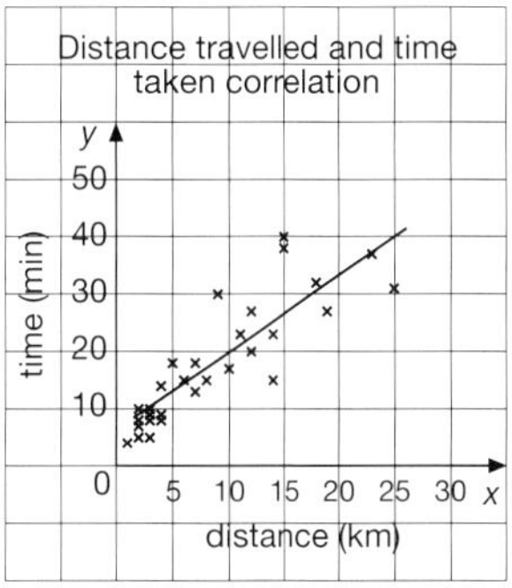

Student Assessment 4 page 318

1. Mean = 86.8 m Median = 90.5 m
Mode = 93 m

2. a) 26
b) i) Mean = 7.7 (1 d.p.) ii) Median = 7.5
iii) Mode = 10

3. Student's own pie chart

4. Student's own histogram

5. Student's own histogram

6. a)

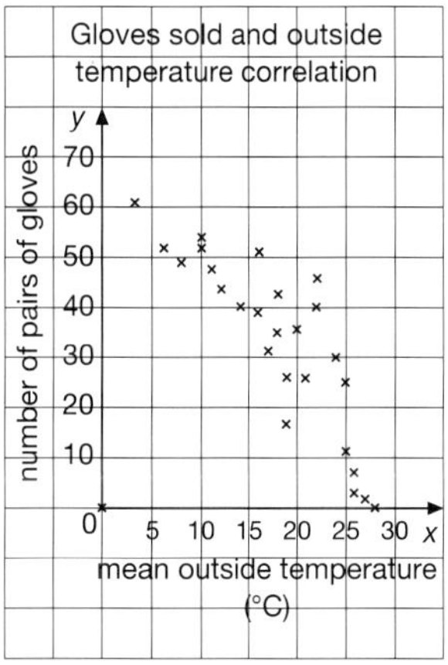

b) Positive correlation
c) Student's own answer

32 Probability

Exercise 32.1 page 320

1. Student's own drawing

2. Student's own drawing

3. Student's own answers

Exercise 32.2 page 321

1. a) $\frac{1}{6}$ **b)** $\frac{5}{6}$ **c)** $\frac{1}{2}$ **d)** $\frac{5}{6}$ **e)** 0 **f)** 1

2. a) i) $\frac{1}{7}$ ii) $\frac{6}{7}$ **b)** Total = 1

3. a) $\frac{1}{250}$ **b)** $\frac{1}{50}$ **c)** 1 **d)** 0

4. a) $\frac{5}{8}$ **b)** $\frac{3}{8}$

5. a) $\frac{1}{13}$ **b)** $\frac{5}{26}$ **c)** $\frac{21}{26}$ **d)** $\frac{3}{26}$

6. $\frac{1}{6}$

7. a) i) $\frac{1}{10}$ ii) $\frac{1}{4}$ **b)** i) $\frac{1}{19}$ ii) $\frac{3}{19}$

8. a) $\frac{1}{37}$ **b)** $\frac{36}{37}$ **c)** $\frac{18}{37}$ **d)** $\frac{1}{37}$ **e)** $\frac{21}{37}$
f) $\frac{12}{37}$ **g)** $\frac{17}{37}$ **h)** $\frac{11}{37}$

9. a) RCA RAC CRA CAR ARC ACR
b) $\frac{1}{6}$ **c)** $\frac{1}{3}$ **d)** $\frac{1}{2}$ **e)** $\frac{1}{24}$

10. a) $\frac{1}{4}$ **b)** $\frac{3}{4}$ **c)** $\frac{1}{13}$ **d)** $\frac{1}{26}$ **e)** $\frac{3}{13}$
f) $\frac{1}{52}$ **g)** $\frac{5}{13}$ **h)** $\frac{4}{13}$

Exercise 32.3 page 323

1. a) $\frac{1}{6}$ **b)** $\frac{5}{6}$ **c)** 0 **d)** $\frac{1}{2}$ **e)** $\frac{1}{2}$ **f)** $\frac{5}{6}$

2. $\frac{1}{2}$

3. a) $\frac{1}{8}$ **b)** $\frac{7}{8}$ **c)** 1 **d)** $\frac{5}{8}$ **e)** $\frac{1}{2}$ **f)** $\frac{1}{2}$
g) $\frac{1}{2}$ **h)** $\frac{3}{4}$

4. a) $\frac{1}{13}$ **b)** $\frac{1}{13}$ **c)** $\frac{1}{13}$ **d)** $\frac{3}{13}$ **e)** $\frac{1}{4}$ **f)** $\frac{1}{2}$
g) $\frac{1}{52}$ **h)** $\frac{3}{52}$ **i)** $\frac{16}{52}$ **j)** $\frac{3}{26}$ **k)** $\frac{15}{52}$ **l)** $\frac{4}{13}$
m) $\frac{11}{26}$ **n)** $\frac{19}{52}$ **o)** $\frac{5}{52}$

5. a) $\frac{1}{200}$ **b)** $\frac{1}{20}$ **c)** $\frac{3}{40}$ **d)** $\frac{1}{8}$ **e)** $\frac{1}{2}$ **f)** $\frac{1}{2}$
g) $\frac{3}{4}$ **h)** 1

6. $f = 40$

7. 25

8. a) $\frac{1}{16}$ **b)** $\frac{7}{16}$ **c)** $\frac{1}{2}$ **d)** $\frac{15}{16}$ **e)** $\frac{11}{16}$

9. a) $\frac{14}{45}$ **b)** $\frac{7}{45}$ **c)** $\frac{13}{30}$ **d)** $\frac{1}{15}$ **e)** $\frac{14}{15}$

10. 35 blue, 28 red, 21 yellow, 49 green, 7 white

11. $t = 300$

12. 14

13. 2

14. 2

Exercise 32.4 page 324

1. Student's graph. The relative frequency approaches 0.5.
2. Student's graph
3. Student's graph
4. Student's explanation

Student Assessment 1 page 327

1. **a)** $\frac{1}{6}$ **b)** $\frac{5}{6}$ **c)** $\frac{2}{3}$ **d)** 0
2. **a)** $\frac{1}{4}$ **b)** $\frac{1}{13}$ **c)** $\frac{1}{2}$ **d)** $\frac{3}{13}$ **e)** $\frac{4}{13}$
3. **a)** $\frac{1}{250}$ **b)** $\frac{1}{50}$ **c)** $\frac{2}{25}$ **d)** $\frac{3}{10}$ **e)** 1
4. $\frac{11}{40}$ **b)** $\frac{1}{8}$ **c)** $\frac{1}{4}$ **d)** $\frac{4}{5}$
5. 160 red, 96 blue, 64 green

Student Assessment 2 page 327

1. **a)** $\frac{1}{8}$ **b)** $\frac{7}{8}$ **c)** $\frac{5}{8}$ **d)** 0
2. **a)** 32 cards in pack
   b i) $\frac{1}{8}$ ii) $\frac{3}{8}$ iii) $\frac{1}{4}$ iv) $\frac{1}{8}$ v) $\frac{11}{32}$
3. **a)** $\frac{1}{180}$ **b)** $\frac{1}{20}$ **c)** $\frac{1}{12}$ **d)** $\frac{2}{9}$ **e)** 1
4. **a)** $\frac{9}{32}$ **b)** $\frac{3}{8}$ **c)** $\frac{21}{32}$
5. 128 red, 80 blue, 112 green
6. 50 000 tickets

ICT Section

Statistics and Probability Activity 1

page 329

1–5. Student's timelines, tables and fully labelled pie charts.

6. Student's comments on the differences between the time spent on activities on a weekday compared with a day at the weekend.

Statistics and Probability Activity 2

page 329

4. Theoretical probability is $\frac{n}{6}$, where n is the number of times a colour is used.

8. The fairness of a spinner will *generally* depend on how close experimental probability is to the theoretical probability. (Note 60 spins is probably not sufficient to reliably determine whether a spinner is fair or not.)

Reviews for Chapters 31 to 32

Review 26 page 331

1. **a)** Mean = 5 Median = 5 Mode = 4
   Range = 7
   b) Mean = 7.75 Median = 7.5 Mode = 4
   Range = 13
   c) Mean = 7.36 Median = 7.75
   Mode = 8.2 Range = 3.1
2. **a)** 20 **b)** 60
3. Student's own pie chart – angles: home 160°, draw 120°, away 80°
4. Student's own bar chart
5.

No. of words	Frequency
1–10	17
11–20	14
21–30	7
31–40	2

6. **a)** $\frac{1}{37}$ **b)** $\frac{18}{37}$ **c)** $\frac{1}{37}$ **d)** $\frac{9}{37}$ **e)** $\frac{11}{37}$
7. **a)** 17 000 **b)** 20 000
8. Student's own diagram
9. $\frac{1}{165}$ **b)** $\frac{1}{11}$ **c)** $\frac{1}{15}$
10. Student's own answer

Review 27 page 332

1. 6.8
2. Student's own pictogram
3. Student's own pie chart – angles: 0–120°, 1–80°, 2–40°, 3–60°, 4–40°, 5–20°

4. Student's own bar chart
5. Student's own diagram
6. a) 5 b) 50 c) 30
7. a) $\frac{1}{3}$ b) $\frac{5}{9}$ c) $\frac{4}{9}$
8. 750
9. Student's own explanations
10. Student's own answers

Review 28 page 333

1. a) Mean = 8 Median = 8 Mode = none
   b) Mean = 5.1 Median = 5 Mode = 4.5
   c) Mean = 19.6 Median = 16 Mode = 8
2. Student's own pictogram
3. Student's own pie chart – angles: crime 56°, sci-fi 90°, romance 64°, other 40°, non 110°
4.

Time	Frequency
$0 \leq t < 10$	7
$10 \leq t < 20$	15
$20 \leq t < 30$	14
$30 \leq t < 40$	15
$40 \leq t < 50$	16
$50 \leq t < 60$	13

Student's own histogram

5. a) 300 b) 66
6. a) $\frac{1}{8}$ b) $\frac{1}{2}$ c) $\frac{1}{2}$
7. a) $\frac{4}{11}$ b) $\frac{4}{11}$ c) $\frac{7}{11}$
8. $\frac{1}{40}$
9. Student's own explanation
10. Student's own explanation

Review 29 page 334

1. Mean = 19.4 Range = 39
2. Student's own pictogram
3. Mean = 40.5 Median = 40 Range = 6
4. Student's own bar chart
5. ART, ATR, RAT, RTA, TAR, TRA
   a) $\frac{1}{6}$ b) $\frac{1}{2}$
6. $\frac{6}{20}(\frac{3}{10})$
7. a) $\frac{18}{30}(\frac{3}{5})$ b) $\frac{12}{30}(\frac{2}{5})$
8. a) $\frac{1}{26}$ b) $\frac{5}{26}$ c) $\frac{4}{26}(\frac{2}{13})$
9. Student's own answer
10. Student's own answer

Review 30 page 335

1. Student's own bar chart
2.

No. of apples	Frequency
0–19	4
20–39	13
40–59	8
60–79	5

Student's own bar chart

3. Student's own pie chart – angles: football 180°, tennis 80°, hockey 40°, basketball 60°
4. Mean = 6.7 Mode = 7
5. a) $\frac{1}{11}$ b) $\frac{2}{11}$ c) $\frac{4}{11}$ d) $\frac{7}{11}$
6. a) $\frac{1}{4}$ b) $\frac{3}{13}$ c) $\frac{4}{13}$
7. $\frac{1}{50}$
8. a) $\frac{1}{7}$ b) $\frac{30}{365}$ c) $\frac{12}{365}$
9. Student's own questionnaire
10. Student's own explanation

General Reviews

General Review 1 page 337

1. a) Tetrahedron/triangular-based pyramid
   b)

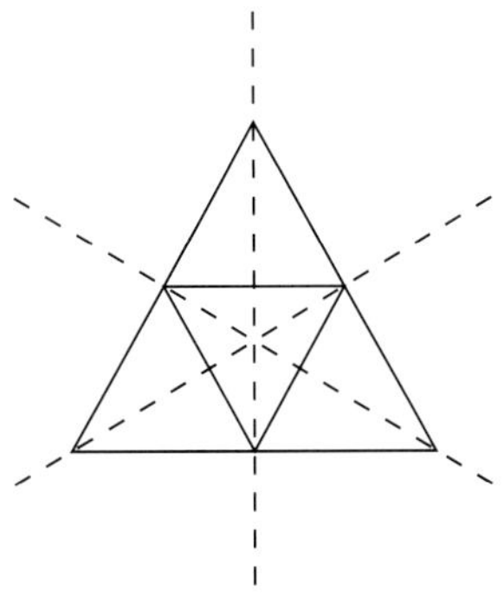

   c) Order 3
   d) 15.6 cm^2 (1 d.p.)

2. **a)** and **b)**

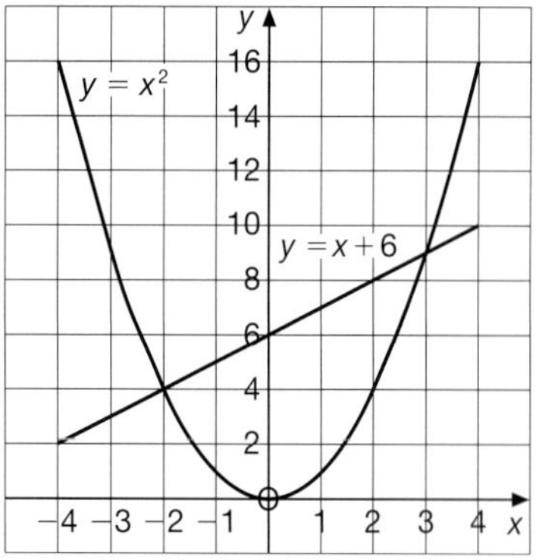

c) $x = -2$ and 3

3. **a)** −3 °C **b)** $H = 150(20 - T)$
c) 1200 m

4. **a)** 57 600 ÷ 12 = $4800 **b)** $3072
c) $62 784

General Review 2 page 337

1. **a)** 4.2 × 4.2 × 12.6 cm
Volume = 222 cm^3 (3 s.f.)
b) 53% (2 s.f.) **c)** 175 cm^3 **d)** 1 : 3

2. **a)**

Test score	0	1	2	3	4	5	6	7	8	9	10	Total
Frequency	3	6	9	6	9	15	12	18	12	3	3	96
Frequency × Score	0	6	18	18	36	75	72	126	96	27	30	504

b) 5.25 **c)** 5.5 **d)** 7
e)

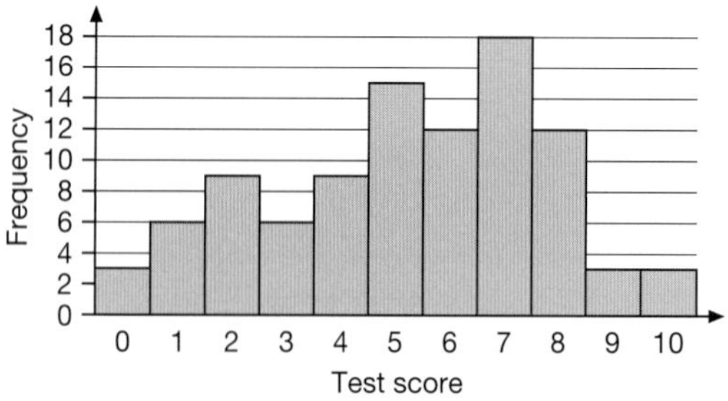

3.

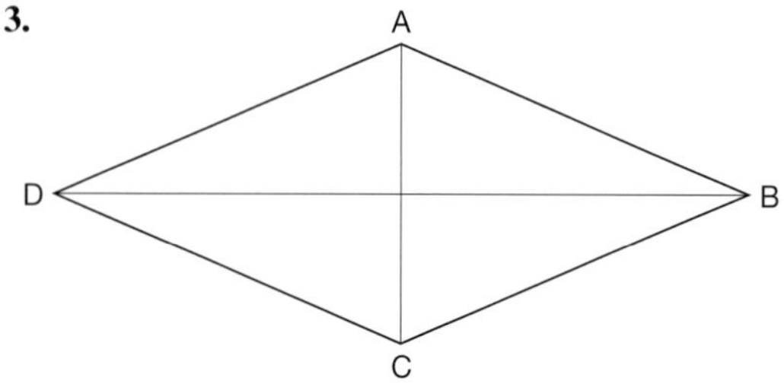

a) Rhombus **b)** Order 2 **c)** AB = 10 cm
d) ∠ABC = 74° (2 s.f.)

4. **a)** $x = 14$ **b)** $x = 7, y = 1$
c) $a = c(d - a) - b$

General Review 3 page 339

1. **a)** Student's own drawing
b)

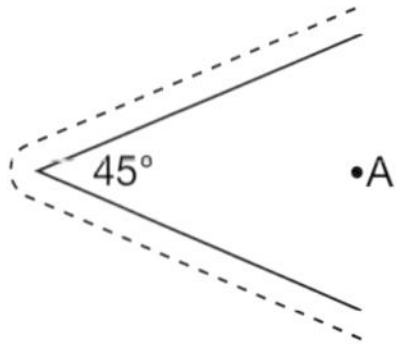

c) 48 m (2 s.f.)

2. **a)** i) 26, 31 ii) 768, 3072
b) i) 24 ii) 9800
c) i) 125, 216 ii) 1 000 000 iii) n^3
d) $n^3 + 5$

3. **a)** If PQ is a tangent, ∠OQP = 90°. Therefore △OQP must follow Pythagoras's theorem.
$12^2 = 13^2 - 5^2$
b) ∠POR = 67° (2 s.f.)
c) 79 cm^2 (2 s.f.)
d) 14.7 cm^2 (1 d.p.)

4. **a)** 24 **b)** i) $\frac{1}{24}$ ii) $\frac{1}{4}$

5. **a)** i) 21.1 (1 d.p.) ii) 20 iii) 17

Score	Frequency
6	H
12	H
13	H
14	H
17	HHH
20	H
21	HH
23	HH
28	H
42	HH

b)

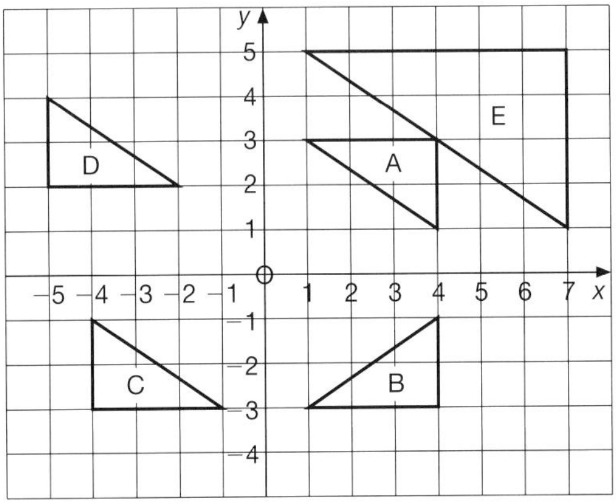

General Review 4 page 340

1. **a) – d)**

2. **a)** 1750 m^2 **b)** 1750 m^2 **c)** 2625 m^2
d) 61 m (2 s.f.) **e)** 35° (2 s.f.)

3. **a)** a **b)** $13xyz(3xy - 2yz + 4xz)$
c) -25 **d)** $p = \frac{3m + n}{r} + q$

4. **a)** Student's own construction
b) Student's own proof
c) 5 **d)** 12

General Review 5 page 341

1. **a)** 14 074 335 m^3 **b)** 562 974
c) £250 million (3 s.f.)

2. **a)** i) $\frac{1}{13}$ ii) $\frac{1}{4}$ iii) $\frac{3}{13}$ iv) $\frac{16}{52}$ or $\frac{4}{13}$
b) i) $\frac{1}{10}$ ii) $\frac{1}{4}$ iii) 0 iv) $\frac{13}{40}$
c) Student's examples

3. **a)** 60° **b)** 11.3 cm (1 d.p.) **c)** 5 cm
d) 22.6° (1 d.p.)
e) To be parallel, ∠ACD must equal ∠BAC and hence be alternate. But ∠BAC = 60° as △ABC is equilateral. Therefore ∠ACD ≠ ∠BAC. Hence AB and CD are not parallel.

4. **a)**

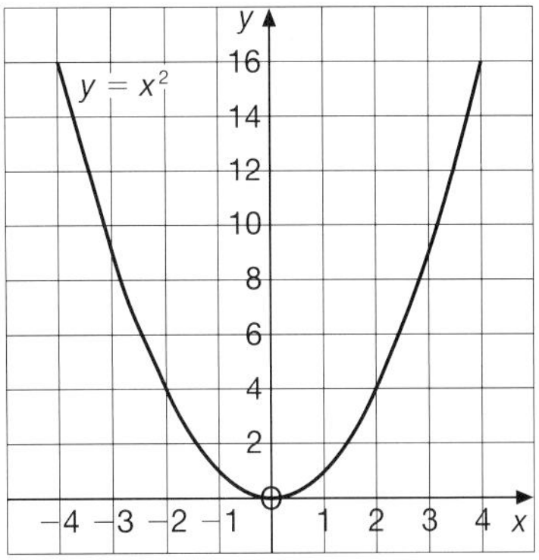

b)

x	0	1	2	3	4	5	6	7
y	18	10	4	0	–2	–2	0	4

c)

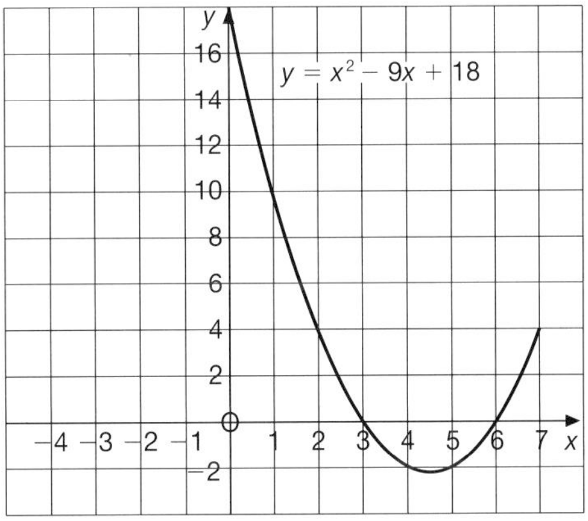

d) $x = 3$ and 6
e) $x = 4$ and 5

INDEX

Page numbers refer to pages where topics are described: mentions on pages devoted wholly to exercises, assessments, review sections, or solutions are not indexed.